![Haynes logo] THE BOOK ®

Citroën XM
Service and Repair Manual

Steve Rendle and Spencer Drayton

Models covered

(3451-320-1AD3)

Citroën XM Hatchback and Estate models, including special/limited editions
4-cylinder 1998 cc 8-valve multi-point fuel injection petrol engines (including turbo)
2088 cc, 2138 cc and 2445 cc diesel engines (including turbo-diesels)

Does NOT cover 16-valve, carburettor, single-point fuel injection or V6 petrol engines

© Haynes Publishing 2000

ABCDE
FGHIJ
KLMNO

Printed in the USA

A book in the **Haynes Service and Repair Manual Series**

ISBN **1 85960 451 X**

British Library Cataloguing in Publication Data

A catalogue record for this book is available from the British Library.

Haynes Publishing
Sparkford, Nr Yeovil, Somerset BA22 7JJ, England

Haynes North America, Inc
861 Lawrence Drive, Newbury Park, California 91320, USA

Editions Haynes S.A.
Tour Aurore - IBC, 18 Place des Reflets,
92975 Paris La Défense 2, Cedex, France

Haynes Publishing Nordiska AB
Box 1504, 751 45 Uppsala, Sverige

Contents

LIVING WITH YOUR CITROËN XM

Roadside Repairs

Weekly Checks

MAINTENANCE

Routine Maintenance and Servicing

Contents

The Citroën XM model range covered by this manual was introduced into the UK in October 1989, initially in 5-door Hatchback form, with a range of petrol engines. Diesel engine models were introduced in May 1990, and 5-door Estate models became available at the end of 1991. The whole model range was revised, with the introduction of a new range of engines, and a facelifted interior, in June 1994.

XM models have been produced with 2.0 litre carburettor, fuel injection and turbo engines, and V6 fuel injection petrol engine; although the 2.0 litre carburettor and V6 engines are not covered by this manual. 2.1 litre and 2.5 litre turbo-diesel engines are also available (a 2.1 litre non-turbo engine was also used, but did not appear in the UK).

Models may be fitted with five-speed manual, or 4-speed automatic transmissions, mounted at the left-hand side of the engine.

All models have front-wheel-drive, and use a Citroën-patented hydropneumatic suspension system, which is electronically-controlled, and adjusts automatically to the prevailing road surface and the driving style. Conventional shock absorbers are replaced by hydraulic units, in which the pressure may be varied to effectively stiffen or soften the suspension.

All models have a high trim level, which is very comprehensive in the upper model range. Driver's and passenger's air bag systems, central locking, electric windows, an electric sunroof, anti-lock brakes, and air conditioning are all available.

Provided that regular servicing is carried out in accordance with the manufacturer's recommendations, the XM should prove reliable, and most of the items requiring frequent attention are easily accessible.

Your Citroën XM Manual

The aim of this manual is to help you get the best value from your vehicle. It can do so in several ways. It can help you decide what work must be done (even should you choose to get it done by a garage), provide information on routine maintenance and servicing, and give a logical course of action and diagnosis when random faults occur. However, it is hoped that you will use the manual by tackling the work yourself. On simpler jobs, it may even be quicker than booking the car into a garage and going there twice, to leave and collect it. Perhaps most important, a lot of money can be saved by avoiding the costs a garage must charge to cover its labour and overheads.

The manual has drawings and descriptions to show the function of the various components,

Citroën XM Turbo SED

Citroën XM Estate

so that their layout can be understood. Then the tasks are described and photographed in a clear step-by-step sequence.

References to 'left' and 'right' of the vehicle are in the sense of a person in the driver's seat facing forwards.

Acknowledgements

Thanks are due to Champion Spark Plug who supplied the illustrations showing spark plug conditions and to Duckhams Oils, who provided lubrication data. Special thanks to Paul Johnson at the XM Centre, and to Olds, Puddletown, who provided the project vehicles used in the origination of this manual. Thanks are also due to Draper Tools Limited, who provided some of the workshop tools, to Harry Boswell who proof read the draft copies, to James Colla for technical assistance, and to all those people at Sparkford who helped in the production of this manual.

We take great pride in the accuracy of information given in this manual, but vehicle manufacturers make alterations and design changes during the production run of a particular vehicle of which they do not inform us. No liability can be accepted by the authors or publishers for loss, damage or injury caused by any errors in, or omissions from, the information given.

Working on your car can be dangerous. This page shows just some of the potential risks and hazards, with the aim of creating a safety-conscious attitude.

General hazards

Scalding

• Don't remove the radiator or expansion tank cap while the engine is hot.
• Engine oil, automatic transmission fluid or power steering fluid may also be dangerously hot if the engine has recently been running.

Burning

• Beware of burns from the exhaust system and from any part of the engine. Brake discs and drums can also be extremely hot immediately after use.

Crushing

• When working under or near a raised vehicle, always supplement the jack with axle stands, or use drive-on ramps. *Never venture under a car which is only supported by a jack.*
• Take care if loosening or tightening high-torque nuts when the vehicle is on stands. Initial loosening and final tightening should be done with the wheels on the ground.

Fire

• Fuel is highly flammable; fuel vapour is explosive.
• Don't let fuel spill onto a hot engine.
• Do not smoke or allow naked lights (including pilot lights) anywhere near a vehicle being worked on. Also beware of creating sparks (electrically or by use of tools).
• Fuel vapour is heavier than air, so don't work on the fuel system with the vehicle over an inspection pit.
• Another cause of fire is an electrical overload or short-circuit. Take care when repairing or modifying the vehicle wiring.
• Keep a fire extinguisher handy, of a type suitable for use on fuel and electrical fires.

Electric shock

• Ignition HT voltage can be dangerous, especially to people with heart problems or a pacemaker. Don't work on or near the ignition system with the engine running or the ignition switched on.

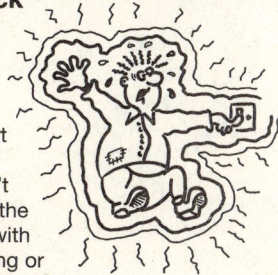

• Mains voltage is also dangerous. Make sure that any mains-operated equipment is correctly earthed. Mains power points should be protected by a residual current device (RCD) circuit breaker.

Fume or gas intoxication

• Exhaust fumes are poisonous; they often contain carbon monoxide, which is rapidly fatal if inhaled. Never run the engine in a confined space such as a garage with the doors shut.
• Fuel vapour is also poisonous, as are the vapours from some cleaning solvents and paint thinners.

Poisonous or irritant substances

• Avoid skin contact with battery acid and with any fuel, fluid or lubricant, especially antifreeze, brake hydraulic fluid and Diesel fuel. Don't syphon them by mouth. If such a substance is swallowed or gets into the eyes, seek medical advice.
• Prolonged contact with used engine oil can cause skin cancer. Wear gloves or use a barrier cream if necessary. Change out of oil-soaked clothes and do not keep oily rags in your pocket.
• Air conditioning refrigerant forms a poisonous gas if exposed to a naked flame (including a cigarette). It can also cause skin burns on contact.

Asbestos

• Asbestos dust can cause cancer if inhaled or swallowed. Asbestos may be found in gaskets and in brake and clutch linings. When dealing with such components it is safest to assume that they contain asbestos.

Special hazards

Hydrofluoric acid

• This extremely corrosive acid is formed when certain types of synthetic rubber, found in some O-rings, oil seals, fuel hoses etc, are exposed to temperatures above 400°C. The rubber changes into a charred or sticky substance containing the acid. *Once formed, the acid remains dangerous for years. If it gets onto the skin, it may be necessary to amputate the limb concerned.*
• When dealing with a vehicle which has suffered a fire, or with components salvaged from such a vehicle, wear protective gloves and discard them after use.

The battery

• Batteries contain sulphuric acid, which attacks clothing, eyes and skin. Take care when topping-up or carrying the battery.
• The hydrogen gas given off by the battery is highly explosive. Never cause a spark or allow a naked light nearby. Be careful when connecting and disconnecting battery chargers or jump leads.

Air bags

• Air bags can cause injury if they go off accidentally. Take care when removing the steering wheel and/or facia. Special storage instructions may apply.

Diesel injection equipment

• Diesel injection pumps supply fuel at very high pressure. Take care when working on the fuel injectors and fuel pipes.

⚠ *Warning: Never expose the hands, face or any other part of the body to injector spray; the fuel can penetrate the skin with potentially fatal results.*

Remember...

DO

• Do use eye protection when using power tools, and when working under the vehicle.
• Do wear gloves or use barrier cream to protect your hands when necessary.
• Do get someone to check periodically that all is well when working alone on the vehicle.
• Do keep loose clothing and long hair well out of the way of moving mechanical parts.
• Do remove rings, wristwatch etc, before working on the vehicle – especially the electrical system.
• Do ensure that any lifting or jacking equipment has a safe working load rating adequate for the job.

DON'T

• Don't attempt to lift a heavy component which may be beyond your capability – get assistance.
• Don't rush to finish a job, or take unverified short cuts.
• Don't use ill-fitting tools which may slip and cause injury.
• Don't leave tools or parts lying around where someone can trip over them. Mop up oil and fuel spills at once.
• Don't allow children or pets to play in or near a vehicle being worked on.

The following pages are intended to help in dealing with common roadside emergencies and breakdowns. You will find more detailed fault finding information at the back of the manual, and repair information in the main chapters.

If your car won't start and the starter motor doesn't turn

- [ ] If it's a model with automatic transmission, make sure the selector lever is in 'P' or 'N'.
- [ ] Open the bonnet and make sure that the battery terminals are clean and tight.
- [ ] Switch on the headlights and try to start the engine. If the headlights go very dim when you're trying to start, the battery is probably flat. Get out of trouble by jump starting (see next page) using a friend's car.

If your car won't start even though the starter motor turns as normal

- [ ] Is there fuel in the tank?
- [ ] Is there moisture on electrical components under the bonnet? Switch off the ignition, then wipe off any obvious dampness with a dry cloth. Spray a water-repellent aerosol product (WD-40 or equivalent) on ignition and fuel system electrical connectors like those shown in the photos. Pay special attention to the ignition coil wiring connector and HT leads. (Note that diesel engines don't normally suffer from damp.)

A Check the condition and security of the battery connections.

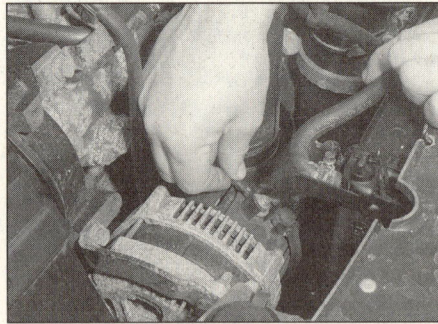

B Check that the alternator wiring is secure.

C Check that the fuel/ignition system wiring connectors are securely connected.

D Check the condition of the fuses, and make sure that none have blown.

Check that electrical connections are secure (with the ignition switched off) and spray them with a water dispersant spray like WD-40 if you suspect a problem due to damp

Jump starting

Jump starting will get you out of trouble, but you must correct whatever made the battery go flat in the first place. There are three possibilities:

1) *The battery has been drained by repeated attempts to start, or by leaving the lights on.*
2) *The charging system is not working properly (alternator drivebelt slack or broken, alternator wiring fault or alternator itself faulty).*
3) *The battery itself is at fault (electrolyte low, or battery worn out).*

When jump-starting a car using a booster battery, observe the following precautions:

✔ Before connecting the booster battery, make sure that the ignition is switched off.

✔ Ensure that all electrical equipment (lights, heater, wipers, etc) is switched off.

✔ Take note of any special precautions printed on the battery case.

✔ Make sure that the booster battery is the same voltage as the discharged one in the vehicle.

✔ If the battery is being jump-started from the battery in another vehicle, the two vehicles MUST NOT TOUCH each other.

✔ Make sure that the transmission is in neutral (or PARK, in the case of automatic transmission).

1 Connect one end of the red jump lead to the positive (+) terminal of the flat battery

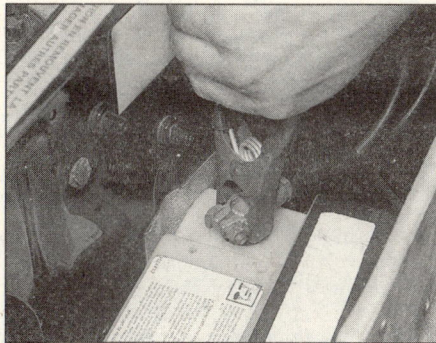

2 Connect the other end of the red lead to the positive (+) terminal of the booster battery

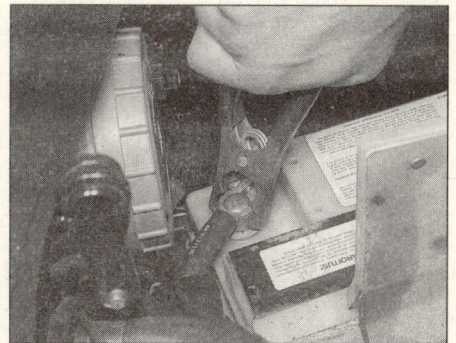

3 Connect one end of the black jump lead to the negative (-) terminal of the booster battery

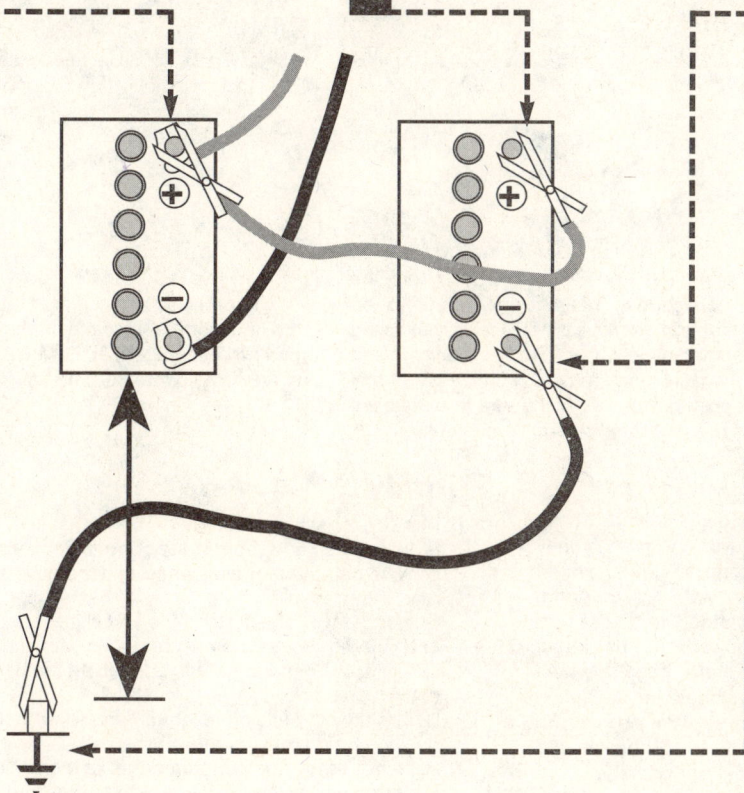

4 Connect the other end of the black jump lead to a bracket on the cylinder head on the vehicle to be started

5 Make sure that the jump leads will not come into contact with the fan, drivebelts or other moving parts of the engine

6 Start the engine using the booster battery and run it at idle speed. Switch on the lights, rear window demister and heater blower motor, then disconnect the jump leads in the reverse order of connection. Turn off the lights etc.

Wheel changing

⚠️ **Warning: Do not change a wheel in a situation where you risk being hit by other traffic. On busy roads, try to stop in a lay-by or gateway. Be wary of passing traffic while changing the wheel - it is easy to become distracted by the job in hand.**

☐ When a puncture occurs, stop as soon as it is safe to do so.

☐ Park on firm level ground, if possible, and well out of the way of other traffic.

☐ Use hazard warning lights if necessary.

☐ If you have one, use a warning triangle to alert other drivers of your presence.

☐ With the engine idling, set the suspension height control lever to the 'Maximum' position. Wait until the suspension has reached maximum height, then stop the engine.

☐ Apply the parking brake and engage first or reverse gear (or 'P' on models with automatic transmission).

☐ Chock the wheel diagonally opposite the one being removed - a couple of large stones will do for this.

☐ If the ground is soft, use a flat piece of wood to spread the load under the foot of the jack.

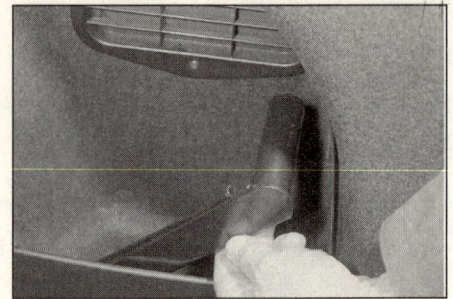

1 The jack and its handle are located in a container inside the spare wheel, under the rear of the vehicle. The wheel brace is located in a holder on the right-hand side of the luggage compartment.

2 The spare wheel is located in a cradle, under the rear of the vehicle. Working in the luggage compartment, pull up the carpet (Hatchback models), or the cover (Estate models), for access to the spare wheel cradle securing screw. Using the wheel brace, loosen, but do not fully unscrew the screw, then, working under the back of the vehicle, unhook the spare wheel cradle from the securing bracket, and lower the cradle until it is resting on the floor. Note that on some models, it may be necessary to unclip a safety cord or hook from the cradle before it can be fully lowered.

3 Lift the spare wheel from the cradle.

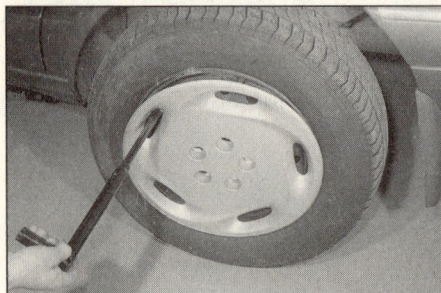

4 Where applicable, prise off the wheel trim from the wheel with the flat tyre, using the end of the wheel brace (extend the wheel brace handle), then loosen each wheel bolt by half a turn.

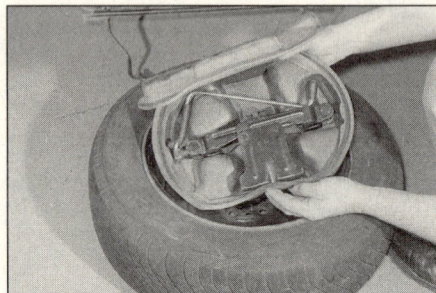

5 Where applicable, prise the cover from the jacking point nearest the wheel to be changed. Remove the jack from its housing in the spare wheel, then engage the jack head with the jacking point (don't jack the car at any other point on the sill). Make sure that the base of the jack is located on firm, level ground.

6 Turn the jack handle clockwise until the wheel is raised clear of the ground. Remove the bolts and lift the wheel clear, then fit the spare wheel. Refit the wheel bolts and tighten moderately with the wheel brace.

7 Lower the car to the ground, then finally tighten the wheel bolts in a diagonal sequence, and where applicable, fit the wheel trim. Note that the wheel bolts should be slackened and retightened to the specified torque at the earliest opportunity.

Finally...

☐ Remove the wheel chocks.

☐ Stow the jack and tools in the correct locations in the vehicle, and the removed wheel in the cradle under the rear of the vehicle.

☐ Check the tyre pressure on the wheel just fitted. If it is low, or if you don't have a pressure gauge with you, drive slowly to the nearest garage and inflate the tyre to the correct pressure.

☐ Return the suspension height control lever to the 'Normal' position before driving the vehicle.

☐ Repair or renew the punctured tyre at the earliest opportunity; set the pressure of the spare tyre at just above that of the highest roadwheel.

Puddles on the garage floor or drive, or obvious wetness under the bonnet or underneath the car, suggest a leak that needs investigating. It can sometimes be difficult to decide where the leak is coming from, especially if the engine bay is very dirty already. Leaking oil or fluid can also be blown rearwards by the passage of air under the car, giving a false impression of where the problem lies.

⚠️ **Warning: Most automotive oils and fluids are poisonous. Wash them off skin, and change out of contaminated clothing, without delay.**

Identifying leaks

HAYNES HiNT *The smell of a fluid leaking from the car may provide a clue to what's leaking. Some fluids are distinctively coloured. It may help to clean the car and to park it over some clean paper as an aid to locating the source of the leak. Remember that some leaks may only occur while the engine is running.*

Sump oil

Engine oil may leak from the drain plug...

Oil from filter

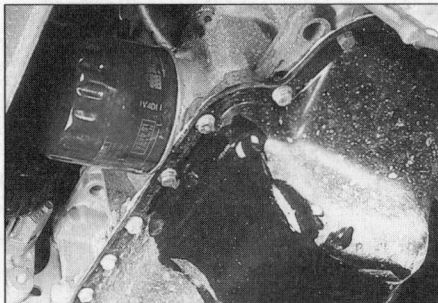

...or from the base of the oil filter.

Gearbox oil

Gearbox oil can leak from the seals at the inboard ends of the driveshafts.

Antifreeze

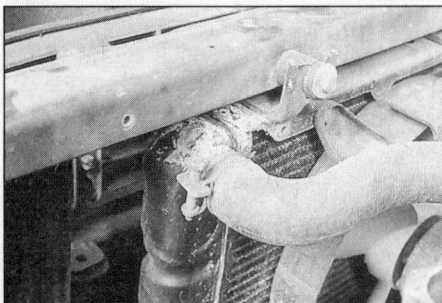

Leaking antifreeze often leaves a crystalline deposit like this.

Brake fluid

A leak occurring at a wheel is almost certainly brake fluid.

Power steering fluid

Power steering fluid may leak from the pipe connectors or seals on the steering rack.

Towing

When all else fails, you may find yourself having to get a tow home – or of course you may be helping somebody else. Long-distance recovery should only be done by a garage or breakdown service.

⚠️ **Warning: To prevent damage to the catalytic converter, a vehicle must not be push-started, or started by towing, when the engine is at operating temperature. Use jump leads (see 'Jump starting').**

Towing an XM

⚠️ **Warning: If the engine is not running, or there is a problem with the hydraulic system, the footbrake will not work. DO NOT attempt to tow using a tow rope unless the engine is running, and the hydraulic system (and braking system) is fully operational.**

⚠️ **Warning: Estate models must not be towed if the suspension is at minimum height.**

☐ If the engine will still run and the hydraulic system is functioning, short-distance DIY towing using another car is easy enough, but observe the following points.
☐ Use a proper tow-rope – they are not expensive. The vehicle being towed must display on ON TOW sign in its rear window.
☐ Have the engine running when being towed, so that the steering is released, and the direction indicators and brake lights will work.
☐ Only attach the tow-rope to the towing eye provided.
☐ If a Hatchback is being towed with front or rear end off the ground and the suspension at minimum height, the wheels must be lifted so that they are only just clear of the ground, and the distance towed must be kept to a minimum.
☐ Before being towed, release the parking brake and select neutral on the transmission.
☐ The driver of the car being towed must keep the tow-rope taut at all times to avoid snatching.
☐ Make sure that both drivers know the route before setting off.
☐ Only drive at moderate speeds and keep the distance towed to a minimum. Drive smoothly and allow plenty of time for slowing down at junctions.
☐ On models with automatic transmission, special procedures apply. If in doubt, do not tow, or transmission damage may result.

The front towing eye is located behind the access cover in the front spoiler

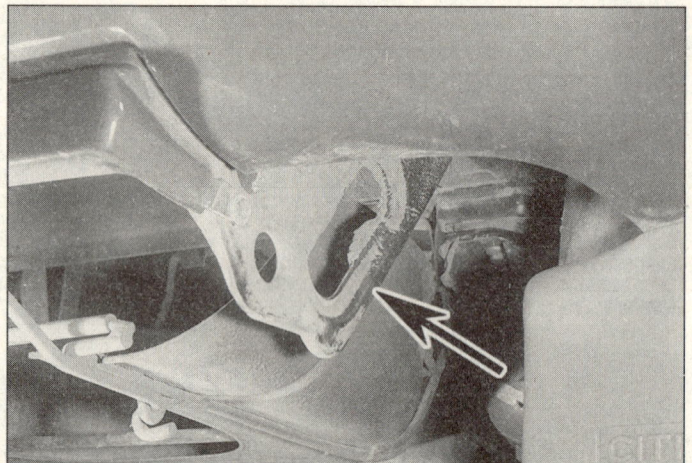

The rear towing eyes are located under the rear bumper

Towing another car

☐ Short-distance DIY towing using another car is easy enough, but observe the following points.
☐ Use a proper tow-rope – they are not expensive. The vehicle being towed must display on ON TOW sign in its rear window.
☐ Turn the ignition key to the 'on' position when being towed, so that the steering is released, and the direction indicators and brake lights will work).

☐ Only attach the tow-rope to the towing eye provided.
☐ Before being towed, release the parking brake and select neutral on the transmission.
☐ Greater-than-usual pedal pressure will be required to operate the brakes, since the vacuum servo unit only works with the engine running.
☐ On models with power steering, greater-than-usual steering effort will also be required.
☐ The driver of the car being towed must

keep the tow-rope taut at all times to avoid snatching.
☐ Make sure that both drivers know the route before setting off.
☐ Only drive at moderate speeds and keep the distance towed to a minimum. Drive smoothly and allow plenty of time for slowing down at junctions.
☐ On models with automatic transmission, special procedures apply. If in doubt, do not tow, or transmission damage may result.

Introduction

There are some very simple checks which need only take a few minutes to carry out, but which could save you a lot on inconvenience and expense.

☐ These *Weekly Checks* require no great skill or special tools, and the small amount of time they take to perform could well prove to be very well spent, for example;

☐ Keeping an eye on tyre condition and pressures, will not only help to stop them wearing out prematurely, but could also save your life.

☐ Many breakdowns are caused by electrical problems. Battery-related faults are particularly common, and a quick check on a regular basis will often prevent the majority of these.

☐ If your car develops a hydraulic fluid leak, the first time you might know about it could be when your brakes don't work properly. Checking the level regularly will give advance warning of this kind of problem.

☐ If the oil or coolant levels run low, the cost of repairing any engine damage will be far greater than fixing the leak.

Underbonnet check points

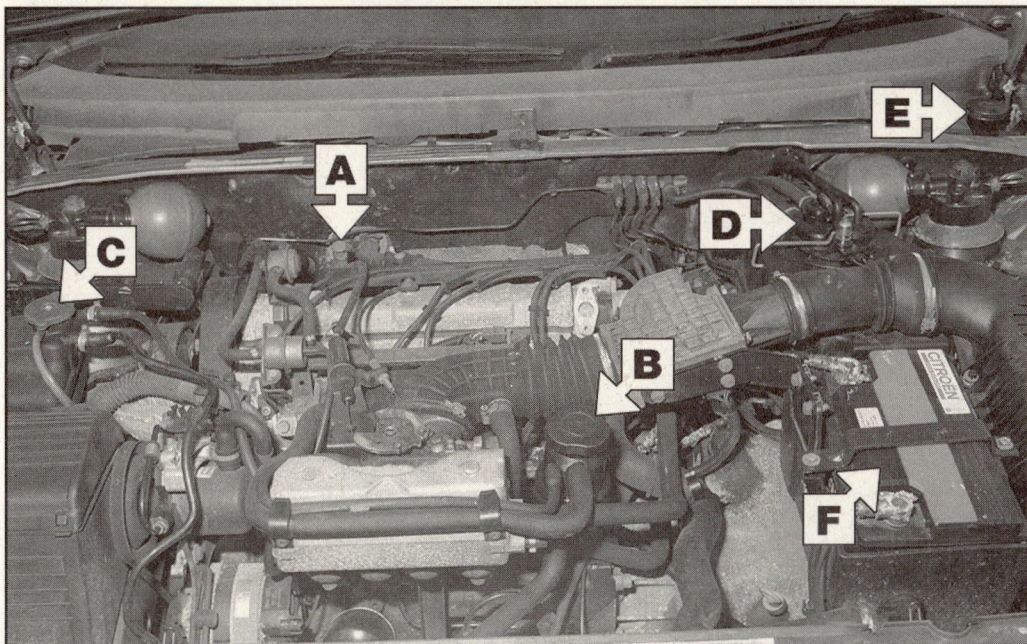

◄ 2.0 litre non-turbo petrol engine

A *Engine oil level dipstick*
B *Engine oil filler cap*
C *Coolant reservoir (expansion tank)*
D *Hydraulic fluid reservoir*
E *Windscreen/tailgate washer fluid reservoir*
F *Battery*

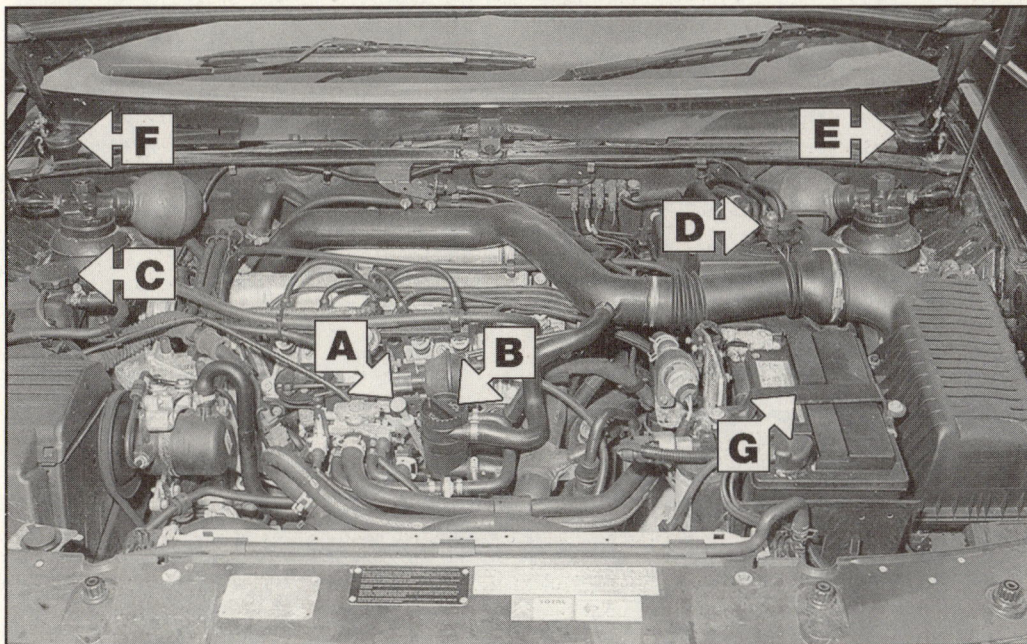

◄ 2.0 litre turbo petrol engine

A *Engine oil level dipstick*
B *Engine oil filler cap*
C *Coolant reservoir (expansion tank)*
D *Hydraulic fluid reservoir*
E *Windscreen/tailgate washer fluid reservoir*
F *Headlight washer fluid reservoir*
G *Battery*

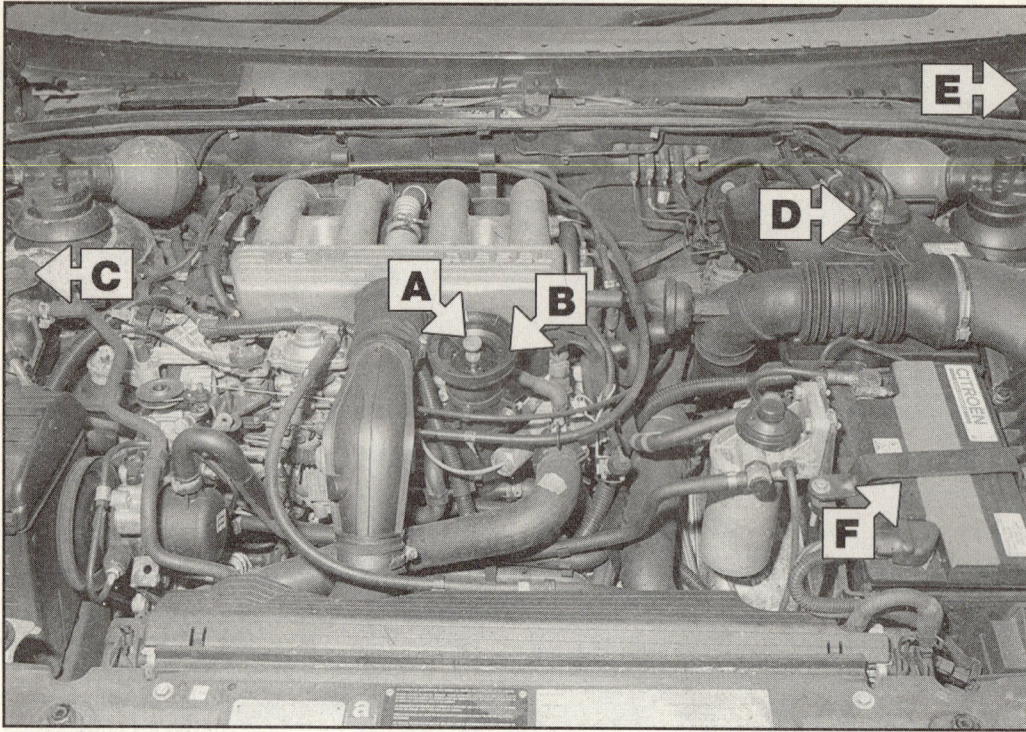

◀ **2.1 litre turbo-diesel engine**

A *Engine oil level dipstick*

B *Engine oil filler cap*

C *Coolant reservoir (expansion tank)*

D *Hydraulic fluid reservoir*

E *Windscreen/tailgate washer fluid reservoir*

F *Battery*

◀ **2.5 litre turbo-diesel engine**

A *Engine oil level dipstick*

B *Engine oil filler cap*

C *Coolant reservoir (expansion tank)*

D *Hydraulic fluid reservoir*

E *Windscreen/tailgate washer fluid reservoir*

F *Headlight washer fluid reservoir*

G *Battery*

Engine oil level

Before you start

✔ Make sure that your car is on level ground.
✔ Check the oil level before the car is driven, or at least 5 minutes after the engine has been switched off.

> **HAYNES HiNT**
> *If the oil level is checked immediately after driving the vehicle, some of the oil will remain in the upper engine components, resulting in an inaccurate reading on the dipstick!*

The correct oil

Modern engines place great demands on their oil. It is very important that the correct oil for your car is used (See "Lubricants, fluids and tyre pressures").

Car care

● If you have to add oil frequently, you should check whether you have any oil leaks. Place some clean paper under the car overnight, and check for stains in the morning. If there are no leaks, then engine may be burning oil (see *Fault Finding*).

● Always maintain the level between the upper and lower dipstick marks (see photo 3). If the level is too low, severe engine damage may occur. Oil seal failure may result if the engine is overfilled by adding too much oil.

1 The dipstick is located in a tube at the front or rear of the engine (see *Underbonnet Check Points* on pages 0•11 and 0•12 for exact location). Withdraw the dipstick.

2 Using a clean rag or paper towel, wipe all the oil from the dipstick. Insert the clean dipstick into the tube as far as it will go, then withdraw it again.

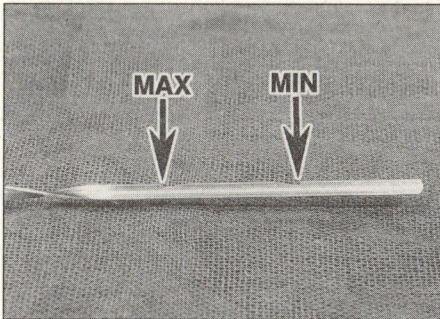

3 Note the oil level on the end of the dipstick, which should be between the upper (MAX) mark and the lower (MIN) mark.

4 Oil is added through the filler cap. Unscrew or pull out the filler cap, then top-up the level. A funnel may help to reduce spillage. Add the oil slowly, checking the level on the dipstick often. Don't overfill.

Washer fluid

Car care

● Screenwash additives not only keep the windscreen clean during bad weather, they also prevent the washer system freezing in cold weather - which is when you are likely to need it most. Don't top up using plain water, as the screenwash will become diluted, and will freeze in cold weather.

⚠ *Warning: On no account use engine coolant antifreeze in the screen washer system - this may damage the paintwork.*

1 The windscreen/tailgate washer fluid reservoir is located at the rear left-hand corner of the engine compartment. If topping-up is necessary, open the cap.

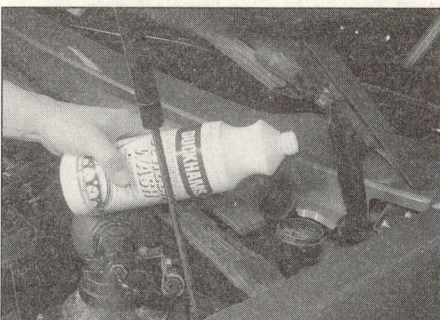

2 When topping-up the reservoir a screenwash additive should be added in the quantities recommended on the bottle.

3 On models with headlight washers, a separate fluid reservoir is located at the rear right-hand corner of the engine compartment.

Hydraulic fluid level

Before you start

✔ Make sure that the car is on level ground.

✔ Cleanliness is of great importance when dealing with the hydraulic system, so take care to clean around the reservoir cap before topping-up. Use only clean LHM hydraulic fluid.

Safety first

● If the reservoir requires repeated topping-up, this is an indication of a fluid leak somewhere in the system, which should be investigated immediately. The XM relies on this main reservoir for the braking, power steering and suspension systems.

● If a leak is suspected, the car should not be driven until the hydraulic systems (brakes, steering and suspension) have been checked. Never take any risks where the hydraulic systems are concerned.

⚠ **Warning: The fluid used in the XM hydraulic system is LHM mineral fluid, which is green in colour. The use of any other type of fluid, including normal brake fluid, will damage the system rubber seals and hoses. Keep LHM fluid carefully sealed in its original container.**

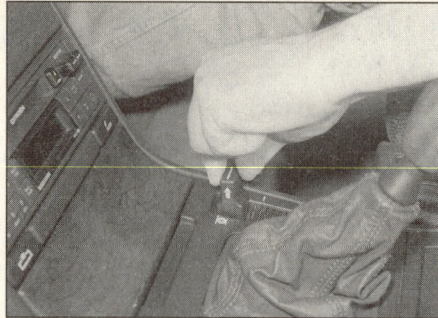

1 With the engine idling, working inside the vehicle, set the suspension height control lever to the 'Maximum' position.

2 The hydraulic fluid reservoir is at the rear of the engine compartment, on the left-hand side. Wipe the area around the filler cap with a clean rag. Locate the level sight-glass on top of the reservoir.

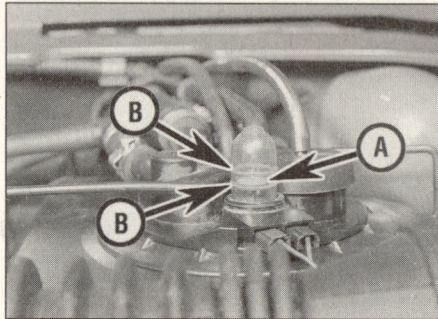

3 The fluid level is indicated by a yellow indicator float in the sight glass. The indicator float (A) must be between the two red rings (B). The level indication is only accurate after the suspension has stabilised at its maximum height.

4 If topping up is necessary, use only genuine green LHM fluid, and make sure that it's clean. The difference between the upper and lower red rings is approximately 0.5 litre. Remove the filler cap and wipe it clean. Add fluid until the indicator reaches the upper mark, then refit the reservoir cap and switch off the engine.

Coolant level

⚠️ **Warning: Do not attempt to remove the expansion tank pressure cap when the engine is hot, as there is a very great risk of scalding. Do not leave open containers of coolant about, as it is poisonous.**

Car care

● With a sealed-type cooling system, adding coolant should not be necessary on a regular basis. If frequent topping-up is required, it is likely there is a leak. Check the radiator, all hoses and joint faces for signs of staining or wetness, and rectify as necessary.

● It is important that antifreeze is used in the cooling system all year round, not just during the winter months. Don't top up with water alone, as the antifreeze will become diluted.

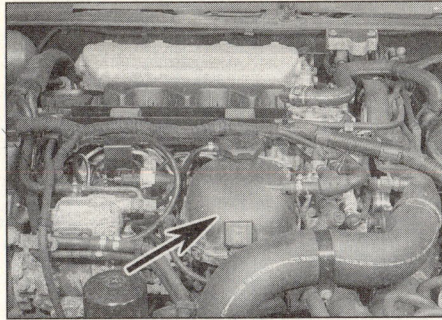

1 The coolant level varies with the temperature of the engine. The level should be checked with the engine cold. The level should be checked in the expansion tank, which is located at the right-hand side of the engine compartment on all except 2.5 litre diesel engine models, and at the front of the engine on 2.5 litre diesel engines. A red level indicator is fitted inside the expansion tank.

3 With a cold engine, the coolant level should be 20 mm above the level indicator.

2 To check the coolant level, wait until the engine is cold, then turn the pressure cap on the expansion tank slowly anti-clockwise, and pause until any pressure remaining in the system is released. Unscrew the cap and lift off.

4 If topping up is necessary, add a mixture of water and antifreeze to the expansion tank until the coolant is up to the specified level. Refit the cap, turning it clockwise as far as it will go until it is secure. Re-check that the cap is securely tightened once the engine is warm.

Wiper blades

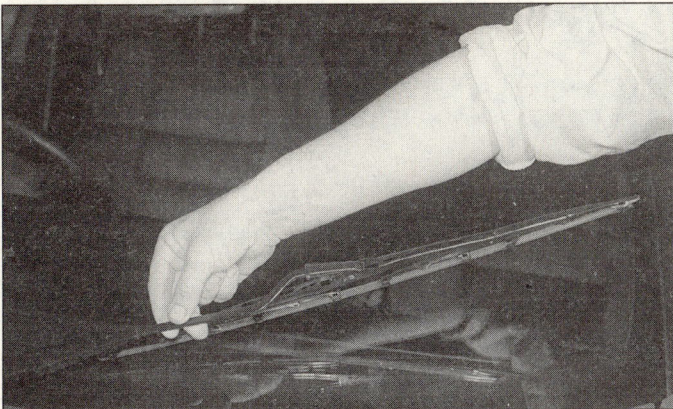

1 Check the condition of the wiper blades; if they are cracked or show any signs of deterioration, or if the glass swept area is smeared, renew them. For maximum clarity of vision, wiper blades should be renewed annually, as a matter of course.

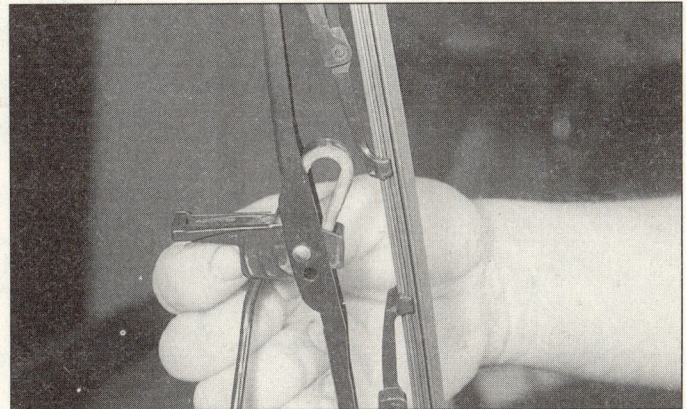

2 To remove a windscreen wiper blade, pull the arm fully away from the screen until it locks. Swivel the blade through 90°, then depress the locking clip at the base of the mounting block, and slide the blade out of the hooked end of the arm. Don't forget to check the tailgate wiper blade as well. The blade can be removed by swivelling the blade through 90°, then sliding the blade from the arm.

Tyre condition and pressure

It is very important that tyres are in good condition, and at the correct pressure - having a tyre failure at any speed is highly dangerous. Tyre wear is influenced by driving style - harsh braking and acceleration, or fast cornering, will all produce more rapid tyre wear. As a general rule, the front tyres wear out faster than the rears. Interchanging the tyres from front to rear ("rotating" the tyres) may result in more even wear. However, if this is completely effective, you may have the expense of replacing all four tyres at once! Remove any nails or stones embedded in the tread before they penetrate the tyre to cause deflation. If removal of a nail does reveal that the tyre has been punctured, refit the nail so that its point of penetration is marked. Then immediately change the wheel, and have the tyre repaired by a tyre dealer.

Regularly check the tyres for damage in the form of cuts or bulges, especially in the sidewalls. Periodically remove the wheels, and clean any dirt or mud from the inside and outside surfaces. Examine the wheel rims for signs of rusting, corrosion or other damage. Light alloy wheels are easily damaged by "kerbing" whilst parking; steel wheels may also become dented or buckled. A new wheel is very often the only way to overcome severe damage.

New tyres should be balanced when they are fitted, but it may become necessary to re-balance them as they wear, or if the balance weights fitted to the wheel rim should fall off. Unbalanced tyres will wear more quickly, as will the steering and suspension components. Wheel imbalance is normally signified by vibration, particularly at a certain speed (typically around 50 mph). If this vibration is felt only through the steering, then it is likely that just the front wheels need balancing. If, however, the vibration is felt through the whole car, the rear wheels could be out of balance. Wheel balancing should be carried out by a tyre dealer or garage.

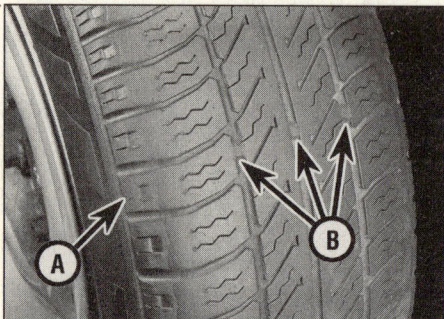

1 Tread Depth - visual check
The original tyres have tread wear safety bands (B), which will appear when the tread depth reaches approximately 1.6 mm. The band positions are indicated by a triangular mark on the tyre sidewall (A).

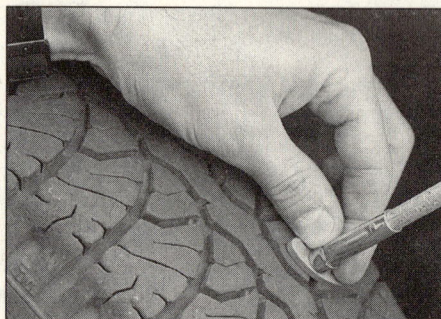

2 Tread Depth - manual check
Alternatively, tread wear can be monitored with a simple, inexpensive device known as a tread depth indicator gauge.

3 Tyre Pressure Check
Check the tyre pressures regularly with the tyres cold. Do not adjust the tyre pressures immediately after the vehicle has been used, or an inaccurate setting will result.

Tyre tread wear patterns

Shoulder Wear

Underinflation (wear on both sides)
Under-inflation will cause overheating of the tyre, because the tyre will flex too much, and the tread will not sit correctly on the road surface. This will cause a loss of grip and excessive wear, not to mention the danger of sudden tyre failure due to heat build-up.
Check and adjust pressures
Incorrect wheel camber (wear on one side)
Repair or renew suspension parts
Hard cornering
Reduce speed!

Centre Wear

Overinflation
Over-inflation will cause rapid wear of the centre part of the tyre tread, coupled with reduced grip, harsher ride, and the danger of shock damage occurring in the tyre casing.
Check and adjust pressures

If you sometimes have to inflate your car's tyres to the higher pressures specified for maximum load or sustained high speed, don't forget to reduce the pressures to normal afterwards.

Uneven Wear

Front tyres may wear unevenly as a result of wheel misalignment. Most tyre dealers and garages can check and adjust the wheel alignment (or "tracking") for a modest charge.
Incorrect camber or castor
Repair or renew suspension parts
Malfunctioning suspension
Repair or renew suspension parts
Unbalanced wheel
Balance tyres
Incorrect toe setting
Adjust front wheel alignment
Note: *The feathered edge of the tread which typifies toe wear is best checked by feel.*

Battery

Caution: Before carrying out any work on the vehicle battery, read the precautions given in 'Safety first' at the start of this manual.

✔ Make sure that the battery tray is in good condition, and that the clamp is tight. Corrosion on the tray, retaining clamp and the battery itself can be removed with a solution of water and baking soda. Thoroughly rinse all cleaned areas with water. Any metal parts damaged by corrosion should be covered with a zinc-based primer, then painted.

✔ Periodically (approximately every three months), check the charge condition of the battery as described in Chapter 5A.

✔ If the battery is flat, and you need to jump start your vehicle, see *Roadside Repairs*.

1 The battery is located at the left-hand side of the engine compartment. The exterior of the battery should be inspected periodically for damage such as a cracked case or cover.

2 Check the tightness of the battery cable clamps to ensure good electrical connections. You should not be able to move them. Also check each cable for cracks and frayed conductors.

HAYNES HINT

Battery corrosion can be kept to a minimum by applying a layer of petroleum jelly to the clamps and terminals after they are reconnected.

3 If corrosion (white fluffy deposits) is evident, remove the cables from the terminals, clean them with a small wire brush, then refit them. Automotive stores sell a useful tool for cleaning the battery post and terminals.

Electrical systems

✔ Check all external lights and the horn. Refer to the appropriate Sections of Chapter 13 for details if any of the circuits are found to be inoperative.

✔ Visually check all accessible wiring connectors, harnesses and retaining clips for security, and for signs of chafing or damage.

HAYNES HINT

If you need to check your brake lights and indicators unaided, back up to a wall or garage door and operate the lights. The reflected light should show if they are working properly.

1 If a single indicator light, brake light or headlight has failed, it is likely that a bulb has blown and will need to be replaced. Refer to Chapter 13 for details. If both brake lights have failed, it is possible that the brake light switch operated by the brake pedal has failed. Refer to Chapter 10 for details.

2 If more than one indicator light or headlight has failed, it is likely that either a fuse has blown or that there is a fault in the circuit (see *Electrical fault finding* in Chapter 13). The main fuses are mounted in a panel located at the lower left-hand side of the facia under a cover. Pull the cover to release it from the facia. Additional fuses are located in the auxiliary fusebox at the front left-hand corner of the engine compartment.

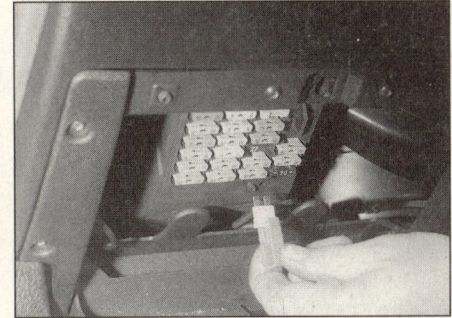

3 To replace a blown fuse, remove it, where applicable, using the plastic tool provided. Fit a new fuse of the same rating, available from car accessory shops. It is important that you find the reason that the fuse blew (see *Electrical fault finding* in Chapter 13).

Lubricants and fluids

Petrol engine . . . . Multigrade engine oil, viscosity SAE 5W/40 to 20W/50, to API SG or better
(Duckhams Hypergrade Petrol Engine Oil, or Duckhams QXR Premium Petrol Engine Oil)

Diesel engine . . . . Multigrade engine oil, viscosity SAE 5W/40 to 20W/50, to API CD or better
(Duckhams Hypergrade Diesel Engine Oil, or Duckhams QXR Premium Diesel Engine Oil)

Cooling system . . . . Ethylene glycol-based antifreeze
(Duckhams Antifreeze and Summer Coolant)

Manual transmission . . . . Synthetic gear oil, viscosity 75W/80
(Duckhams Hypoid Gear Oil 75W-80W GL5)

Automatic transmission . . . . Dexron II type automatic transmission fluid
(Duckhams ATF Autotrans III)

Hydraulic system . . . . LHM mineral fluid *(Duckhams LHM Fluid)*

Choosing your engine oil

Engines need oil, not only to lubricate moving parts and minimise wear, but also to maximise power output and to improve fuel economy. By introducing a simplified and improved range of engine oils, Duckhams has taken away the confusion and made it easier for you to choose the right oil for your engine.

HOW ENGINE OIL WORKS

• Beating friction

Without oil, the moving surfaces inside your engine will rub together, heat up and melt, quickly causing the engine to seize. Engine oil creates a film which separates these moving parts, preventing wear and heat build-up.

• Cooling hot-spots

Temperatures inside the engine can exceed 1000° C. The engine oil circulates and acts as a coolant, transferring heat from the hot-spots to the sump.

• Cleaning the engine internally

Good quality engine oils clean the inside of your engine, collecting and dispersing combustion deposits and controlling them until they are trapped by the oil filter or flushed out at oil change.

OIL CARE - FOLLOW THE CODE

To handle and dispose of used engine oil safely, always:

• **Avoid skin contact with used engine oil.** Repeated or prolonged contact can be harmful.
• **Dispose of used oil and empty packs in a responsible manner in an authorised disposal site.** Call 0800 663366 to find the one nearest to you. Never tip oil down drains or onto the ground.

OIL CARE
FOLLOW THE CODE
OIL BANK LINE
0800 66 33 66

DUCKHAMS ENGINE OILS

For the driver who demands a premium quality oil for complete reassurance, we recommend synthetic formula **Duckhams QXR Premium Engine Oils**.
For the driver who requires a straightforward quality engine oil, we recommend **Duckhams Hypergrade Engine Oils**.

For further information and advice, call the Duckhams UK Helpline on 0800 212988.

DUCKHAMS

Tyre pressures

Note: *Pressures given here are a guide only, and apply to original-equipment tyres – the recommended pressures may vary if any other make or type of tyre is fitted; check with the vehicle handbook, or the tyre manufacturer or supplier for latest recommendations.*

	Front	Rear
Hatchback models:		
All except turbo petrol engine models	33 psi (2.3 bars)	28 psi (1.9 bars)
Turbo petrol engine models	35 psi (2.4 bars)	29 psi (2.0 bars)
Estate models:		
2.1 litre diesel engine models:		
Manual transmission models	35 psi (2.4 bars)	35 psi (2.4 bars)
Automatic transmission models	33 psi (2.3 bars)	33 psi (2.3 bars)
2.5 litre diesel engine models	33 psi (2.3 bars)	33 psi (2.3 bars)
Petrol engine models except turbo	35 psi (2.4 bars)	35 psi (2.4 bars)
Turbo petrol engine models	33 psi (2.3 bars)	33 psi (2.3 bars)

Chapter 1 Part A:
Routine maintenance and servicing - petrol engine models

Contents

Degrees of difficulty

Easy, suitable for novice with little experience	**Fairly easy,** suitable for beginner with some experience	**Fairly difficult,** suitable for competent DIY mechanic	**Difficult,** suitable for experienced DIY mechanic	**Very difficult,** suitable for expert DIY or professional

Lubricants and fluids

Refer to the end of *Weekly checks*

Capacities (approximate)

Engine oil

Non-turbo models . . . . 5.40 litres

Turbo models:

    Models without air conditioning . . . . 4.75 litres

    Models with air conditioning . . . . 4.50 litres

Cooling system

Non-turbo models . . . . 7.30 litres

Turbo models . . . . 10.80 or 11.30 litres (according to equipment)

Transmission

Manual transmission . . . . 1.90 litres

Automatic transmission:

    From dry . . . . 7.50 litres

    Fluid change . . . . 2.40 litres

Hydraulic LHM fluid

All models . . . . 5.40 litres

Fuel tank

All models . . . . 80.00 litres

Engine

Oil filter:

    Models up to June 1994 . . . . Champion F104

    Models from July 1994 . . . . Champion F118

Cooling system

Antifreeze mixture:

    28% antifreeze . . . . Protection down to -15°C (5°F)

    50% antifreeze . . . . Protection down to -30°C (-22°F)

Note: *Refer to antifreeze manufacturer's recommendations*

Fuel system

Air filter element . . . . Champion U561

Fuel filter . . . . Champion L206

Ignition system

Spark plugs:	Type	Electrode gap*
Non-turbo models without catalytic converter	Champion RC7YCC	0.8 mm
Non-turbo models with catalytic converter	Champion RC9YCC	0.8 mm
Turbo models, engine code RGY	Champion RC7BMC	Not adjustable
Turbo models, engine code RGX	Champion RC8DMC	Not adjustable

Ignition HT lead set:

    Models with distributor ignition system . . . . Champion LS-55

    Models with distributorless ignition system . . . . Champion LS-51

*The spark plug gap quoted is that recommended by Champion for their specified plugs listed above. If spark plugs of any other type are to be fitted, refer to their manufacturer's recommendations

Brakes

Brake pad friction material minimum thickness:

    Front pads . . . . 3.0 mm

    Rear pads . . . . 2.0 mm

Wiper blades

Front . . . . Champion X55

Rear:

    Hatchback models . . . . Champion X45

    Estate models . . . . Champion X41

Clutch

Clutch pedal travel . . . . 45 -0+10 mm

Tyre pressures

See end of *Weekly checks*

Torque wrench settings

	Nm	lbf ft
Roadwheel bolts	90	66
Spark plugs	25	18
Manual transmission oil filler/level plug	20	15
Manual transmission oil drain plug	30	22

1 The maintenance intervals in this manual are provided with the assumption that you, not the dealer, will be carrying out the work. These are the minimum maintenance intervals recommended by us for vehicles driven daily. If you wish to keep your vehicle in peak condition at all times, you may wish to perform some of these procedures more often. We encourage frequent maintenance, because it enhances the efficiency, performance and resale value of your vehicle.

2 If the vehicle is driven in very dusty areas, used to tow a trailer, spends long periods with the engine idling, is driven frequently at slow speeds (eg in heavy traffic) or is used mainly for short journeys, then *more frequent* maintenance intervals are recommended.

3 When the vehicle is new, it should be serviced by a factory-authorised dealer service department, in order to preserve the factory warranty.

Every 250 miles (400 km) or weekly

☐ Refer to *Weekly checks*.

Every 9000 miles or 12 months, whichever comes first

☐ Renew the engine oil and filter (Section 3)
☐ Renew the automatic transmission fluid - where applicable (Section 4)
☐ Check the clutch pedal height – models with cable-operated clutch (Section 5)
☐ Check all components, pipes and hoses for fluid leaks (Section 6)
☐ Check the condition and security of the steering and suspension components (Section 7)
☐ Check the condition of the driveshafts (Section 8)
☐ Lubricate all hinges and locks (Section 9)
☐ Check the fault diagnosis system memory for fault codes (Section 10)

Every 18 000 miles

☐ Renew the air filter element - Turbo models (Section 11)
☐ Renew the spark plugs (Section 12)
☐ Check the condition of the front and rear brake pads and discs (Section 13)

Every 2 years, regardless of mileage

☐ Renew the coolant (Section 14)

Every 36 000 miles

☐ Renew the air filter element - normally-aspirated models (Section 15)
☐ Renew the hydraulic fluid and clean the hydraulic fluid return filters (Section 16)
☐ Renew the fuel filter (Section 17)
☐ Check the manual transmission oil level (Section 18)
☐ Check the condition of the auxiliary drivebelts (Section 19)
☐ Clean the automatic transmission fluid strainer - where applicable (Section 20)
☐ Carry out a road test (Section 21)

Every 72 000 miles

☐ Renew the timing belt (Section 22)

Note: *Although this is the normal interval for timing belt renewal, it is strongly recommended that the interval is reduced on vehicles which are subjected to intensive use, ie, mainly short journeys or a lot of stop-start driving. The actual belt renewal is therefore very much up to the individual owner, but bear in mind that severe engine damage may result if the belt breaks*

1A

Underbonnet view of a 2.0 litre turbo petrol engine

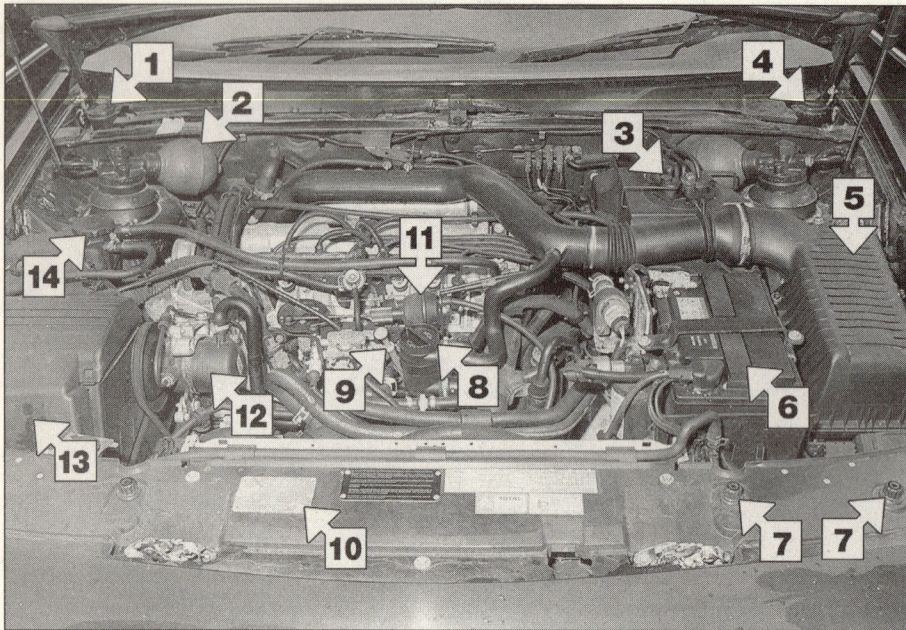

1 Headlight washer fluid reservoir
2 Air suspension hydraulic fluid reservoir bulb
3 Hydraulic fluid reservoir
4 Windscreen/tailgate washer fluid reservoir
5 Air cleaner housing
6 Battery
7 Headlight beam adjustment screws
8 Engine oil filler cap
9 Engine oil level dipstick
10 VIN plate
11 Cruise control vacuum actuator
12 Hydraulic fluid pump
13 Electronic control unit housing
14 Coolant reservoir (expansion tank)

Underbonnet view of a 2.0 litre non-turbo petrol engine

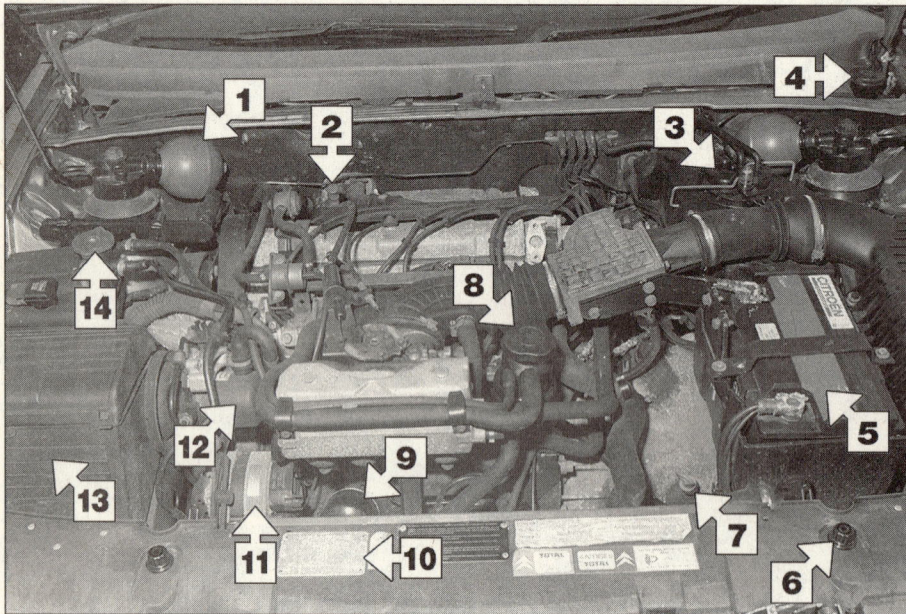

1 Air suspension hydraulic fluid reservoir bulb
2 Engine oil level dipstick
3 Hydraulic fluid reservoir
4 Windscreen/tailgate washer fluid reservoir
5 Battery
6 Headlight beam adjustment screw
7 Automatic transmission fluid level dipstick
8 Engine oil filler cap
9 Engine oil filter
10 VIN plate
11 Alternator
12 Hydraulic fluid pump
13 Electronic control unit housing
14 Coolant reservoir (expansion tank)

Front underbody view (diesel engine model shown, petrol engine model similar)

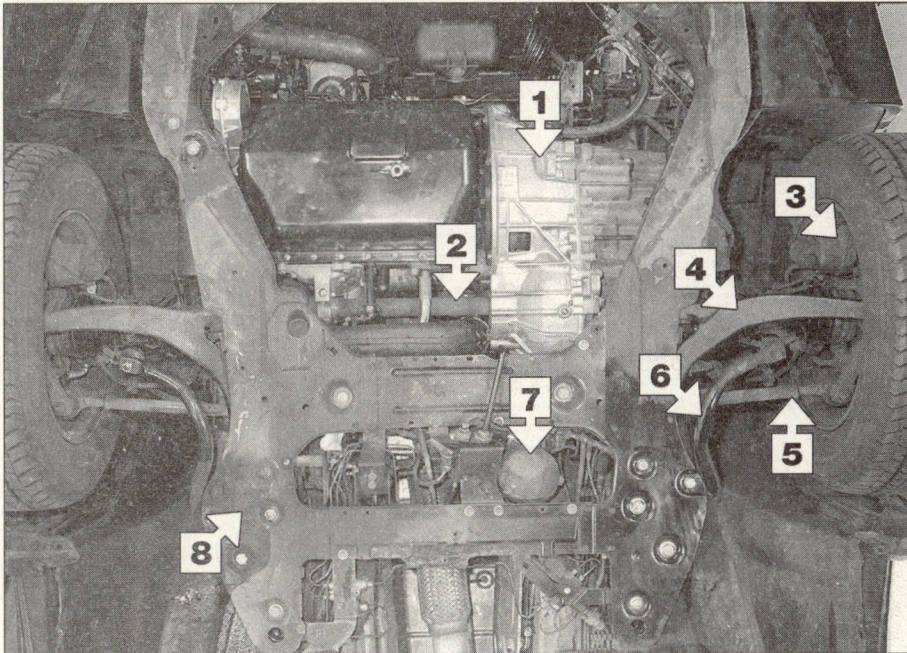

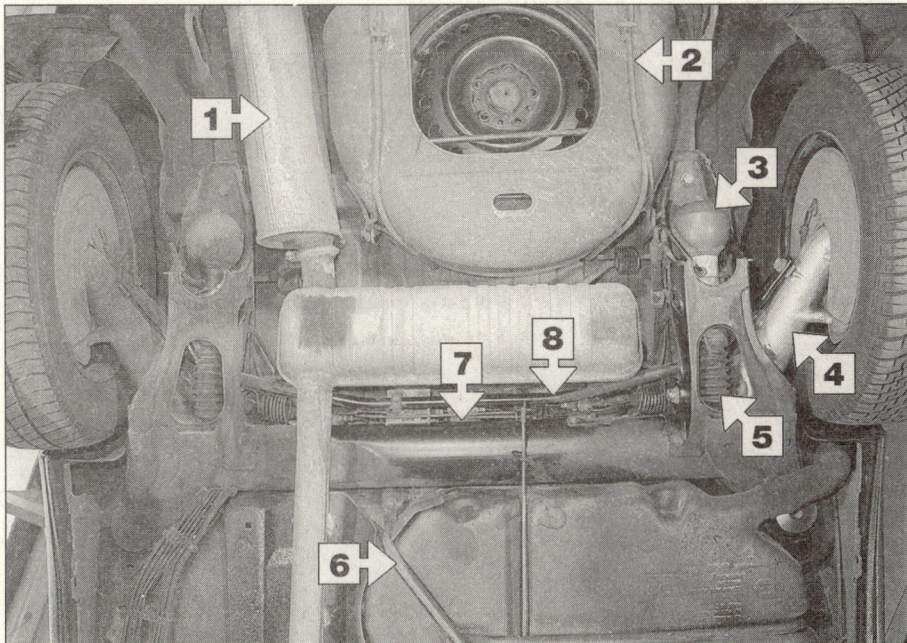

1 Manual transmission
2 Driveshaft
3 Brake caliper
4 Front suspension lower arm
5 Track-rod
6 Anti-roll bar
7 Suspension soft-setting hydraulic fluid reservoir bulb
8 Front suspension subframe

Rear underbody view (diesel engine model shown, petrol engine model similar)

1 Exhaust expansion box
2 Spare wheel carrier
3 Hydraulic fluid reservoir bulb
4 Rear suspension trailing arm
5 Rear suspension hydraulic unit
6 Fuel tank
7 Rear suspension crosstube
8 Rear anti-roll bar

1A

1 Introduction

1 This Chapter is designed to help the home mechanic maintain his/her vehicle for safety, economy, long life and peak performance.

2 The Chapter contains a maintenance schedule, followed by Sections dealing specifically with each task in the schedule. Visual checks, adjustments, component renewal and other helpful items are included. Refer to the accompanying illustrations of the engine compartment and the underside of the vehicle for the locations of the various components.

3 Servicing your vehicle in accordance with the above recommendations and the following Sections will provide a planned maintenance programme, which should result in a long and reliable service life. This is a comprehensive plan, so maintaining some items but not others at the specified service intervals, will not produce the same results.

4 As you service your vehicle, you will discover that many of the procedures can - and should - be grouped together, because of the particular procedure being performed, or because of the proximity of two otherwise-unrelated components to one another. For example, if the vehicle is raised for any reason, the exhaust can be inspected at the same time as the suspension and steering components.

5 The first step in this maintenance programme is to prepare yourself before the actual work begins. Read through all the Sections relevant to the work to be carried out, then make a list and gather all the parts and tools required. If a problem is encountered, seek advice from a parts specialist, or a dealer service department.

2 Regular maintenance

1 If, from the time the vehicle is new, the routine maintenance schedule is followed closely, and frequent checks are made of fluid levels and high-wear items, as suggested throughout this manual, the engine will be kept in relatively good running condition, and the need for additional work will be minimised.

2 It is possible that there will be times when the engine is running poorly due to the lack of regular maintenance. This is even more likely if a used vehicle, which has not received regular and frequent maintenance checks, is purchased. In such cases, additional work may need to be carried out, outside of the regular maintenance intervals.

3 If engine wear is suspected, a compression test (refer to the relevant Part of Chapter 2) will provide valuable information regarding the overall performance of the main internal components. Such a test can be used as a basis to decide on the extent of the work to be carried out. If, for example, a compression test indicates serious internal engine wear, conventional maintenance as described in this Chapter will not greatly improve the performance of the engine, and may prove a waste of time and money, unless extensive overhaul work is carried out first.

4 The following series of operations are those most often required to improve the performance of a generally poor-running engine:

Primary operations

a) Clean, inspect and test the battery (See Weekly checks and Chapter 5A).
b) Check all the engine-related fluids (See Weekly checks).
c) Check the condition and tension of the auxiliary drivebelt(s) (Section 19).
d) Renew the spark plugs (Section 12).
e) Check the condition of the air filter, and renew if necessary (Sections 11 and 15).
f) Check the fuel filter (Section 17).
g) Check the condition of all hoses, and check for fluid leaks (Section 6).

5 If the above operations do not prove fully effective, carry out the following secondary operations:

Secondary operations

All items listed under Primary operations, plus the following:

a) Check the charging system (Chapter 5A).
b) Check the ignition system (see relevant Part of Chapter 5).
c) Check the fuel system (see relevant Part of Chapter 4).

Every 9000 miles or 12 months, whichever comes first

3 Engine oil and filter renewal

Note: A suitable square-section wrench may be required to undo the sump drain plug on some models. These wrenches can be obtained from most motor factors or your Citroën dealer.

1 Frequent oil and filter changes are the most important preventative maintenance procedures which can be undertaken by the DIY owner. As engine oil ages, it becomes diluted and contaminated, which leads to premature engine wear.

2 Before starting this procedure, gather together all the necessary tools and materials. Make sure that you have plenty of clean rags and newspapers handy, to mop up any spills, and a container of suitable size to drain the oil into. Ideally, the engine oil should be warm, as it will drain better, and more built-up sludge will be removed with it. Take care, however, not to touch the exhaust or any other hot parts of the engine when working under the vehicle. To avoid any possibility of scalding, and to protect yourself from possible skin irritants and other harmful contaminants in used engine oils, it is advisable to wear gloves when carrying out this work. Access to the underside of the vehicle will be greatly improved if it can be raised on a lift, driven onto ramps, or jacked up and supported on axle stands (see Jacking and Vehicle Support). Whichever method is chosen, make sure that the vehicle remains level, or if it is at an angle, that the drain plug is at the lowest point. Where necessary remove the splash guard from under the engine.

3 Slacken the drain plug about half a turn; on some models, a square-section wrench may be needed to slacken the plug. Position the draining container under the drain plug, then remove the plug completely. If possible, try to keep the plug pressed into the sump while unscrewing it by hand the last couple of turns (see illustration and Haynes Hint).

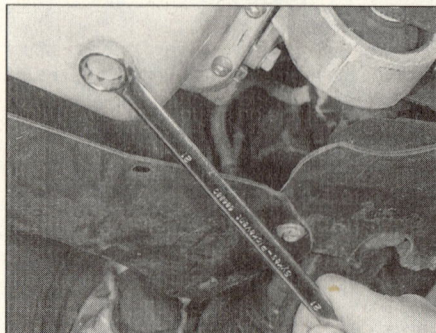

3.3 Slackening the sump drain plug

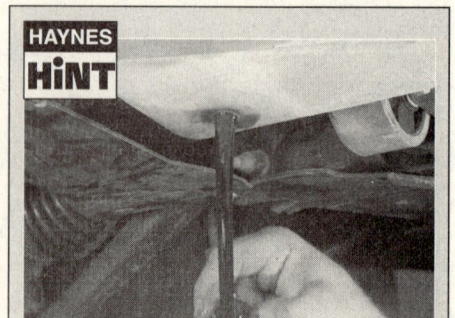

HAYNES HiNT

As the drain plug releases from the threads, move it away sharply, so that the stream of oil issuing from the sump runs into the container - not down your arm.

3.8a Using an oil filter removal tool, slacken the filter . . .

3.8b . . . then unscrew it by hand the rest of the way

4 Recover the sealing ring from the drain plug.

5 Allow some time for the old oil to drain, noting that it may be necessary to reposition the container as the oil flow slows to a trickle.

6 After all the oil has drained, wipe off the drain plug with a clean rag, and fit a new sealing washer. Clean around the drain plug opening, then refit and tighten the plug.

7 If the filter is also to be renewed, move the container into position under the oil filter. On all models, the filter is located on the front side of the cylinder block.

8 Using an oil filter removal tool if necessary, slacken the filter initially, then unscrew it by hand the rest of the way **(see illustrations)**. If any oil remains in the old filter, empty it into the container.

9 Use a clean rag to remove all oil, dirt and sludge from the filter sealing area on the engine. Check the old filter to make sure that the rubber sealing ring hasn't stuck to the engine. If it has, carefully remove it.

10 Apply a light coating of engine oil to the sealing ring on the new filter, then screw it onto the engine. Tighten the filter firmly by hand only - **do not** use any tools. If necessary, refit the splash guard under the engine.

11 Remove the old oil and all tools from under the car, then lower the car to the ground (if applicable).

12 Remove the dipstick, then unscrew the oil filler cap from the cylinder head cover. Fill the engine, using the correct grade and type of oil (*see Weekly checks*). An oil can spout or funnel may help to reduce spillage. Pour in half the specified quantity of oil first, then wait a few minutes for the oil to fall to the sump. Continue adding oil a small quantity at a time until the level is up to the lower mark on the dipstick. Adding approximately 1.5 litres will bring the level up to the upper mark on the dipstick. Refit the filler cap.

13 Start the engine and run it for a few minutes; check for leaks around the oil filter seal and the sump drain plug. There may be a delay of a few seconds before the oil pressure warning light goes out when the engine is first started, as the oil circulates through the engine oil galleries and the new oil filter (where fitted) before the pressure builds up.

14 Switch off the engine, and wait a few minutes for the oil to settle in the sump once more. With the new oil circulated and the filter completely full, recheck the level on the dipstick, and add more oil as necessary.

15 Dispose of the used engine oil safely, with reference to *General Repair Procedures*.

4 Automatic transmission fluid renewal

1 Take the vehicle on a short run, to warm the transmission up to normal operating temperature.

2 Park the car on level ground, then switch off the ignition and apply the parking brake firmly. For improved access, chock the rear wheels then jack up the front of the car and support it securely on axle stands (see *Jacking and Vehicle Support*). When refilling and checking the fluid level, the car must be lowered to the ground, and level, to ensure accuracy.

3 Remove the dipstick, then position a suitable container under the transmission. The transmission unit has one drain plug, located at the bottom of the differential housing.

⚠ *Warning: If the fluid is hot, take precautions against scalding.*

4 Unscrew the drain plug, and allow the fluid to drain completely into the container. Clean the drain plug, being especially careful to wipe any metallic particles off the magnetic insert. Discard the original sealing washers; these should be renewed whenever they are disturbed.

5 When the fluid has finished draining, clean the drain plug threads and those of the transmission casing. Fit a new sealing washer to the drain plug, and refit the plug to the transmission, tightening it securely. If the car was raised for the draining operation, now lower it to the ground. Make sure that the car is level (front-to-rear and side-to-side).

6 Refilling the transmission is an awkward operation, adding the specified type of fluid to the transmission a little at a time via the dipstick tube. Alternatively, use the filler cap (breather) having cleaned around the area first. Use a funnel with a fine-mesh gauze, to avoid spillage, and to ensure that no foreign matter enters the transmission. Allow plenty of time for the fluid level to settle properly.

7 Once the level is up to the MAX mark on the dipstick, refit the dipstick. Start the engine, and allow it to idle for a few minutes in P, then recheck the level, topping-up if necessary. Take the car on a short run to fully distribute the new fluid around the transmission, then recheck the fluid level.

5 Clutch pedal height check – models with cable-operated clutch

1 The clutch adjustment is checked by measuring the clutch pedal travel. If a new cable has been fitted, settle it in position by depressing the clutch pedal at least thirty times.

2 Ensure that there are no obstructions beneath the clutch pedal then measure the distance (L1) from the centre of the clutch pedal pad to the base of the steering wheel with the pedal in the at-rest position. Depress the clutch pedal fully to the floor, and measure the distance (L2) from the centre of the clutch pedal pad to the base of the steering wheel **(see illustration)**.

3 Subtract the first measurement from the second to obtain the clutch pedal travel. If this is not with the range given in the Specifications at the start of this Chapter, adjust the clutch as follows.

4 The clutch cable is adjusted by means of the adjuster nut on the transmission end of the cable. Access to the locknut is limited and, if required, the air cleaner duct or housing component can be removed or disconnected to improve access (refer to Chapter 4A). Access can be further

1A

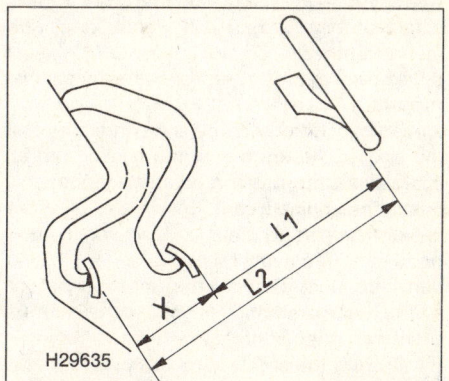

5.2 Clutch pedal height measurement

L1 Pedal height at rest
L2 Pedal height fully depressed
X   Clutch pedal travel

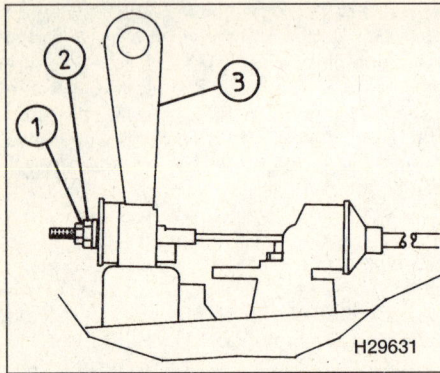

5.4 Clutch cable adjustment details

1	Locknut	3	Clutch release
2	Adjustment nut		arm

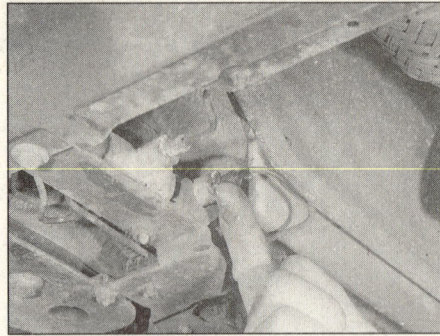

6.2 Check the hydraulic fluid pipes for leaks

A leak in the cooling system will usually show up as white or rust-coloured deposits on the area adjoining the leak

impoved by removing the battery and its holder (refer to Chapter 5A) (**see illustration**).

5 Working in the engine compartment, slacken the locknut from the end of the clutch cable. Adjust the position of the adjuster nut, then depress the clutch pedal ten times and re-measure the clutch pedal travel. Repeat this procedure until the clutch pedal travel is as specified.

6 Once the adjuster nut is correctly positioned, and the pedal travel is correctly set, securely tighten the cable locknut. Where necessary, refit any disturbed air cleaner duct/housing components (see Chapter 4A).

6 Hose and fluid leak check

1 Visually inspect the engine joint faces, gaskets and seals for any signs of water or oil leaks. Pay particular attention to the areas around the cylinder head cover, cylinder head, oil filter and sump joint faces. Bear in mind that, over a period of time, some very slight seepage from these areas is to be expected - what you are really looking for is any indication of a serious leak. Should a leak be found, renew the offending gasket or oil seal by referring to the appropriate Chapters in this manual.

2 Also check the security and condition of all the engine-related pipes and hoses, and all hydraulic and braking system pipes and hoses (**see illustration**). Ensure that all cable ties or securing clips are in place, and in good condition. Clips which are broken or missing can lead to chafing of the hoses, pipes or wiring, which could cause more serious problems in the future.

3 Carefully check the radiator hoses and heater hoses along their entire length . Renew any hose which is cracked, swollen or deteriorated. Cracks will show up better if the hose is squeezed. Pay close attention to the hose clips that secure the hoses to the

cooling system components. Hose clips can pinch and puncture hoses, resulting in cooling system leaks. If the crimped-type hose clips are used, it may be a good idea to replace them with standard worm-drive clips.

4 Inspect all the cooling system components (hoses, joint faces, etc) for leaks (**see Haynes Hint**).

5 Where any problems are found on system components, renew the component or gasket with reference to Chapter 3.

6 With the vehicle raised, inspect the fuel tank and filler neck for punctures, cracks and other damage. The connection between the filler neck and tank is especially critical. Sometimes a rubber filler neck or connecting hose will leak due to loose retaining clamps or deteriorated rubber.

7 Carefully check all rubber hoses and metal fuel lines leading away from the fuel tank. Check for loose connections, deteriorated hoses, crimped lines, and other damage. Pay particular attention to the vent pipes and hoses, which often loop up around the filler neck and can become blocked or crimped. Follow the lines to the front of the vehicle, carefully inspecting them all the way. Renew damaged sections as necessary. Similarly, whilst the vehicle is raised, take the opportunity to inspect all underbody brake fluid pipes and hoses.

8 From within the engine compartment,

check the security of all fuel, vacuum and brake hose attachments and pipe unions, and inspect all hoses for kinks, chafing and deterioration.

9 Check the condition of the power steering fluid pipes and hoses and, where applicable, the automatic transmission fluid cooler pipes and hoses.

7 Steering and suspension check

Suspension and steering check

1 Raise the front of the vehicle, and securely support it on axle stands (see *Jacking and Vehicle Support*).

2 Visually inspect the balljoint dust covers and the steering gear gaiters for splits, chafing or deterioration (**see illustrations**). Any wear of these components will cause loss of lubricant, together with dirt and water entry, resulting in rapid deterioration of the balljoints or steering gear.

3 Check the hydraulic fluid hoses and pipes for chafing or deterioration, and the pipe and hose unions for fluid leaks. Also check for signs of fluid leakage under pressure from the steering gear rubber gaiters, which would

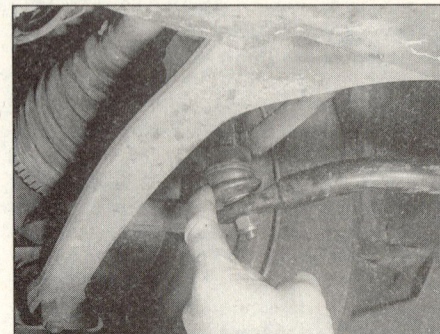

7.2a Inspect the balljoint dust covers . . .

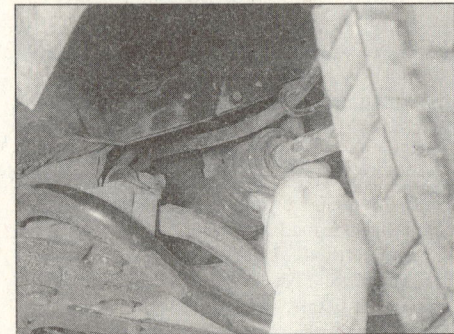

7.2b . . . and the steering gear gaiters

7.4 Check for wear in the hub bearings by grasping the wheel and trying to rock it

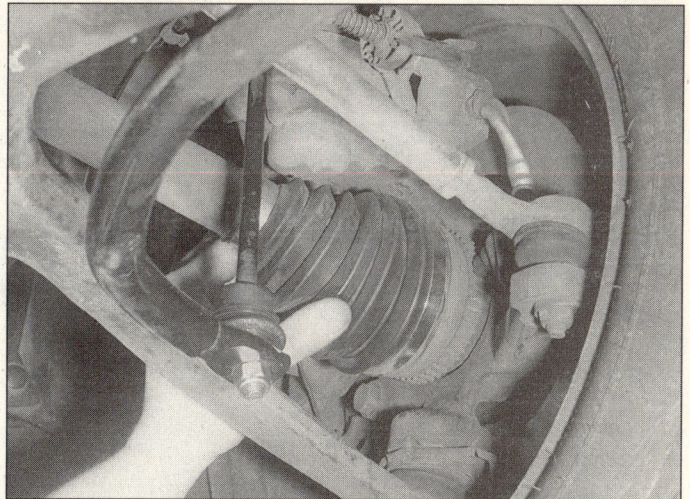

8.1 Check the condition of the driveshaft gaiters

indicate failed fluid seals within the steering gear.

4 Grasp the roadwheel at the 12 o'clock and 6 o'clock positions, and try to rock it **(see illustration)**. Very slight free play may be felt, but if the movement is appreciable, further investigation is necessary to determine the source. Continue rocking the wheel while an assistant depresses the footbrake. If the movement is now eliminated or significantly reduced, it is likely that the hub bearings are at fault. If the free play is still evident with the footbrake depressed, then there is wear in the suspension joints or mountings.

5 Now grasp the wheel at the 9 o'clock and 3 o'clock positions, and try to rock it as before. Any movement felt now may again be caused by wear in the hub bearings or the steering track-rod balljoints. If the outer balljoint is worn, the visual movement will be obvious. If the inner joint is suspect, it can be felt by placing a hand over the rack-and-pinion rubber gaiter and gripping the track-rod. If the wheel is now rocked, movement will be felt at the inner joint if wear has taken place.

6 Using a large screwdriver or flat bar, check for wear in the suspension mounting bushes by levering between the relevant suspension component and its attachment point. Some movement is to be expected, as the mountings are made of rubber, but excessive wear should be obvious. Also check the condition of any visible rubber bushes, looking for splits, cracks or contamination of the rubber.

7 With the car standing on its wheels, have an assistant turn the steering wheel back and forth, about an eighth of a turn each way. There should be very little, if any, lost movement between the steering wheel and roadwheels. If this is not the case, closely observe the joints and mountings previously described. In addition, check the steering

column universal joints for wear, and also check the rack-and-pinion steering gear itself.

Suspension strut/hydraulic unit check

8 Check for any signs of fluid leakage around the suspension strut/hydraulic unit body, or from the rubber gaiter around the piston rod. Should any fluid be noticed, the suspension strut/hydraulic unit may be defective internally, and may need renewing (although leaky connectors could be the cause).

9 Check the condition and security of all hydraulic pipes and hoses. If any signs of leakage are found, investigate the cause. Pipe and hose renewal is described in Chapter 9.

10 Similarly, check around all hydraulic valves and connectors for signs of leakage.

11 If it is suspected that there is a fault in the operation of the suspension, the system can be checked by a Citroën dealer using specialist test equipment.

8 Driveshaft check

1 With the vehicle raised and securely supported on stands, turn the steering onto full lock then slowly rotate the roadwheel. Inspect the condition of the outer constant velocity (CV) joint rubber gaiters while squeezing the gaiters to open out the folds **(see illustration)**. Check for signs of cracking, splits or deterioration of the rubber which may allow the grease to escape and lead to water and grit entry into the joint. Also check the security and condition of the retaining clips. Repeat these checks on the inner CV joints. If any damage or deterioration is found, the gaiters should be renewed as described in Chapter 8.

2 At the same time check the general condition

of the CV joints themselves by first holding the driveshaft and attempting to rotate the wheel. Repeat this check by holding the inner joint and attempting to rotate the driveshaft. Any appreciable movement indicates wear in the joints, wear in the driveshaft splines or loose driveshaft retaining nut.

9 Hinge and lock lubrication

1 Work around the vehicle and lubricate the hinges of the bonnet, doors and tailgate with a light machine oil.

2 Lightly lubricate the bonnet release mechanism and exposed section of inner cable with a smear of grease.

3 Check carefully the security and operation of all hinges, latches and locks, adjusting them where required. Check the operation of the central locking system (if fitted).

4 Check the condition and operation of the tailgate and bonnet struts, renewing them if there is any evidence of leakage, or if the struts are no longer able to support the tailgate or bonnet securely when raised.

10 Fault diagnosis system memory check

The vehicle fault diagnosis systems (for the engine management system, suspension control system, and ABS, where applicable) can be analysed and checked for stored fault codes, which will indicate any faults which have occurred in the relevant systems.

This check must be carried out by a Citroën dealer using the appropriate specialist diagnostic equipment.

1A

11.1 Slacken the retaining screws around the edge of the air cleaner

11.2 Lift the filter element from the air cleaner casing

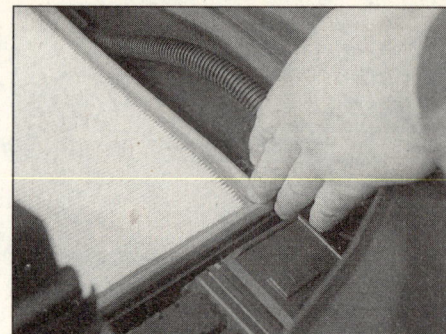

11.4 Fit the new element in position in the air cleaner, pressing the rubber seal firmly into the recess at the edge of the casing

Every 18 000 miles

11 Air filter element renewal - Turbo models

1 Slacken and withdraw the retaining screws around the edge of the air cleaner (**see illustration**), then remove the cover.
2 Lift the filter element from the air cleaner casing (**see illustration**). Check that the replacement element is the same, before discarding the old one.
3 Remove any dirt or debris from the inside of the air cleaner casing, using a brush or rag.
4 Fit the new element in position in the air cleaner, pressing the rubber edging firmly into the recess in the air cleaner casing (**see illustration**).
5 Refit the air cleaner cover, securing it in position with its retaining screws.
6 Reconnect the intake duct to the rear of the air cleaner cover, and securely tighten its retaining clip.

12 Spark plug renewal

1 The correct functioning of the spark plugs is vital for the correct running and efficiency of the engine. It is essential that the plugs fitted are appropriate for the engine (a suitable type is specified at the end of this Chapter). If this type is used and the engine is in good condition, the spark plugs should not need attention between scheduled replacement intervals. Spark plug cleaning is rarely necessary, and should not be attempted unless specialised equipment is available, as damage can easily be caused to the firing ends.
2 If the marks on the original-equipment spark plug (HT) leads cannot be seen, mark the leads '1' to '4', to correspond to the cylinder the lead serves (No 1 cylinder is at the transmission end of the engine). Pull the leads from the plugs by gripping the end fitting, not the lead, otherwise the lead connection may be fractured (**see illustration**).
3 It is advisable to remove the dirt from the spark plug recesses using a clean brush, vacuum cleaner or compressed air before removing the plugs, to prevent dirt dropping into the cylinders.
4 Unscrew the plugs using a spark plug spanner, suitable box spanner or a deep socket and extension bar . Keep the socket aligned with the spark plug - if it is forcibly moved to one side, the ceramic insulator may be cracked or broken off. As each plug is removed, examine it as follows.
5 Examination of the spark plugs will give a good indication of the condition of the engine. If the insulator nose of the spark plug is clean and white, with no deposits, this is indicative of a weak mixture or too hot a plug (a hot plug transfers heat away from the electrode slowly, a cold plug transfers heat away quickly).
6 If the tip and insulator nose are covered with powdered, black-looking deposits, then this is indicative that the mixture is too rich. Should the plug be black and oily, then it is possible that the either the valve guides, piston rings, or turbocharger (where applicable) are worn.
7 If the insulator nose is covered with light tan to greyish-brown deposits, then the mixture is correct and it is likely that the engine is in good condition.
8 The spark plug electrode gap is of considerable importance as, if it is too large or too small, the size of the spark and its efficiency will be seriously impaired. The gap should be set to the value given in the Specifications at the beginning of this Chapter.
9 To set the gap, measure it with a feeler blade, and then bend the outer plug electrode until the correct gap is achieved (**see illustration**). The centre electrode should never be bent, as this may crack the insulator and cause plug failure, if nothing worse. If using feeler blades, the gap is correct when the appropriate-size blade is a firm sliding fit.
10 Special spark plug electrode gap adjusting tools are available from most motor accessory shops, or from some spark plug manufacturers.
11 Before fitting the spark plugs, check that the threaded connector sleeves are tight, and that the plug exterior surfaces and threads are clean. It is very often difficult to insert spark plugs into their holes without cross-threading them. To avoid this possibility, fit a short

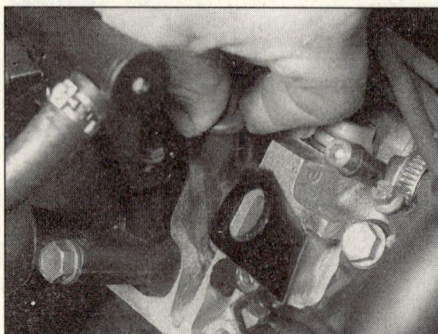

12.2 Pull the HT leads from the spark plugs using the end fittings, not the lead itself

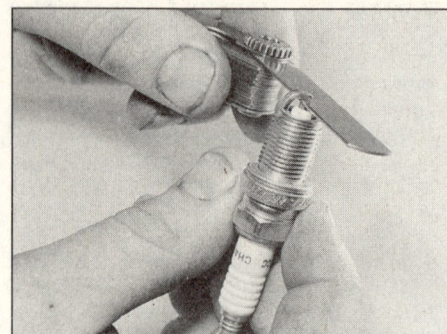

12.9 Measuring the spark plug electrode gap using a feeler gauge

It is often very difficult to insert spark plugs into their holes without cross-threading them. To avoid this, fit a short length of 5/16" internal diameter rubber hose over the end of the spark plug. The flexible hose acts as a universal joint to help align the plug with the plug hole. Should the plug begin to cross-thread, the hose will slip on the spark plug preventing thread damage to the cylinder head.

length of hose over the end of the spark plug **(see Haynes Hint).**
12 Remove the rubber hose (if used), and tighten the plug to the specified torque (see

12.12 Tighten the spark plug to the specified torque wrench setting

Specifications) using the spark plug socket and a torque wrench. Refit the remaining plugs in the same way **(see illustration)**.
13 Connect the HT leads in the correct order, and refit any components removed for access.

13 Front and rear brake pad and disc check

1 Apply the parking brake, then jack up the front and rear of the car and support it

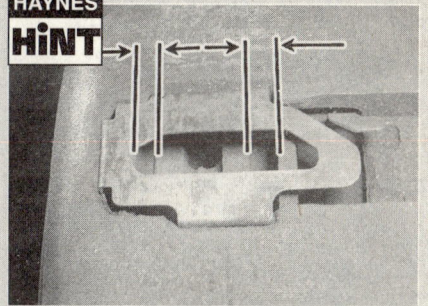

For a quick check, the thickness of friction material remaining on each brake pad can be measured through the aperture in the caliper body

securely on axle stands (see *Jacking and Vehicle Support*). Remove the roadwheels.
2 For a comprehensive check, the brake pads should be removed and cleaned. The operation of the caliper can then also be checked, and the condition of the brake disc itself can be fully examined on both sides. Refer to Chapter 10 for further information.
3 On completion refit the roadwheels and lower the car to the ground.

1A

Every 2 years, regardless of mileage

14 Coolant renewal

Cooling system draining

Warning: Wait until the engine is cold before starting this procedure. Do not allow antifreeze to come in contact with your skin, or with the painted surfaces of the vehicle. Rinse off spills immediately with plenty of water. Never leave antifreeze lying around in an open

container, or in a puddle in the driveway or on the garage floor. Children and pets are attracted by its sweet smell, but antifreeze can be fatal if ingested.

1 With the engine completely cold, remove the expansion tank filler cap. Turn the cap anti-clockwise until it reaches the first stop. Wait until any pressure remaining in the system is released, then push the cap down, turn it anti-clockwise to the second stop, and lift it off.
2 Position a suitable container beneath the coolant drain outlet at the lower left-hand side of the radiator.
3 Loosen the drain plug (there is no need to remove it completely) and allow the coolant to drain into the container **(see Haynes Hint)**.

4 To assist draining, open the cooling system bleed screws. These may be located in the heater matrix inlet hose, on the top of the thermostat housing or hoses, and in the top of the radiator, depending on model **(see illustrations)**.
5 If the coolant has been drained for a reason other than renewal, then provided it is clean and less than two years old, it can be re-used, though this is not recommended.
6 Tighten the radiator drain plug on completion of draining.

Cooling system flushing

7 To avoid the possibility of the cooling system losing efficiency, as the coolant passages become restricted due to rust, scale

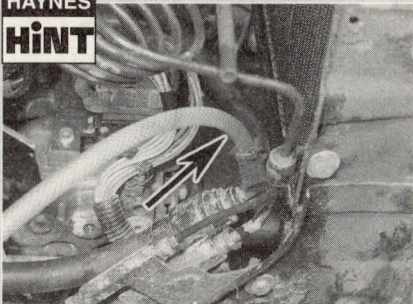

A hose can be attached to the outlet on the radiator drain plug to direct the flow of coolant

14.4a Coolant bleed screw (arrowed) in coolant hose

14.4b Coolant bleed screw (arrowed) in radiator

deposits, and other sediment, the cooling system should be flushed as a matter of course whenever the coolant is renewed. The cooling system efficiency can be restored by flushing the system clean.

8 The radiator should be flushed independently of the engine, to avoid unnecessary contamination.

Radiator flushing

9 To flush the radiator, first tighten the radiator drain plug, and the radiator bleed screw, where applicable.

10 Disconnect the top and bottom hoses and any other relevant hoses from the radiator, with reference to Chapter 3.

11 Insert a garden hose into the radiator top inlet. Direct clean water through the radiator, and continue flushing until clean water emerges from the radiator bottom outlet.

12 If after a reasonable period, the water still does not run clear, the radiator can be flushed with a good proprietary cleaning agent. It is important that their manufacturer's instructions are followed carefully. If the contamination is particularly bad, insert the hose in the radiator bottom outlet, and reverse-flush the radiator.

Engine flushing

13 To flush the engine, first tighten the cooling system bleed screws.

14 Remove the thermostat (Chapter 3), then temporarily refit the thermostat cover.

15 With the top and bottom hoses disconnected from the radiator, insert a length of garden hose into the radiator top hose. Direct a flow of clean water through the engine, and continue flushing until clean water emerges from the radiator bottom hose.

16 On completion of flushing, refit the thermostat and reconnect the hoses with reference to Chapter 3.

Cooling system filling

17 Before filling the cooling system, make

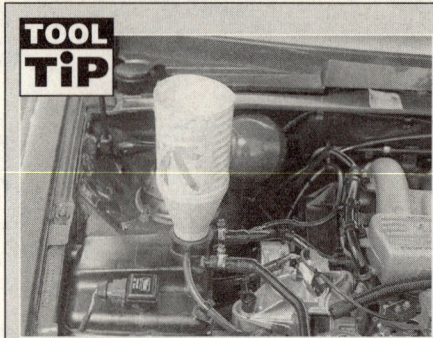

TOOL TIP

Although Citroën dealers use a special header tank, the same effect can be achieved by using a plastic bottle with the bottom cut out, and a seal between the bottle and the expansion tank. The seal (a suitable O-ring can be used) must be as airtight as possible

sure that all hoses and clips are in good condition, and that the clips are tight. Note that an antifreeze mixture must be used all year round, to prevent corrosion of the engine components (see following sub-Section). Also check that the radiator drain plug is in place and tight.

18 Remove the expansion tank filler cap.

19 Open all the cooling system bleed screws (see paragraph 4).

20 Some of the cooling system hoses are positioned at a higher level than the top of the radiator expansion tank. It is therefore necessary to use a 'header tank' when refilling the cooling system, to reduce the possibility of air being trapped in the system **(see Tool Tip)**.

21 Fit the 'header tank' to the expansion tank and slowly fill the system. Coolant will emerge from each of the bleed screws in turn, starting with the lowest screw. As soon as coolant free from air bubbles emerges from the lowest screw, tighten that screw, and watch the next

bleed screw in the system. Repeat the procedure until the coolant is emerging from the highest bleed screw in the cooling system and all bleed screws are securely tightened.

22 Ensure that the 'header tank' is full. Start the engine, and run it at a fast idle speed (do not exceed 2000 rpm) until the cooling fan cuts in, and then cuts out. Stop the engine. **Note:** *Take great care not to scald yourself with the hot coolant during this operation.*

23 Allow the engine to cool then remove the 'header tank'.

24 When the engine has cooled, check the coolant level with reference to *Weekly checks*. Top-up the level if necessary and refit the expansion tank cap.

Antifreeze mixture

25 The antifreeze should always be renewed at the specified intervals. This is necessary not only to maintain the antifreeze properties, but also to prevent corrosion which would otherwise occur as the corrosion inhibitors become progressively less effective.

26 Always use an ethylene-glycol based antifreeze which is suitable for use in mixed-metal cooling systems. The quantity of antifreeze and levels of protection are given in the Specifications at the end of this Chapter.

27 Before adding antifreeze, the cooling system should be completely drained, preferably flushed, and all hoses checked for condition and security.

28 After filling with antifreeze, a label should be attached to the expansion tank, stating the type and concentration of antifreeze used, and the date installed. Any subsequent topping-up should be made with the same type and concentration of antifreeze.

29 Do not use engine antifreeze in the windscreen/tailgate washer system, as it will cause damage to the vehicle paintwork. A screenwash additive should be added to the washer system in the quantities stated on the bottle.

Every 36 000 miles

15 Air filter element renewal - normally-aspirated models

Refer to the information given in Section 11.

16 Hydraulic fluid renewal and fluid return filter cleaning

⚠️ **Warning:** *The fluid used in the XM hydraulic system is LHM mineral fluid, which is green in*

colour. The use of any other type of fluid will damage the system rubber seals and hoses. Keep the fluid carefully sealed in its original container.

1 Remove and empty the hydraulic fluid reservoir, as described in Chapter 9.

2 Also drain the fluid from the high pressure hose connecting the reservoir centre section to the fluid pump.

3 Remove the metal clip securing the two filters to the bottom of the reservoir centre section **(see illustration)**.

4 Pull the semi-circular filter from the centre section, then twist the round filter to release it **(see illustrations)**.

16.3 Remove the metal clip securing the filters

16.4a Pull out the semi-circular filter . . .

16.4b . . . then twist the round filter to remove it

17.1 Fuel filter location

5 Clean the filters and the reservoir using clean petrol, then dry the components. Ideally, the components should be blown through using compressed air.

⚠ **Warning: Wear eye protection when using compressed air.**

6 Refit the filters to the reservoir centre section, then refit the reservoir (Chapter 9).
7 Refill the reservoir with fresh LHM fluid.
8 Prime the hydraulic fluid circuit if necessary, as described in Chapter 9.
9 On completion, check and top-up the hydraulic fluid level (see *Weekly checks*).

17 Fuel filter renewal

⚠ **Warning: Before carrying out the following operation, refer to the precautions given in Safety first! at the beginning of this manual, and follow them implicitly.**

1 The fuel filter is situated underneath the rear of the vehicle, on the right-hand side of the fuel tank **(see illustration)**. To gain access to the filter, apply the parking brake, then jack up the rear of the vehicle and support it on axle stands (see *Jacking and Vehicle Support*).
2 Depressurise the fuel system (Chapter 4A).
3 Unscrew the screw and release the retaining clamp **(see illustration)**.
4 Noting the direction of the arrow marked on the filter body, release the retaining clips and disconnect the fuel hoses from the filter. Where the original crimped-type clips are still fitted, cut and discard them; replace with standard worm-drive hose clips on refitting.
5 Remove the filter from the car. Dispose of the old filter safely; it will be highly flammable, and may explode if thrown on a fire.
6 Connect the new filter to the hoses and tighten the retaining clips. Make sure that the arrow on the filter points in the correct direction (ie towards the hose which leads to the engine compartment) **(see illustration)**.
7 Locate the filter in the retaining strap, then insert and tighten the screw.

17.3 Unscrew the screw and release the filter retaining clamp

8 Start the engine, check the filter hose connections for leaks, then lower the vehicle to the ground.

18 Manual transmission oil level check

Note: *A suitable square-section wrench may be required to undo the transmission filler/level plug on some models. These wrenches can be obtained from most motor factors or your Citroën dealer.*

1 Park the car on a level surface. The oil level must be checked before the car is driven, or at least 5 minutes after the engine has been switched off. If the oil is checked immediately after driving the car, some of the oil will

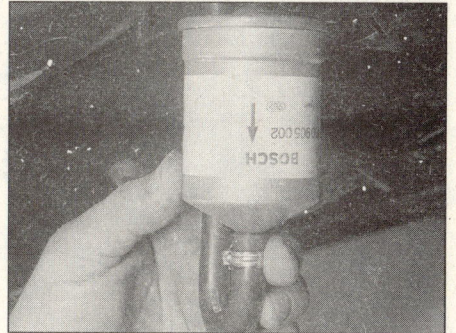
18.3a Slacken the filler/level plug with a spanner . . .

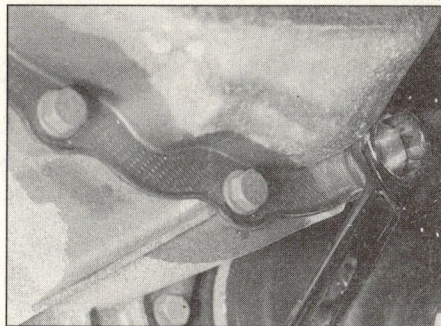
17.6 When fitting the new filter, ensure that the arrow on the filter casing points in the correct direction

remain distributed around the transmission components, resulting in an inaccurate level reading.
2 Prise out the retaining clips and remove the access cover from the left-hand wheelarch liner. On some models it may be necessary to remove the splash guard from under the engine.
3 Wipe clean the area around the filler/level plug, which is situated on the left-hand end of the transmission. Slacken the plug with a spanner or socket and wrench, then unscrew it by hand and clean it; discard the sealing washer **(see illustrations)**.
4 The oil level should reach the lower edge of the filler/level hole. A certain amount of oil will have gathered behind the filler/level plug, and will trickle out when it is removed; this does **not** necessarily indicate that the level is correct. To ensure that a true level is

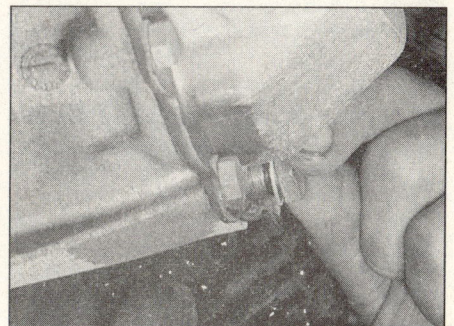
18.3b . . . then unscrew and remove it

1A

18.4 Topping up the transmission oil level

established, wait until the initial trickle has stopped, then add oil as necessary until a trickle of new oil can be seen emerging **(see illustration)**. The level will be correct when the flow ceases; use only good-quality oil (see *Lubricants and Fluids*).

5 Filling the transmission with oil is an extremely awkward operation; above all, allow plenty of time for the oil level to settle properly before checking it. If a large amount is added to the transmission, and a large amount flows out on checking the level, refit the filler/level plug and take the vehicle on a short journey so that the new oil is distributed fully around the transmission components, then recheck the level when it has settled again.

6 If the transmission has been overfilled so that oil flows out as soon as the filler/level plug is removed, check that the car is completely level (front-to-rear and side-to-side), and allow the surplus to drain off into a suitable container.

7 When the level is correct, fit a new sealing washer to the filler/level plug. Refit the plug, tightening it to the specified torque wrench setting (see *Specifications*). Wash off any spilt oil then refit the access cover securing it in position with the retaining clips.

19 Auxiliary drivebelt check

Note: *The auxiliary drivebelt configuration varies considerably depending on model. Removal and refitting should be self-explanatory. If in doubt, consult a Citroën dealer or specialist for advice on the correct tensioning procedure.*

Alternator/hydraulic pump/air conditioning pump drivebelt(s)

Checking

1 The alternator/hydraulic pump/air conditioning compressor drivebelt(s) is/are located on the right-hand end of the engine.

2 The number and configuration of drivebelts fitted varies considerably depending on model, and whether the model is equipped with air conditioning.

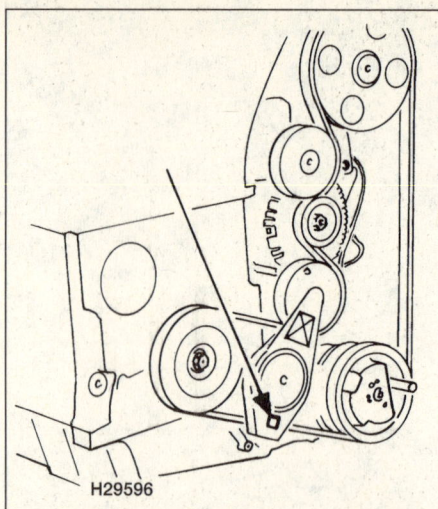

19.7a Auxiliary drivebelt arrangement – later turbo models with air conditioning

Insert a square-drive extension into the hole (arrowed) and lever to release the belt tension

3 Due to their function and material makeup, drivebelts are prone to failure after a period of time and should therefore be inspected, and if necessary adjusted periodically.

4 Since the drivebelt(s) is/are located very close to the right-hand side of the engine compartment, it may be possible to gain better access by raising the front of the vehicle and removing the right-hand wheel, then removing the wheel arch liner.

5 With the engine switched off, inspect the full length of the drivebelt(s) for cracks and separation of the belt plies. It will be necessary to turn the engine in order to move the belt(s) from the pulleys so that the belt(s) can be inspected thoroughly. Twist the belt(s) between the pulleys so that both sides can be viewed. Also check for fraying, and glazing which gives the belt(s) a shiny appearance. Check the pulleys for nicks, cracks, distortion and corrosion.

Tensioning

6 Citroën technicians use a special electronic gauge to measure the tension of the auxiliary drivebelt(s). In the absence of this equipment, an approximate setting must be used. As an approximation, the belt(s) should be tensioned to that, under firm thumb pressure, there is approximately 5.0 mm of free movement at the mid-point between the pulleys on the longest belt run. Ideally, the tension should be re-checked using the special gauge at the earliest opportunity.

7 If adjustment is necessary, where applicable loosen the alternator pivot bolt first, then loosen the adjustment bolt(s). Alternatively, on models equipped with a separate belt tensioner/adjuster mechanism, loosen the tensioner bolt(s) and move or turn the tensioner (as applicable) to relieve the tension in the belt **(see illustrations)**.

8 To apply tension to the belt, on models

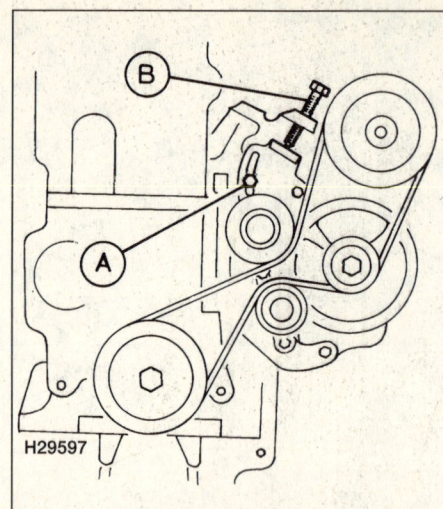

19.7b Auxiliary drivebelt arrangement – later turbo models without air conditioning

Slacken the bolt (A) and turn the bolt (B) to adjust the tension

without a separate belt tensioner/adjuster mechanism, turn the adjuster screw as necessary, and move the alternator to tension the belt. Tighten the adjustment bolt(s) and the pivot bolt.

9 To apply tension to the belt, on models with a separate belt tensioner/adjuster mechanism, turn or reposition the tensioner (as applicable) to achieve the correct belt tension. Tighten the tensioner bolt(s) securely on completion.

10 Run the engine for about 5 minutes, then recheck the tension.

Removal and refitting

11 To remove a belt, slacken the belt tension fully as described previously. Slip the belt off the pulleys, noting its routing to aid refitting, then fit the new belt ensuring that it is routed correctly. Note that on models with two drivebelts, it will be necessary to remove the front drivebelt for access to the rear belt.

12 With the belt in position, adjust the tension as previously described.

20 Automatic transmission fluid strainer cleaning

Refer to the information given in Chapter 7B.

21 Road test

Instruments and electrical equipment

1 Check the operation of all instruments and electrical equipment.

2 Make sure that all instruments read correctly, and switch on all electrical equipment in turn, to check that it functions properly.

Steering and suspension

3 Check for any abnormalities in the steering, suspension, handling or road 'feel'.
4 Drive the vehicle, and check that there are no unusual vibrations or noises.
5 Check that the steering feels positive, with no excessive 'sloppiness', or roughness, and check for any suspension noises when cornering and driving over bumps.

Drivetrain

6 Check the performance of the engine, clutch, transmission and driveshafts.

7 Listen for any unusual noises from the engine, clutch and transmission.
8 Make sure that the engine runs smoothly when idling, and that there is no hesitation when accelerating.
9 Check that the clutch action is smooth and progressive, that the drive is taken up smoothly, and that the pedal travel is not excessive. Also listen for any noises when the clutch pedal is depressed.
10 Check that all gears can be engaged smoothly without noise, and that the gear lever action is not abnormally vague or 'notchy'.
11 Listen for a metallic clicking sound from the front of the vehicle, as the vehicle is driven slowly in a circle with the steering on full-lock. Carry out this check in both directions. If a clicking noise is heard, this indicates wear in a driveshaft joint (see Chapter 8).

Check the operation and performance of the braking system

12 Make sure that the vehicle does not pull to one side when braking, and that the wheels do not lock prematurely when braking hard.
13 Check that there is no vibration through the steering when braking.
14 Check that the parking brake operates correctly, and that it holds the vehicle stationary on a slope.

Every 72 000 miles

22 Timing belt renewal

Refer to the relevant Part of Chapter 2.

1A

Notes

Chapter 1 Part B:
Routine maintenance and servicing - diesel engine models

Contents

Degrees of difficulty

Easy, suitable for novice with little experience	**Fairly easy,** suitable for beginner with some experience	**Fairly difficult,** suitable for competent DIY mechanic	**Difficult,** suitable for experienced DIY mechanic	**Very difficult,** suitable for expert DIY or professional

Lubricants and fluids
Refer to the end of *Weekly checks*

Capacities (approximate)
Engine oil
Non-turbo models .... 6.00 litres
2.1 litre turbo models:
   Early models .... 6.00 litres
   Later models:
      Models without air conditioning .... 6.25 litres
      Models with air conditioning .... 6.00 litres
2.5 litre turbo models .... 8.00 litres

Cooling system
Non-turbo models .... 9.60 litres
2.1 litre turbo models:
   Early models (P8A engine) .... 10.00 litres
   Later models (P8B and P8C engines) .... 11.40 or 12.00 litres (according to equipment)
2.5 litre turbo models .... 13.20 litres

Transmission
Manual transmission .... 1.90 litres
Automatic transmission:
   From dry .... 7.80 litres
   Fluid change .... 2.70 litres

Hydraulic LHM fluid
All models .... 5.40 litres

Fuel tank
All models .... 80.00 litres

Engine
Oil filter:
   2.1 litre models:
      Models up to June 1995 .... Champion F104
      Models from July 1995 .... Champion F132
   2.5 litre turbo models .... Champion F104

Cooling system
Antifreeze mixture:
   28% antifreeze .... Protection down to -15°C (-5°C)
   50% antifreeze .... Protection down to -30°C (-22°C)
Note: *Refer to antifreeze manufacturer's recommendations*

Fuel system
Air filter element .... Champion U561
Fuel filter:
   2.1 litre models:
      Models with Lucas/CAV fuel filter housing .... Champion L132
      Models with Bosch fuel filter housing .... Champion L135
   2.5 litre turbo models .... Champion L135

Engine electrical system
Glow plugs:
   2.1 litre models, up to June 1994 .... Champion CH68
   2.1 litre models, July 1994 onwards, and 2.5 litre models .... Champion CH163

Brakes
Brake pad friction material minimum thickness:
   Front pads .... 3.0 mm
   Rear pads .... 2.0 mm

Wiper blades
Front .... Champion X55
Rear:
   Hatchback models .... Champion X45
   Estate models .... Champion X41

Clutch
Clutch pedal travel .... 45 -0+10 mm

Tyre pressures
See end of *Weekly checks*

Torque wrench settings

	Nm	lbf ft
Roadwheel bolts	90	66
Manual transmission oil filler/level plug	20	15
Manual transmission oil drain plug	30	22

1 The maintenance intervals in this manual are provided with the assumption that you, not the dealer, will be carrying out the work. These are the minimum maintenance intervals recommended by us for vehicles driven daily. If you wish to keep your vehicle in peak condition at all times, you may wish to perform some of these procedures more often. We encourage frequent maintenance, because it enhances the efficiency, performance and resale value of your vehicle.

2 If the vehicle is driven in very dusty areas, used to tow a trailer, spends long periods with the engine idling, is driven frequently at slow speeds (eg in heavy traffic) or is used mainly for short journeys, then *more frequent* maintenance intervals are recommended.

3 When the vehicle is new, it should be serviced by a factory-authorised dealer service department, in order to preserve the factory warranty.

Every 250 miles (400 km) or weekly
☐ Refer to *Weekly checks*.

Every 6000 miles or 12 months, whichever comes first
☐ Renew the engine oil and filter (Section 3)
☐ Check the clutch pedal height (Section 4)
☐ Check all components, pipes and hoses for fluid leaks (Section 5)
☐ Check the condition and security of the steering and suspension components (Section 6)
☐ Check the condition of the driveshafts (Section 7)
☐ Lubricate all hinges and locks (Section 8)
☐ Check the fault diagnosis system memory for fault codes (Section 9)

Every 18,000 miles
☐ Renew the automatic transmission fluid - where applicable (Section 10)
☐ Check the condition of the front and rear brake pads and discs (Section 11)
☐ Renew the fuel filter (Section 12)

Every 2 years, regardless of mileage
☐ Renew the coolant (Section 13)

Every 36,000 miles
☐ Renew the air filter element (Section 14)
☐ Renew the hydraulic fluid and clean the hydraulic fluid return filters (Section 15)
☐ Check the manual transmission oil level (Section 16)
☐ Check the condition of the auxiliary drivebelts (Section 17)
☐ Clean the automatic transmission fluid strainer - where applicable (Section 18)
☐ Carry out a road test (Section 19)

Every 48 000 miles
☐ Renew the timing belt on models up to 1994 (Section 20)
Note: *Although this is the normal interval for timing belt renewal, it is strongly recommended that the interval is reduced on vehicles which are subjected to intensive use, ie, mainly short journeys or a lot of stop-start driving. The actual belt renewal is therefore very much up to the individual owner, but bear in mind that severe engine damage may result if the belt breaks*

Every 72, 000 miles
☐ Renew the timing belt on models from 1994-on (Section 20)
Note: *Although this is the normal interval for timing belt renewal, it is strongly recommended that the interval is reduced on vehicles which are subjected to intensive use, ie, mainly short journeys or a lot of stop-start driving. The actual belt renewal is therefore very much up to the individual owner, but bear in mind that severe engine damage may result if the belt breaks*

Underbonnet view of a 2.1 litre turbo-diesel engine

1 Air suspension unit hydraulic fluid reservoir bulb
2 Hydraulic fluid reservoir
3 Windscreen/tailgate washer fluid reservoir
4 Air cleaner housing
5 Battery
6 Headlight beam adjustment screw
7 Fuel filter
8 Engine oil filler cap
9 Engine oil level dipstick
10 VIN plate
11 Fuel injection pump
12 Hydraulic fluid pump
13 Electronic control unit housing
14 Coolant reservoir (expansion tank)

Underbonnet view of a 2.5 litre turbo-diesel engine

1 Headlight washer fluid reservoir
2 Air suspension unit hydraulic fluid reservoir bulb
3 Windscreen/tailgate washer fluid reservoir
4 Air cleaner housing
5 Hydraulic fluid reservoir
6 Battery cover
7 Fuel filter housing
8 Coolant reservoir (expansion tank)
9 Engine oil filler cap
10 Engine oil level dipstick
11 Engine oil filter
12 Alternator
13 Fuel injection pump
14 Electronic control unit housing

Front underbody view – 2.5 litre diesel engine model

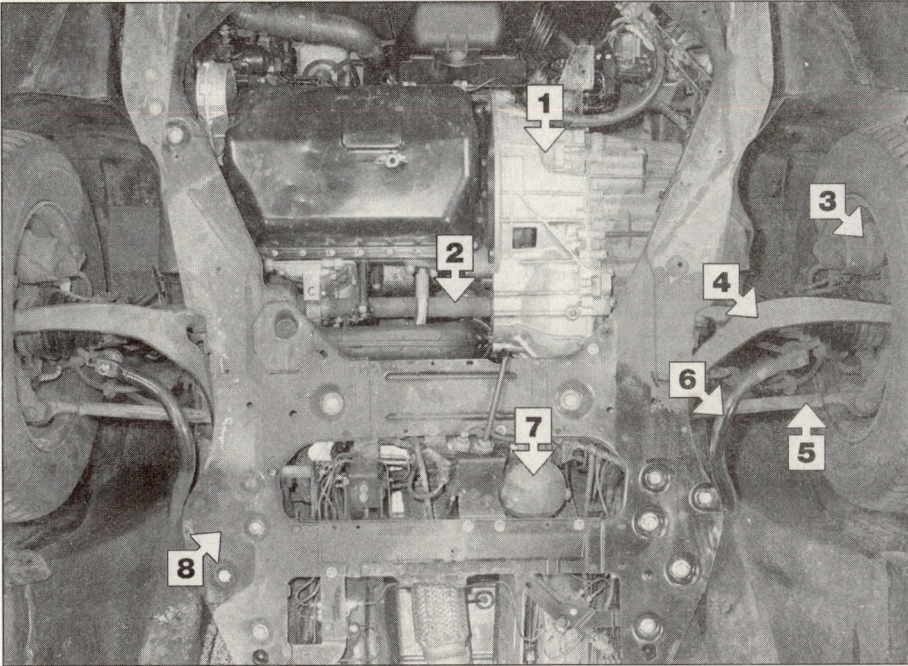

1 Manual transmission
2 Driveshaft
3 Brake caliper
4 Front suspension lower arm
5 Track-rod
6 Anti-roll bar
7 Suspension soft-setting hydraulic fluid reservoir bulb
8 Front suspension subframe

Rear underbody view

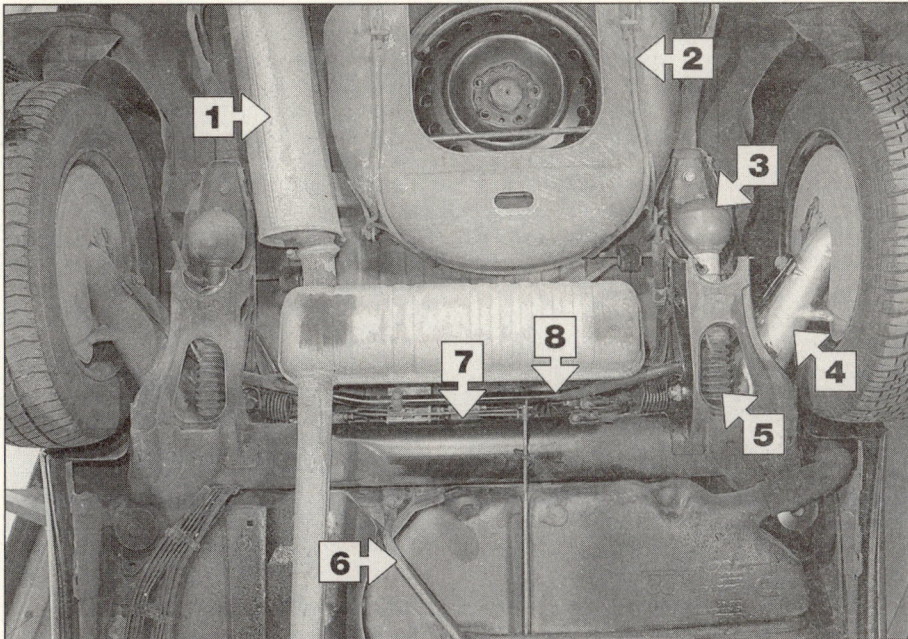

1 Exhaust expansion box
2 Spare wheel carrier
3 Hydraulic fluid reservoir bulb
4 Rear suspension trailing arm
5 Rear suspension hydraulic unit
6 Fuel tank
7 Rear suspension crosstube
8 Rear anti-roll bar

1B

1 Introduction

1 This Chapter is designed to help the home mechanic maintain his/her vehicle for safety, economy, long life and peak performance.

2 The Chapter contains a maintenance schedule, followed by Sections dealing specifically with each task in the schedule. Visual checks, adjustments, component renewal and other helpful items are included. Refer to the accompanying illustrations of the engine compartment and the underside of the vehicle for the locations of the various components.

3 Servicing your vehicle in accordance with the above recommendations and the following Sections will provide a planned maintenance programme, which should result in a long and reliable service life. This is a comprehensive plan, so maintaining some items but not others at the specified service intervals, will not produce the same results.

4 As you service your vehicle, you will discover that many of the procedures can - and should - be grouped together, because of the particular procedure being performed, or because of the proximity of two otherwise-unrelated components to one another. For example, if the vehicle is raised for any reason, the exhaust can be inspected at the same time as the suspension and steering components.

5 The first step in this maintenance programme is to prepare yourself before the actual work begins. Read through all the Sections relevant to the work to be carried out, then make a list and gather all the parts and tools required. If a problem is encountered, seek advice from a parts specialist, or a dealer service department.

2 Regular maintenance

1 If, from the time the vehicle is new, the routine maintenance schedule is followed closely, and frequent checks are made of fluid levels and high-wear items, as suggested throughout this manual, the engine will be kept in relatively good running condition, and the need for additional work will be minimised.

2 It is possible that there will be times when the engine is running poorly due to the lack of regular maintenance. This is even more likely if a used vehicle, which has not received regular and frequent maintenance checks, is purchased. In such cases, additional work may need to be carried out, outside of the regular maintenance intervals.

3 If engine wear is suspected, a compression test (refer to the relevant Part of Chapter 2) will provide valuable information regarding the overall performance of the main internal components. Such a test can be used as a basis to decide on the extent of the work to be carried out. If, for example, a compression test indicates serious internal engine wear, conventional maintenance as described in this Chapter will not greatly improve the performance of the engine, and may prove a waste of time and money, unless extensive overhaul work is carried out first.

4 The following series of operations are those most often required to improve the performance of a generally poor-running engine:

Primary operations

a) *Clean, inspect and test the battery (See Weekly checks).*
b) *Check all the engine-related fluids (See Weekly checks).*
c) *Check the condition and tension of the auxiliary drivebelt(s) (Section 17).*
d) *Check the condition of the air filter, and renew if necessary (Section 14).*
e) *Check the fuel filter (Section 12).*
f) *Check the condition of all hoses, and check for fluid leaks (Section 5).*

5 If the above operations do not prove fully effective, carry out the following secondary operations:

Secondary operations

All items listed under *Primary operations*, plus the following:

a) *Check the charging system (Chapter 5A).*
b) *Check the fuel system (see relevant Part of Chapter 4).*
c) *Check the preheating system (see Chapter 5C).*

Every 6000 miles or 12 months, whichever comes first

3 Engine oil and filter renewal

Note: *A suitable square-section wrench may be required to undo the sump drain plug on some models. These wrenches can be obtained from most motor factors or your Citroën dealer.*

1 Frequent oil and filter changes are the most important preventative maintenance procedures which can be undertaken by the DIY owner. As engine oil ages, it becomes diluted and contaminated, which leads to premature engine wear.

2 Before starting this procedure, gather together all the necessary tools and materials. Make sure that you have plenty of clean rags and newspapers handy, to mop up any spills, and a container of suitable size to drain the oil into. Ideally, the engine oil should be warm, as it will drain better, and more built-up sludge will be removed with it. Take care, however, not to touch the exhaust or any other hot parts of the engine when working under the vehicle. To avoid any possibility of scalding, and to protect yourself from possible skin irritants and other harmful contaminants in used engine oils, it is advisable to wear gloves when carrying out this work. Access to the underside of the vehicle will be greatly improved if it can be raised on a lift, driven onto ramps, or jacked up and supported on axle stands (see *Jacking and Vehicle Support*). Whichever method is chosen, make sure that the vehicle remains level, or if it is at an angle, that the drain plug is at the lowest point. Where necessary remove the splash guard from under the engine.

3 Slacken the drain plug about half a turn; on some models, a square-section wrench may be needed to slacken the plug. Position the draining container under the drain plug, then remove the plug completely. If possible, try to keep the plug pressed into the sump while unscrewing it by hand the last couple of turns **(see illustration and Haynes Hint)**.

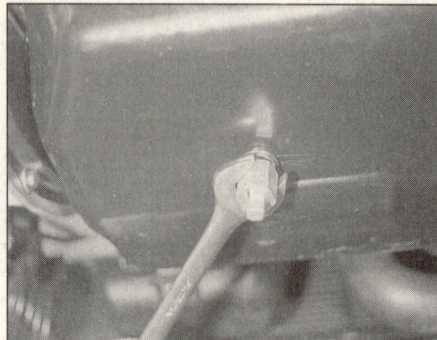

3.3 Slacken the sump drain plug about half a turn using a square-section wrench

HAYNES HINT

As the drain plug releases from the threads, move it away sharply, so that the stream of oil issuing from the sump runs into the container - not down your arm

3.8a Using an oil filter removal tool, slacken the filter . . .

3.8b . . . then unscrew it by hand the rest of the way

3.10 Screw the new filter into position - 2.5 litre engine shown

4 Recover the sealing ring from the drain plug.

5 Allow some time for the old oil to drain, noting that it may be necessary to reposition the container as the oil flow slows to a trickle.

6 After all the oil has drained, wipe off the drain plug with a clean rag, and fit a new sealing washer. Clean around the drain plug opening, then refit and tighten the plug.

7 If the filter is also to be renewed, move the container into position under the oil filter. On 2.1 litre models, the filter is located on the front side of the cylinder block. On 2.5 litre models, the filter is mounted vertically on a housing, which incorporates the oil cooler.

8 Using an oil filter removal tool if necessary, slacken the filter initially, then unscrew it by hand the rest of the way **(see illustrations)**. If any oil remains in the old filter, empty it into the container.

9 Use a clean rag to remove all oil, dirt and sludge from the filter sealing area on the engine. Check the old filter to make sure that the rubber sealing ring hasn't stuck to the engine. If it has, carefully remove it.

10 Apply a light coating of engine oil to the sealing ring on the new filter, then screw it onto the engine **(see illustration)**. Tighten the filter firmly by hand only - **do not** use any tools. If necessary, refit the splash guard under the engine.

11 Remove the old oil and all tools from under the car, then lower the car to the ground (if applicable).

12 Remove the dipstick, then unscrew the oil filler cap from the cylinder head cover. Fill the engine, using the correct grade and type of oil (see Weekly checks) **(see illustrations)**. An oil can spout or funnel may help to reduce spillage. Pour in half the specified quantity of oil first, then wait a few minutes for the oil to fall to the sump. Continue adding oil a small quantity at a time until the level is up to the lower mark on the dipstick. Adding approximately 1.5 litres will bring the level up to the upper mark on the dipstick. Refit the filler cap.

13 Start the engine and run it for a few minutes; check for leaks around the oil filter seal and the sump drain plug. There may be a delay of a few seconds before the oil pressure warning light goes out when the engine is first started, as the oil circulates through the engine oil galleries and the new oil filter (where fitted) before the pressure builds up.

14 Switch off the engine, and wait a few minutes for the oil to settle in the sump once more. With the new oil circulated and the filter completely full, recheck the level on the dipstick, and add more oil as necessary **(see illustrations)**.

15 Dispose of the used engine oil safely, with reference to General Repair Procedures.

1B

3.12a Remove the dipstick . . .

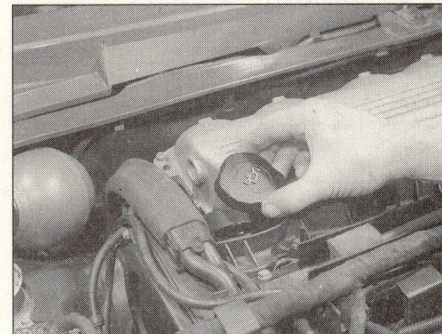

3.12b . . . then unscrew the oil filler cap from the cylinder head cover - 2.5 litre engine shown

3.12c Fill the engine with the correct quantity and grade of oil

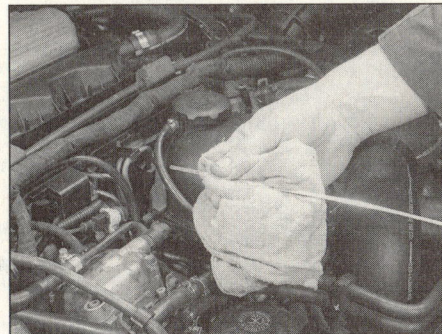

3.14a To obtain a true level reading, remove the dipstick, wipe the dipstick with a clean cloth, then reinsert it. . .

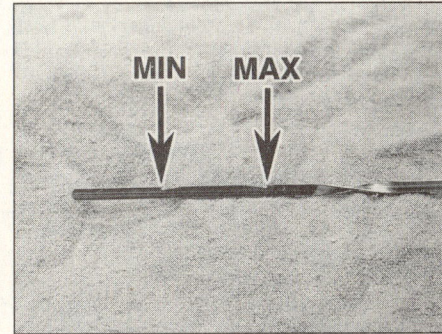

3.14b . . . withdraw the dipstick again check that the oil level is up to the MAX marking

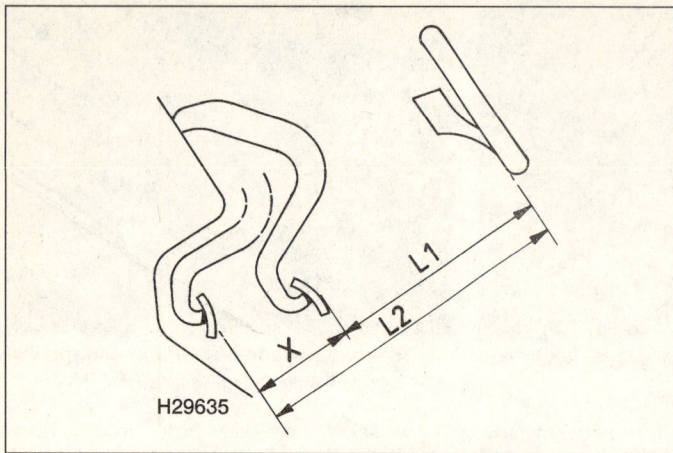

4.2 Clutch pedal height measurement

L1 Pedal height at rest
L2 Pedal height fully depressed

X Clutch pedal travel

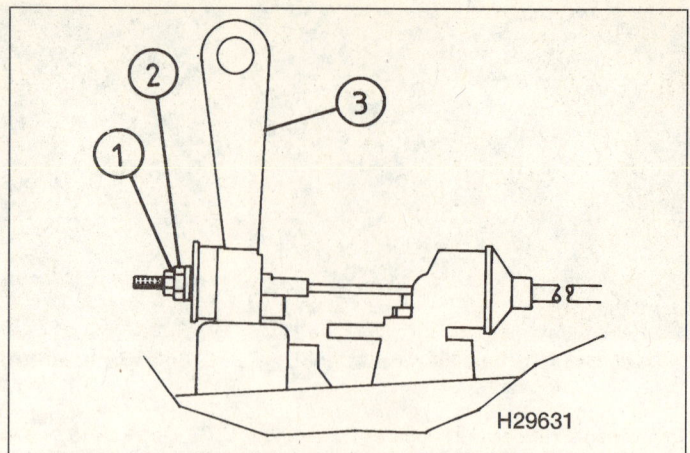

4.4 Clutch cable adjustment details

1 Locknut   2 Adjustment nut   3 Clutch release arm

4 Clutch pedal height check – models with cable-operated clutch

Note: *This procedure applies to 2.1 litre turbo and non-turbo models only. On 2.5 litre models with hydraulic clutch control, there is no pedal height adjustment facility.*

1 The clutch adjustment is checked by measuring the clutch pedal travel. If a new cable has been fitted, settle it in position by depressing the clutch pedal at least thirty times.

2 Ensure that there are no obstructions beneath the clutch pedal then measure the distance (L1) from the centre of the clutch pedal pad to the base of the steering wheel with the pedal in the at-rest position. Depress the clutch pedal fully to the floor, and measure the distance (L2) from the centre of the clutch pedal pad to the base of the steering wheel (see illustration).

3 Subtract the first measurement from the second to obtain the clutch pedal travel. If this is not with the range given in the Specifications at the start of this Chapter, adjust the clutch as follows.

4 The clutch cable is adjusted by means of the adjuster nut on the transmission end of the cable. Access to the locknut is limited and, if required, the air cleaner duct or housing component can be removed or disconnected to improve access (refer to Chapter 4A). Access can be further improved by removing the battery and its holder (refer to Chapter 5A) (see illustration).

5 Working in the engine compartment, slacken the locknut from the end of the clutch cable. Adjust the position of the adjuster nut, then depress the clutch pedal ten times and re-measure the clutch pedal travel. Repeat this procedure until the clutch pedal travel is as specified.

6 Once the adjuster nut is correctly positioned, and the pedal travel is correctly set, securely tighten the cable locknut. Where necessary, refit any disturbed air cleaner duct/housing components (see Chapter 4).

5 Hose and fluid leak check

1 Visually inspect the engine joint faces, gaskets and seals for any signs of water or oil leaks. Pay particular attention to the areas around the cylinder head cover, cylinder head, oil filter and sump joint faces. Bear in mind that, over a period of time, some very slight seepage from these areas is to be expected - what you are really looking for is any indication of a serious leak. Should a leak be found, renew the offending gasket or oil seal by referring to the appropriate Chapters in this manual.

2 Also check the security and condition of all the engine-related pipes and hoses, and all hydraulic and braking system pipes and hoses (see illustration). Ensure that all cable ties or securing clips are in place, and in good condition. Clips which are broken or missing can lead to chafing of the hoses, pipes or wiring, which could cause more serious problems in the future.

3 Carefully check the radiator hoses and heater hoses along their entire length . Renew any hose which is cracked, swollen or deteriorated. Cracks will show up better if the hose is squeezed. Pay close attention to the hose clips that secure the hoses to the cooling system components. Hose clips can pinch and puncture hoses, resulting in cooling system leaks. If the crimped-type hose clips are used, it may be a good idea to replace them with standard worm-drive clips.

4 Inspect all the cooling system components (hoses, joint faces, etc) for leaks (see Haynes Hint).

5 Where any problems are found on system components, renew the component or gasket with reference to Chapter 3.

6 With the vehicle raised, inspect the fuel tank and filler neck for punctures, cracks and other damage. The connection between the filler neck and tank is especially critical.

5.2 Check the hydraulic fluid pipes for leaks

HAYNES HINT

A leak in the cooling system will usually show up as white or rust-coloured deposits on the area adjoining the leak

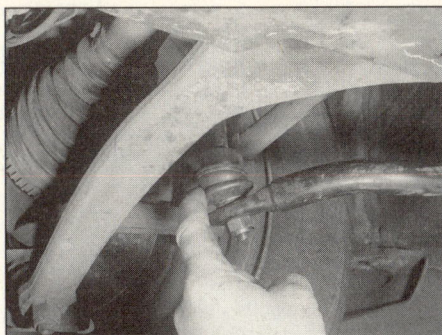
6.2a Inspect the balljoint dust covers . . .

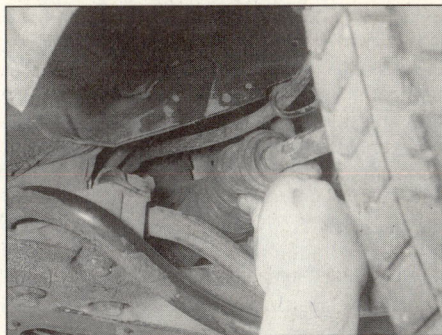
6.2b . . . and the steering gear gaiters

6.4 Check for wear in the hub bearings by grasping the wheel and trying to rock it

Sometimes a rubber filler neck or connecting hose will leak due to loose retaining clamps or deteriorated rubber.

7 Carefully check all rubber hoses and metal fuel lines leading away from the fuel tank. Check for loose connections, deteriorated hoses, crimped lines, and other damage. Pay particular attention to the vent pipes and hoses, which often loop up around the filler neck and can become blocked or crimped. Follow the lines to the front of the vehicle, carefully inspecting them all the way. Renew damaged sections as necessary. Similarly, whilst the vehicle is raised, take the opportunity to inspect all underbody brake fluid pipes and hoses.

8 From within the engine compartment, check the security of all fuel, vacuum and brake hose attachments and pipe unions, and inspect all hoses for kinks, chafing and deterioration.

9 Check the condition of the power steering fluid pipes and hoses and, where applicable, the automatic transmission fluid cooler pipes and hoses.

6 Steering and suspension check

Suspension and steering check

1 Raise the front of the vehicle, and securely support it on axle stands (see *Jacking and Vehicle Support*).

2 Visually inspect the balljoint dust covers and the steering gear gaiters for splits, chafing or deterioration **(see illustrations)**. Any wear of these components will cause loss of lubricant, together with dirt and water entry, resulting in rapid deterioration of the balljoints or steering gear.

3 Check the hydraulic fluid hoses and pipes for chafing or deterioration, and the pipe and hose unions for fluid leaks. Also check for signs of fluid leakage under pressure from the steering gear rubber gaiters, which would indicate failed fluid seals within the steering gear.

4 Grasp the roadwheel at the 12 o'clock and

6 o'clock positions, and try to rock it **(see illustration)**. Very slight free play may be felt, but if the movement is appreciable, further investigation is necessary to determine the source. Continue rocking the wheel while an assistant depresses the footbrake. If the movement is now eliminated or significantly reduced, it is likely that the hub bearings are at fault. If the free play is still evident with the footbrake depressed, then there is wear in the suspension joints or mountings.

5 Now grasp the wheel at the 9 o'clock and 3 o'clock positions, and try to rock it as before. Any movement felt now may again be caused by wear in the hub bearings or the steering track-rod balljoints. If the outer balljoint is worn, the visual movement will be obvious. If the inner joint is suspect, it can be felt by placing a hand over the rack-and-pinion rubber gaiter and gripping the track-rod. If the wheel is now rocked, movement will be felt at the inner joint if wear has taken place.

6 Using a large screwdriver or flat bar, check for wear in the suspension mounting bushes by levering between the relevant suspension component and its attachment point. Some movement is to be expected, as the mountings are made of rubber, but excessive wear should be obvious. Also check the condition of any visible rubber bushes, looking for splits, cracks or contamination of the rubber.

7 With the car standing on its wheels, have an assistant turn the steering wheel back and forth, about an eighth of a turn each way.

7.1 Check the condition of the driveshaft gaiters

There should be very little, if any, lost movement between the steering wheel and roadwheels. If this is not the case, closely observe the joints and mountings previously described. In addition, check the steering column universal joints for wear, and also check the rack-and-pinion steering gear itself.

Suspension strut/hydraulic unit check

8 Check for any signs of fluid leakage around the suspension strut/hydraulic unit body, or from the rubber gaiter around the piston rod. Should any fluid be noticed, the suspension strut/hydraulic unit is defective internally, and should be renewed.

9 Check the condition and security of all hydraulic pipes and hoses. If any signs of leakage are found, investigate the cause. Pipe and hose renewal is described in Chapter 9.

10 Similarly, check around all hydraulic valves and connectors for signs of leakage.

11 If it is suspected that there is a fault in the operation of the suspension, the system can be checked by a Citroën dealer using specialist test equipment.

7 Driveshaft check

1 With the vehicle raised and securely supported on stands, turn the steering onto full lock then slowly rotate the roadwheel. Inspect the condition of the outer constant velocity (CV) joint rubber gaiters while squeezing the gaiters to open out the folds **(see illustration)**. Check for signs of cracking, splits or deterioration of the rubber which may allow the grease to escape and lead to water and grit entry into the joint. Also check the security and condition of the retaining clips. Repeat these checks on the inner CV joints. If any damage or deterioration is found, the gaiters should be renewed as described in Chapter 8.

2 At the same time check the general condition of the CV joints themselves by first holding the driveshaft and attempting to rotate the wheel. Repeat this check by holding

1B

the inner joint and attempting to rotate the driveshaft. Any appreciable movement indicates wear in the joints, wear in the driveshaft splines or loose driveshaft retaining nut.

8 Hinge and lock lubrication

1 Work around the vehicle and lubricate the hinges of the bonnet, doors and tailgate with a light machine oil.

2 Lightly lubricate the bonnet release mechanism and exposed section of inner cable with a smear of grease.

3 Check carefully the security and operation of all hinges, latches and locks, adjusting them where required. Check the operation of the central locking system (if fitted).

4 Check the condition and operation of the tailgate and bonnet struts, renewing them if there is any evidence of leakage, or if the struts are no longer able to support the tailgate or bonnet securely when raised.

9 Fault diagnosis system memory check

The vehicle fault diagnosis systems (for the engine management system, suspension control system, and ABS, where applicable) can be analysed and checked for stored fault codes, which will indicate any faults which have occurred in the relevant systems.

This check must be carried out by a Citroën dealer using the appropriate specialist diagnostic equipment.

Every 18 000 miles

10 Automatic transmission fluid renewal

1 Take the vehicle on a short run, to warm the transmission up to normal operating temperature.

2 Park the car on level ground, then switch off the ignition and apply the parking brake firmly. For improved access, chock the rear wheels then jack up the front of the car and support it securely on axle stands (see *Jacking and Vehicle Support*). When refilling and checking the fluid level, the car must be lowered to the ground, and level, to ensure accuracy.

3 Remove the dipstick, then position a suitable container under the transmission. The 4HP18 transmission unit has one drain plug, located at the bottom of the differential housing.

⚠️ *Warning: If the fluid is hot, take precautions against scalding.*

4 Unscrew the drain plug, and allow the fluid to drain completely into the container. Clean the drain plug, being especially careful to wipe any metallic particles off the magnetic insert. Discard the original sealing washer; these should be renewed whenever they are disturbed.

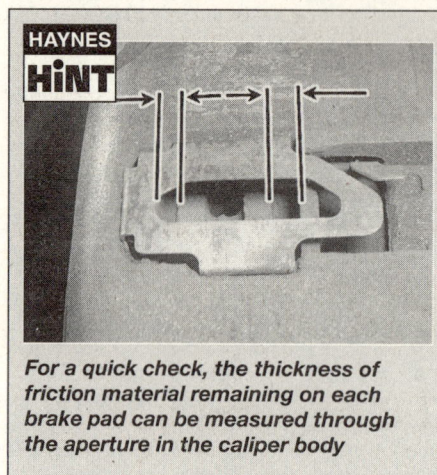

For a quick check, the thickness of friction material remaining on each brake pad can be measured through the aperture in the caliper body

5 When the fluid has finished draining, clean the drain plug threads and those of the transmission casing. Fit a new sealing washer to the drain plug, and refit the plug to the transmission, tightening it securely. If the car was raised for the draining operation, now lower it to the ground. Make sure that the car is level (front-to-rear and side-to-side).

6 Refilling the transmission is an awkward operation, adding the specified type of fluid to the transmission a little at a time via the dipstick tube. Alternatively, use the filler cap

(breather) having cleaned around the area first. Use a funnel with a fine-mesh gauze, to avoid spillage, and to ensure that no foreign matter enters the transmission. Allow plenty of time for the fluid level to settle properly.

7 Once the level is up to the MAX mark on the dipstick, refit the dipstick. Start the engine, and allow it to idle for a few minutes in P, then recheck the level, topping-up if necessary. Take the car on a short run to fully distribute the new fluid around the transmission, then recheck the fluid level.

11 Front and rear brake pad and disc check

1 Apply the parking brake, then jack up the front and rear of the car and support it securely on axle stands (see *Jacking and Vehicle Support*). Remove the roadwheels.

2 For a comprehensive check, the brake pads should be removed and cleaned. The operation of the caliper can then also be checked, and the condition of the brake disc itself can be fully examined on both sides. Refer to Chapter 10 for further information.

3 On completion refit the roadwheels and lower the car to the ground.

12 Fuel filter renewal

1 The combined fuel filter/priming pump is located at the front left hand corner of the engine compartment, attached to the battery holder. Access can be greatly improved by unbolting the unit from the battery holder (**see illustration**).

2 Where applicable, cover the clutch bellhousing with a piece of plastic sheeting to protect the clutch from fuel spillage.

3 Unplug the water sensor wiring from the filter canister at the connector (**see illustration**).

12.1 Access to the fuel filter/priming pump can be greatly improved by unbolting the unit from the battery holder

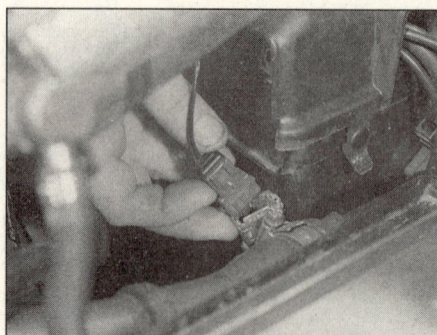

12.3 Unplug the water sensor wiring from the filter canister at the connector

12.4 Open the drain screw at the base of the filter housing and allow the fuel to drain completely

12.5 Slacken and withdraw the centre bolt, thenlower the filter canister and filter element away from the housing

12.6 Place the new filter in the canister

4 Position a suitable container under the end of the fuel filter drain hose. Open the drain screw at the base of the filter housing and allow the fuel to drain completely (**see illustration**).

5 Using a suitable Allen key or hexagon bit, slacken the vertical centre bolt from the bottom of the fuel filter canister. Withdraw the bolt and lower the filter canister and filter element away from the housing. Be prepared for some fuel spillage as you do this (**see illustration**).

6 Place the new filter in the canister, then offer canister up to the housing (**see illustration**).

7 Fit a new seal to the housing mating surface (**see illustration**).

8 Coat the threads of the canister centre bolt with thread-locking compound, then insert the bolt and tighten it securely.

9 Reconnect the water sensor wiring and close the filter drain screw. Mount the housing on the battery holder, then insert the bolts and tighten them securely.

10 Prime and bleed the fuel system as described in Chapter 4B.

11 Start the engine and check around the top of the filter canister for fuel leakage.

12.7 Fit a new seal to the housing mating surface

Every 2 years, regardless of mileage

13 Coolant renewal

Cooling system draining

⚠ **Warning: Wait until the engine is cold before starting this procedure. Do not allow antifreeze to come in contact** with your skin, or with the painted surfaces of the vehicle. Rinse off spills immediately with plenty of water. Never leave antifreeze lying around in an open container, or in a puddle in the driveway or on the garage floor. Children and pets are attracted by its sweet smell, but antifreeze can be fatal if ingested.

1 With the engine completely cold, remove the expansion tank filler cap. Turn the cap anti-clockwise until it reaches the first stop. Wait until any pressure remaining in the system is released, then push the cap down, turn it anti-clockwise to the second stop, and lift it off.

2 Position a suitable container beneath the coolant drain outlet at the lower left-hand side of the radiator.

3 Loosen the drain plug (there is no need to remove it completely) and allow the coolant to drain into the container (**see Haynes Hint**).

4 To assist draining, open the cooling system bleed screws. These may be located on the top of the thermostat housing, and in the heater matrix inlet hose, depending on model (**see illustrations**).

HAYNES HiNT

A hose can be attached to the outlet on the radiator drain plug to direct the flow of coolant

13.4a Coolant bleed screw (arrowed) in coolant hose – 2.5 litre engine model

13.4b Coolant bleed screw (arrowed) in radiator – 2.5 litre engine model

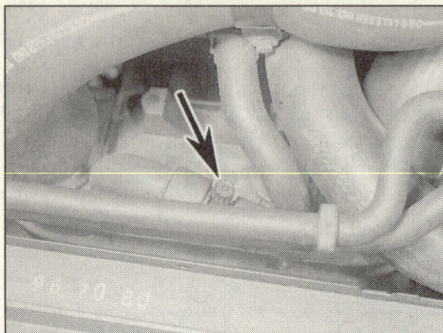

13.4c Coolant bleed screw (arrowed) in coolant bypass hose – 2.5 litre engine model

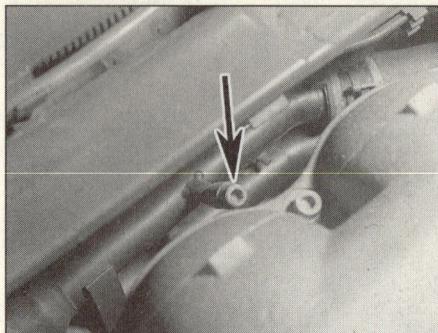

13.4d Coolant bleed screw (arrowed) in coolant pipe at rear of engine compartment – 2.5 litre engine model

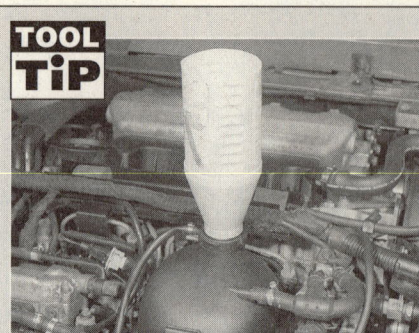

Although Citroën dealers use a special header tank, the same effect can be achieved by using a plastic bottle with the bottom cut out, and a seal between the bottle and the expansion tank. The seal (a suitable O-ring can be used) must be as airtight as possible

5 On 2.5 litre engine models, proceed as follows.
a) *When the flow of coolant stops, reposition the container below the cylinder block drain plug at the front of the cylinder block, to the right of the oil filter.*
b) *Remove the drain plug, and allow the coolant to drain into the container.*

6 If the coolant has been drained for a reason other than renewal, then provided it is clean and less than two years old, it can be re-used, though this is not recommended.

7 Refit and tighten the radiator and cylinder block drain plugs on completion of draining.

Cooling system flushing

8 To avoid the possibility of the cooling system losing efficiency, as the coolant passages become restricted due to rust, scale deposits, and other sediment, the cooling system should be flushed as a matter of course whenever the coolant is renewed. The cooling system efficiency can be restored by flushing the system clean.

9 The radiator should be flushed independently of the engine, to avoid unnecessary contamination.

Radiator flushing

10 To flush the radiator, first tighten the radiator drain plug, and the radiator bleed screw, where applicable.

11 Disconnect the top and bottom hoses and any other relevant hoses from the radiator, with reference to Chapter 3.

12 Insert a garden hose into the radiator top inlet. Direct clean water through the radiator, and continue flushing until clean water emerges from the radiator bottom outlet.

13 If after a reasonable period, the water still does not run clear, the radiator can be flushed with a good proprietary cleaning agent. It is important that their manufacturer's instructions are followed carefully. If the contamination is particularly bad, insert the hose in the radiator bottom outlet, and reverse-flush the radiator.

Engine flushing

14 To flush the engine, first make sure that the cylinder block drain plug is tight, and tighten the cooling system bleed screws.

15 Remove the thermostat (Chapter 3), then temporarily refit the thermostat cover.

16 With the top and bottom hoses disconnected from the radiator, insert a length of garden hose into the radiator top hose. Direct a flow of clean water through the engine, and continue flushing until clean water emerges from the radiator bottom hose.

17 On completion of flushing, refit the thermostat and reconnect the hoses with reference to Chapter 3.

Cooling system filling

18 Before filling the cooling system, make sure that all hoses and clips are in good condition, and that the clips are tight. Note that an antifreeze mixture must be used all year round, to prevent corrosion of the engine components (see following sub-Section). Also check that the radiator drain plug is in place and tight.

19 Remove the expansion tank filler cap.

20 Open all the cooling system bleed screws (see paragraph 4).

21 Some of the cooling system hoses are positioned at a higher level than the top of the radiator expansion tank. It is therefore necessary to use a 'header tank' when refilling the cooling system, to reduce the possibility of air being trapped in the system **(see Tool Tip)**.

22 Fit the header tank to the expansion tank

13.22 Filling the cooling system through the 'header tank'

and slowly fill the system **(see illustration)**. Ensure that the header tank is kept full throughout the filling procedure. Coolant will emerge from each of the bleed screws in turn, starting with the lowest screw. As soon as coolant free from air bubbles emerges from the lowest screw, tighten that screw, and watch the next bleed screw in the system. Repeat the procedure until the coolant is emerging from the highest bleed screw in the cooling system and all bleed screws are securely tightened.

23 Proceed as follows according to engine type.

2.1 litre engines

24 Ensure that the header tank is full, then start the engine and run it at 2000 rpm for 2 minutes. If necessary, top up the header tank so that the coolant level remains visible.

25 Stop the engine, then remove the header tank, and refit the expansion tank cap.

26 Start the engine, and run it until the cooling fan cuts in, and then cuts out.

27 Stop the engine and allow it to cool. When the engine has cooled, check the coolant level with reference to *Weekly checks*. Top-up the level if necessary and refit the expansion tank cap.

2.5 litre engines

28 Ensure that the header tank is full, then start the engine, and run it at a fast idle speed (do not exceed 1500 rpm) until the cooling fan cuts in, and then cuts out three times. Stop the engine. **Note:** *Take great care not to scald yourself with the hot coolant during this operation.*

29 Allow the engine to cool then remove the 'header tank'.

30 When the engine has cooled, check the coolant level with reference to *Weekly checks*. Top-up the level if necessary and refit the expansion tank cap.

Antifreeze mixture

31 The antifreeze should always be renewed

at the specified intervals. This is necessary not only to maintain the antifreeze properties, but also to prevent corrosion which would otherwise occur as the corrosion inhibitors become progressively less effective.

32 Always use an ethylene-glycol based antifreeze which is suitable for use in mixed-metal cooling systems. The quantity of antifreeze and levels of protection are given in the Specifications at the end of this Chapter.

33 Before adding antifreeze, the cooling system should be completely drained, preferably flushed, and all hoses checked for condition and security.

34 After filling with antifreeze, a label should be attached to the expansion tank, stating the type and concentration of antifreeze used, and the date installed. Any subsequent topping-up should be made with the same type and concentration of antifreeze.

35 Do not use engine antifreeze in the windscreen/tailgate washer system, as it will cause damage to the vehicle paintwork. A screenwash additive should be added to the washer system in the quantities stated on the bottle.

Every 36 000 miles

14 Air filter element renewal

1 Slacken and withdraw the retaining screws around the edge of the air cleaner (**see illustration**), then remove the cover.

2 Lift the filter element from the air cleaner casing (**see illustration**). Check that the replacement element is the same, before discarding the old one.

3 Remove any dirt or debris from the inside of the air cleaner casing, using a brush or rag.

4 Fit the new element in position in the air cleaner, pressing the rubber edging firmly into the recess in the air cleaner casing (**see illustration**).

5 Refit the air cleaner cover, securing it in position with its retaining screws.

6 Reconnect the intake duct to the rear of the air cleaner cover, and securely tighten its retaining clip.

15 Hydraulic fluid renewal and fluid return filter cleaning

⚠ *Warning: The fluid used in the XM hydraulic system is LHM mineral fluid, which is green in colour. The use of any other type of fluid will damage the system rubber seals and hoses. Keep the fluid carefully sealed in its original container.*

1 Remove and empty the hydraulic fluid reservoir, as described in Chapter 9.

2 Also drain the fluid from the high pressure hose connecting the reservoir centre section to the fluid pump.

3 Remove the metal clip securing the two filters to the bottom of the reservoir centre section (**see illustration**).

4 Pull the semi-circular filter from the centre section, then twist the round filter to release it (**see illustrations**).

5 Clean the filters and the reservoir using clean petrol, then dry the components. Ideally, the components should be blown through using compressed air.

⚠ *Warning: Wear eye protection when using compressed air.*

6 Refit the filters to the reservoir centre section, then refit the reservoir (Chapter 9).

7 Refill the reservoir with fresh LHM fluid.

8 Prime the hydraulic fluid circuit as described in Chapter 9.

9 On completion, check and top-up the hydraulic fluid level (see *Weekly checks*).

1B

14.1 Slacken the retaining screws around the edge of the air cleaner

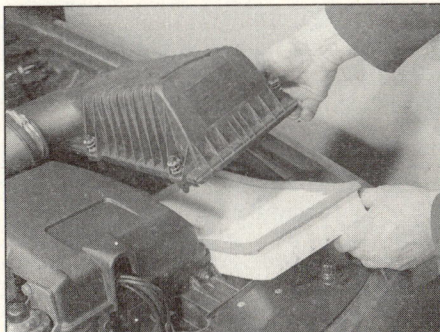
14.2 Lift the filter element from the air cleaner casing

14.4 Fit the new element in position in the air cleaner, pressing the rubber seal firmly into the recess at the edge of the casing

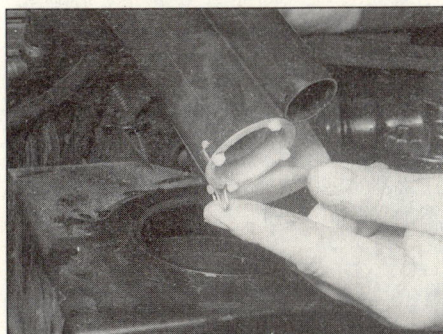
15.3 Remove the metal clip securing the filters

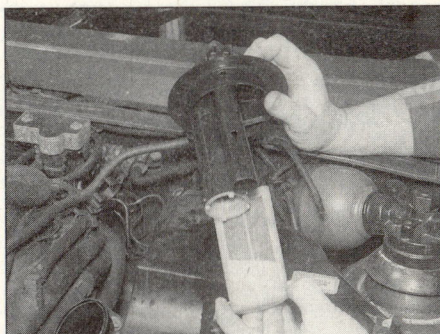
15.4a Pull out the semi-circular filter . . .

15.4b . . . then twist the round filter to remove it

16.3a Slacken the filler/level plug with a spanner . . .

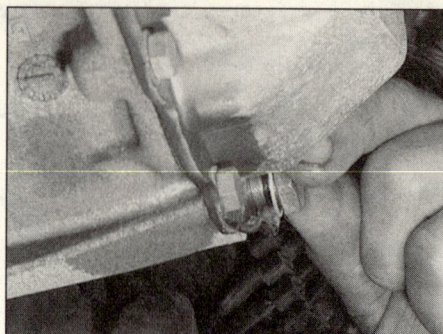
16.3b . . . then unscrew and remove it.

16.4 Topping up the transmission oil level

16 Manual transmission oil level check

Note: *A suitable square-section wrench may be required to undo the transmission filler/level plug on some models. These wrenches can be obtained from most motor factors or your Citroën dealer.*

1 Park the car on a level surface. The oil level must be checked before the car is driven, or at least 5 minutes after the engine has been switched off. If the oil is checked immediately after driving the car, some of the oil will remain distributed around the transmission components, resulting in an inaccurate reading.
2 Prise out the retaining clips and remove the access cover from the left-hand wheelarch liner. On some models it may be necessary to remove the splash guard from under the engine.
3 Wipe clean the area around the filler/level plug, which is situated on the left-hand end of the transmission. Slacken the plug with a spanner or socket and wrench, then unscrew it by hand and clean it; discard the sealing washer **(see illustrations)**.
4 The oil level should reach the lower edge of the filler/level hole. A certain amount of oil will have gathered behind the filler/level plug, and will trickle out when it is removed; this does **not** necessarily indicate that the level is correct. To ensure that a true level is established, wait until the initial trickle has stopped, then add oil as necessary until a trickle of new oil can be seen emerging **(see illustration)**. The level will be correct when the flow ceases; use only good-quality oil (see *Lubricants and Fluids*).
5 Filling the transmission with oil is an extremely awkward operation; above all, allow plenty of time for the oil level to settle properly before checking it. If a large amount is added to the transmission, and a large amount flows out on checking the level, refit the filler/level plug and take the vehicle on a short journey so that the new oil is distributed fully around the transmission components, then recheck the level when it has settled again.
6 If the transmission has been overfilled so that oil flows out as soon as the filler/level plug is removed, check that the car is completely level (front-to-rear and side-to-side), and allow the surplus to drain off into a suitable container.
7 When the level is correct, fit a new sealing washer to the filler/level plug. Refit the plug, tightening it to the specified torque wrench setting (see *Specifications*). Wash off any spilt oil then refit the access cover securing it in position with the retaining clips.

17 Auxiliary drivebelt check

Note: *The auxiliary drivebelt configuration varies considerably depending on model. Removal and refitting should be self-explanatory. If in doubt, consult a Citroën dealer or specialist for advice on the correct tensioning procedure.*

Alternator/hydraulic pump/air conditioning pump drivebelt(s)
Checking

1 The alternator/hydraulic pump/air conditioning compressor drivebelt(s) is/are located on the right-hand end of the engine.
2 The number and configuration of drivebelts fitted varies considerably depending on model, and whether the model is equipped with air conditioning.
3 Due to their function and material makeup, drivebelts are prone to failure after a period of time and should therefore be inspected, and if necessary adjusted periodically.
4 Since the drivebelt(s) is/are located very close to the right-hand side of the engine compartment, it may be possible to gain better access by raising the front of the vehicle and removing the right-hand wheel, then removing the wheel arch liner.
5 With the engine switched off, inspect the full length of the drivebelt(s) for cracks and separation of the belt plies. It will be necessary to turn the engine in order to move the belt(s) from the pulleys so that they can be inspected thoroughly. Twist the belt(s) between the pulleys so that both sides can be viewed. Also check for fraying, and glazing which gives the belt(s) a shiny appearance. Check the pulleys for nicks, cracks, distortion and corrosion.

Tensioning

6 Citroën technicians use a special electronic gauge to measure the tension of the auxiliary drivebelt(s). In the absence of this equipment, an approximate setting must be used. As an approximation, the belt(s) should be tensioned to that, under firm thumb pressure, there is approximately 5.0 mm of free movement at the mid-point between the pulleys on the longest belt run. Ideally, the tension should be re-checked using the special gauge at the earliest opportunity.
7 If adjustment is necessary, where applicable loosen the alternator pivot bolt first, then loosen the adjustment bolt(s). Alternatively, on models equipped with a separate belt tensioner/adjuster mechanism, loosen the locknut (where applicable) and the tensioner bolt(s) and move or turn the tensioner (as applicable) to relieve the tension in the belt **(see illustrations)**.

17.7a Slackening the auxiliary drivebelt tensioner pulley bolt – 2.5 litre engine

17.7b Slacken the tensioner locknut (arrowed) . . .

17.7c . . . then turn the tensioner adjuster bolt to adjust the belt tension – 2.5 litre engine

17.11 Removing the auxiliary drivebelt – 2.5 litre engine

17.19 Turning the drivebelt adjuster screw - 2.5 litre engine (viewed with engine removed for clarity)

Tensioner pulley locking screw (arrowed)

8 To apply tension to the belt, on models without a separate belt tensioner/adjuster mechanism, turn the adjuster screw as necessary, and move the alternator to tension the belt. Tighten the adjustment bolt(s) and the pivot bolt.

9 To apply tension to the belt, on models with a separate belt tensioner/adjuster mechanism, turn or reposition the tensioner (as applicable) to achieve the correct belt tension. Tighten the tensioner bolt(s) securely on completion.

10 Run the engine for about 5 minutes, then recheck the tension.

Removal and refitting

11 To remove a belt, slacken the belt tension fully as described previously. Slip the belt off the pulleys, noting its routing to aid refitting, then fit the new belt ensuring that it is routed correctly **(see illustration)**. Note that on models with two drivebelts, it will be necessary to remove the front drivebelt for access to the rear belt.

12 With the belt in position, adjust the tension as previously described.

Coolant pump drivebelt (2.5 litre engine)

Checking

13 The coolant pump drivebelt is driven by a pulley fitted to the end of the camshaft at the left-hand end of the engine.

14 For access to the drivebelt, proceed as described in paragraphs 21 to 35, ignoring the references to draining the cooling system and disconnecting the coolant hose.

15 Check the condition of the drivebelt as described in paragraph 5.

16 Refit the components removed for access to the drivebelt, with reference to paragraph 37.

Tensioning

17 For access to the drivebelt, proceed as described in paragraphs 21 to 35, ignoring the references to draining the cooling system and disconnecting the coolant hose.

18 Citroën technicians use a special electronic gauge to measure the tension of the coolant pump drivebelt. In the absence of this equipment, an approximate setting must be used. As an approximation, the belt should be tensioned to that, under firm thumb

pressure, there is approximately 5.0 mm of free movement at the mid-point between the pulleys on the longest belt run. Ideally, the tension should be re-checked using the special gauge at the earliest opportunity.

19 To adjust the belt tension, slacken the tensioner pulley locking screw (at the centre of the pulley), then turn the tensioner adjuster screw as necessary to adjust the belt tension **(see illustration)**.

20 On completion of adjustment, tighten the tensioner pulley locking screw.

Removal

21 Drain the cooling system as described in Section 13.

22 Release the securing clips, and withdraw the battery cover.

23 Unclip the plastic cover from the top of the engine.

24 Remove the battery as described in Chapter 5A.

25 Remove the air cleaner assembly as described in Chapter 4.

26 Remove the hydraulic fluid reservoir as described in Chapter 9.

27 Unbolt the following components, and move them to one side to enable access to the coolant pump.

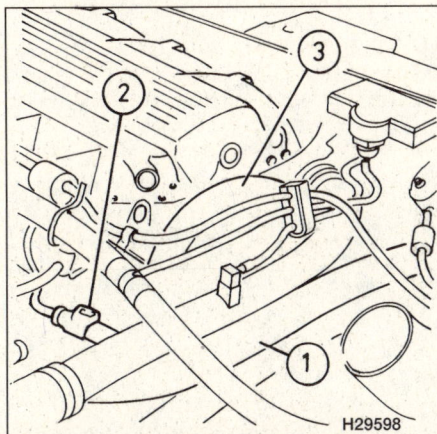

17.34 Disconnect the coolant hose (1), remove the air temperature sensor (2), and unclip the hoses and wiring harnesses from the turbocharger pipe/belt cover (3) – 2.5 litre engine (left-hand-drive shown)

a) *Auxiliary fusebox.*
b) *Pre-heating control unit.*
c) *Fuel priming pump/fuel filter assembly.*
d) *Dehydrator reservoir (models with air conditioning).*
e) *Battery tray.*

28 Release the wiring harnesses from the battery tray.

29 Release the fusebox(es) from their mountings.

30 Remove the MAP sensor as described in Chapter 4.

31 On left-hand-drive models, remove the securing screw from the engine compartment bulkhead, and tilt the throttle actuator support bracket clear of the working area.

32 Disconnect the coolant hose from the housing on the cylinder head.

33 Remove the air temperature sensor as described in Chapter 4.

34 Unclip the hoses and wiring harnesses from the turbocharger pipe/belt cover **(see illustration)**.

35 Remove the four turbocharger pipe/belt cover securing screws, then disconnect the trunking from the turbocharger pipe/belt cover, and remove the pipe/belt cover.

36 Slacken the bolt securing the coolant pump drivebelt tensioner pulley, then turn the tensioner pulley adjustment screw until the drivebelt can be removed from the pulleys **(see illustration)**.

17.36 Removing the coolant pump drivebelt – 2.5 litre engine (viewed with engine removed for clarity)

1B

Refitting

37 Refitting is a reversal of removal, bearing in mind the following points.

a) *Before tightening the drivebelt tensioner pulley bolt, tension the drivebelt as described previously in this Section.*

b) *On completion, refill and bleed the cooling system as described in Section 13.*

18 Automatic transmission fluid strainer cleaning

Refer to the information given in Chapter 7B.

19 Road test

Instruments and electrical equipment

1 Check the operation of all instruments and electrical equipment.

2 Make sure that all instruments read correctly, and switch on all electrical equipment in turn, to check that it functions properly.

Steering and suspension

3 Check for any abnormalities in the steering, suspension, handling or road 'feel'.

4 Drive the vehicle, and check that there are no unusual vibrations or noises.

5 Check that the steering feels positive, with no excessive 'sloppiness', or roughness, and check for any suspension noises when cornering and driving over bumps.

Drivetrain

6 Check the performance of the engine, clutch, transmission and driveshafts.

7 Listen for any unusual noises from the engine, clutch and transmission.

8 Make sure that the engine runs smoothly when idling, and that there is no hesitation when accelerating.

9 Check that the clutch action is smooth and progressive, that the drive is taken up smoothly, and that the pedal travel is not excessive. Also listen for any noises when the clutch pedal is depressed.

10 Check that all gears can be engaged smoothly without noise, and that the gear lever action is not abnormally vague or 'notchy'.

11 Listen for a metallic clicking sound from the front of the vehicle, as the vehicle is driven slowly in a circle with the steering on full-lock. Carry out this check in both directions. If a clicking noise is heard, this indicates wear in a driveshaft joint (see Chapter 8).

Check the operation and performance of the braking system

12 Make sure that the vehicle does not pull to one side when braking, and that the wheels do not lock prematurely when braking hard.

13 Check that there is no vibration through the steering when braking.

14 Check that the parking brake operates correctly, and that it holds the vehicle stationary on a slope.

Every 48 000 miles or 72 000 miles (depending on year of vehicle)

20 Timing belt renewal

Refer to the relevant Part of Chapter 2.

Chapter 2 Part A:
Petrol engine in-car repair procedures

Contents

Degrees of difficulty

Easy, suitable for novice with little experience	**Fairly easy,** suitable for beginner with some experience	**Fairly difficult,** suitable for competent DIY mechanic	**Difficult,** suitable for experienced DIY mechanic	**Very difficult,** suitable for expert DIY or professional

2A

Specifications

Engine (general)

Note: *At the time of writing, no definitive manufacturers specifications were available for the 2.0 litre turbocharged engine; refer to your Citroën dealer for the latest details regarding this engine. Figures given here relate to the 2.0 litre injection, normally aspirated engine.*

Designation XU10
Engine codes*:
   1998 cc fuel injected engine . . . . R6A, RFZ or RDZ (XU10 J2/Z/M)
   1998 cc fuel injected, turbocharged engine . . . . RGY or RGX (XU10 J2T/Z/L/L3)
Bore . . . . 86.00 mm
Stroke . . . . 86.00 mm
Direction of crankshaft rotation . . . . Clockwise (viewed from the right-hand side of vehicle)
No 1 cylinder location . . . . At transmission end of block
Compression ratio:
   1998 cc engine . . . . 8.8 : 1
   1998 cc turbo engine . . . . 7.9 : 1

The engine code is stamped directly onto the front face of the cylinder block (just to the left of the oil filter) on 1998 cc engines. This is the code most often used by Citroën. The code given in brackets is the factory identification number, and is not often referred to by Citroën or this manual.

Camshaft

Drive . . . . Toothed belt
No of bearings . . . . 5 (No 1 at transmission end of engine)
Camshaft bearing journal diameter (external diameter):
   No 1 journal . . . . 26.980 to 26.959 mm
   No 2 journal . . . . 27.480 to 27.459 mm
   No 3 journal . . . . 27.980 to 27.959 mm
   No 4 journal . . . . 28.480 to 28.459 mm
   No 5 journal . . . . 35.975 to 35.950 mm
Cylinder head bearing journal diameter (internal diameter):
   No 1 journal . . . . 27.000 to 27.033 mm
   No 2 journal . . . . 27.500 to 27.533 mm
   No 3 journal . . . . 28.000 to 28.033 mm
   No 4 journal . . . . 28.500 to 28.533 mm
   No 5 journal . . . . 36.000 to 36.039 mm

Valve clearances
Inlet . . . . 0.20 ± 0.05 mm
Exhaust . . . . 0.40 ± 0.05 mm

Lubrication system
Oil pump type . . . . Gear-type, chain-driven off the crankshaft right-hand end
Minimum oil pressure at 80°C:
  R6A, RFZ, RDZ . . . . 4.0 bars at 4000 rpm
  RGX, RGY . . . . 6.4 bars at 4000 rpm
Oil pressure warning switch operating pressure . . . . 0.5 bars

Torque wrench settings

	Nm	lbf ft
Cylinder head cover nuts/bolts	10	7
Timing belt cover bolts	8	6
Crankshaft pulley retaining bolt	110	81
Timing belt tensioner pulley bolt	20	15
Camshaft sprocket retaining bolt	35	26
Camshaft bearing cap nuts/bolts	16	12
Cylinder head bolts:		
Stage 1	35	26
Stage 2	70	52
Stage 3	Angle-tighten a further 160°	
Sump retaining bolts	19	14
Alternator bracket bolts	22	16
Oil pump retaining bolts	16	12
Flywheel/driveplate retaining bolts	50	37
Big-end bearing cap nuts:		
Stage 1	40	30
Fully slacken all nuts, then tighten to:		
Stage 2	20	15
Stage 3	Angle-tighten a further 70°	
Main bearing cap bolts	70	52
Piston oil jet spray tube bolt	10	7
Front oil seal carrier bolts	16	12
Engine/transmission right-hand mounting:		
Mounting bracket retaining nuts	45	33
Curved retaining plate	20	15
Engine/transmission left-hand mounting:		
Mounting rubber-to-body bolts	20	15
Mounting stud	50	37
Centre nut	65	48
Engine/transmission rear mounting:		
Mounting assembly-to-block bolts	45	33
Mounting link-to-mounting bolt	50	37
Mounting link-to-subframe bolt	70	52

1 General information

How to use this Chapter

This Part of Chapter 2 describes those repair procedures that can reasonably be carried out on the XU series petrol engine, while it remains in the car. If the engine has been removed from the car and is being dismantled as described in Part C, any preliminary dismantling procedures can be ignored.

Note that, while it may be possible physically to overhaul items such as the piston/connecting rod assemblies while the engine is in the car, such tasks are not usually carried out as separate operations. Usually, several additional procedures (not to mention the cleaning of components and oilways) have to be carried out. For this reason, all such tasks are classed as major overhaul procedures, and are described in Part C of this Chapter.

Part C describes the removal of the engine/transmission unit from the vehicle, and the full overhaul procedures that can then be carried out.

XU series engine description

The XU series engine is a well-proven engine which has been fitted to many previous Citroën vehicles. The engine is of the in-line four-cylinder type, mounted transversely at the front of the car. The clutch and transmission are attached to its left-hand end.

The crankshaft runs in five main bearings. Thrustwashers are fitted to No 2 main bearing cap, to control crankshaft endfloat.

The connecting rods rotate on horizontally-split bearing shells at their big-ends. The pistons are attached to the connecting rods by gudgeon pins. The gudgeon pins are an interference fit in the connecting rod small-end eyes. The aluminium alloy pistons are fitted with three piston rings - two compression rings and an oil control ring.

The engine is of the conventional 'dry-liner' type. The cylinder block is cast in iron, and no separate bore liners are fitted.

On all models, the camshaft is driven by a toothed timing belt, and it operates the eight valves via followers located beneath each cam lobe. The valve clearances are adjusted by shims, positioned between the followers and the tip of the valve stem. The camshaft runs in bearing caps which are bolted to the top of the cylinder head. The inlet and exhaust valves are each closed by coil springs, and operate in guides pressed into the cylinder head. Both the valve seats and guides can be renewed separately if worn.

The water pump is driven by the timing belt and located in the right-hand end of the cylinder block.

Lubrication is by means of an oil pump which is driven (via a chain and sprocket) off the crankshaft right-hand end. It draws oil through a strainer located in the sump, and then forces it through an externally-mounted filter into galleries in the cylinder block/crankcase. From there, the oil is distributed to the crankshaft (main bearings) and camshaft. The big-end bearings are supplied with oil via internal drillings in the crankshaft; the camshaft bearings also receive a pressurised supply. The camshaft lobes and valves are lubricated by splash, as are all other engine components. An oil cooler is fitted to some models to keep the oil temperature constant under severe operating conditions - it is mounted behind the oil filter. The oil cooler is supplied with coolant from the engine cooling system.

Throughout the manual, it is often necessary to identify the engines not only by their cubic capacity, but also by their engine code. The engine code consists of three letters (eg. R6A.) and is stamped directly onto the front face of the cylinder block, on the machined surface located just to the left of the oil filter (next to the crankcase vent hose union).

Repair operations possible with the engine in the car

The following work can be carried out with the engine in the car:
a) Compression pressure - testing.
b) Cylinder head cover - removal and refitting.
c) Crankshaft pulley - removal and refitting.
d) Timing belt covers - removal and refitting.
e) Timing belt - removal, refitting and adjustment.
f) Timing belt tensioner and sprockets - removal and refitting.
g) Camshaft oil seal - renewal.
h) Camshaft and followers - removal, inspection and refitting.
i) Valve clearances - checking and adjustment.
j) Cylinder head - removal and refitting.
k) Cylinder head and pistons - decarbonising.
l) Sump - removal and refitting.
m) Oil pump - removal, overhaul and refitting.
n) Crankshaft oil seals - renewal.

o) Engine/transmission mountings - inspection and renewal.
p) Flywheel/driveplate - removal, inspection and refitting.

2 Compression test - description and interpretation

1 When engine performance is down, or if misfiring occurs which cannot be attributed to the ignition or fuel systems, a compression test can provide diagnostic clues as to the engine's condition. If the test is performed regularly, it can give warning of trouble before any other symptoms become apparent.
2 The engine must be fully warmed-up to normal operating temperature, the battery must be fully charged, and all the spark plugs must be removed (Chapter 1). The aid of an assistant will also be required.
3 Disable the ignition system by disconnecting the LT wiring connector from the ignition HT coil(s), referring to Chapter 5 for further information.
4 Fit a compression tester to the No 1 cylinder spark plug hole - the type of tester which screws into the plug thread is to be preferred.
5 Have the assistant hold the throttle wide open, and crank the engine on the starter motor; after one or two revolutions, the compression pressure should build up to a maximum figure, and then stabilise. Record the highest reading obtained.
6 Repeat the test on the remaining cylinders, recording the pressure in each.
7 All cylinders should produce very similar pressures; a difference of more than 2 bars between any two cylinders indicates a fault. Note that the compression should build up quickly in a healthy engine; low compression on the first stroke, followed by gradually-increasing pressure on successive strokes, indicates worn piston rings. A low compression reading on the first stroke, which does not build up during successive strokes, indicates leaking valves or a blown head gasket (a cracked head could also be the cause). Deposits on the undersides of the valve heads can also cause low compression.
8 Citroën do not specify exact compression pressures. As a guide, any cylinder pressure of below 10 bars can be considered as less than healthy. Refer to a Citroën dealer or other specialist if in doubt as to whether a particular pressure reading is acceptable.
9 If the pressure in any cylinder is low, carry out the following test to isolate the cause. Introduce a teaspoonful of clean oil into that cylinder through its spark plug hole, and repeat the test.
10 If the addition of oil temporarily improves the compression pressure, this indicates that bore or piston wear is responsible for the pressure loss. No improvement suggests that

leaking or burnt valves, or a blown head gasket, may be to blame.
11 A low reading from two adjacent cylinders is almost certainly due to the head gasket having blown between them; the presence of coolant in the engine oil will confirm this.
12 If one cylinder is about 20 percent lower than the others and the engine has a slightly rough idle, a worn camshaft lobe could be the cause.
13 If the compression reading is unusually high, the combustion chambers are probably coated with carbon deposits. If this is the case, the cylinder head should be removed and decarbonised.
14 On completion of the test, refit the spark plugs and reconnect the ignition system.

3 Engine assembly/valve timing holes - general information and usage

Note: Do not attempt to rotate the engine whilst the crankshaft/camshaft are locked in position. If the engine is to be left in this state for a long period of time, it is a good idea to place suitable warning notices inside the vehicle, and in the engine compartment. This will reduce the possibility of the engine being accidentally cranked on the starter motor, which is likely to cause damage with the locking pins in place.

1 On all models, timing holes are drilled in the camshaft sprocket and crankshaft pulley. The holes are used to align the crankshaft and camshaft, to prevent the possibility of the valves contacting the pistons when refitting the cylinder head, or when refitting the timing belt. When the holes are aligned with their corresponding holes in the cylinder head and cylinder block (as appropriate), suitable diameter pins can be inserted to lock both the camshaft and crankshaft in position. Proceed as follows:
2 Jack up the front of the car and support it on axle stands (see Jacking and Vehicle Support). Remove the right-hand front roadwheel.
3 From underneath the front of the car, unscrew the bolts and prise out the clips securing the plastic cover to the inner wing valance. Remove the cover to gain access to the crankshaft pulley bolt. The crankshaft can then be turned using a suitable socket and extension bar fitted to the pulley bolt. Note that the crankshaft must always be turned in a clockwise direction (viewed from the right-hand side of vehicle).
4 Remove the timing belt upper cover with reference to Section 6.
5 Rotate the crankshaft pulley until the timing hole in the camshaft sprocket is aligned with its corresponding hole in the cylinder head. Note that the holes are aligned when the sprocket hole is in the 8 o'clock position, when viewed from the right-hand end of the engine.

2A

3.6 8 mm diameter drill inserted through the crankshaft pulley timing hole

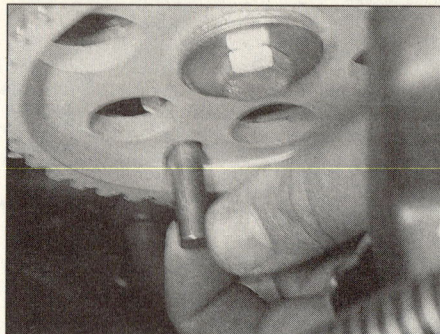

3.7 9.5 mm diameter drill inserted through the camshaft pulley timing hole

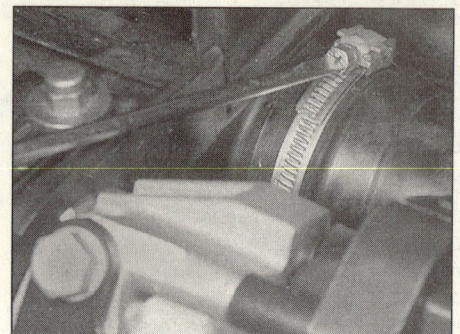

4.3 Disconnecting the intake duct

6 With the camshaft sprocket timing hole correctly positioned, insert an 8 mm diameter bolt or drill through the timing (8 mm diameter) hole in the crankshaft pulley, and locate it in the corresponding hole in the end of the cylinder block **(see illustration)**. Note that it may be necessary to rotate the crankshaft slightly, to get the holes to align.

7 Once the crankshaft pulley is locked in position, insert an 9.5 mm diameter bolt or drill through the camshaft sprocket hole and locate it in the cylinder head **(see illustration)**.

8 The crankshaft and camshaft are now locked in position, preventing rotation.

4 Cylinder head cover - removal and refitting

Removal

1 Disconnect the battery negative lead.

2 Slacken the retaining clips, and disconnect the breather hoses from the front right-hand end of the cover. Where the original crimped-type Citroën hose clips are still fitted, cut them off and discard them; use standard worm-drive hose clips on refitting.

3 Slacken the retaining clip, and disconnect the air cleaner-to-throttle housing duct from

the front of the cylinder head cover. Also remove the intake duct from the left-hand side of the head cover **(see illustration)**.

4 Release the two retaining clips, then undo the two retaining screws located at the front, and remove the air cleaner element cover from the cylinder head cover. Remove the air cleaner element, and store it with the cover.

5 Evenly and progressively unscrew the ten cylinder head cover retaining nuts, lift off the cylinder head cover, and remove it along with its rubber seal **(see illustration)**. Examine the seal for signs of damage and deterioration, and if necessary, renew it.

Refitting

6 Clean the cylinder head and cover mating surfaces, and remove all traces of oil.

7 Locate the rubber seal in the cover groove, ensuring that it is correctly located along its entire length.

8 Apply a smear of suitable sealant to each camshaft end bearing cap around the area where the cap contacts the cylinder head mating surface **(see illustration)**.

9 Carefully refit the cylinder head cover to the engine, taking great care not to displace the rubber seal.

10 Check that the seal is correctly located, then refit the cover retaining nuts and, working in a spiral sequence, starting at the centre of the cover, tighten them evenly and

progressively to the specified torque (see *Specifications*).

11 Refit the air cleaner element, and install the element cover. Securely tighten the cover retaining screws, and secure it in position with the retaining clips.

12 Reconnect the breather hoses, intake duct and throttle housing duct to the cover, tightening their retaining clips securely. Reconnect the battery.

5 Crankshaft pulley - removal and refitting

Removal

1 Remove the auxiliary drivebelt (Chapter 1).

2 To prevent the crankshaft turning whilst the pulley retaining bolt is being slackened, select 4th gear and have an assistant apply the brakes firmly. If the engine has been removed from the vehicle, lock the flywheel ring gear using the arrangement shown **(see illustration)**. *Do not* attempt to lock the pulley by inserting a bolt/drill through the timing hole. If the locking pin is in position, temporarily remove it prior to slackening the pulley bolt, then refit it once the bolt has been slackened.

3 Unscrew the retaining bolt and washer, then slide the pulley off the end of the

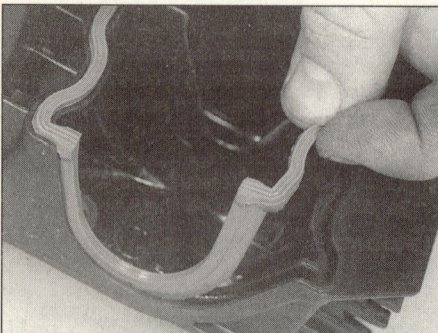

4.5 Removing the cylinder head cover seal

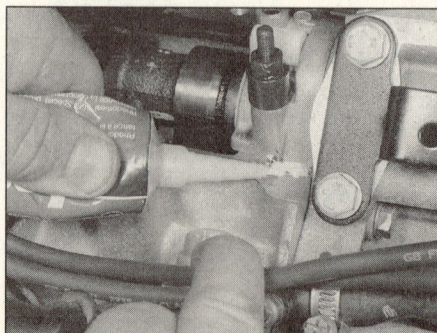

4.8 Applying sealant to the camshaft end bearing caps

5.2 Use a fabricated tool like this one to lock the flywheel ring gear and prevent crankshaft rotation

crankshaft (see illustrations). If the pulley locating roll pin or Woodruff key (as applicable) is a loose fit, remove it and store it with the pulley for safe-keeping. If the pulley is tight fit, it can be drawn off the crankshaft using a suitable puller.

Refitting

4 Ensure the Woodruff key is correctly located in its crankshaft groove, or that the roll pin is in position (as applicable). Refit the pulley to the end of the crankshaft, aligning its locating groove or hole with the Woodruff key or pin.
5 Thoroughly clean the threads of the pulley retaining bolt, then apply a coat of locking compound to the bolt threads. Citroën recommend Loctite (available from your Citroën dealer); in the absence of this, any good-quality locking compound may be used.
6 Refit the crankshaft pulley retaining bolt and washer. Tighten the bolt to the specified torque (see Specifications), preventing the crankshaft from turning using the method employed on removal.
7 Refit and tension the auxiliary drivebelt as described in Chapter 1.

6 Timing belt covers - removal and refitting

Removal

Upper cover

1 Release the retaining clip, and free the fuel hoses from the top of the timing belt cover.
2 Slacken and remove the two cover retaining bolts, then lift the upper cover upwards and out of the engine compartment (see illustrations).

Lower cover

3 Remove the crankshaft pulley (Section 5).
4 Slacken and remove the three retaining bolts, then remove the lower timing belt cover from the engine. Note that on some models it may be necessary to unbolt the auxiliary

5.3a Removing the crankshaft pulley retaining bolt

5.3b Removing the crankshaft pulley from the end of the crankshaft

drivebelt tensioner assembly and remove it from the engine in order to allow the cover to be removed (see illustrations).

Refitting

5 Refitting is a reversal of the relevant removal procedure, ensuring that each cover section is correctly located, and that the cover retaining nuts and/or bolts are securely tightened to the specified torque, where given (see Specifications).

7 Timing belt - general information, removal and refitting

Note: Citroën specify the use of a special electronic tool (SEEM 4122-T) to correctly set

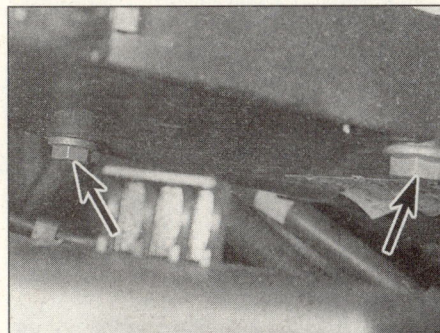
6.2a Upper timing belt cover retaining bolts (arrowed)

the timing belt tension. If access to this equipment cannot be obtained, an approximate setting can be achieved using the method described below. If the method described is used, the tension must be checked using the special electronic tool at the earliest possible opportunity. Do not drive the vehicle over large distances, or use high engine speeds, until the belt tension is known to be correct. Refer to a Citroën dealer for advice.

General information

1 The timing belt drives the camshaft and coolant pump from a toothed sprocket on the front of the crankshaft. If the belt breaks or slips in service, the pistons are likely to hit the valve heads, resulting in extensive (and expensive) damage.

2A

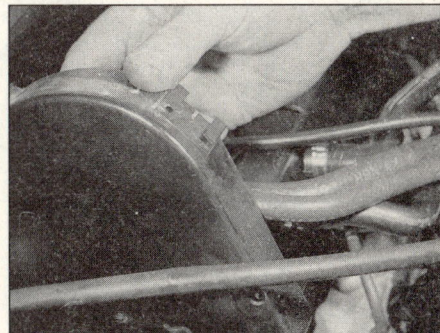
6.2b Removing the upper timing belt cover

6.4a Removing the auxiliary drivebelt tensioner assembly

6.4b Unscrew the retaining bolts ...

6.4c ... and remove the lower timing cover

2 The timing belt should be renewed at the specified intervals (see Chapter 1), or earlier if it is contaminated with oil, or if it is at all noisy in operation (a 'scraping' noise due to uneven wear).

3 If the timing belt is being removed, it is a wise precaution to check the condition of the coolant pump at the same time (check for signs of coolant leakage). This may avoid the need to remove the timing belt again at a later stage, should the coolant pump fail.

Removal

4 Disconnect the battery negative terminal.
5 Jack up the front of the vehicle and support it on axle stands (see *Jacking and Vehicle Support*). Remove the right-hand front wheel.
6 Prise out the clips and unbolt the inner splash guard.
7 Remove the auxiliary drivebelt as described in Chapter 1. Also unbolt and remove the auxiliary drivebelt tensioner.
8 Unbolt and remove the upper timing belt cover as described in Section 6.
9 Align the engine assembly/valve timing holes as described in Section 3, and lock the camshaft sprocket and crankshaft pulley in position. *Do not* attempt to rotate the engine whilst the pins are in position.
10 Remove the crankshaft pulley as described in Section 5.
11 Unbolt and remove the lower timing belt cover (refer to Section 6 if necessary).
12 Loosen the timing belt tensioner pulley retaining bolt. Pivot the pulley in a clockwise direction, using a suitable square-section key fitted to the hole in the pulley hub, then securely retighten the retaining bolt.
13 If the timing belt is to be re-used, use white paint or chalk to mark the direction of rotation on the belt (if markings do not already exist), then slip the belt off the sprockets **(see illustration)**. Note that the crankshaft must not be rotated whilst the belt is removed.
14 Check the timing belt carefully for any signs of uneven wear, splitting, or oil contamination. Pay particular attention to the roots of the teeth. Renew it if there is the slightest doubt about its condition. If the engine is undergoing an overhaul, and has covered more than 36 000 miles (60 000 km) with the existing belt fitted, renew the belt as a

7.13 Removing the timing belt from the water pump sprocket

matter of course, regardless of its apparent condition. The cost of a new belt is nothing compared with the cost of repairs, should the belt break in service. If signs of oil contamination are found, trace the source of the oil leak and rectify it. Wash down the engine timing belt area and all related components, to remove all traces of oil.

Refitting

15 Before refitting, thoroughly clean the timing belt sprockets. Check that the tensioner pulley rotates freely, without any sign of roughness. If necessary, renew the tensioner pulley as described in Section 8.
16 Ensure that the camshaft sprocket locking pin is still in position. Temporarily refit the crankshaft pulley, and insert the locking pin through the pulley timing hole to ensure that the crankshaft is still correctly positioned.
17 Remove the crankshaft pulley. Manoeuvre the timing belt into position, ensuring that any arrows on the belt are pointing in the direction of rotation (clockwise when viewed from the right-hand end of the engine).
18 Do not twist the timing belt sharply while refitting it. Fit the belt over the crankshaft and camshaft sprockets. Ensure that the belt 'front run' is taut - ie, any slack should be on the tensioner pulley side of the belt. Fit the belt over the water pump sprocket and tensioner pulley. Ensure that the belt teeth are seated centrally in the sprockets.
19 Temporarily refit the crankshaft pulley at this stage and tighten the bolt moderately, then refit the locking pin. **Note:** *The timing belt is tensioned with the timing covers removed, then the pulley is removed to fit the covers and finally refitted.*
20 Loosen the tensioner pulley retaining bolt. Using the square-section key, pivot the pulley anti-clockwise to remove all free play from the timing belt.
21 If the special belt tension measuring equipment is available, it should be fitted to the 'front run' of the timing belt. The tensioner roller should be adjusted so that the initial belt tension is 16 ± 2 units .
22 Remove the locking pins, then rotate the crankshaft through two complete rotations in a clockwise direction (viewed from the right-hand end of the engine). Realign the camshaft and crankshaft engine assembly/valve timing holes (see Section 3). *Do not* at any time rotate the crankshaft anti-clockwise. Both camshaft and crankshaft timing holes should be aligned so that the locking pins can be easily inserted. This indicates that the valve timing is correct. If all is well, remove the pins.
23 If the timing holes are not correctly positioned, repeat the fitting procedure so far.
24 If the tension is being set without using the special measuring tool, proceed as follows. Check that, under moderate pressure from the thumb and forefinger, the belt can just be twisted through 90° at the mid-point of the 'front run' of the belt. Note that this method is only an initial setting, and the belt tension *must*

be checked at the earliest available opportunity using the special measuring tool. Failure to do so could lead to the belt breaking (through over-tightening) or 'jumping a tooth' (through slackness), resulting in serious engine damage. If necessary, readjust the tensioner pulley position as required. Tighten its retaining bolt to the specified torque (see *Specifications*) on completion.
25 If the special measuring tool is being used, rotate the crankshaft two more turns without turning backwards and refit the camshaft locking pin, then check that the final belt tension on the taut 'front run' of the belt is 44 ± 2 units. If not, repeat the complete fitting procedure.
26 With the belt tension correctly set remove the camshaft locking pin, then remove the crankshaft pulley and refit the timing cover(s).
27 Refit the crankshaft pulley but this time apply locking fluid to the threads of the bolt before inserting it. Tighten the bolt to the specified torque (see *Specifications*) and refer to Section 5 if necessary.
28 Refit the auxiliary drivebelt tensioner then refit and tension the drivebelt with reference to Chapter 1.
29 Refit the inner splash guard and front right-hand wheel, then lower the vehicle to the ground.
30 Reconnect the battery negative terminal.

---

8 Timing belt tensioner and sprockets - removal, inspection and refitting

Note: *This Section describes the removal and refitting of the components concerned as individual operations - if more than one is to be removed at the same time, start by removing the timing belt as described in Section 7; remove the actual component as described below, ignoring the preliminary dismantling steps.*

Removal

1 Disconnect the battery negative lead.
2 Align the engine assembly/valve timing holes as described in Section 3, locking the camshaft sprocket and the crankshaft pulley in position, and proceed as described under the relevant sub-heading. *Do not* attempt to rotate the engine whilst the pins are in position.

Camshaft sprocket

3 Remove the upper timing belt cover as described in Section 6.
4 Loosen the timing belt tensioner pulley retaining bolt. Rotate the pulley in a clockwise direction, using a suitable square-section key fitted to the hole in the pulley hub, then retighten the retaining bolt.
5 Remove the locking pin from the camshaft sprocket. Disengage the timing belt from the sprocket and position it clear, taking care not to bend or twist the belt sharply.

6 Slacken the camshaft sprocket retaining bolt and remove it, along with its washer. To prevent the camshaft rotating as the bolt is slackened, a sprocket holding tool will be required. In the absence of the special Citroën tool, an acceptable substitute can be fabricated from two lengths of steel strip (one long, the other short) and three nuts and bolts, as follows. One nut and bolt form the pivot of a forked tool, with the remaining two nuts and bolts at the tips of the 'forks' to engage with the sprocket spokes **(see Tool Tip)**. *Do not* attempt to use the sprocket locking pin to prevent the sprocket from rotating whilst the bolt is slackened.

Using a home-made tool to retain the camshaft sprocket whilst the sprocket retaining bolt is tightened

7 With the retaining bolt removed, slide the sprocket off the end of the camshaft. If the locating peg is a loose fit in the rear of the sprocket, remove it for safe-keeping. Examine the camshaft oil seal for signs of oil leakage and, if necessary, renew it (see Section 9).

Crankshaft sprocket

8 Remove the upper, centre and/or lower timing belt cover(s) (as applicable) as described in Section 6.
9 Loosen the timing belt tensioner pulley retaining bolt. Rotate the pulley in a clockwise direction, using a suitable square-section key fitted to the hole in the pulley hub, then retighten the retaining bolt.
10 Disengage the timing belt from the crankshaft sprocket, and slide the sprocket off the end of the crankshaft. Remove the Woodruff key from the crankshaft, and store it with the sprocket for safe-keeping. Where necessary, also slide the spacer (where fitted) off the end of the crankshaft.
11 Examine the crankshaft oil seal for signs of oil leakage and, if necessary, renew it as described in Section 16.

Tensioner pulley

12 Remove the upper and where necessary the centre timing belt covers (see Section 6).
13 Slacken and remove the timing belt tensioner pulley retaining bolt, and slide the pulley off its mounting stud. Examine the mounting stud for signs of damage and if necessary, renew it.

Inspection

14 Clean the camshaft/crankshaft sprockets thoroughly, and renew any that show signs of wear, damage or cracks.
15 Clean the tensioner assembly, but do not use any strong solvent which may enter the pulley bearing. Check that the pulley rotates freely on the backplate, with no sign of stiffness or free play. Renew the assembly if there is any doubt about its condition, or if there are any signs of wear or damage.

Refitting

Camshaft sprocket

16 Refit the locating peg (where removed) to the rear of the sprocket. Locate the sprocket on the end of the camshaft, ensuring that the locating peg is correctly engaged with the cutout in the camshaft end.
17 Refit the sprocket retaining bolt and washer, and tighten it to the specified torque (see *Specifications*). Retain the sprocket with the tool used on removal **(see Tool Tip)**.
18 Realign the hole in the camshaft sprocket with the corresponding hole in the cylinder head, and refit the locking pin. Check that the crankshaft pulley locking pin is still in position.
19 Refit the timing belt to the camshaft sprocket. Ensure that the 'front run' of the belt is taut - ie, that any slack is on the tensioner pulley side of the belt. Do not twist the belt sharply while refitting it, and ensure that the belt teeth are seated centrally in the sprockets.
20 With the timing belt correctly engaged on the sprockets, tension the belt (see Section 7).
21 Once the belt is correctly tensioned, refit the timing belt covers (refer to Section 6).

Crankshaft sprocket

22 Slide the spacer (where fitted) into position, taking great care not to damage the crankshaft oil seal, and refit the Woodruff key to its slot in the crankshaft end.
23 Slide on the crankshaft sprocket, aligning its slot with the Woodruff key.
24 Ensure that the camshaft sprocket locking pin is still in position. Temporarily refit the crankshaft pulley, and insert the locking pin through the pulley timing hole, to ensure that the crankshaft is still correctly positioned.
25 Remove the crankshaft pulley. Engage the timing belt with the crankshaft sprocket. Ensure the belt 'front run' is taut - ie, that any slack is on the tensioner pulley side of the belt. Fit the belt over the water pump sprocket and tensioner pulley. Do not twist the belt sharply while refitting it, and ensure the belt teeth are seated centrally in the sprockets.
26 Tension the timing belt (see Section 7).
27 Remove the crankshaft pulley, then refit the timing belt cover(s) (refer to Section 6).
28 Refit the crankshaft pulley as described in Section 5, and reconnect the battery negative terminal.

Tensioner pulley

29 Refit the tensioner pulley to its mounting stud, and fit the retaining bolt.

30 Ensure that the 'front run' of the belt is taut - ie, that any slack is on the pulley side of the belt. Check that the belt is centrally located on all its sprockets. Rotate the pulley anti-clockwise to remove all free play from the timing belt, and securely tighten the pulley retaining nut.
31 Tension the belt (see Section 7).
32 Once the belt is correctly tensioned, refit the timing belt covers (refer to Section 6).

9 Camshaft oil seal(s) - renewal

Note: *If the camshaft oil seal is to be renewed with the timing belt still in place, check first that the belt is free from oil contamination. (Renew the belt as a matter of course if signs of oil contamination are found; see Section 7). Cover the belt, to protect it from contamination by oil, while work is in progress. If the timing belt is removed, ensure that all traces of oil are removed from the area before the belt is refitted.*

1 Remove the camshaft sprocket as described in Section 8.
2 Punch or drill two small holes opposite each other in the oil seal. Screw a self-tapping screw into each, and pull on the screws with pliers to extract the seal.
3 Clean the seal housing, and polish off any burrs or raised edges, which may have caused the seal to fail in the first place.
4 Lubricate the lips of the new seal with clean engine oil, and drive it into position until it seats on its locating shoulder. Use a suitable tubular drift, such as a socket, which bears only on the hard outer edge of the seal. Take care not to damage the seal lips during fitting. Note that the seal lips should face inwards.
5 Refit the camshaft sprocket as described in Section 8.

10 Camshaft and followers - removal, inspection and refitting

Removal

1 Disconnect the battery negative terminal, and remove the cylinder head cover as described in Section 4.
2 Remove the camshaft sprocket as described in Section 8.
3 Remove the ignition HT coil as described in Chapter 5B.
4 With the coil removed, slacken the upper bolt securing the thermostat housing to the left-hand end of the cylinder head. Remove the bolt, along with its sealing washer. This is necessary since the bolt screws into the left-hand (No 1) camshaft bearing cap.
5 Carefully ease the oil supply pipe out from the top of the camshaft bearing caps, and remove it. Note the O-ring seals fitted to each

2A

10.5 Removing the oil supply pipe from the camshaft bearing caps

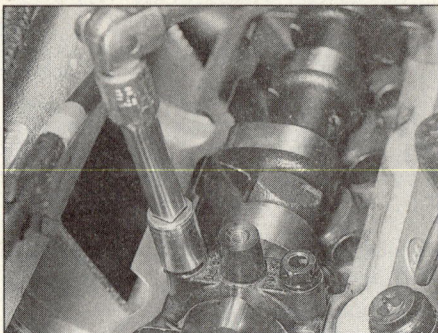

10.7 Working as described in the text, unscrew the retaining nuts . . .

of the pipe unions **(see illustration)**. Also note the position of the adapters at each end of the supply tube.

6 The camshaft bearing caps should be numbered 1 to 5, number 1 being at the transmission end of the engine. If not, make identification marks on the caps, using white paint or a suitable marker pen. Also mark each cap in some way to indicate its correct fitted orientation. This will avoid the possibility of installing the caps the wrong way around on refitting.

7 Evenly and progressively slacken the camshaft bearing cap retaining nuts by one turn at a time. This will relieve the valve spring pressure on the bearing caps gradually and evenly. Once the pressure has been relieved, the nuts can be fully unscrewed and removed **(see illustration)**.

8 Note the correct fitted orientation of the bearing caps, then remove them from the cylinder head **(see illustration)**.

9 Lift the camshaft away from the cylinder head, and slide the oil seal off the camshaft end **(see illustration)**.

10 Obtain eight small, clean plastic containers, and number them 1 to 8; alternatively, divide a larger container into eight compartments. Using a rubber sucker, withdraw each follower in turn, and place it in its respective container. Do not interchange the cam followers, or the rate of wear will be much-increased. If necessary, also remove the shim from the top of the valve stem, and store it with its respective follower. Note that

the shim may stick to the inside of the follower as it is withdrawn. If this happens, take care not to allow it to drop out as the follower is removed.

Inspection

11 Examine the camshaft bearing surfaces and cam lobes for signs of wear ridges and scoring. Renew the camshaft if any of these conditions are apparent. Examine the condition of the bearing surfaces, both on the camshaft journals and in the cylinder head/bearing caps. If the head bearing surfaces are worn excessively, the cylinder head will need to be renewed. If suitable measuring equipment is available, camshaft bearing journal wear can be checked by direct measurement (where the necessary specifications have been quoted by Citroën), noting that No 1 journal is at the transmission end of the head.

12 Examine the cam follower bearing surfaces which contact the camshaft lobes for wear ridges and scoring. Renew any follower on which these conditions are apparent. If a follower bearing surface is badly scored, also examine the corresponding lobe on the camshaft for wear, as it is likely that both will be worn. Renew worn components as necessary.

Refitting

13 Where removed, refit each shim to the top of its original valve stem. *Do not* interchange the shims, as this will upset the valve clearances (see Section 11).

14 Liberally oil the cylinder head cam follower bores and the followers. Carefully refit the followers to the cylinder head, ensuring that each follower is refitted to its original bore. Some care will be required to enter the followers squarely into their bores.

15 Liberally oil the camshaft bearings and lobes, then refit the camshaft to the cylinder head. Temporarily refit the sprocket to the end of the shaft, and position it so that the sprocket timing hole is aligned with the corresponding cutout in the cylinder head. Also ensure that the crankshaft is still locked in position (see Section 3).

16 Ensure that the bearing cap and head mating surfaces are completely clean, unmarked, and free from oil. Apply a smear of sealant to the thermostat housing mating surface of the left-hand (No 1) bearing cap, then refit all the caps, using the identification marks noted on removal to ensure that each is installed correctly and in its original location.

17 Evenly and progressively tighten the camshaft bearing cap nuts by one turn at a time until the caps touch the cylinder head. Then go round again and tighten all the nuts to the specified torque setting (see *Specifications*). Work only as described, to impose the pressure of the valve springs gradually and evenly on the bearing caps.

18 Examine the oil supply pipe union O-rings for signs of damage or deterioration, and renew as necessary. Apply a smear of clean engine oil to the O-rings. Ease the pipe into position in the top of the bearing caps, taking great care not to displace the O-rings.

19 Examine the sealing washer for signs of damage or deterioration, and renew it if necessary. Refit and tighten the thermostat housing upper retaining bolt.

20 Refit the ignition HT coil as described in Chapter 5B.

21 Fit a new camshaft oil seal, using the information given in Section 9, then refit the camshaft sprocket as described in Section 8.

22 Check the valve clearances as described in Section 11.

23 Refit the cylinder head cover as described in Section 4, and reconnect the battery negative terminal.

11 Valve clearances - checking and adjustment

Checking

1 The importance of having the valve clearances correctly adjusted cannot be overstressed, as they vitally affect the performance of the engine. Checking should not be regarded as a routine operation, however. It should only be necessary when the valve gear has become noisy, after engine overhaul, or when trying to trace the cause of power loss. The clearances are checked as

10.8 . . . and remove the camshaft bearing caps . . .

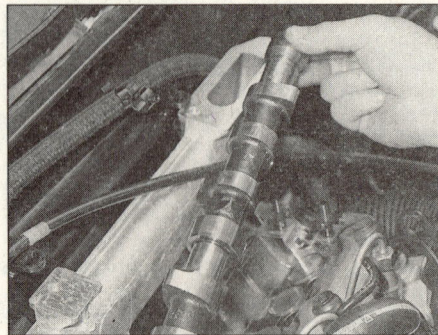

10.9 . . . then lift the camshaft away from the cylinder head

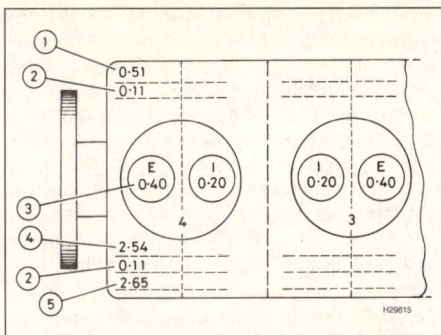

11.6 Example of valve shim thickness calculation

I   Inlet
E   Exhaust
1   Measured clearance
2   Difference between 1 and 3
3   Specified clearance
4   Thickness of original shim fitted
5   Thickness of new shim required

follows. The engine must be cold for the check to be accurate.

2 Apply the parking brake, then jack up the front of the car and support it on axle stands (see *Jacking and Vehicle Support*). Remove the right-hand front roadwheel.

3 From underneath the front of the car, prise out the retaining clips and unscrew the bolts, and remove the plastic cover from the wing valance to gain access to the crankshaft sprocket bolt. Where necessary, unclip the coolant hoses from the bracket to improve access further.

4 The engine can now be turned over using a suitable socket and extension bar fitted to the crankshaft pulley bolt.

5 Remove the cylinder head cover as described in Section 4.

6 Draw the outline of the engine on a piece of paper, numbering the cylinders 1 to 4, with No 1 cylinder at the transmission end of the engine. Show the position of each valve, together with the specified valve clearance (see paragraph 10). Above each valve, draw two lines for noting (1) the actual clearance and (2) the amount of adjustment required **(see illustration)**.

7 Turn the crankshaft until the inlet valve of No 1 cylinder (nearest the transmission end) is fully closed, with the tip of the cam facing directly away from the cam follower.

8 Using feeler gauges, measure the clearance between the base of the cam and the follower **(see illustration)**. Record the clearance on line (1).

9 Repeat the measurement for the other seven valves, turning the crankshaft as necessary so that the cam lobe in question is always facing directly away from the relevant follower.

10 Calculate the difference between each measured clearance and the desired value, and record it on line (2). Since the clearance is different for inlet and exhaust valves, make sure that you are aware which valve you are dealing with. The valve sequence from either end of the engine is:

Ex - In - In - Ex - Ex - In - In - Ex

11 If all the clearances are within tolerance, refit the cylinder head cover with reference to Section 4. Clip the coolant hoses into position (if removed) and refit the plastic cover to the wing valance. Refit the roadwheel, and lower the vehicle to the ground.

12 If any clearance measured is outside the specified tolerance, adjustment must be carried out as described in the following paragraphs.

Adjustment

13 Remove the camshaft as described in Section 10.

14 Withdraw the first follower from the cylinder head, and recover the shim from the top of the valve stem. Note that the shim may stick to the inside of the follower as it is withdrawn. If this happens, take care not to allow it to drop out as the follower is removed. Remove all traces of oil from the shim, and measure its thickness with a micrometer **(see illustrations)**. The shims usually carry thickness markings, but wear may have reduced the original thickness.

15 Refer to the clearance recorded for the valve concerned. If the clearance was more than that specified, the shim thickness must be *increased* by the difference recorded (2). If the clearance was less than that specified, the

thickness of the shim must be *decreased* by the difference recorded (2).

16 Draw three more lines beneath each valve on the calculation paper, as shown in illustration 11.6. On line (4), note the measured thickness of the shim, then add or deduct the difference from line (2) to give the final shim thickness required on line (5).

17 Shims are available in thicknesses between 2.225 mm and 3.550 mm, in steps of 0.025 mm. Clean new shims before measuring or fitting them.

18 Repeat the procedure in paragraphs 14 to 16 on the remaining valves, keeping each follower identified for position.

19 When reassembling, oil the shim, and fit it on the valve stem with the size marking face downwards. Alternatively, if the shims have chamfered sides ensure that these sides face the followers. Oil the follower, and lower it onto the shim. Do not raise the follower after fitting, as the shim may become dislodged.

20 When all the followers are in position, complete with their shims, refit the camshaft as described in Section 10. Recheck the valve clearances before refitting the cylinder head cover, to make sure they are correct.

12 Cylinder head - removal and refitting

Removal

1 Disconnect the battery negative lead.

2 Drain the cooling system (Chapter 1).

3 Align the engine assembly/valve timing holes as described in Section 3, locking both the camshaft sprocket and crankshaft pulley in position, and proceed as described under the relevant sub-heading. *Do not* attempt to rotate the engine whilst the pins are in position.

4 Remove the cylinder head cover as described in Section 4.

5 Remove the air cleaner-to-throttle housing duct as described in Chapter 4A.

6 Note that the following text assumes that the cylinder head will be removed with both inlet and exhaust manifolds attached; this is

2A

11.8 Measuring a valve clearance using a feeler blade

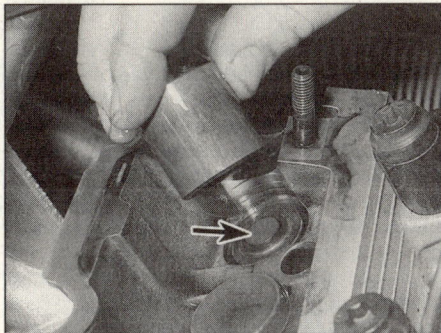
11.14a Lift out the follower and remove the shim (arrowed)

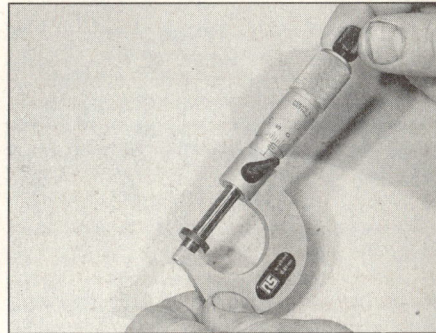
11.14b Using a micrometer to measure shim thickness

easier, but makes it a bulky and heavy assembly to handle. If it is wished first to remove the manifolds, proceed as described in Chapter 4A.

7 Working as described in Chapter 4, disconnect the exhaust system front pipe from the manifold. Where necessary, disconnect or release the lambda sensor wiring, so that it is not strained by the weight of the exhaust.

8 Carry out the following operations as described in Chapter 4:
a) *Depressurise the fuel system, and disconnect the fuel feed and return hoses. Plug all openings, to prevent loss of fuel and the entry of dirt into the system.*
b) *Disconnect the accelerator cable.*
c) *Disconnect all the other relevant vacuum/breather hoses, from the inlet manifold and throttle housing. Release the hoses from the retaining clips on the manifold.*
d) *Disconnect all the electrical connector plugs from the throttle housing.*
e) *Disconnect the wiring connectors from the fuel injectors, and free the wiring loom from the manifold.*

9 Slacken the retaining clips, and disconnect the coolant hoses from the thermostat housing (left-hand end of the cylinder head).

10 Depress the retaining clip(s), and disconnect the wiring connector(s) from the electrical switch(es) and/or sensor(s) screwed into the thermostat housing, or the left-hand end of the cylinder head (as appropriate).

11 Slacken and remove the bolt securing the engine oil dipstick tube to the inlet manifold and withdraw the tube from the cylinder block.

12 Disconnect the wiring connector from the ignition HT coil. If the cylinder head is to be dismantled for overhaul, remove the ignition HT coil as described in Chapter 5B. Note that the HT leads should be disconnected from the spark plugs instead of the coil, and the coil and leads removed as an assembly. If the cylinder numbers are not already marked on the HT leads, number each lead, to avoid the possibility of the leads being incorrectly connected on refitting.

13 Remove the engine mounting bracket from the right-hand end of the cylinder head. Release the timing belt tensioner and disengage the timing belt from the camshaft sprocket as described in Section 8.

14 Working in the *reverse* of the sequence shown in illustration 12.37, progressively slacken the ten cylinder head bolts by half a turn at a time, until all bolts can be unscrewed by hand.

15 Remove all the bolts, along with their washers, and check them as described in paragraph 24.

16 With all the cylinder head bolts removed, lift the cylinder head away. Seek assistance if possible, as it is a heavy assembly.

17 Remove the gasket from the top of the block, noting the two locating dowels. If the locating dowels are a loose fit, remove them and store them with the head for safe-keeping.

18 If the cylinder head is to be dismantled for overhaul, remove the camshaft as described in Section 10, then refer to the relevant Sections of Part C of this Chapter.

Preparation for refitting

19 The mating faces of the cylinder head and cylinder block/crankcase must be perfectly clean before refitting the head. Use a hard plastic or wooden scraper to remove all traces of gasket and carbon; also clean the piston crowns. Make sure that the carbon is not allowed to enter the oil and water passages - this is particularly important for the lubrication system, as carbon could block the oil supply to the engine's components. Using adhesive tape and paper, seal the water, oil and bolt holes in the cylinder block/crankcase. To prevent carbon entering the gap between the pistons and bores, smear a little grease in the gap. After cleaning each piston, use a small brush to remove all traces of grease and carbon from the gap, then wipe away the remainder with a clean rag. Clean all the pistons in the same way.

20 Check the mating surfaces of the cylinder block/crankcase and the cylinder head for nicks, deep scratches and other damage. If slight, they may be removed carefully with a file, but if excessive, machining may be the only alternative to renewal.

21 If warpage of the cylinder head gasket surface is suspected, use a straight-edge to check it for distortion. Refer to Part C of this Chapter if necessary.

22 When purchasing a new cylinder head gasket, it is essential that a gasket of the correct thickness is obtained. On some models only one thickness of gasket is available, so this is not a problem. However on other models, there are two different thicknesses available - the standard gasket which is fitted at the factory, and a slightly thicker 'repair' gasket (+ 0.2 mm), for use once the head gasket face has been machined. If the cylinder head has been machined, it should have the letter 'R' stamped adjacent to the No 3 exhaust port, and the gasket should also have the letter 'R' stamped adjacent to No 3 cylinder on its front upper face. The gaskets can also be identified as described in the following paragraph, using the cut-outs on the left-hand end of the gasket.

23 With the gasket fitted the correct way up on the cylinder block, there will be either a single hole, or a series of holes, punched in the tab on the left-hand end of the gasket. The standard (1.2 mm) gasket has only one hole punched in it; the slightly thicker (1.4 mm) gasket has either two or three holes punched in it, depending on its manufacturer. Identify the gasket type, and ensure the new gasket obtained is of the correct thickness. If there is any doubt as to which gasket is fitted, take the old gasket along to your Citroën dealer, and have the dealer confirm the gasket type.

24 Check the condition of the cylinder head bolts, and particularly their threads, whenever they are removed. Wash the bolts in a suitable solvent, and wipe them dry. Check each bolt for any sign of visible wear or damage, renewing them if necessary. Measure the length of each bolt (without the washer fitted) from the underside of its head to the end of the bolt. If all bolts are less than 122 mm, they may be re-used. However, if any one bolt is longer than the specified length, *all* of the bolts should be renewed as a complete set. Considering the stress which the cylinder head bolts are under, it is recommended that they are renewed, regardless of their apparent condition.

Refitting

25 Wipe clean the mating surfaces of the cylinder head and cylinder block/crankcase. Check that the two locating dowels are in position at each end of the cylinder block/crankcase surface. Where applicable, remove the cylinder liner clamps.

26 Position a new gasket on the cylinder block/crankcase surface, ensuring that its identification holes are at the left-hand end of the gasket.

27 Check that the crankshaft pulley and camshaft sprocket are still locked in position with their respective pins. With the aid of an assistant, carefully refit the cylinder head assembly to the block, aligning it with the locating dowels.

28 Apply a smear of grease to the threads, and to the underside of the heads, of the cylinder head bolts. Citroën recommend the use of Molykote G Rapid Plus (available from your Citroën dealer); in the absence of the specified grease, any good-quality high-melting-point grease may be used.

29 Carefully enter each bolt and washer into its relevant hole (*do not drop it in*) and screw it in finger-tight.

30 Working progressively and in the sequence shown **(see illustration),** tighten the cylinder head bolts to their stage 1 torque setting, using a torque wrench and a suitable socket (see *Specifications*).

31 Once all the bolts are tightened to their stage 1 torque setting, tighten all bolts to their stage 2 torque setting, again following the specified sequence (see *Specifications*).

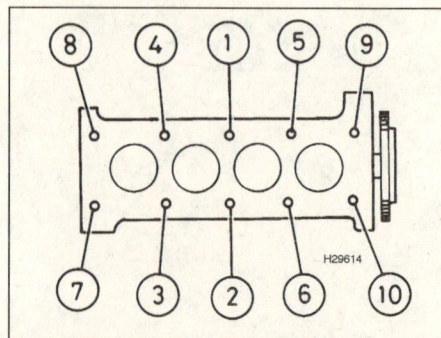

12.30 Cylinder head bolt tightening sequence

32 Working in the specified sequence, angle-tighten the bolts through the specified stage 3 angle, using a socket and extension bar. If an angle tightening gauge is not available, a dab of paint on the bolt head can be used to check that the bolt has been turned through the specified angle. Use a protractor (obtainable from a stationers) to ensure that the angle measurement is accurate.

33 Once the cylinder head bolts are correctly tightened, reconnect the wiring connector to the ignition HT coil. Otherwise, if the head was stripped for overhaul, refit the HT coil, as described in Chapter 5B.

34 Refit the timing belt to the camshaft sprocket as described in Section 8, and tension the belt as described in Section 7.

35 The remainder of the refitting procedure is a reversal of removal, noting the following points:

a) *Ensure that all wiring is correctly routed, and that all connectors are securely reconnected to the correct components.*

b) *Ensure that the coolant hoses are correctly reconnected, and that their retaining clips are securely tightened.*

c) *Ensure that all vacuum/breather hoses are correctly reconnected.*

d) *Refit the cylinder head cover as described in Section 4.*

e) *Reconnect the exhaust system to the manifold, refit the air cleaner housing and ducts, and adjust the accelerator cable, as described in Chapter 4A. If the manifolds were removed, refit these as described in Chapter 4A.*

f) *On completion, refill the cooling system as described in Weekly Checks, and reconnect the battery.*

13 Sump -
removed and refitting

Removal

1 Disconnect the battery negative lead.

2 Chock the rear wheels, jack up the front of the vehicle and support it on axle stands (see *Jacking and Vehicle Support*).

3 Drain the engine oil (Chapter 1), then clean and refit the engine oil drain plug, tightening it securely. If the engine is nearing its service interval when the oil and filter are due for renewal, it is recommended that the filter is also removed, and a new one fitted. After reassembly, the engine can then be refilled with fresh oil - refer to *Weekly Checks*.

4 Where necessary, disconnect the wiring connector from the oil temperature sender unit, which is screwed into the sump.

5 Move the pressure regulator accumulator to one side and remove the auxiliary drivebelt as described in Chapter 1.

6 On models without air conditioning, unbolt the hydraulic high pressure pump and pipe. Tie it to one side without disconnecting the hydraulic lines. Also remove the alternator (Chapter 5A) and unbolt the bracket from the side of the sump.

7 On models with air conditioning, where the compressor is located on the side of the sump, unbolt the compressor and position it clear of the sump. Support the weight of the compressor by tying it to the vehicle, to prevent any excess strain being placed on the compressor lines. *Do not* disconnect the refrigerant lines from the compressor (refer to the warnings given in Chapter 3).

8 On models with air conditioning, unbolt the transmission lower cover plate.

9 Progressively slacken and remove all the sump retaining bolts. Since the sump bolts vary in length, remove each bolt in turn, and store it in its correct fitted order by pushing it through a clearly-marked cardboard template. This will avoid the possibility of installing the bolts in the wrong locations on refitting.

10 Break the joint by striking the sump with the palm of your hand. Lower the sump, and withdraw it from underneath the vehicle. Remove the gasket (where fitted), and discard it; a new one must be used on refitting. While the sump is removed, take the opportunity to

check the oil pump pick-up/strainer for signs of clogging or splitting. If necessary, remove the pump as described in Section 14, and clean or renew the strainer.

11 On some models, a large spacer plate is fitted between the sump and the base of the cylinder block/crankcase. If this plate is fitted, undo the two retaining screws from diagonally-opposite corners of the plate. Remove the plate from the base of the engine, noting which way round it is fitted.

Refitting

12 Clean all traces of sealant/gasket from the mating surfaces of the cylinder block/crankcase and sump, then use a clean rag to wipe out the sump and the engine's interior.

13 Where a spacer plate is fitted, remove all traces of sealant/gasket from the spacer plate, then apply a thin coating of suitable sealant (see paragraph 14) to the plate upper mating surface. Offer up the plate to the base of the cylinder block/crankcase, and securely tighten its retaining screws **(see illustrations)**.

14 On models where the sump was fitted without a gasket, ensure that the sump mating surfaces are clean and dry, then apply a thin coating of sealant to the sump mating surface.

15 On models where the sump was fitted with a gasket, ensure that all traces of the old gasket have been removed, and that the sump mating surfaces are clean and dry. Position the new gasket on the top of the sump, using a dab of grease to hold it in position.

16 Offer up the sump to the cylinder block/crankcase. Refit its retaining bolts, ensuring that each is screwed into its original location. Tighten the bolts evenly and progressively to the specified torque setting (see *Specifications*).

17 On models with air conditioning, refit the transmission lower cover plate and tighten the bolts securely. Refit the compressor to the side of the sump and tighten the bolts to the specified torque (see *Specifications*).

18 On models without air conditioning, refit

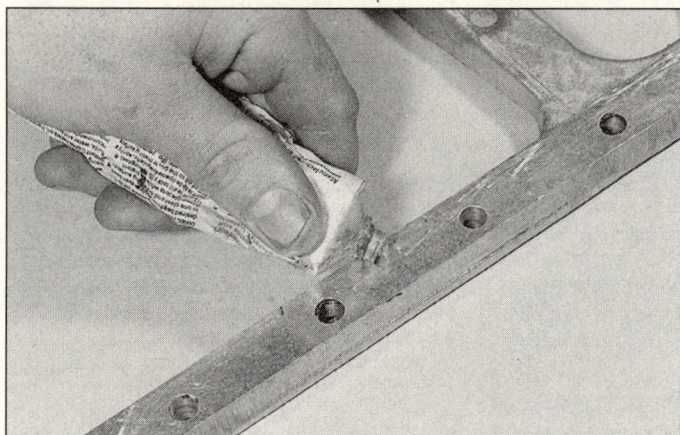

13.13a If a sump spacer plate is fitted, apply sealant to the plate upper surface . . .

13.13b . . . then refit the plate to the base of the cylinder block/crankcase

14.3 Removing the oil pump

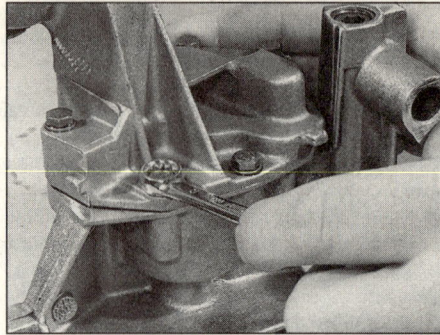

14.5a Remove the oil pump cover retaining bolts . . .

14.5b . . . then lift off the cover and remove the spring . . .

the alternator bracket, alternator and hydraulic high pressure pump and pipe.
19 Refit the auxiliary drivebelt (see Chapter 1) and the pressure regulator accumulator.
20 Reconnect the wiring connector to the oil temperature sensor (where fitted).
21 Lower the vehicle to the ground, then refill the engine with oil (see *Weekly Checks*) and reconnect the battery negative lead.

14 Oil pump - removal, inspection and refitting

Removal

1 Remove the sump (see Section 13).
2 Undo the two screws, and slide the sprocket cover off the front of the oil pump.
3 Slacken and remove the three bolts securing the oil pump to the base of the cylinder block/crankcase. Disengage the pump sprocket from the chain, and remove the oil pump **(see illustration)**. Where necessary, also remove the spacer plate which is fitted behind the oil pump.

Inspection

4 Examine the oil pump sprocket for signs of damage and wear, such as chipped or missing teeth. If the sprocket is worn, the pump assembly must be renewed - the sprocket is not available separately. It is recommended that the chain and drive

sprocket, fitted to the crankshaft, be renewed at the same time. To renew the chain and drive sprocket, first remove the crankshaft timing belt sprocket (see Section 8). Unbolt the oil seal carrier from the cylinder block. The sprocket, spacer (where fitted) and chain are then slid off the end of the crankshaft. See Part C for more information.
5 Slacken and remove the bolts (along with the baffle plate, where fitted) securing the strainer cover to the pump body. Lift off the strainer cover, and take off the relief valve piston and spring, noting which way round they are fitted **(see illustrations)**.
6 Examine the pump rotors and body for signs of wear ridges or scoring. If worn, the complete pump assembly must be renewed.
7 Examine the relief valve piston for signs of wear or damage, and renew if necessary. The condition of the relief valve spring can only be measured by comparing it with a new one; if there is any doubt about its condition, it should also be renewed. Both the piston and spring are available individually.
8 Thoroughly clean the oil pump strainer with a suitable solvent, and check it for signs of clogging or splitting. If the strainer is damaged, the strainer and cover assembly must be renewed.
9 Locate the relief valve spring and piston in the strainer cover. Refit the cover to the pump body, aligning the relief valve piston with its bore in the pump. Refit the baffle plate (where fitted) and the cover retaining bolts, and tighten them securely.

Refitting

10 Offer up the spacer plate (where fitted), then locate the pump sprocket with its drive chain. Seat the pump on the base of the cylinder block/crankcase. Refit the pump retaining bolts, and tighten them to the specified torque setting (see *Specifications*).
11 Where necessary, slide the sprocket cover into position on the pump. Refit its retaining bolts, tightening them securely.
12 Refit the sump as described in Section 13.
13 Before starting the engine, prime the oil pump as follows. Disconnect the fuel injection ECU, then spin the engine on the starter until the oil pressure light goes out. Reconnect the ECU on completion.

15 Oil cooler - removal and refitting

Removal

1 Firmly apply the parking brake, then jack up the front of the vehicle and support it on axle stands (see *Jacking and Vehicle Support*).
2 Drain the cooling system as described in Chapter 1. Alternatively, clamp the oil cooler coolant hoses directly above the cooler, and be prepared for some coolant loss as the hoses are disconnected.
3 Position a suitable container beneath the oil filter. Unscrew the filter using an oil filter removal tool if necessary, and drain the oil into the container. If the oil filter is damaged or distorted during removal, it must be renewed. Given the low cost of a new oil filter relative to the cost of repairing the damage which could result if a re-used filter springs a leak, it is a good idea to renew the filter in any case.
4 Release the hose clips, and disconnect the coolant hoses from the oil cooler.
5 Unscrew the oil cooler/oil filter bolt from the cylinder block, and withdraw the cooler. Note the locating notch in the cooler flange, which fits over the lug on the cylinder block **(see illustration)**. Discard the oil cooler sealing ring; a new one must be used on refitting.

14.5c . . . and relief valve piston, noting which way round it is fitted

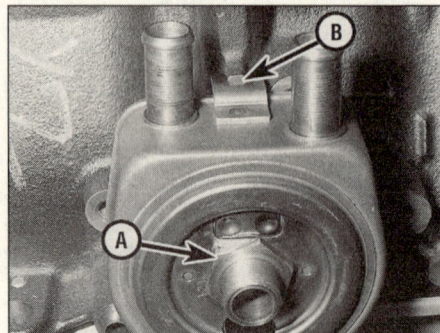

15.5 Oil cooler/oil filter mounting bolt (A) and locating notch (B)

16.2 Using a self-tapping screw and pliers to remove the crankshaft oil seal

Refitting

6 Fit a new sealing ring to the recess in the rear of the cooler, then offer the cooler to the cylinder block.

7 Ensure that the locating notch in the cooler flange is correctly engaged with the lug on the cylinder block, then refit the mounting bolt and tighten it securely.

8 Fit the oil filter, then lower the vehicle to the ground. Top-up the engine oil level as described in *Weekly Checks*.

9 Refill or top-up the cooling system as described in *Weekly Checks* (as applicable). Start the engine, and check the oil cooler for signs of leakage.

16 Crankshaft oil seals - renewal

Right-hand oil seal

1 Remove the crankshaft sprocket and (where fitted) spacer (see Section 8). Secure the timing belt clear of the working area, so that it cannot be contaminated with oil. Make a note of the correct fitted depth of the seal in its housing.

2 Punch or drill two small holes opposite each other in the seal. Screw a self-tapping screw into each, and pull on the screws with pliers to extract the seal **(see illustration)**. Alternatively, the seal can be levered out of position. Use a flat-bladed screwdriver, and take great care not to damage the crankshaft shoulder or seal housing.

3 Clean the seal housing, and polish off any burrs or raised edges, which may have caused the seal to fail in the first place.

4 Lubricate the lips of the new seal with clean engine oil, and carefully locate the seal on the end of the crankshaft. Note that its sealing lip must be facing inwards. Take care not to damage the seal lips during fitting.

5 Fit the new seal using a suitable tubular drift, such as a socket, which bears only on the hard outer edge of the seal. Tap the seal into position, to the same depth in the housing as the original was prior to removal.

6 Wash off any traces of oil, then refit the crankshaft sprocket (see Section 8).

Left-hand oil seal

7 Remove the flywheel/driveplate as described in Section 17. Make a note of the correct fitted depth of the seal in its housing.

8 Punch or drill two small holes opposite each other in the seal. Screw a self-tapping screw into each, and pull on the screws with pliers to extract the seal.

9 Clean the seal housing, and polish off any burrs or raised edges, which may have caused the seal to fail in the first place.

10 Lubricate the lips of the new seal with clean engine oil, and carefully locate the seal on the end of the crankshaft.

11 Fit the new seal using a suitable tubular drift, which bears only on the hard outer edge of the seal. Drive the seal into position, to the same depth in the housing as the original was prior to removal.

12 Wash off any traces of oil, then refit the flywheel/driveplate (see Section 17).

17 Flywheel/driveplate - removal, inspection and refitting

Removal

Flywheel (manual transmission models)

1 Remove the transmission as described in Chapter 7A, then remove the clutch assembly as described in Chapter 6.

2 Prevent the flywheel from turning by locking the ring gear teeth with a similar arrangement to that shown in illustration 5.2. Alternatively, bolt a strap between the flywheel and the cylinder block/crankcase. *Do not* attempt to lock the flywheel in position using the crankshaft pulley locking pin described in Section 3.

3 Slacken and remove the flywheel bolts, and remove the flywheel from the end of the crankshaft. Take care not to drop it; it is heavy. If the flywheel locating dowel is loose in the crankshaft end, remove and store it with the flywheel for safe-keeping. Discard the flywheel bolts; new ones must be used on refitting.

Driveplate (automatic transmission models)

4 Remove the transmission as described in Chapter 7B. Lock the driveplate as described in paragraph 2. Mark the relationship between the torque converter plate and the driveplate, and slacken all the driveplate retaining bolts.

5 Remove the retaining bolts, along with the torque converter plate and the two shims (one fitted on each side of the torque converter plate). Note that the shims are of different thickness, the thicker one being on the outside of the torque converter plate. Discard the driveplate retaining bolts; new ones must be used on refitting.

6 Remove the driveplate from the end of the crankshaft. If the locating dowel is a loose fit in the crankshaft end, remove it and store it with the driveplate for safe-keeping.

Inspection

7 On models with manual transmission, examine the flywheel for scoring of the clutch face, and for wear or chipping of the ring gear teeth. If the clutch face is scored, the flywheel may be surface-ground, but renewal is preferable. Seek the advice of a Citroën dealer or engine specialist to see if machining is possible. If the ring gear is worn or damaged, the flywheel must be renewed, as it is not possible to renew the ring gear separately.

8 On models with automatic transmission, check the torque converter driveplate carefully for signs of distortion. Look for any hairline cracks around the bolt holes or radiating outwards from the centre, and inspect the ring gear teeth for signs of wear or chipping. If any sign of wear or damage is found, the driveplate must be renewed.

Refitting

Flywheel - models with manual transmission

9 Clean the mating surfaces of the flywheel and crankshaft. Remove any locking compound from the threads of the crankshaft holes, using the correct-size tap, if available.

> **HAYNES HINT** *If a suitable tap is not available, cut two length-wise slots into the threads of an old flywheel bolt and use the bolt to remove the locking compound from the threads.*

10 If the new flywheel retaining bolts are not supplied with their threads already pre-coated, apply a suitable thread-locking compound to the threads of each bolt **(see illustration)**.

11 Ensure the locating dowel is in position. Offer up the flywheel, locating it on the dowel, and fit the new retaining bolts.

12 Lock the flywheel using the method employed on dismantling, and tighten the

17.10 If the new flywheel bolt threads are not supplied with their threads pre-coated, apply a suitable locking compound . . .

2A

17.12 . . . then refit the flywheel, and tighten the bolts to the specified torque

retaining bolts to the specified torque (see *Specifications*) **(see illustration)**.

13 Refit the clutch as described in Chapter 6. Remove the flywheel locking tool, and refit the transmission as described in Chapter 7A.

Driveplate - models with automatic transmission

14 Carry out the operations described above in paragraphs 9 and 10, substituting 'driveplate' for all references to the flywheel.

15 Locate the driveplate on its locating dowel.

16 Offer up the torque converter plate, with the thinner shim positioned behind the plate and the thicker shim on the outside, and align the marks made prior to removal.

17 Fit the new retaining bolts, then lock the driveplate using the method employed on dismantling. Tighten the retaining bolts to the specified torque wrench setting (see *Specifications*).

18 Remove the driveplate locking tool, and refit the transmission (refer to Chapter 7B).

18 Engine/transmission mountings - inspection and renewal

Inspection

1 If improved access is required, raise the front of the car and support it securely on axle stands (see *Jacking and Vehicle Support*).

2 Check the mounting rubber to see if it is cracked, hardened or separated from the metal at any point; renew the mounting if any such damage or deterioration is evident.

3 Check that all the mounting's fasteners are securely tightened; use a torque wrench to check if possible.

4 Using a large screwdriver or a crowbar, check for wear in the mounting by levering against it to check for free play. Where this is not possible, enlist the aid of an assistant to move the engine/transmission unit back and forth, or from side to side, while you watch the mounting. While some free play is to be expected even from new components,

excessive wear should be obvious. If excessive free play is found, check first that the fasteners are correctly secured, then renew any worn components as described below.

Renewal

Right-hand mounting

5 Disconnect the battery negative lead. Release all the relevant hoses and wiring from their retaining clips. Place the hoses/wiring clear of the mounting so that the removal procedure is not hindered.

6 Place a jack beneath the engine, with a block of wood on the jack head. Raise the jack until it is supporting the weight of the engine.

7 Undo the two bolts securing the curved mounting retaining plate to the body. Lift off the plate, and withdraw the rubber damper from the top of the mounting bracket.

8 Slacken and remove the two nuts and two bolts securing the right-hand engine/ transmission mounting bracket to the engine. Remove the single nut securing the bracket to the mounting rubber, and lift off the bracket.

9 Lift the rubber buffer plate off the mounting rubber stud, then unscrew the mounting rubber from the body and remove it from the vehicle. If necessary, the mounting bracket can be unbolted and removed from the front of the cylinder block.

10 Check all components carefully for signs of wear or damage, and renew as necessary.

11 On reassembly, screw the mounting rubber into the vehicle body, and tighten it securely. Where removed, refit the mounting bracket to the front of the cylinder head, and securely tighten its retaining bolts.

12 Refit the rubber buffer plate to the mounting rubber stud, and install the mounting bracket.

13 Tighten the mounting bracket retaining nuts to the specified torque setting (see *Specifications*), and remove the jack from underneath the engine.

14 Refit the rubber damper to the top of the mounting bracket, and refit the curved retaining plate. Tighten the retaining plate bolts to the specified torque (see *Specifications*), and reconnect the battery.

Left-hand mounting

15 Remove the battery and battery tray, as described in Chapter 5A.

16 Place a jack beneath the transmission, with a block of wood on the jack head. Raise the jack until it is supporting the transmission.

17 Slacken and remove the centre nut and washer from the left-hand mounting, then undo the nuts securing the mounting in position and remove it from the engine compartment.

18 If necessary, slide the spacer (where fitted) off the mounting stud, then unscrew the

stud from the top of the transmission housing, and remove it along with its washer. If the mounting stud is tight, a universal stud extractor can be used to unscrew it.

19 Check all components carefully for signs of wear or damage, and renew as necessary.

20 Clean the threads of the mounting stud, and apply a coat of thread-locking compound to its threads. Refit the stud and washer to the top of the transmission, and tighten it to the specified torque setting (see *Specifications*).

21 Slide the spacer (where fitted) onto the mounting stud, then refit the rubber mounting. Tighten both the mounting-to-body bolts and the mounting centre nut to their specified torque settings (see *Specifications*), and remove the jack from below the transmission.

22 Refit the battery support plate, tightening its retaining bolts securely, then refit the battery as described in Chapter 5A.

Rear mounting

23 If not already done, chock the rear wheels, then jack up the front of the vehicle and support it securely on axle stands (see *Jacking and Vehicle Support*).

24 Unscrew and remove the bolt securing the rear mounting link to the mounting on the rear of the cylinder block **(see illustration)**.

25 Remove the bolt securing the rear mounting link to the bracket on the underbody. Withdraw the link.

26 To remove the mounting assembly it will first be necessary to remove the right-hand driveshaft as described in Chapter 8.

27 With the driveshaft removed, undo the retaining bolts and remove the mounting from the rear of the cylinder block.

28 Check carefully for signs of wear or damage on all components, and renew them where necessary.

29 On reassembly, fit the rear mounting assembly to the rear of the cylinder block, and tighten its retaining bolts to the specified torque (see *Specifications*). Refit the driveshaft as described in Chapter 8.

30 Refit the rear mounting link, and tighten both its bolts to their specified torque settings (see *Specifications*).

31 Lower the vehicle to the ground.

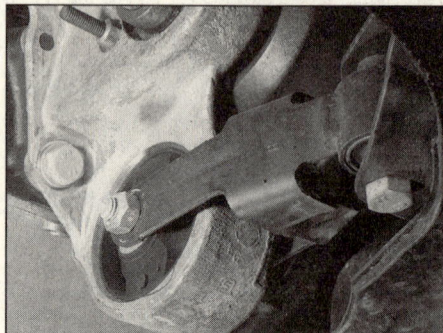

18.24 Rear engine mounting and link

Chapter 2 Part B:
Diesel engine in-car repair procedures

Contents

Degrees of difficulty

Easy, suitable for novice with little experience	**Fairly easy,** suitable for beginner with some experience	**Fairly difficult,** suitable for competent DIY mechanic	**Difficult,** suitable for experienced DIY mechanic	**Very difficult,** suitable for expert DIY or professional

Specifications

General

Designation:
2.1 litre (2088 cc) turbo engine	XUD11ATE/BTE
2.1 litre (2138 cc) non-turbo engine	XUD11A
2.5 litre (2445 cc) turbo engine	DK5

Engine codes*:
2.1 litre turbo engine	PHZ, P8A/B/C
2.1 litre non-turbo engine	P9A, PJZ
2.5 litre turbo engine	THY

Bore:
2.1 litre turbo engine	85.00 mm
2.1 litre non-turbo engine	86.00 mm
2.5 litre engine	92.00 mm

Stroke:
2.1 litre turbo engine	92.00 mm
2.1 litre non-turbo engine	92.00 mm
2.5 litre engine	92.00 mm
Direction of crankshaft rotation	Clockwise (viewed from right-hand side of vehicle)
No 1 cylinder location	At the transmission end of block

Compression ratio:
2.1 litre turbo engine	21.5 : 1
2.1 litre non-turbo	22.5 : 1
2.5 litre engine	22.0 : 1

The engine code is stamped on a plate attached to the front of the cylinder block. This is the code most often used by Citroën. The code given in brackets is the factory identification number, and is not often referred to by Citroën or this manual.

Camshaft

Drive	Toothed belt

No of bearings:
2.1 litre engine	5
2.5 litre engine	5

Endfloat:
2.1 litre engine	Not available at time of writing
2.5 litre engine	Not available at time of writing

2B

Lubrication system

Oil pump type	Gear-type, chain-driven off the crankshaft right-hand end
Oil pressure:	
2.1 litre turbo engine	2.5 bar at 2000 rpm, 100°C
2.1 litre non-turbo engine	2.7 bar at 2000 rpm, 100°C
2.5 litre turbo engine	3.0 bar at 2000 rpm, 90°C
Oil pressure warning switch operating pressure	0.5 bars

Camshaft timing belt

Belt tension(2.5 litre engine only)*:	
New belt:	
Pre-tension	800 N (107 SEEM units)
Final tension	300 N (58 SEEM units)
Used belt:	
Pre-tension	500 N (80 SEEM units)
Final tension	250 N (51 SEEM units)

On 2.1 litre engines, the camshaft timing belt tension is set by the automatic tensioner.

Balance shaft timing belt (2.5 litre engines only)

Belt tension:	
New belt	
Pre-tension	400 N (70 SEEM units)
Final tension	120 N (31 SEEM units)
Used belt:	
Pre-tension	250 N (51 SEEM units)
Final tension	90 N (26 SEEM units)

Torque wrench settings

2.5 litre engines	Nm	lbf ft
Auxiliary drivebelt tensioner pulley	43	32
Camshaft carrier-to-cylinder head nuts	20	15
Camshaft sprocket bolts:		
Stage 1	10	7
Stage 2	25	18
Camshaft sprocket centre nut	43	32
Camshaft oil seal housing bolts	12	9
Crankshaft sprocket bolt:		
Stage 1	70	52
Stage 2	Angle tighten through 51°	
Crankshaft pulley bolts	20	15
Cylinder head bolts:		
M10 bolts:		
Stage 1	35	26
Stage 2	Angle tighten through 120°	
M12 bolts:		
Stage 1	50	37
Stage 2	Angle tighten through 120°	
Cylinder head cover bolts	8	6
Cylinder head blanking plate bolts	15	11
Flywheel/driveplate bolts	50	37
Injection pump bolts:		
Stage 1	10	7
Stage 2	25	18
Left-hand engine/transmission mounting:		
Mounting bracket-to-body	30	22
Rubber mounting-to-bracket bolts	20	15
Mounting stud-to-transmission	50	37
Centre nut	65	48
Rear engine mounting/torque control mechanism:		
Torque arm to mounting bracket	50	37
Mounting bracket to engine	55	41
Right hand engine mounting:		
Mounting bracket to chassis bolts	50	37
Through-bolt	90	66
Lower torque control mechanism:		
Pushrod mounting bolts	110	81
Torque arm mounting bracket-to-chassis bolts	50	37
Torque arm-to-mounting bracket bolts (front)	110	81
Torque arm-to-mounting bracket bolts (rear)	60	44

Torque wrench settings

	Nm	lbf ft
2.5 litre engines (continued)		
Oil pump mounting bolts	9	7
Oil pressure switch	22	16
Piston oil jet spray tube bolt	10	7
Sump bolts	8	6
Timing belt tensioner centre nut	45	33
Balance shaft drivebelt tensioner centre nut	45	33
Balance shaft drivebelt idler pulley centre nut	45	33
2.1 litre engines*		
Camshaft carrier bolts	25	18
Camshaft sprocket bolt	50	37
Crankshaft front oil seal housing bolts	16	12
Crankshaft pulley bolt:		
Stage 1	40	30
Stage 2	Tighten through a further 60°	
Cylinder head cover bolts	8	6
Cylinder head bolts (new bolts with guide bosses)†:		
Stage 1	20	15
Stage 2	60	44
Stage 3	Angle tighten through 180°	

† Note: *Where new cylinder head bolts are fitted, there is no requirement to re-tighten the bolts after the engine has been started and warmed up for the first time.*

	Nm	lbf ft
Flywheel/driveplate bolts	50	37
Injection pump sprocket puller retaining screws	10	7
Injection pump sprocket nut	50	37
Left-hand engine/transmission mounting:		
Mounting bracket-to-body	30	22
Rubber mounting-to-bracket bolts	30	22
Mounting stud-to-transmission	60	44
Centre nut	65	48
Lower engine movement limiter-to-driveshaft intermediate bearing		
housing	50	37
Lower engine movement limiter-to-subframe	85	62
Oil pump mounting bolts	13	10
Piston oil jet spray tube bolt	10	7
Right-hand engine/transmission mounting:		
Mounting bracket-to-engine nuts	45	33
Mounting bracket-to-rubber mounting nut	45	33
Rubber mounting-to-body nut	40	29
Upper engine movement limiter bolts	50	37
Sump bolts	16	12
Timing belt idler pulley	37	27
Timing belt tensioner nut/bolt	10	7

***Note:** At the time of writing, a definitive set of torque figures was not available for the 2.1 litre engine. The above settings are given for guidance only.*

1 General information

How to use this Chapter

This Part of Chapter 2 describes the repair procedures that can reasonably be carried out on the engine while it remains in the vehicle. If the engine has been removed from the vehicle and is being dismantled as described in Part C, any preliminary dismantling procedures can be ignored.

Note that, while it may be possible physically to overhaul items such as the piston/connecting rod assemblies while the engine is in the car, such tasks are not usually carried out as separate operations. Usually, several additional procedures are required (not to mention the cleaning of components and oilways); for this reason, all such tasks are classed as major overhaul procedures, and are described in Part C of this Chapter.

Part C describes the removal of the engine/transmission from the car, and the full overhaul procedures that can then be carried out.

Engine description

The engine is a four-cylinder overhead camshaft design, mounted transversely, with the transmission mounted on the left-hand side.

An aluminium alloy cylinder head is fitted, incorporating twelve valves (two inlet and one exhaust per cylinder). The valve clearances are self-adjusting by means of hydraulic tappets fitted to the cam followers.

The camshaft is supported by five bearings within a separate carrier. A toothed timing belt drives the camshaft, fuel injection pump and coolant pump.

The crankshaft runs in five main bearings of the usual shell type. Endfloat is controlled by thrustwashers either side of No 2 main bearing.

The pistons are selected to be of matching weight, and incorporate fully-floating gudgeon pins retained by circlips. The underside of each piston crown has a cooling channel, which receives oil sprayed from jets that are mounted at the lower end of each cylinder.

The oil pump is chain-driven from the front of the crankshaft. An oil cooler is fitted to all engines.

The 2.5 litre DK5 engine is fitted with two balance shafts; one mounted either side of the cylinder block. The balance shafts are driven

2B

from the crankshaft pulley via a dedicated toothed belt that runs behind the camshaft timing belt. Eccentrically-mounted weights on each balance shaft rotate at twice crankshaft speed. The synchronisation between the two balance shafts and the crankshaft is such that the lateral forces generated by the rotation of the balance shafts tends to cancel out the forces generated by the movement of the crankshaft and pistons. The result is a substantial reduction in engine vibration.

Throughout the manual, it is often necessary to identify the engines not only by their cubic capacity, but also by their engine code. The engine code consists of three characters (eg: THY). The code is stamped on a plate attached to the front of the cylinder block.

Repair operations and precautions

Both the 2.1 and 2.5 litre engines are complex units with numerous accessories and ancillary components. The design of the XM engine compartment is such that every conceivable space has been utilised, and access to virtually all of the engine components is very limited. In many cases, ancillary components will have to be removed, or moved to one side, and wiring, pipes and hoses will have to be disconnected or removed from various cable clips and support brackets.

When working on these engines, read through the entire procedure first, look at the

car and engine at the same time, and establish whether you have the necessary tools, equipment, skill, patience and time to proceed. Allow considerable time for any operation, and be prepared for the unexpected. Any major work on these engines is not for the faint-hearted!

Because of the limited access, many of the photographs appearing in this Chapter were, by necessity, taken with the engine removed from the vehicle.

Repair operations possible with the engine in the vehicle

The following operations can be carried out without having to remove the engine from the vehicle:

2.1 litre models

a) *Removal and refitting of the timing belt and sprockets.*
b) *Removal and refitting of the camshaft and hydraulic tappets.*
c) *Removal and refitting of the sump.*
d) *Removal and refitting of the oil pump.*
e) *Renewal of the engine/transmission mountings.*
f) *Removal and refitting of the flywheel/driveplate.*

2.5 litre models

a) *Removal and refitting of the camshaft timing belt and sprockets.*
b) *Removal and refitting of the camshaft and hydraulic tappets.*

c) *Removal of the cylinder head.*
d) *Removal and refitting of the balance shaft drivebelt (Chapter 2C).*
e) *Removal and refitting of the balance shafts (Chapter 2C).*
f) *Removal and refitting of the sump.*
g) *Removal and refitting of the oil pump.*
h) *Renewal of the engine/transmission mountings.*
i) *Removal and refitting of the flywheel/driveplate.*

2 Camshaft and followers - removal, inspection and refitting

Removal

2.1 litre engine

1 Remove the camshaft cover as described in Section 3.
2 Remove the camshaft sprocket as described in Section 15.
3 Remove the braking system vacuum pump. Recover the vacuum pump oil feed tube from the end of the camshaft **(see illustration)**.
4 Refer to Chapter 4B and remove the fuel supply and leak-off pipes from the fuel injectors.
5 Disconnect the oil return hose from the front of the camshaft carrier.
6 Working in a spiral sequence, progressive slacken and remove the camshaft carrier retaining bolts.
7 Lift the camshaft carrier, complete with camshaft, upwards off the locating dowels **(see illustration)**.
8 Extract the oil seal from the end of the camshaft carrier.
9 Undo the two bolts securing the camshaft thrust plate, and carefully slide the camshaft out of the carrier **(see illustration)**.
10 Obtain twelve clean plastic containers, and number them inlet 1 to 8, and exhaust 1 to 4; alternatively, divide a larger container into twelve compartments.
11 Lift off the rockers and their guides and place them in their respective containers **(see illustrations)**. Withdraw each hydraulic tappet

2.3 Recover the vacuum pump oil feed tube (2.1 litre models)

2.7 On 2.1 litre models, lift the camshaft carrier, complete with camshaft, upwards off the locating dowels

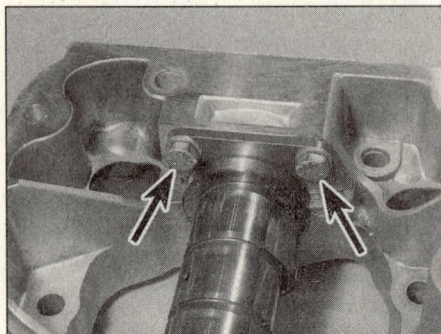
2.9 Undo the two bolts securing the camshaft thrust plate (arrowed)

2.11a Lift off the rockers . . .

2.11b . . . and their guides and place them in their respective containers

2.12 **Remove the oil filter tube from its cylinder head location**

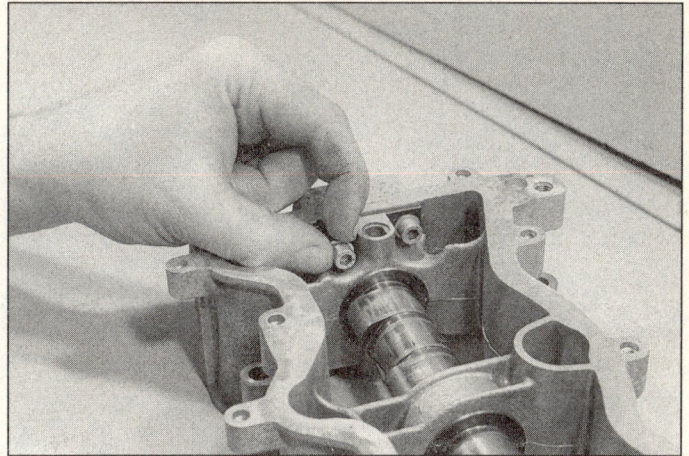

2.25a **Slacken and withdraw the thrust plate screws . . .**

in turn, and place it in its respective container. Do not interchange the tappets, or the rate of wear will be much-increased.

12 Remove the oil filter tube from its location in the cylinder head (**see illustration**).

2.5 litre engine

13 Refer to Chapter 3 and carry out the following:
a) *Remove the coolant pump drivebelt.*
b) *Unbolt the coolant pump pulley from the end of the camshaft.*

14 Refer to Section 4 and remove the auxiliary drivebelt(s).

15 Undo the securing bolts and remove the auxiliary drive belt roller tensioner.

16 With reference to Section 6, remove the camshaft timing belt cover.

17 Working from Section 12, set the engine to TDC on cylinder No 4 and lock it in position as described, using the timing belt fuel injection pump sprocket and flywheel locking pins.

18 Release the camshaft sprocket from the end of the camshaft, as described in Section 15.

19 Unscrew the nut, lift off the washer and remove the camshaft sprocket plate from the end of the camshaft.

20 Unbolt and remove the upper section of the inlet manifold, as described in Chapter 4B.

21 Refer to Section 3 and remove the camshaft cover.

22 Working in a spiral sequence, progressive slacken and remove the camshaft carrier retaining bolts.

23 Lift the camshaft carrier, complete with camshaft, upwards off the locating dowels.

24 Extract the oil seals from both ends of the camshaft carrier (Section 5).

25 Undo the bolts securing the camshaft thrust plate, and carefully slide the camshaft out of the carrier (**see illustrations**).

26 Obtain twelve clean plastic containers, and number them inlet 1 to 8, and exhaust 1 to 4; alternatively, divide a larger container into twelve compartments.

27 Lift off the rockers and their guides and place them in their respective containers. Withdraw each hydraulic tappet in turn, and place it in its respective container. Do not interchange the tappets, or the rate of wear will be much-increased.

28 Remove the oil filter tube from its location in the cylinder head (see Section 9).

Inspection

29 Examine the camshaft bearing surfaces and cam lobes for signs of wear ridges and scoring. Renew the camshaft if any of these conditions are apparent. Examine the condition of the bearing surfaces, both on the camshaft journals and in the cylinder head/camshaft carrier/bearing caps. If the bearing surfaces are worn excessively, the cylinder head/camshaft carrier will need to be renewed.

30 Examine the cam follower/tappet bearing surfaces which contact the camshaft lobes or rockers for wear ridges and scoring. Renew any component on which these conditions are apparent. If a follower/tappet bearing surface is badly scored, also examine the corresponding rocker or lobe on the camshaft for wear, as it is likely that both will be worn. Renew worn components as necessary.

Refitting

All engines

31 Liberally lubricate the camshaft and the camshaft bearing journals in the carrier and slide the camshaft into the carrier. Refit the thrust plate and secure with the two bolts.

32 Liberally lubricate each hydraulic tappet and place it in its respective bore (**see illustration**).

33 Lubricate the guides and rockers and place all twelve over their respective valves

2.25b **. . . remove the thrust plate . . .**

2.25c **. . . and withdraw the camshaft**

2.32 **Lubricate each hydraulic tappet and place it in its bore**

2B

2.33a Lubricate the ends of each valve stem . . .

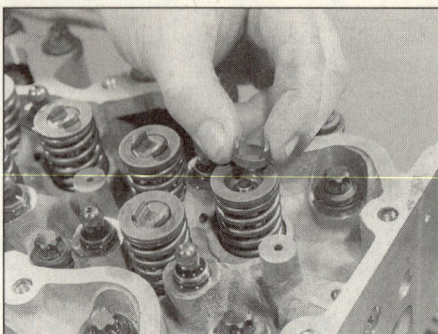

2.33b . . . then position the guides . . .

2.33c . . . and rockers over their respective valves

2.35a Lubricate the lips of a new camshaft oil seal and fit the seal to the camshaft carrier . . .

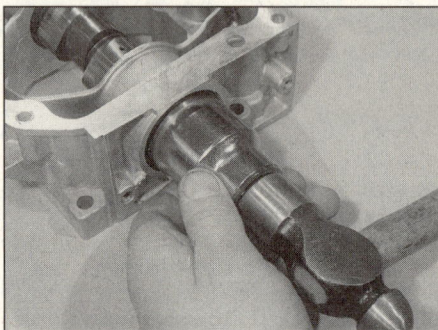

2.35b . . . tap the seal into position using a suitable socket

(see illustrations). Ensure that the guides are fitted with their slots facing upwards, and that the rockers engage with the guide slots.

34 Insert a new oil filter tube to its bore in the cylinder head.

35 Liberally lubricate the lips of a new camshaft oil seal and fit the seal to the camshaft carrier (note that on 2.5 litre engines there is a seal at each end of the camshaft; see Section 5). Tap the seal into position using a socket of suitable diameter, or the old seal (see illustrations).

36 On 2.5 litre engines, ensure that the Woodruff key is still in place, then refit the camshaft sprocket plate and secure it in position with the washer and nut. Insert a bolt through the timing holes in the plate and camshaft carrier to ensure that the camshaft is correctly positioned then tighten the plate securing nut to the specified torque (see illustrations).

37 Apply a bead of silicone sealant to the space between the groove and the outer edge of the camshaft carrier (see illustration). Ensure that the sealant is applied all around the two bolt holes at the timing belt end of the carrier.

38 Locate the assembled camshaft carrier on the cylinder head, taking care not to dislodge the rockers and guides.

39 Refit the retaining bolts, then working in a spiral sequence from the centre outward, progressively tighten the camshaft carrier retaining bolts to the specified torque (see illustrations).

2.36a On 2.5 litre engines, refit the camshaft sprocket plate

2.36b Insert a bolt (arrowed) through the timing holes in the plate and camshaft carrier

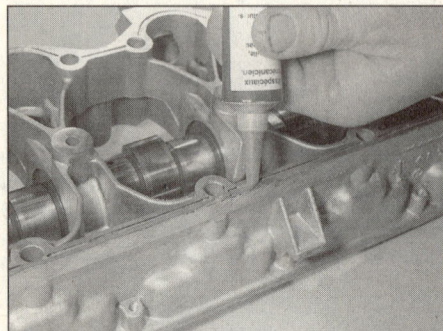

2.37 Apply a bead of silicone sealant to the space between the groove and the outer edge of the camshaft carrier

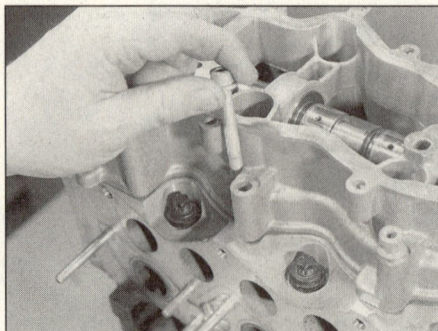

2.39a Refit the retaining bolts . . .

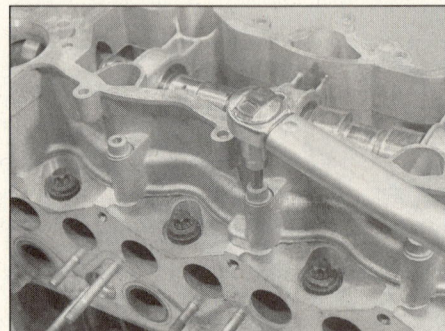

2.39b . . . then working in a spiral sequence, tighten the camshaft carrier retaining bolts

40 Reconnect the oil return hose to the front of the carrier.
41 The remainder of the refitting procedure is a reversal of removal, noting the following points:
a) *Observe the correct torque settings, where specified.*
b) *On 2.1 litre engines, refit the fuel supply and leak-off pipes to the fuel injectors with reference to Chapter 4B.*
c) *On 2.1 litre engines, refit the braking system vacuum pump.*
d) *Refit the camshaft sprocket as described in Section 15.*
e) *Refit the timing belt and cover as described in Section 6.*
f) *Refit the cylinder head cover as described in Section 3.*
g) *On 2.5 litre engines, refit the auxiliary drivebelt(s) as described in Section 4.*
h) *On 2.5 litre engines, refit the coolant pump pulley and drivebelt with reference to Chapter 3.*

3 Camshaft cover - removal and refitting

2.5 litre models

Removal

1 Unclip the fuel supply and return hoses from the top of the camshaft timing belt outer cover.
2 With reference to Chapter 4B, unbolt and remove the upper section of the inlet manifold. It will be necessary to first unbolt the EGR solenoid valve bracket from the side of the manifold.
3 Slacken the clip and disconnect the breather hose from the port on the cylinder head cover.
4 Note the locations of any brackets secured by the cylinder head cover retaining bolts, then unscrew the bolts in a progressive spiral sequence **(see illustration)**.
5 Carefully move any hoses clear of the cylinder head cover.

3.11a Removing the fuel hose bracket from the cylinder head cover

3.4 Unscrew the cylinder head cover retaining bolts

6 Lift off the cover, and recover the rubber seal **(see illustration)**. Examine the seal for signs of damage and deterioration, and if necessary, renew it.

Refitting

7 Refitting is a reversal of removal. Ensure that the camshaft cover seal is firmly pressed into its recess. Observe the correct torque when tightening the camshaft cover bolts - do not overtighten them, or the seal may be distorted, causing leakage.

2.1 litre models

Removal

8 Remove the timing belt upper cover as described in Section 6.
9 Remove the upper section of the inlet manifold as described in Chapter 4B.
10 Disconnect the breather hose from the front of the cylinder head cover.
11 Note the locations of any brackets secured by the cylinder head cover retaining bolts, then unscrew the bolts in a progressive spiral sequence **(see illustrations)**.
12 Carefully move any hoses clear of the cylinder head cover.
13 Lift off the cover, and recover the rubber seal **(see illustration)**. Examine the seal for signs of damage and deterioration, and if necessary, renew it.

Refitting

14 Refitting is a reversal of removal, bearing in mind the following points:

3.11b Remove the cylinder head cover retaining bolts and washers

3.6 Lift off the cylinder head cover , and recover the rubber seal (2.5 litre model)

a) *Refit any brackets in their original positions noted before removal.*
b) *Refit the inlet manifold and air inlet ducts described in Chapter 4B.*

4 Auxiliary drivebelt - removal and refitting

Note: *Depending on model and equipment fitted, access to the auxiliary drivebelt can be extremely limited. Where necessary, greater working clearance can be gained by removing the diesel injection electronic control unit (ECU) and its mounting box as described in Chapter 4B. If working on the 2.1 litre model, the help of an assistant will also be beneficial.*
Note: *On models with a manually adjusted drivebelt, Citroën specify the use of a special electronic tool (SEEM C105.5) to correctly set the drivebelt tension. If access to this equipment cannot be obtained, an approximate setting can be achieved as described in the text. If this method is used, the tension must be checked using the special electronic tool at the earliest opportunity.*
1 All models covered by this manual are equipped with either one or two multi-ribbed auxiliary drivebelts. The belt tension is adjusted manually on models without air conditioning, and automatically (once an initial setting procedure has been carried out), by means of a spring-loaded tensioner, on models with air conditioning.

2B

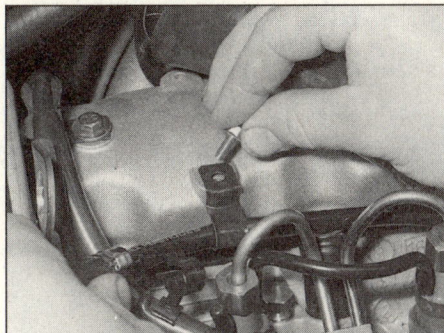
3.13 Lifting off the cylinder head cover on 2.1 litre models

Condition check

2 Apply the handbrake, then jack up the front of the car and support it on axle stands. Remove the right-hand front roadwheel.

3 Release the screws and clips and remove the wheel arch liner from under the right-hand front wing for access to the crankshaft pulley bolt. Where fitted, also remove the splash guard from under the front of the engine.

4 Using a suitable socket and bar fitted to the crankshaft pulley bolt, rotate the crankshaft so that the entire length of the drivebelt can be examined. Examine the drivebelt for cracks, splitting, fraying or damage. Check also for signs of glazing (shiny patches) and for separation of the belt plies. Renew the belt if worn or damaged.

5 If the condition of the belt is satisfactory, check the drivebelt tension as described below.

2.1 litre models without air conditioning

Removal

6 If not already done, proceed as described in paragraphs 2 and 3.

7 Disconnect the battery negative lead.

8 Slacken the two bolts securing the tensioning pulley assembly to the engine (see illustration).

9 Rotate the adjuster bolt to move the tensioner pulley away from the drivebelt until there is sufficient slack for the drivebelt to be removed from the pulleys.

Refitting

10 Fit the drivebelt around the pulleys in the following order:
a) Power steering pump.
b) Crankshaft.
c) Alternator.
d) Tensioner pulley.

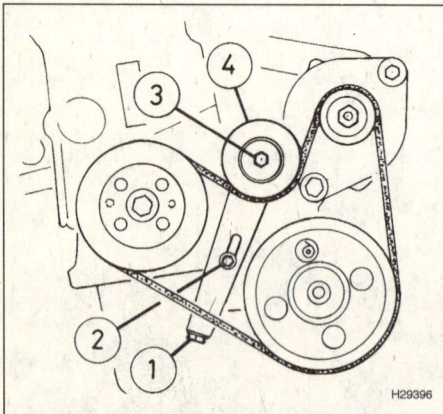

4.8 Auxiliary drivebelt adjustment details (2.1 litre models)

1 Adjuster bolt
2 Tensioner pulley assembly - lower securing bolt
3 Tensioner pulley assembly - upper securing bolt
4 Tensioner pulley

11 Ensure that the ribs on the belt are correctly engaged with the grooves in the pulleys, and that the drivebelt is correctly routed. Take all the slack out of the belt by turning the tensioner pulley adjuster bolt. Tension the belt as follows.

Tensioning

12 If not already done, proceed as described in paragraphs 2 and 3.

13 Correct tensioning of the drivebelt will ensure that it has a long life. A belt which is too slack will slip and perhaps squeal. Beware, however, of overtightening, as this can cause wear in the alternator bearings.

14 The belt should be tensioned so that, under firm thumb pressure, there is approximately 5.0 mm of free movement at the mid-point between the pulleys on the longest belt run.

15 To adjust the tension, with the tensioner pulley assembly retaining bolts slackened, rotate the adjuster bolt until the correct tension is achieved.

16 Once the belt is correctly tensioned, rotate the crankshaft four complete revolutions in the normal direction of rotation and recheck the tension.

17 When the belt is correctly tensioned, tighten the tensioner pulley assembly retaining bolts to the specified torque, then reconnect the battery negative lead.

18 Refit the wheel arch liner and, where fitted, the engine splash guard. Refit the roadwheel, and lower the vehicle to the ground.

2.5 litre models without air conditioning

Removal

19 If not already done, proceed as described in paragraphs 2 and 3.

20 Disconnect the battery negative lead.

21 Relieve all tension from the drivebelt tensioner by slackening the tensioner adjuster bolt.

22 Reach through the right hand wheel arch and release the belt from the tensioner pulley. Slide the belt from the alternator, hydraulic pump and crankshaft pulleys and remove it from the vehicle (see illustration).

4.22 Removing the auxiliary drivebelt via the right hand wheel arch (2.5 litre models without air conditioning)

Refitting and tensioning

23 Fit the drivebelt around the pulleys in the following order:
a) Crankshaft.
b) Hydraulic pump.
c) Alternator.
d) Tensioner pulley.

24 Adjust the drivebelt tension by tightening the adjuster bolt. The belt should be tensioned so that, under firm thumb pressure, there is approximately 5.0 mm of free movement at the mid-point between the pulleys on the longest belt run.

25 Once the belt is correctly tensioned, rotate the crankshaft four complete revolutions in the normal direction of rotation and recheck the tension.

26 When the belt is correctly tensioned, reconnect the battery negative lead, refit the wheel arch liner and, where fitted, the engine splash guard/undertray. Refit the roadwheel, and lower the vehicle to the ground.

2.5 litre models with air conditioning

Removal

27 If not already done, proceed as described in paragraphs 2 and 3.

28 Disconnect the battery negative lead.

29 Working under the wheel arch, slacken the retaining bolt located in the centre of the eccentric tensioner pulley (see illustration).

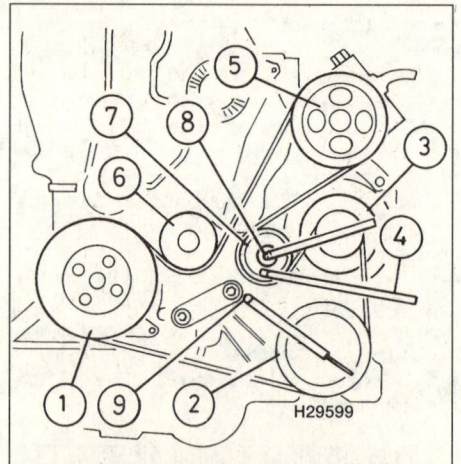

4.29 Auxiliary drivebelt tensioning details (2.5 litre models with air conditioning)

1 Crankshaft pulley
2 Air con compressor pulley
3 Alternator pulley
4 Cranked, square section bar
5 Hydraulic pump pulley
6 Automatic tensioner pulley
7 Eccentric tensioner pulley
8 Eccentric tensioner pulley retaining bolt (with socket and wrench attached)
9 Setting tool inserted in 8 mm hole in arm of automatic tensioner pulley

30 Insert a cranked, square section bar (a quarter inch square drive socket bar for example) into the square hole on the front face of the eccentric tensioner pulley.

31 Using the bar, turn the eccentric tensioner pulley until the hole in the arm of the automatic tensioner pulley is aligned with the hole in the mounting bracket behind. When the holes are aligned, slide a suitable setting tool (a bolt, or cranked length of rod of approximately 8.0 mm diameter) through the hole in the arm and into the mounting bracket.

32 With the automatic tensioner locked, turn the eccentric tensioner pulley until the drivebelt tension is released sufficiently to enable the belt to be removed.

Refitting and tensioning

33 Fit the drivebelt around the pulleys in the following order:
a) Air conditioning compressor.
b) Crankshaft.
c) Automatic tensioner pulley.
d) Hydraulic pump/power steering pump.
e) Alternator.
f) Eccentric tensioner pulley.

34 Ensure that the ribs on the belt are correctly engaged with the grooves in the pulleys.

35 Turn the eccentric tensioner pulley to apply tension to the drivebelt, until the load is released from the setting bolt. Without altering the position of the eccentric tensioner pulley, tighten its retaining bolt to the specified torque.

36 Remove the setting bolt from the automatic tensioner arm, then rotate the crankshaft four complete revolutions in the normal direction of rotation.

37 Check that the holes in the automatic adjuster arm and the mounting bracket are still aligned by re-inserting the setting bolt. If the bolt will not slide in easily, repeat the tensioning procedure from paragraph 35 onward.

38 On completion, reconnect the battery negative lead, refit the wheel arch liner and, where fitted, the engine splash guard. Refit the roadwheel, and lower the vehicle to the ground.

5 Camshaft oil seal(s) - renewal

Camshaft right-hand oil seal

1 Remove the camshaft sprocket from end of the camshaft as described in Section 15.

2 On 2.5 litre engines, remove the nut and slide the sprocket plate off the end of the camshaft (see Section 2).

3 Pull the oil seal from the housing using a hooked instrument. Alternatively, drill a small hole in the oil seal and use a self-tapping screw and a pair of pliers to remove it. Take great care to avoid drilling into the camshaft or carrier sealing surfaces.

4 Clean the oil seal housing and the camshaft sealing surface.

5 Smear the new oil seal with clean engine oil, then fit it over the end of the camshaft, open end first **(see illustration)**. A piece of thin plastic or tape wound around the front of the camshaft is useful to prevent damage to the oil seal as it is fitted.

6 Press the seal partially into the housing. Use a nut (screwed into the end of the camshaft), packing washers and a suitable tube or socket to drive the seal into the housing, until it is flush with the end face of the cylinder head.

7 On 2.5 litre models, refit the sprocket plate (see Section 2). On all models, refit the camshaft sprocket as described in Section 15.

8 If oil leakage from the old seal has contaminated the timing belt, fit a new belt as described in Section 6.

Camshaft left-hand oil seal

2.1 litre models

9 No oil seal is fitted to the left-hand end of the camshaft. The sealing is provided by an O-ring fitted to the end plate flange. The O-ring can be renewed after unbolting the plate from the cylinder head.

2.5 litre models

10 Refer to Chapter 3 and carry out the following:

a) Remove the coolant pump drive belt
b) Unbolt the coolant pump drivebelt pulley from the end of the of the camshaft.

11 Pull the oil seal from the housing using a hooked instrument. Alternatively, drill a small hole in the oil seal and use a self-tapping screw and a pair of pliers to remove it. Take great care to avoid drilling into the camshaft or carrier sealing surfaces.

12 Clean the oil seal housing and the camshaft sealing surface.

13 Smear the new oil seal with clean engine oil, then fit it over the end of the camshaft, open end first. A piece of thin plastic or tape wound around the front of the camshaft is useful to prevent damage to the oil seal as it is fitted **(see illustration)**.

14 Press the seal partially into the housing, then using a mallet and a length of tube or a socket as a drift, tap the seal squarely into position. until it is flush with the end face of the cylinder head **(see illustration)**.

15 Refit the coolant pump drivebelt pulley as described in Chapter 3.

16 If oil leakage from the old seal has contaminated the coolant pump drive belt, fit a new belt as described in Chapter 3.

6 Camshaft timing belt and outer covers - removal and refitting

General

1 The timing belt drives the camshaft, injection pump, (and on 2.1 litre engines, the coolant pump) from a toothed sprocket on the front of the crankshaft. If the belt breaks or slips in service, the pistons are likely to hit the valve heads, resulting in expensive damage.

2 The timing belt should be renewed at the specified intervals, or earlier if it is contaminated with oil, or at all noisy in operation (a 'scraping' noise due to uneven wear).

3 If the timing belt is being removed on a 2.1 litre engine, it is a wise precaution to check the coolant pump for signs of coolant leakage at the same time. This may avoid the need to remove the timing belt again at a later stage, should the coolant pump fail.

2B

5.5 Fitting a new right hand camshaft oil seal

5.13 Fitting a new left hand camshaft oil seal

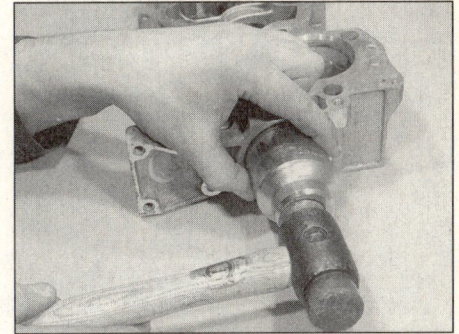
5.14 Drive the seal squarely into its housing

6.5 Undo the single retaining bolt, located in the centre of the cover (2.1 litre models)

6.6 Turn the upper fastener a quarter of a turn clockwise to release the locking peg (2.1 litre models)

6.9 Remove the centre cover from the front of the injection pump (2.1 litre models)

4 Due to the design of the fuel injection system fitted to 2.5 litre models, it is possible to remove the fuel injection pump sprocket from the pump shaft without removing the timing belt from the sprocket. The same is true for the removal of the camshaft sprocket - refer to Section 15 for more detail.

Timing belt covers

2.1 litre engine upper cover - removal

5 Undo the single retaining bolt, located in the centre of the cover (see illustration).
6 Turn the upper fastener a quarter of a turn clockwise to release the locking peg (see illustration).
7 Manipulate the cover up and off the front of the engine.

6.13 Removing the timing belt lower cover (2.1 litre models)

2.1 litre engine centre cover - removal

8 Remove the auxiliary drivebelt as described in Section 4.
9 Undo the two bolts and remove the centre cover from the front of the injection pump (see illustration).

2.1 litre engine lower cover - removal

10 Remove the crankshaft pulley as described in Chapter 2A, Section 5.
11 Remove the right-hand engine mounting assembly as described in Section 19.
12 Remove both upper covers as described previously.
13 Slacken and remove the retaining bolts, and remove the lower cover (see illustration).

2.5 litre engine upper cover - removal

14 Disconnect the battery negative cable.
15 Withdraw each of the ECU's from the casing at the front right hand corner of the engine compartment. Unbolt the casing and remove it from the engine compartment.
16 Disconnect the plastic housing from the side of the upper cover.
17 Remove the auxiliary drivebelt as described in Section 4.
18 Unbolt the auxiliary drivebelt tensioner mechanism and remove it from the engine.
19 Undo the securing screws and lift off the upper cover (see illustration).

2.5 litre engine lower cover - removal

20 Remove the upper cover as described in the previous sub-Section.

21 Remove the four bolts which secure the crankshaft pulley, followed by the pulley itself. There is no need to remove the centre bolt.
22 Undo the securing screws and lift off the lower cover (see illustration).

Refitting

23 Refitting is a reversal of the relevant removal procedure, ensuring that each cover section is correctly located.

Timing belt

Removal - 2.1 litre engines

24 Remove the timing belt covers as described at the beginning of this Section.
25 Align the engine assembly/valve timing holes as described in Section 12, and lock the camshaft sprocket, injection pump sprocket and flywheel in position. Do not attempt to rotate the engine whilst the pins are in position. Disconnect the battery negative terminal.
26 Remove the remaining timing belt covers as described earlier in this Section.
27 Slacken the timing belt tensioner pulley retaining nut, situated just to the left of the engine mounting carrier bracket.
28 Using a 5 mm Allen key inserted through the hole in the engine mounting carrier bracket, slacken the timing belt tensioner locking bolt (see illustration).
29 Using a 10 mm socket or box spanner inserted through the same hole, retract the

6.19 Removing the upper timing belt cover (2.5 litre engine)

6.22 Removing the lower timing belt cover (2.5 litre engine)

6.28 On 2.1 litre models, slacken the timing belt tensioner locking bolt using a 5 mm Allen key

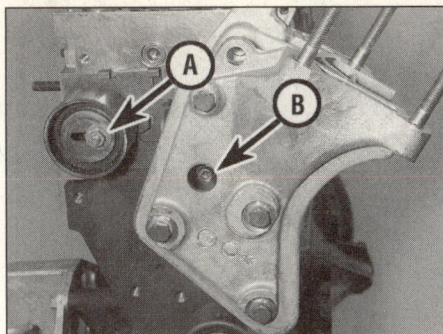

6.29a Timing belt tensioner pulley retaining nut (A) and locking bolt (B) on 2.1 litre models

6.29b Timing belt tensioner arrangement on 2.1 litre models showing tensioner 10 mm shaft (arrowed)

6.30 Removing the timing belt from the sprockets

tensioner by turning its shaft clockwise to the extent of its travel (see illustrations).

30 Mark the timing belt with an arrow to indicate its running direction, if it is to be re-used. Remove the belt from the sprockets (see illustration).

Removal - 2.5 litre engines

31 Jack up the front of the car and support it securely on axle stands. Remove the front right hand roadwheel.

32 Remove the fixings and detach the front section of the right hand wheel arch inner liner.

33 Remove the timing belt covers as described earlier in this Section.

34 Refer to Section 12 and lock the engine at TDC on cylinder No 4 using the appropriate locking tools at the flywheel, camshaft sprocket and fuel injection pump sprocket.

35 Slacken the timing belt tensioner pulley retaining nut, to relieve the tension on the timing belt.

36 Support the engine at the lifting eyelet, using a lifting beam or a engine crane.

37 Working via the right hand wheel arch and with reference to Section 10, unbolt the right hand lower engine mounting from the chassis, together with its tie-rod. Note that there is no need to remove the large centre bolt from the cylinder block engine mounting.

38 Lower the engine slightly, using the crane or lifting beam, until there is a gap of about 20-30 mm between the chassis and the right hand engine mounting (see illustration).

39 Mark the timing belt with an arrow to indicate its running direction, if it is to be re-used. Remove the belt from the crankshaft sprocket, pass it between the right hand engine mounting and the chassis and then remove it from the camshaft and fuel injection pump sprockets.

Inspection - all engines

40 Check the timing belt carefully for any signs of uneven wear, split or oil contamination. Pay particular attention to the roots of the teeth. Renew it if there is the slightest doubt about its condition. If the engine is undergoing an overhaul, and has covered more than 36 000 miles (60 000 km)

with the existing belt fitted, renew the belt as a matter of course, regardless of its apparent condition. The cost of a new belt is nothing compared with the cost of repairs, should the belt break in service. If signs of oil contamination are found, trace the source of the oil leak and rectify it.

41 Wash down the engine timing belt area and all related components, to remove all traces of oil. Check that the tensioner and idler pulleys rotate freely, without any sign of roughness.

2.1 litre models- refitting

Note: The final tension of the timing belt should ideally be checked using a dedicated Citroën electronic tester. This procedure describes how to achieve an approximate belt tension setting, but it is recommended that you have the tension checked accurately at the earliest opportunity.

42 Commence refitting by ensuring that the locking tools are still fitted to the camshaft and fuel injection pump sprockets, and that the rod/drill is positioned in the timing hole in the flywheel.

43 Ensure that the timing belt tensioner is still retracted, then tighten the tensioner pulley retaining nut. Using the 10 mm socket or box spanner, release the tensioner by turning it anti-clockwise to the extent of its travel.

44 Locate the timing belt on the crankshaft

6.38 Lower the engine slightly, using the crane or lifting beam, until there is a gap of about 20 to 30 mm between the chassis and the right hand engine mounting

sprocket, making sure that, where applicable, the direction of rotation arrow is facing the correct way.

45 Engage the timing belt with the crankshaft sprocket, hold it in position, then feed the belt over the remaining sprockets in the following order:

a) Idler roller.
b) Fuel injection pump.
c) Camshaft.
d) Coolant pump.
e) Tensioner roller.

46 Be careful not to kink or twist the belt. To ensure correct engagement, locate only a half-width on the injection pump sprocket before feeding the timing belt onto the camshaft sprocket, keeping the belt taut and fully engaged with the crankshaft sprocket. Locate the timing belt fully onto the sprockets.

47 Slacken the tensioner pulley retaining nut to allow the tensioner to tension the belt.

48 Unscrew and remove the bolts from the camshaft and fuel injection pump sprockets and remove the rod/drill from the timing hole in the flywheel.

49 Rotate the crankshaft through two complete turns in the normal running direction (clockwise). Do not rotate the crankshaft backwards, as the timing belt must be kept tight between the crankshaft, fuel injection pump and camshaft sprockets.

50 Tighten the tensioner pulley retaining nut, then rotate the crankshaft through a further two complete turns in the normal running direction, stopping at the timing setting position.

51 Slacken the tensioner pulley retaining nut one turn to allow the tensioner to finally tension the belt. Tighten the tensioner pulley retaining nut and the timing belt tensioner locking bolt to the specified torque.

52 Check that the timing holes are all correctly positioned by reinserting the sprocket locking bolts and the rod/drill in the flywheel timing hole, as described in Section 12. If the timing holes are not correctly positioned, the timing belt has been incorrectly fitted (possibly one tooth out on one of the sprockets) - in this case, repeat the refitting procedure from the beginning.

2B

6.55a Rotate the camshaft sprocket fully clockwise, so that the securing bolts reach the end of their slotted mounting holes -2.5 litre model (arrowed bolt is slackened to show positioning)

53 The remaining refitting procedure is a reversal of removal.

2.5 litre models - refitting

Note: *The tension of the timing belt must be set using a dedicated Citroën electronic tester.*

Note: *Timing belt tensioning is a complicated, two-stage operation that involves pre-tensioning the belt, turning the engine by hand, and then resetting the belt tension to a final value. Read through the text carefully before starting work.*

54 Check that the engine locking tools are still in place at the flywheel, camshaft sprocket and fuel injection pump sprocket (see Section 12).

55 With reference to Section 15, slacken off the camshaft sprocket securing bolts. Re-tighten them by hand, so that the sprocket is only just held in position. Repeat this procedure at the fuel injection pump sprocket. Rotate both sprockets fully clockwise, so that the securing bolts reach the ends of the slotted mounting holes (see illustrations).

56 Locate the timing belt on the crankshaft sprocket, making sure that, where applicable, the direction of rotation arrow is facing the correct way.

57 Engage the timing belt with the crankshaft sprocket, hold it in position, then feed the belt

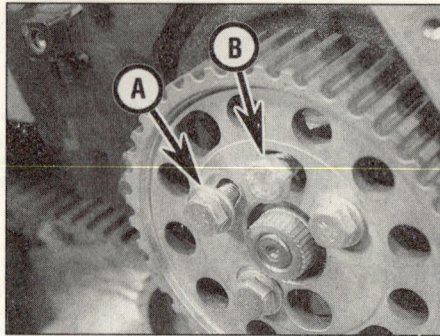

6.55b Fuel injection pump sprocket rotated fully clockwise, so that the securing bolts reach the end of their slotted mounting holes - 2.5 litre model

A   Securing bolt (slackened to show positioning)
B   Locking tool inserted through sprocket timing hole

over the remaining sprockets in the following order (Be careful not to kink or twist the belt) **(see illustrations)**:
a)   Idler roller.
b)   Fuel injection pump.
c)   Camshaft.
d)   Tensioner roller.

58 Adjust the position of the camshaft and fuel injection pump sprockets by rotating them anticlockwise as necessary, until the teeth of the timing belt engage with those on the sprockets. The rotation needed to engage the belt with the sprockets should not exceed the width of one sprocket tooth. After fitting the belt, the sprocket bolts should now be positioned near the centre of their slotted mounting holes **(see illustration)**.

59 Tension the belt by fitting a wrench to the square hole in the side of the tensioner pulley. Turn the pulley with the wrench until the correct **pre-tension** value for a new or used belt (as applicable) is attained. Keep the tensioner pulley in this position, and use a second wrench to tighten the tensioner centre nut to the specified torque **(see illustration)**.

60 Remove the engine locking tools from the flywheel, camshaft and fuel injection pump sprockets.

6.57a Fitting the camshaft timing belt - 2.5 litre model

61 Using a socket and bar on the crankshaft sprocket, turn the engine **in its normal direction of rotation** through at least ten revolutions.

62 Refit the flywheel locking tool, as described in Section 12.

63 Slacken the tensioner pulley centre nut, so that the tension on the belt is completely removed. Slacken off the camshaft and fuel injection pump sprocket securing bolts again, as described in paragraph 55, then refit the engine locking tools to both sprockets, as described in Section 12.

64 Tension the belt by fitting a wrench to the square hole in the side of the tensioner pulley. Turn the pulley with the wrench until the correct **final-tension** value for a new or used belt (as applicable) is attained. Keep the tensioner pulley in this position, and use a second wrench to tighten the tensioner centre nut to the specified torque.

65 Tighten the camshaft and fuel injection pump securing bolts, as described in Section 15.

66 Using a socket and bar on the crankshaft sprocket, turn the engine **in its normal direction of rotation** through two revolutions.

67 Check that the timing holes are all correctly positioned by reinserting the engine locking tools in the flywheel, and the camshaft and fuel injection pump sprockets, as described in Section 12. If the timing holes are not correctly positioned, the timing belt has

6.57b Correctly fitted camshaft timing belt - 2.5 litre model

6.58 After fitting the belt, the sprocket bolts should now be positioned near the centre of their slotted mounting holes (arrowed) - 2.5 litre models

6.59 Tensioning the timing belt - 2.5 litre engine

7.5a Fitting a new right hand crankshaft oil seal . . .

7.5b . . . and driving it into the housing using a suitable tube or socket - 2.5 litre model shown

7.9 Fitting a new left hand crankshaft oil seal - 2.5 litre model shown

been incorrectly fitted (possibly one tooth out on one of the sprockets) - in this case, repeat the refitting procedure from the beginning.
68 The remaining refitting procedure is a reversal of removal.

7 Crankshaft oil seals - renewal

Crankshaft right-hand oil seal

1 Remove the crankshaft sprocket as described in Section 15.
2 Measure and note the fitted depth of the oil seal.
3 Pull the oil seal from the housing using a hooked instrument. Alternatively, drill a small hole in the oil seal, and use a self-tapping screw and a pair of pliers to remove it.
4 Clean the oil seal housing and the crankshaft sealing surface.
5 Press it into the housing (open end first) to the previously-noted depth, using a suitable tube or socket (see illustrations). A piece of thin plastic or tape wound around the front of the crankshaft is useful to prevent damage to the oil seal as it is fitted. Do not lubricate the seal with oil as this may lead to leakage.
6 Where applicable, remove the plastic or tape from the end of the crankshaft.
7 Refit the timing belt crankshaft sprocket as described in Section 8.

8.2 Using a diesel engine compression gauge

Crankshaft left-hand oil seal

8 Remove the flywheel/driveplate, as described in Section 11.
9 Proceed as described in paragraphs 2 to 6, noting that when fitted, the outer lip of the oil seal must point outwards; if it is pointing inwards, use a piece of bent wire to pull it out. Take care not to damage the oil seal (see illustration).
10 Refit the flywheel/driveplate, as described in Section 11.

8 Cylinder compression and leakdown test

Compression test

Note: A compression tester specifically designed for diesel engines must be used for this test.

1 When engine performance is down, or if misfiring occurs which cannot be attributed to the fuel system, a compression test can provide diagnostic clues as to the engine's condition. If the test is performed regularly, it can give warning of trouble before any other symptoms become apparent.
2 A compression tester specifically intended for diesel engines must be used, because of the higher pressures involved. The tester is connected to an adapter which screws into the glow plug or injector hole. On these models, an adapter suitable for use in the injector holes will be required, due to the limited access to the glow plug holes (see illustration). It is unlikely to be worthwhile buying such a tester for occasional use, but it may be possible to borrow or hire one - if not, have the test performed by a garage.
3 Unless specific instructions to the contrary are supplied with the tester, observe the following points:
a) The battery must be in a good state of charge, the air filter must be clean, and the engine should be at normal operating temperature.
b) All the injectors or glow plugs should be removed before starting the test. If

removing the injectors, also remove the flame shield washers, otherwise they may be blown out.
c) It should be sufficient to disconnect the fuel injection multi-function relay located in the ECU module box, to prevent the engine from firing during cranking, but the advice of a dealer should be sought first.
4 There is no need to hold the accelerator pedal down during the test, because the diesel engine air inlet is not throttled.
5 The actual compression pressures measured are not so important as the balance between cylinders. The manufacturer does not supply specific figures, but as a rough guide, the difference in pressure between cylinders should be no greater than 5 bar.
6 The cause of poor compression is less easy to establish on a diesel engine than on a petrol one. The effect of introducing oil into the cylinders ('wet' testing) is not conclusive, because there is a risk that the oil will sit in the swirl chamber or in the recess on the piston crown instead of passing to the rings. However, the following can be used as a rough guide to diagnosis.
7 All cylinders should produce very similar pressures; any difference greater than that specified indicates the existence of a fault. Note that the compression should build up quickly in a healthy engine; low compression on the first stroke, followed by gradually-increasing pressure on successive strokes, indicates worn piston rings. A low compression reading on the first stroke, which does not build up during successive strokes, indicates leaking valves or a blown head gasket (a cracked head could also be the cause). Deposits on the undersides of the valve heads can also cause low compression.
8 A low reading from two adjacent cylinders is almost certainly due to the head gasket having blown between them; the presence of coolant in the engine oil will confirm this.
9 If the compression reading is unusually high, the cylinder head surfaces, valves and pistons are probably coated with carbon deposits. If this is the case, the cylinder head should be removed and decarbonised (see Part C).

2B

Leakdown test

10 A leakdown test measures the rate at which compressed air fed into the cylinder is lost. It is an alternative to a compression test, and in many ways it is better, since the escaping air provides easy identification of where pressure loss is occurring (piston rings, valves or head gasket).

11 The equipment needed for leakdown testing is unlikely to be available to the home mechanic. If poor compression is suspected, have the test performed by a suitably-equipped garage.

9 Cylinder head - removal and refitting

2.1 litre engine

Removal

1 Removal of the cylinder head from the 2.1 litre engine with the engine in situ is unfeasible due to the angle at which the engine is inclined towards the rear of the engine compartment. This makes it extremely difficult to fully withdraw the cylinder head bolts as they are obstructed by the engine compartment bulkhead. The description given here assumes that the engine and transmission assembly have been removed from the vehicle.

2 Support the engine and transmission assembly securely on a stand or a workbench.

3 Remove the camshaft and followers as described in Section 2 of this chapter.

4 Refer to Chapter 4B and remove the inlet manifold lower part, and the exhaust manifold.

5 Remove the fuel injection pump sprocket as described in Section 15 of this chapter.

6 Undo the bolts and remove the engine mounting attachment bracket from the front of the engine **(see illustration)**.

7 Disconnect the remaining wiring, hoses, support brackets and connections at the cylinder head.

8 Progressively unscrew the cylinder head bolts, following the tightening sequence **(see illustration 9.22a)** in **reverse**.

9 Lift out the bolts and recover the spacers.

10 Release the cylinder head from the cylinder block and location dowel by rocking it. The Citroën tool for doing this consists simply of two metal rods with 90-degree angled ends. Do not prise between the mating faces of the cylinder head and block, as this may damage the gasket faces.

11 Lift the cylinder head from the block, and recover the gasket.

Preparation for refitting

12 The mating faces of the cylinder head and cylinder block/crankcase must be perfectly clean before refitting the head. Use a hard

9.6 Undo the bolts and remove the engine mounting attachment bracket - 2.1 litre models

plastic or wooden scraper to remove all traces of gasket and carbon; also clean the piston crowns. Make sure that the carbon is not allowed to enter the oil and water passages - this is particularly important for the lubrication system, as carbon could block the oil supply to the engine's components. Using adhesive tape and paper, seal the water, oil and bolt holes in the cylinder block/crankcase. To prevent carbon entering the gap between the pistons and bores, smear a little grease in the gap. After cleaning each piston, use a small brush to remove all traces of grease and carbon from the gap, then wipe away the remainder with a clean rag. Clean all the pistons in the same way.

13 Check the mating surfaces of the cylinder block/crankcase and the cylinder head for nicks, deep scratches and other damage. If slight, they may be removed carefully with a file, but if excessive, machining may be the only alternative to renewal. If warpage of the cylinder head gasket surface is suspected, use a straight-edge to check it for distortion.

14 When purchasing a new cylinder head gasket, it is essential that a gasket of the correct thickness is obtained. Modifications to the cylinder head gasket material, type, and manufacturer are constantly taking place; seek the advice of a Citroën dealer as to the latest recommendations.

15 The cylinder head bolts are designed to stretch in use. They may be re-used provided that they have not stretched beyond the specified maximum length. Refer to your Citroën dealer for details of the maximum length of used cylinder head bolts.

9.18 . . . then lower the cylinder head into position

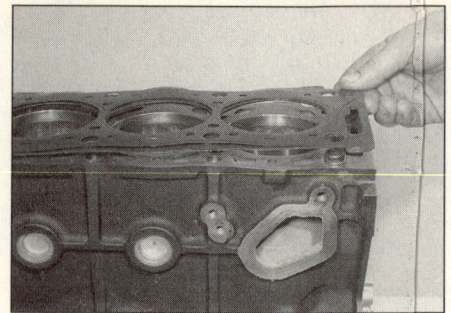
9.17 Position a new gasket on the cylinder block . . .

Refitting

16 Wipe clean the mating surfaces of the cylinder head and cylinder block/crankcase. Check that the two locating dowels are in position at each end of the cylinder block/crankcase surface.

17 Position a new gasket on the cylinder block/crankcase surface, ensuring that its identification holes or the projecting tongue are at the left-hand end of the gasket **(see illustration)**.

18 Lower the cylinder head onto the block **(see illustration)**.

19 Apply a smear of grease to the threads, and to the underside of the heads, of the cylinder head bolts. Citroën recommend the use of Molykote G Rapid Plus (available from your Citroën dealer); in the absence of the specified grease, any good-quality high-melting-point grease may be used **(see illustration)**.

20 Carefully enter each bolt into its relevant hole (*do not drop it in*) and screw it in finger-tight.

21 Two types of cylinder head bolt may be fitted to the 2.1 litre engine. Early engines were fitted with bolts that have a plain threaded section, bolts fitted to later engines (and bolts supplied as new replacement parts) have an unthreaded guide boss at the end of the threaded section. The following tightening sequence assumes that new bolts (with guide bosses) have been fitted.

22 Working progressively and in the sequence shown, tighten the cylinder head bolts to their Stage 1 torque setting, using a torque wrench and suitable socket. See

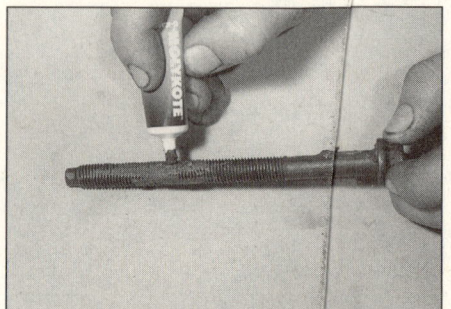
9.19 Apply a smear of grease to the threads and underside of the heads of the cylinder head bolts

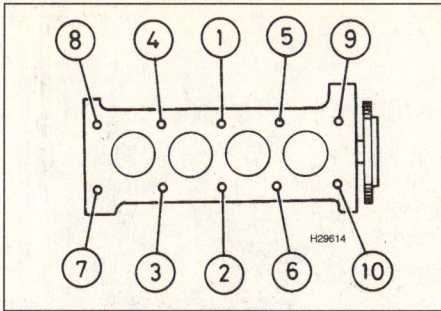

9.22A Cylinder head bolt tightening sequence

9.22B Tighten the cylinder head bolts to their Stage 1 torque setting, using a torque wrench

Specifications for the relevant torque wrench settings **(see illustrations)**.

23 With all the bolts tightened to their Stage 1 setting, tighten the cylinder head bolts to their Stage 2 torque setting, using a torque wrench and suitable socket. See Specifications for the relevant torque wrench settings.

24 With all the bolts tightened to their Stage 2 setting, working again in the specified sequence, angle-tighten the bolts through the specified Stage 3 angle using a socket and extension bar. It is recommended that an angle-measuring gauge is used during this stage of tightening, to ensure accuracy **(see illustration)**. If a gauge is not available, use white paint to make alignment marks between the bolt head and cylinder head prior to tightening; the marks can then be used to check that the bolt has rotated sufficiently.

25 The remainder of refitting is a direct reversal of the removal procedure.

26 Refit the engine and transmission assembly as described in Chapter 2C. Note that if new cylinder head bolts with guide bosses have been fitted, there is no requirement to re-tighten the head bolts after the engine has been warmed up for the first time.

2.5 litre engine

Removal

27 Jack up the front of the vehicle and rest it securely on axle stands. Remove the right hand front road wheel.

28 Disconnect both cables from the battery terminals (negative first). Remove the battery from its holder.

9.24 Angle-tighten the bolts through the specified Stage 3 angle using a socket and extension bar and angle measuring gauge

29 Slacken and withdraw the screws, then remove the engine compartment undertray(s).

30 Working in the right hand wheel arch, remove the screws and detach the centre section of the plastic wheel arch liner.

31 Refer to Chapter 7A and drain the transmission oil from the gearbox.

32 Refer to Chapter 3 and drain the cooling system.

33 Refer to Chapter 8 and remove the right hand driveshaft/intermediate driveshaft assembly (this is optional but gives much improved access to the turbocharger and its mountings).

34 With reference to Chapter 4C, remove the turbocharger from the exhaust manifold.

35 Slacken the hose clip and detach the ducting from the rear of the air cleaner. Remove the securing screws and lift the air cleaner and all its associated intake ducting from the engine compartment with reference to Chapter 4B.

36 Refer to Chapter 9 and remove the cover from the hydraulic fluid reservoir. Place the cover (with the hoses still attached) in a container to prevent fluid contamination and to protect the exposed pipe ends. Cover the open fluid reservoir to prevent entry of foreign materials.

37 Working at the front left hand corner of the engine compartment carry out the following:
a) *Remove the screws, detach the fusebox and glow plug control unit from the battery holder and move them to one side, without disconnecting the wiring.*
b) *Slacken and withdraw the securing bolts, then remove the fuel filter/priming pump assembly and move it to one side, without disconnecting the fuel hoses.*
c) *Slacken and withdraw the securing bolts, then remove the air conditioning dehydrator unit (where fitted) and move it to one side, without disconnecting the hoses.*
d) *Remove the securing screws and lift the battery holder from the engine compartment.*

38 Refer to Chapter 3 and carry out the following:
a) *Remove the coolant pump drive belt.*
b) *Slacken the hose clips, detach the coolant hoses and unplug the wiring from the coolant expansion tank.*

c) *Undo the securing strap clamp screw and remove the coolant expansion tank from the engine.*

39 Slacken the clips and disconnect the fuel hoses from the fuel heater ports, at the left hand end of the cylinder head. Be prepared for some fuel spillage.

40 Trace the engine wiring harness back to the connectors at the left hand end of the cylinder head. Unplug the harness at the connectors, then release the securing clips and separate the harness from the support bracket that runs along the front of the cylinder head. Remove the screws and detach the support bracket from the cylinder head.

41 Refer to Chapter 4B and carry out the following:
a) *Slacken the unions and disconnect the fuel supply pipes from the fuel injectors and the fuel injection pump.*
b) *Disconnect the fuel return hoses from the each of the fuel injectors.*
c) *Unplug the injector lift sensor wiring from injector number three at the connector.*

42 With reference to Chapter 5C, disconnect the glow plug supply wiring at the connector(s).

43 Working in the front right hand corner of the engine compartment, remove the cover from the ECU casing. Remove the ECUs from the casing, unplug the wiring connectors and place the ECU's to one side, away from the work area.

44 Undo the securing screws, detach the ECU casing from the bodywork and remove it from the engine compartment.

45 Unclip the fuel supply and return hoses from the bracket at the top of the timing belt cover.

46 Remove the securing screws, unplug the wiring and detach the EGR solenoid valve from the right hand end of the cylinder head cover.

47 Referring to the relevant Sections in this Chapter, carry out the following:
a) *Remove the auxiliary drivebelt, then unbolt the drivebelt roller tensioner from the engine.*
b) *Unbolt the timing belt outer cover, then remove the timing belt.*
c) *Remove the camshaft timing belt sprocket and the sprocket-to-camshaft mounting flange.*
d) *Unbolt and remove the eccentric roller tensioner.*

48 Remove the securing screws and prise the timing belt inner cover away from the cylinder head.

49 Working at the rear of the cylinder head, slacken the clip and detach the EGR air hose from the inlet manifold upper section.

50 Unbolt and remove the upper section of the inlet manifold, with reference to Chapter 4B.

51 Referring to the relevant sections in this Chapter, carry out the following:
a) *Remove the cylinder head cover.*
b) *Unbolt the rear engine mounting/torque mechanism from the rear of the engine; extract the mounting from socket at the suspension mounting turret.*

2B

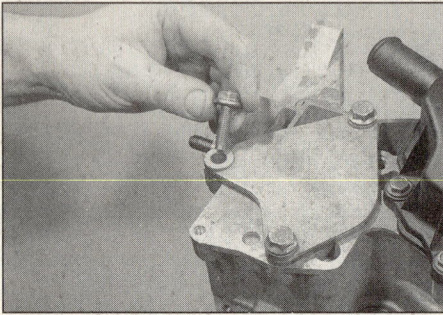

9.52a Remove the blanking plate securing bolts . . .

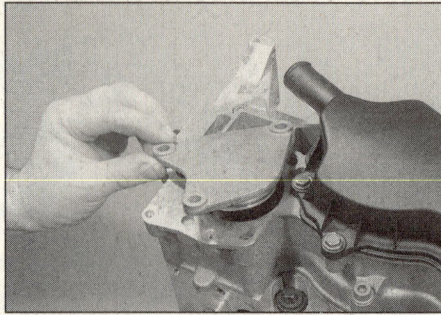

9.52b . . . and lift off the plate to expose the last cylinder head bolt

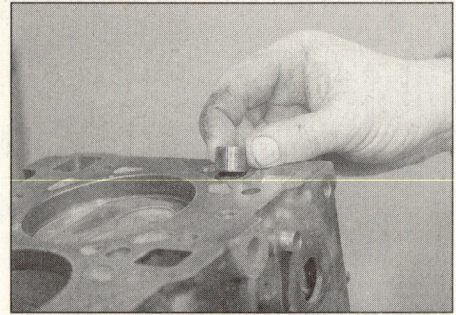

9.58a Check that the cylinder head locating dowels are in place

52 Unbolt the blanking plate from the top left hand side of the cylinder head, to expose the final cylinder head bolt. If required, the camshaft carrier and followers may be unbolted and removed separately; refer to Section 2 for details **(see illustrations).**

53 Progressively unscrew the cylinder head bolts, following the given tightening sequence in **reverse**. Lift out the bolts together with the spacers.

54 Release the cylinder head from the cylinder block and location dowel by rocking it. The Citroën tool for doing this consists simply of two metal rods with 90-degree angled ends. Do not prise between the mating faces of the cylinder head and block, as this may damage the gasket faces.

55 Lift the cylinder head from the block, and recover the gasket.

Preparation for refitting

56 Refer to the information given in the previous sub-Section.

Refitting

57 Check the protrusion of the swirl chambers above the cylinder head mating surface, using a DTI (Dial Test Indicator) gauge and compare the measurement with the Specifications. Renew the swirl chambers if they appear excessively cracked or are damaged in any way.

58 Wipe clean the mating surfaces of the cylinder head and cylinder block/crankcase. Check that the two locating dowels are in position at each end of the cylinder block/crankcase surface **(see illustrations).**

59 Position a new gasket on the cylinder block/crankcase surface, ensuring that it is correctly seated **(see illustration).**

60 Lower the cylinder head onto the block . If

the swirl chambers are loose, hold them in position with stout elastic bands whilst the head is refitted. Cut through the bands and withdraw them when the cylinder head is in position, but ensure that no part of the band is left underneath the head **(see illustrations).**

61 Apply a smear of grease to the threads, and to the underside of the heads, of the cylinder head bolts. Citroën recommend the use of Molykote G Rapid Plus (available from your Citroën dealer); in the absence of the specified grease, any good-quality high-melting-point grease may be used **(see illustration).**

62 Carefully enter each bolt into its relevant hole (*do not drop it in*) and screw it in finger-tight **(see illustration).**

63 Citroën state that the cylinder head bolts fitted to the 2.5 litre engine may be re-used,

9.58b Location of cylinder head dowels - 2.5 litre engine

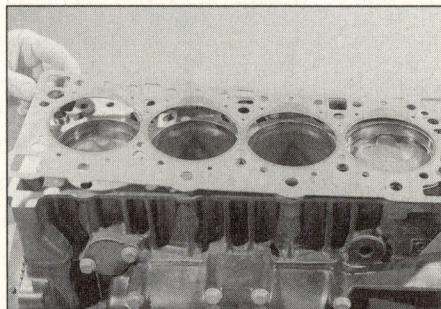

9.59 Position a new gasket on the cylinder block/crankcase surface

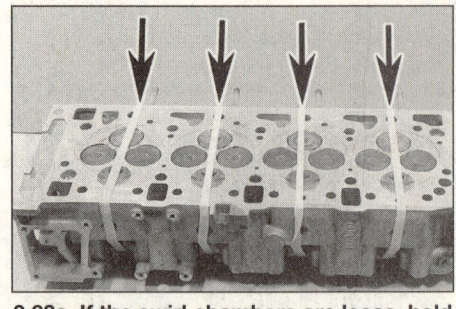

9.60a If the swirl chambers are loose, hold them in position with stout elastic bands whilst the head is refitted

9.60b Lower the cylinder head onto the block

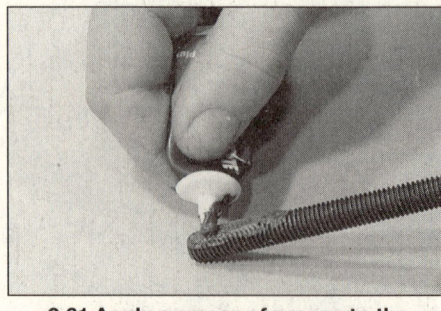

9.61 Apply a smear of grease to the threads, and to the underside of the heads, of the cylinder head bolts

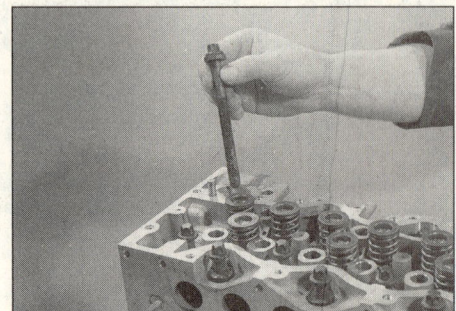

9.62 Carefully enter each cylinder head bolt into its relevant hole

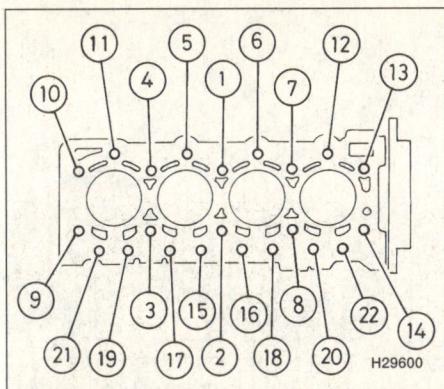

9.64a Cylinder head bolt location and tightening sequence - 2.5 litre engine

1 to 14   12 mm bolts
15 to 22   10 mm bolts

provided they have not stretched beyond their maximum specified length. Refer to your Citroën dealer for details of the maximum length of your cylinder head bolts.

64 Working progressively and in the sequence shown, tighten the 12 mm diameter cylinder head bolts (Nos 1 to 14) to their Stage 1 torque setting, using a torque wrench and suitable socket. See Specifications for the relevant torque wrench settings **(see illustrations)**.

65 Repeat the operation in previous paragraph on the 10 mm diameter cylinder head bolts (Nos 15 to 22), tightening them to their Stage 1 torque setting, using a torque wrench and suitable socket.

66 With all the bolts tightened to their Stage 1 setting, working again in the specified sequence, angle-tighten all the bolts (Nos 1 to 22) through the specified Stage 2 angle using a socket and extension bar. It is recommended that an angle-measuring gauge is used during this stage of tightening, to ensure accuracy **(see illustration)**. If a gauge is not available, use white paint to make alignment marks between the bolt head and cylinder head prior to tightening; the marks can then be used to check that the bolt has rotated sufficiently.

67 The remainder of the cylinder head refitting procedure is a direct reversal of the

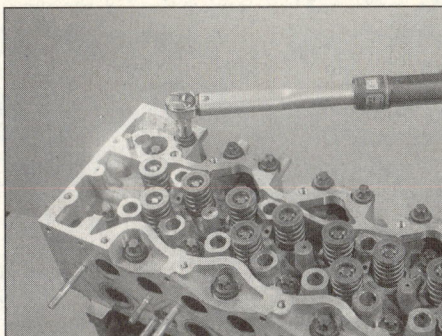

9.64b Tighten the cylinder head bolts to the specified Stage 1 setting using a torque wrench

removal procedure. Note that the design of the cylinder head bolts is such that re-tightening (after the engine has been started and warmed up for the first time) is **not** required **(see illustration)**.

10 Engine mountings - inspection and renewal

2.1 litre engines

Refer to the information given in Chapter 2A.

2.5 litre engines

General information

1 The engine/transmission assembly is supported longitudinally on two mountings; one at left hand side of the engine, above the transmission casing and one at the timing belt end of the engine. Axial rotation of the engine is controlled by two torque control mechanisms: one mounted at the front right hand corner of the engine block (bolted to the suspension subframe) that is linked by means of a pushrod to the right hand engine mounting, and one bolted to the rear right hand corner of the cylinder head.

Transmission mounting

2 Remove the air cleaner assembly, as described in Chapter 4B. Remove the battery and battery tray as described in Chapter 5A.

9.66 Angle-tighten the cylinder head bolts to the specified Stage 2 setting, using an extension bar and an angle measuring gauge

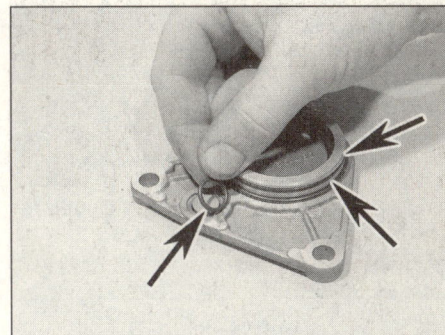

9.67 Fit a new set of seals (arrowed) to the cylinder head blanking plate before refitting it - 2.5 litre engines

3 Place a jack beneath the transmission, with a block of wood on the jack head. Raise the jack until it is supporting the weight of the transmission.

4 Remove the rubber cap, then slacken and remove the centre nut and washer from the left-hand mounting. Undo the bolts securing the mounting in position and remove it from the engine compartment **(see illustration)**.

5 If required, the mounting bracket may be unbolted from the chassis and removed **(see illustration)**.

6 Check all components carefully for signs of wear or damage, and renew as necessary.

7 Clean the threads of the mounting stud, and apply a coat of thread-locking compound to its threads **(see illustration)**.

2B

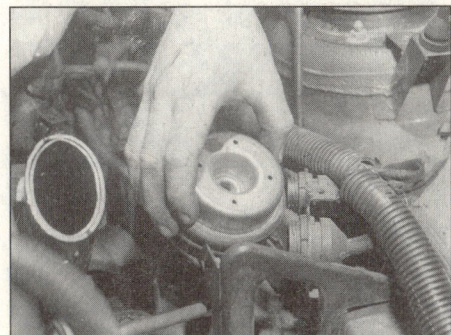

10.4 Removing the left hand engine/transmission mounting

10.5 Removing the left hand engine/transmission mounting bracket

10.7 Apply a coat of thread-locking compound to the left hand engine/transmission mounting stud threads

10.13 Slacken and withdraw the bolts (arrowed) at either end of the torque control mechanism pushrod

10.14 Unbolt the front torque mechanism from the subframe

10.16 Using improvised tools (arrowed) to support the right hand engine mounting

8 Slide the spacer (where fitted) onto the mounting stud, then refit the rubber engine mounting. Tighten both the engine mounting securing bolts and the mounting centre nut to their specified torque settings, then refit the rubber cap.

9 Remove the jack from underneath the transmission, then refit the air cleaner assembly (Chapter 4B). On completion refit the battery (Chapter 5A).

Front torque mechanism/right hand engine mounting

10 If not already done, chock the rear wheels, then jack up the front of the vehicle and support it securely on axle stands.

11 Remove the right hand front roadwheel, then extract the fixings and remove the centre section of the inner wheel arch liner.

12 Remove the rear torque mechanism, as described in the next sub-Section.

13 Slacken and withdraw the bolts at either end of the torque control mechanism pushrod **(see illustration)**. Remove the pushrod from the engine mounting.

14 Unbolt the front torque mechanism from the subframe **(see illustration)** and extract the coupling shaft together with its rubber buffer from the circular housing bracket mounted on the engine. Note that the front torque mechanism bolt holes are slotted to allow adjustment on refitting.

15 Place a jack beneath the engine, with a block of wood on the jack head. Raise the jack until it is supporting the weight of the engine. Do not jack under the sump.

16 Slacken and withdraw the first of the two bolts that secure the right hand engine mounting bracket to the chassis. Obtain a section of threaded rod of the same diameter as the bolt just removed, approximately 200 mm long. Fit three nuts and a large washer to the rod and screw it in place of the engine mounting bolt, until the end of the rod protrudes through the bottom of the mounting. Lock two of the nuts together and use them to adjust the protrusion of the rod. Repeat this procedure on the second engine mounting bolt **(see illustration)**.

17 Lower the engine slightly using the jack, until the head of the right hand engine mounting through-bolt can be seen. Ensure that the washers and nuts at the top of the threaded rods are now resting on the chassis **(see illustration)**.

18 The next step is to slacken and withdraw the through-bolt, to separate the right hand engine mounting from the engine. The threaded rods will brace the engine mounting as you do this, to avoid stressing the rubber material at the centre **(see illustration)**.

19 Slacken and withdraw the through bolt from the right hand engine mounting, then remove the temporarily-fitted threaded rods

and remove the right hand engine mounting from the engine bay **(see illustration)**.

20 Refitting is a reversal of removal, but note that the right hand engine mounting must be braced (as described during removal), to avoid stressing the rubber components, when the through bolt is refitted and tightened to its specified torque. Note also that the right hand engine mounting is dowelled to ensure correct positioning on the chassis member.

Note: *Details of the removal and refitting of the right hand engine mounting rubber coupling are given in Chapter 2C)*

Rear torque mechanism

21 If not already done, chock the rear wheels, then jack up the front of the vehicle and support it securely on axle stands.

22 Place a jack beneath the engine, with a block of wood on the jack head. Raise the jack until it is supporting the weight of the engine. Do not jack under the sump.

23 Where applicable, remove the screws and lift the EGR solenoid valve from the end of the inlet manifold. Unclip the fuel supply and return hoses from the support bracket at top end of the timing belt casing.

24 Remove the three bolts that secure the flexible rubber coupling to the cylinder head bracket.

25 Extract the flexible rubber coupling from the bodywork socket and remove it from the

10.17 Lower the engine using the jack, until the head of the right hand engine mounting through-bolt (arrowed) can be seen

10.18 Slacken and withdraw the right hand engine mounting through-bolt

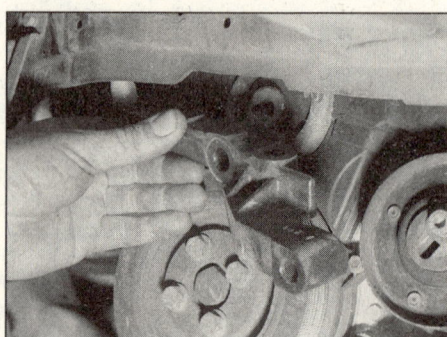

10.19 Removing the right hand engine mounting

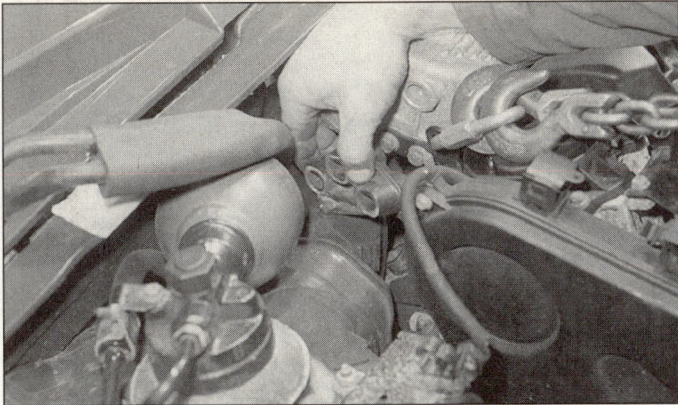

10.25 Extract the flexible rubber coupling from the bodywork socket and remove it from the engine bay

10.26 The rear torque mechanism bolt holes (arrowed) are slotted to allow adjustment

engine bay **(see illustration)**

26 Refitting is a reversal of removal. Note that the engine mounting bolt holes are slotted to allow adjustment **(see illustration)**.

11 Flywheel/driveplate - removal, inspection and refitting

Refer to the information given in Chapter 2A.

12 Engine assembly and valve timing holes - general information and usage

Note: *Do not attempt to rotate the engine whilst the crankshaft/camshaft/injection pump are locked in position. If the engine is to be left in this state for a long period of time, it is a good idea to place suitable warning notices inside the vehicle, and in the engine compartment. This will reduce the possibility of the engine being accidentally cranked on the starter motor, which is likely to cause damage with the locking pins in place.*

1 On all models, timing holes are provided in the camshaft sprocket, injection pump sprocket and flywheel. The holes are used to align the crankshaft, camshaft and injection pump and to prevent the possibility of the

valves contacting the pistons when refitting the cylinder head, or when refitting the timing belt. When the holes are aligned with their corresponding holes in the cylinder head and cylinder block (as appropriate), suitable diameter bolts/pins can be inserted to lock both the camshaft, injection pump and crankshaft in position, preventing them from rotating unnecessarily **(see illustration)**. Proceed as follows. **Note:** *With the timing holes aligned, No 4 cylinder is at TDC on its compression stroke.*

2 Remove the upper timing belt covers as described in Section 6.

3 The crankshaft must now be turned until the bolt holes in the camshaft and injection pump sprockets (one hole in the camshaft sprocket, one or two holes in the injection pump sprocket) are aligned with the corresponding holes in the engine front plate. The crankshaft can be turned by using a spanner on the pulley bolt. To gain access to the pulley bolt, from underneath the front of the car, prise out the retaining clips and remove the screws, then withdraw the plastic wheel arch liner from the wing valance, to gain access to the crankshaft pulley bolt. Where necessary, unclip the coolant hoses from the bracket, to improve access further. The crankshaft can then be turned using a suitable spanner or socket and extension bar fitted to the pulley bolt. Note that the crankshaft must

always be turned in a clockwise direction (viewed from the right-hand side of the vehicle).

4 Insert a cranked rod of approximately 8 mm in diameter (or a long bolt of the same diameter) through the hole in the left-hand flange of the cylinder block by the starter motor. Access is very restricted, and it may be easier to remove the starter motor (see Chapter 5A) to be able to locate the hole. On 2.5 litre models, there is an alternative hole at the rear of the engine (accessible from underneath the car), drilled into the mating surface between the cylinder block and the transmission bellhousing **(see illustrations)**

5 Insert one 8 mm bolt through the guides at

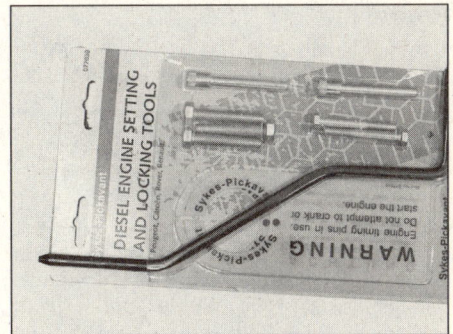

12.1 Typical kit of engine locking tools

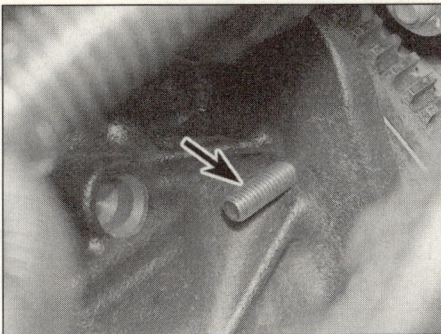

12.4a Crankshaft locking tool in place - 2.1 litre engine

12.4b Crankshaft locking tool in place - 2.5 litre engine

12.4c Crankshaft locking tool in place (alternative hole at rear of engine block) - 2.5 litre engine

2B

12.5a Camshaft sprocket locking tool in use - 2.1 litre engine

12.5b Fuel injection pump sprocket locking tool in use - 2.1 litre engine

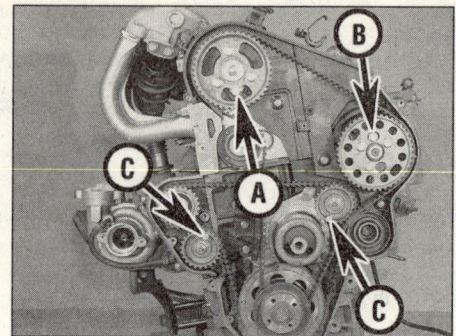

12.5c Complete set of sprocket locking tools in use - 2.5 litre engine

A *Camshaft sprocket locking tool*
B *Fuel injection pump sprocket locking tool*
C *Balance shaft sprocket locking tools*

the base of the camshaft sprocket hub, and another through the guides at the base of the fuel injection pump sprocket hub. Screw the bolts into the engine by hand to ensure that they can't fall out **(see illustrations)**.

6 The crankshaft, camshaft and injection pump are now locked in position, preventing unwanted rotation.

7 Note that on 2.5 litre engines, there are also timing holes that allow the balance shaft sprockets to be locked in position. Refer to Section 17 for details.

13 Oil pump and pickup - removal, inspection and refitting

Removal

1 Remove the sump (see Section 14). Where applicable, unbolt and remove the oil baffle plate **(see illustration)**.

2 On 2.1 litre engines, undo the two retaining screws, and slide the sprocket cover off the front of the oil pump.

3 Slacken and remove the three bolts securing the oil pump to the base of the cylinder block/crankcase. Disengage the pump sprocket from the chain, and remove the oil pump **(see illustration)**. Where necessary, also remove the spacer plate which is fitted behind the oil pump.

On 2.5 litre engines, recover the locating dowel, if it is loose.

Inspection

4 Examine the oil pump sprocket for signs of damage and wear, such as chipped or missing teeth. If the sprocket is worn, the pump assembly must be renewed, since the sprocket is not available separately. It is also recommended that the chain and drive sprocket, fitted to the crankshaft, be renewed at the same time. To renew the chain and drive sprocket, first remove the crankshaft timing belt sprocket as described in Section 15. Unbolt the oil seal carrier from the cylinder block. The sprocket, spacer (where fitted) and chain can then be slid off the end of the crankshaft. Refer to Part C for further information.

5 Slacken and remove the bolts (along with the baffle plate, where fitted) securing the strainer cover to the pump body. Lift off the strainer cover, and take off the relief valve piston and spring, noting which way round they are fitted.

6 Examine the pump rotors and body for signs of wear ridges or scoring. If worn, the complete pump assembly must be renewed.

7 Examine the relief valve piston for signs of wear or damage, and renew if necessary. The condition of the relief valve spring can only be measured by comparing it with a new one; if there is any doubt about its condition, it should also be renewed. Both the piston and spring are available individually.

8 Thoroughly clean the oil pump strainer with a suitable solvent, and check it for signs of clogging or splitting. If the strainer is damaged, the strainer and cover assembly must be renewed.

9 Locate the relief valve spring and piston in the strainer cover. Refit the cover to the pump body, aligning the relief valve piston with its bore in the pump. Refit the baffle plate (where fitted) and the cover retaining bolts, and tighten them securely.

Refitting

10 Offer up the spacer plate (where fitted), then locate the pump sprocket with its drive chain. Ensure that the locating dowel is in place on 2.5 litre engines, then seat the pump on the base of the cylinder block/crankcase. Refit the pump retaining bolts, and tighten them to the specified torque setting **(see illustration)**

11 Where applicable, slide the sprocket cover into position on the pump. Refit its retaining bolts, tightening them securely.

12 Refit the sump as described in Section 14.

13 Before starting the engine, prime the oil pump as follows. Disconnect the stop solenoid wiring or injection system relay (as applicable), then crank the engine on the starter until the oil pressure light goes out. Reconnect the wiring/relay on completion.

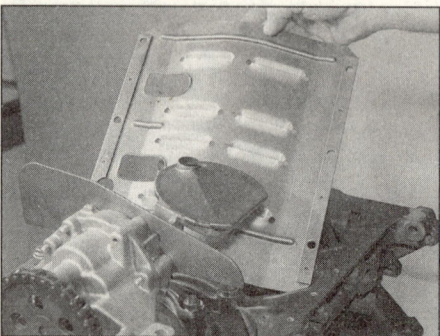

13.1 Unbolt and remove the oil baffle plate

13.3 Disengage the pump sprocket from the drive chain, and remove the oil pump

13.10 Refit the pump retaining bolts, and tighten them to the specified torque setting

14 Sump -
removal and refitting

Removal

1 Disconnect the battery negative terminal.
2 Chock the rear wheels, jack up the front of the vehicle and support it on axle stands.
3 Drain the engine oil as described in Chapter 1B, then clean and refit the engine oil drain plug, tightening it securely. If the engine is nearing its service interval when the oil and filter are due for renewal, it is recommended that the filter is also removed, and a new one fitted. After reassembly, the engine can then be refilled with fresh oil. Refer to Chapter 1B for further information.
4 Where necessary on 2.1 litre models, disconnect the wiring connector from the oil temperature sender unit, which is screwed into the sump.
5 Remove the auxiliary drivebelt as described in Section 4. On 2.1 litre models with air conditioning, where the compressor is located on the side of the sump, unbolt the compressor and position it clear of the sump. Support the weight of the compressor by tying it to the vehicle, to prevent any excess strain being placed on the compressor lines. *Do not* disconnect the refrigerant lines from the compressor (refer to the warnings given in Chapter 3).
6 On 2.5 litre models, carry out the following:
a) *Refer to Section 10 and remove the right hand engine mounting.*
b) *Unbolt the auxiliary belt roller tensioner from the engine block.*
c) *Unbolt the right hand engine mounting/torque control mechanism bracket from the from the engine block.*
d) *Remove the screws and lift off the flywheel protection plate.*
7 Progressively slacken and remove all the sump retaining bolts. Since the sump bolts vary in length, remove each bolt in turn, and store it in its correct fitted order by pushing it through a clearly-marked cardboard template. This will avoid the possibility of installing the bolts in the wrong locations on refitting.

8 Break the joint by striking the sump with the palm of your hand. Lower the sump, and withdraw it from underneath the vehicle. Remove the gasket (where fitted), and discard it; a new one must be used on refitting. While the sump is removed, take the opportunity to check the oil pump pick-up/strainer for signs of clogging or splitting. If necessary, remove the pump as described in Section 13, and clean or renew the strainer.
9 On some models, a large spacer plate is fitted between the sump and the base of the cylinder block/crankcase. If this plate is fitted, undo the two retaining screws from diagonally-opposite corners of the plate. Remove the plate from the base of the engine, noting which way round it is fitted.

Refitting

10 Clean all traces of sealant/gasket from the mating surfaces of the cylinder block/crankcase and sump, then use a clean rag to wipe out the sump and the engine's interior.
11 Where a spacer plate is fitted, remove all traces of sealant/gasket from the spacer plate, then apply a thin coating of suitable sealant to the plate upper mating surface. Offer up the plate to the base of the cylinder block/crankcase, and securely tighten its retaining screws.
12 On models where the sump was fitted without a gasket, ensure that the sump mating surfaces are clean and dry, then apply a thin coating of suitable sealant to the sump mating surface **(see illustration)**
13 On models where the sump was fitted with a gasket, ensure that all traces of the old gasket have been removed, and that the sump mating surfaces are clean and dry. Position the new gasket on the top of the sump, using a dab of grease to hold it in position.
14 Offer up the sump to the cylinder block/crankcase **(see illustration)**. Refit its retaining bolts, ensuring that each is screwed into its original location. Tighten the bolts evenly and progressively to the specified torque setting.
15 The remainder of the refitting procedure is a reversal of removal. On completion, lower the vehicle to the ground and refill with oil as described in Chapter 1B.

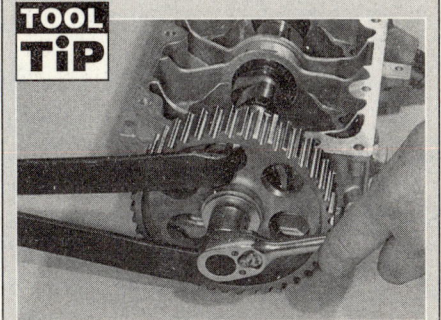

TOOL TiP

A sprocket holding tool can be made from two lengths of steel strip bolted together to form a forked end. Bend the ends of the strip through 90° to form the fork "prongs"

15 Timing belt sprockets -
removal and refitting

Camshaft sprocket (2.1 litre models)

Removal

1 Remove the timing belt (see Section 6).
2 Slacken the camshaft sprocket retaining bolt and remove it, along with its washer. To prevent the camshaft rotating as the bolt is slackened, a sprocket holding tool will be required **(see Tool Tip)**. *Do not* attempt to use the sprocket locking pin to prevent the sprocket from rotating whilst the bolt is slackened.
3 Remove the camshaft sprocket retaining bolt and washer.
4 Unscrew and remove the locking bolt from the camshaft sprocket.
5 With the retaining bolt removed, slide the sprocket off the end of the camshaft **(see illustration)**. Recover the Woodruff key from the end of the camshaft if it is loose. Examine the camshaft oil seal for signs of oil leakage and, if necessary, renew it (see Section 5).

Refitting

6 Refit the Woodruff key to the end of the

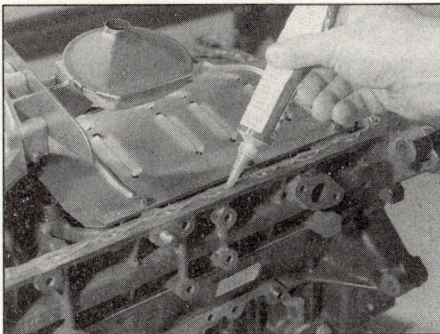

14.12 Apply a bead of sealant to the sump mating surface

14.14 Refitting the sump

15.5 Removing the camshaft sprocket - 2.1 litre engine

2B

15.13 Removing the crankshaft sprocket - 2.1 litre engine

15.21 Using a home made tool to prevent the fuel injection pump sprocket from turning - 2.1 litre engine

15.23 Home-made puller fitted to the fuel injection pump sprocket - 2.1 litre engine

camshaft, then refit the camshaft sprocket. Note that the sprocket will only fit one way round (with the protruding centre boss against the camshaft), as the end of the camshaft is tapered.

7 Refit the sprocket retaining bolt and washer. Tighten the bolt to the specified torque, preventing the camshaft from turning as during removal.

8 Where applicable, refit the cylinder head cover as described in Section 3.

9 Align the holes in the camshaft sprocket and the engine front plate, and refit the 8 mm bolt to lock the camshaft in position.

10 Refit the timing belt as described in Section 6.

11 Refit the timing belt covers as described in Section 6.

Crankshaft sprocket (2.1 litre models)

Note: *On vehicles built before build code 6176, a new version of the crankshaft sprocket Woodruff key must be retro-fitted; refer to your Citroën dealer for details.*

Removal

12 Remove the timing belt (see Section 6).

13 Slide the sprocket off the end of the crankshaft **(see illustration)**.

14 Remove the Woodruff key from the crankshaft, and store it with the sprocket for safe-keeping.

15 Examine the crankshaft oil seal for signs of oil leakage and, if necessary, renew it as described in Section 7.

Refitting

16 Refit the Woodruff key to the end of the crankshaft, then refit the crankshaft sprocket (with the flange nearest the cylinder block).

17 Refit the timing belt as described in Section 6.

Fuel injection pump sprocket (2.1 litre models)

Removal

18 Remove the timing belt as described in Section 6.

19 Remove the locking bolt securing the fuel injection pump sprocket in the TDC position (See section 12).

20 On certain models, the sprocket may be fitted with a built-in puller, which consists of a plate bolted to the sprocket. The plate contains a captive nut (the sprocket securing nut), which is screwed onto the fuel injection pump shaft. On models not fitted with the built-in puller, a suitable puller can be made up using a short length of bar, and two M7 bolts screwed into the holes provided in the sprocket.

21 The fuel injection pump shaft must be prevented from turning as the sprocket nut is unscrewed, and this can be achieved using a tool similar to that shown **(see illustration)**. Use the tool to hold the sprocket stationary by means of the holes in the sprocket.

22 On models with a built-in puller, unscrew the sprocket securing nut until the sprocket is freed from the taper on the pump shaft, then withdraw the sprocket. Recover the Woodruff key from the end of the pump shaft if it is loose. If desired, the puller assembly can be removed from the sprocket by removing the two securing screws and washers.

23 On models not fitted with a built-in puller, partially unscrew the sprocket securing nut, then fit the improvised puller, and tighten the two bolts (forcing the bar against the sprocket nut), until the sprocket is freed from the taper on the pump shaft **(see illustration)**. Withdraw the sprocket and recover the Woodruff key from the end of the pump shaft if it is loose. Remove the puller from the sprocket.

Refitting

24 Refit the Woodruff key to the pump shaft, ensuring that it is correctly located in its groove.

25 Where applicable, if the built-in puller assembly has been removed from the sprocket, refit it, and tighten the two securing screws securely ensuring that the washers are in place.

26 Refit the sprocket, then tighten the securing nut to the specified torque, preventing the pump shaft from turning as during removal.

27 Make sure that the 8 mm locking bolts are fitted to the camshaft and fuel injection pump sprockets, and that the rod/drill is positioned in the flywheel timing hole.

28 Fit the timing belt around the fuel injection pump sprocket, ensuring that the marks made on the belt and sprocket before removal are aligned.

29 Tension the timing belt as described in Section 6.

30 Refit the upper timing belt covers as described in Section 6.

Camshaft sprocket (2.5 litre engine)

Note: *If the camshaft sprocket is being removed to allow the removal of the camshaft, the timing belt can be secured to the sprocket with cable ties, so that the valve timing is preserved.*

Removal

31 Remove the timing belt cover as described in Section 6, then set the engine to TDC on cylinder No 4 and lock it in this position, using the information given in Section 12.

32 With reference to Section 6, relieve the tension on the camshaft timing belt, then slide the belt off the camshaft sprocket.

33 Slacken and remove the three camshaft sprocket retaining bolts. To prevent the camshaft rotating as the nut/bolts are slackened, a sprocket holding tool will be required **(see Tool Tip at the beginning of this section)**. *Do not* attempt to use the sprocket locking pin to prevent the sprocket from rotating whilst the bolt is slackened.

34 Remove the sprocket from its mounting flange.

Refitting

35 Ensure that the flywheel and fuel injection pump locking tools are still in position before proceeding.

36 Offer the camshaft sprocket up to its mounting flange at the end of the camshaft

15.36 Offer the camshaft sprocket up to its mounting flange, then insert the sprocket securing bolts and tighten them by hand - 2.5 litre engine

15.47 Refit the fuel injection pump sprocket, then insert the sprocket securing bolts and tighten them by hand - 2.5 litre engine

15.53 Remove the crankshaft sprocket securing bolt - 2.5 litre engine

(see illustration). At the same time, engage the camshaft sprocket locking tool with the timing hole at the bottom of the sprocket hub.

37 Insert the sprocket securing bolts and tighten them by hand.

38 With reference to Section 6, refit and tension the timing belt. On completion, tighten the sprocket securing bolts to the specified torque remove the engine locking tools.

39 Using a socket and bar of the crankshaft pulley, rotate the engine in its normal direction of rotation through two crankshaft revolutions.

40 Re-check the engine valve timing by fitting the engine locking tools, as described in Section 12. If the valve timing holes do not line up, the fuel injection pump sprocket has been refitted incorrectly - it will be necessary to remove the timing belt and set the valve timing from scratch - refer to Section 6 for details.

41 On completion, refit the timing outer cover.

Fuel injection pump sprocket (2.5 litre engine)

Note: *If the fuel injection pump sprocket is being removed to allow the removal of the fuel injection pump, the timing belt can (if required) be secured to the sprocket with cable ties, so that the valve/pump timing is preserved.*

Removal

42 Remove the timing belt cover as described in Section 6, then set the engine to TDC on cylinder No 4 and lock it in this position, using the information given in Section 12.

43 With reference to Section 6, relieve the tension on the timing belt and slide it from the fuel injection pump sprocket.

44 Unscrew and remove the three fuel injection pump sprocket retaining bolts. To prevent the pump shaft rotating as the nut/bolts are slackened, a sprocket holding tool will be required (see Tool Tip at the beginning of this section). *Do not* attempt to use the sprocket locking pin to prevent the sprocket from rotating whilst the bolt is slackened.

45 Remove the sprocket from the pump shaft.

Refitting

46 Ensure that the flywheel and camshaft sprocket locking tools are still in position before proceeding.

47 Offer the fuel injection pump sprocket up to the end of the pump shaft. At the same time, engage the sprocket locking tool with the timing hole in the sprocket hub (see illustration).

48 Insert the sprocket securing bolts and hand tighten them.

49 With reference to Section 6, refit and tension the timing belt. On completion, tighten the sprocket securing bolts and remove the engine locking tools.

50 Using a socket and bar of the crankshaft pulley, rotate the engine in its normal direction

of rotation through two crankshaft revolutions. Re-check the engine valve timing by fitting the engine locking tools, as described in Section 12. If the valve timing holes so not line up, the fuel injection pump sprocket has been refitted incorrectly - it will be necessary to remove the timing belt and set the valve timing from scratch - refer to Section 6 for details.

51 On completion, refit the timing belt outer covers.

Crankshaft sprocket (2.5 litre engine)

Removal

52 Remove the timing belt (see Section 6).

53 Remove the sprocket securing bolt then slide the sprocket off the end of the crankshaft (see illustration).

54 Remove the Woodruff key from the crankshaft, and store it with the sprocket for safe-keeping.

55 Examine the crankshaft oil seal for signs of oil leakage and, if necessary, renew it as described in Section 7.

Refitting

56 Refit the Woodruff key to the end of the crankshaft, then refit the crankshaft sprocket, ensuring that the sprocket cut-out engages with the woodruff key . Fit a new sprocket retaining bolt and tighten it to the specified stage 1 and 2 torque settings (see illustrations).

57 Refit the timing belt as described in Section 6.

16 Oil cooler - removal and refitting

Refer to the information given in Chapter 2A.

17 Balance shaft drivebelt (DK5 engine) - removal and refitting

Removal

1 Disconnect the negative cable from the battery terminal.

2 Raise the front of the car and support it

15.56a Ensure that the crankshaft sprocket cut-out engages with the Woodruff key - 2.5 litre engine

15.56b Angle tightening the crankshaft sprocket securing bolt

17.7 Remove the camshaft timing belt guide roller and its mountings from the engine bracket

17.10 Balance shaft sprocket locking tools (arrowed) in place

17.12 Tensioning the balance shaft belt

securely on axle stands. Remove the front right hand roadwheel.

3 Extract the fixings and remove the engine bay undertray. Remove the screws and lift off the wheel arch inner plastic liner.

4 Refer to Section 4 and remove the auxiliary drivebelt.

5 Refer to Section 6 and remove the camshaft timing belt. Unbolt the timing belt guide roller assembly from the engine.

6 Lock the balance shafts in position by inserting a suitable bolt through the timing hole in each balance shaft sprocket. Thread the bolts into the holes in the cylinder block, behind the balance shaft sprockets, to prevent them from falling out.

7 Slacken and withdraw the securing bolts(s), then remove the guide roller and its mountings from the engine **(see illustration)**.

17.16 Correctly fitted balance shaft belt, with sprocket locking tools in place

8 Slacken the tensioner roller centre nut to release the tension on the balance shaft belt.

9 Remove the balance shaft belt from the sprockets.

Refitting

Caution: Do not try to fit the balance shaft drivebelt without the sprocket locking tools in place. Incorrect balance shaft timing will result in severe engine vibration that may lead to damage.

Note: *The final tension of the balance shaft belt should must be checked using a dedicated Citroën electronic tester.*

Note: *Balance belt tensioning is a complicated, two-stage operation that involves pre-tensioning the belt, turning the engine by hand, and then resetting the belt tension to a final value. Read through the text carefully before starting work.*

10 Check that the balance shaft sprocket locking tools are still in place before proceeding **(see illustration)**.

11 Pass the drivebelt over the sprockets, tensioner roller and guide roller and ensure that it seats correctly.

12 Tension the belt by fitting a wrench to the square hole in the side of the tensioner pulley. Turn the pulley with the wrench until the correct **pre-tension** value for a new or used belt (as applicable) is attained. Keep the tensioner pulley in this position, and use a second wrench to tighten the tensioner centre nut to the specified torque **(see illustration)**.

13 Refer to Section 6 and refit the timing belt;

ensure that the two stage tensioning procedure is carried out correctly.

14 Slacken the balance shaft belt tensioner pulley centre nut, so that the tension on the balance shaft belt is completely removed.

15 Tension the belt by fitting a wrench to the square hole in the side of the tensioner pulley. Turn the pulley with the wrench until the correct **final-tension** value for a new or used belt (as applicable) is attained. Keep the tensioner pulley in this position, and use a second wrench to tighten the tensioner centre nut to the specified torque.

16 Using a socket and bar on the crankshaft sprocket, turn the engine **in its normal direction of rotation** through two revolutions. Bring the engine around to TDC on cylinder No 4 and lock it in position again, using the engine locking tools as described in Section 12. Re-insert the balance shaft sprocket locking tools (as described in *Removal*) **(see illustration)**.

17 If all the timing holes are all correctly positioned, it should be possible to fit the engine locking tools in the flywheel, and the camshaft, fuel injection pump and balance shaft sprockets without difficulty or misalignment. If the timing holes are not correctly positioned, the balance shaft belt has been incorrectly fitted (possibly one tooth out on one of the sprockets) - in this case, repeat the refitting procedure from the beginning.

18 The remaining refitting procedure is a reversal of removal.

Chapter 2 Part C:
General engine overhaul procedures

Contents

Degrees of difficulty

Easy, suitable for novice with little experience	Fairly easy, suitable for beginner with some experience	Fairly difficult, suitable for competent DIY mechanic	Difficult, suitable for experienced DIY mechanic	Very difficult, suitable for expert DIY or professional

2C

Specifications

Note: *At the time of writing, no manufacturer's specifications were available for the 2.0 litre turbo petrol engine. The petrol engine figures given here relate to the 2.0 fuel injected, non-turbo engines (XU10J2 series - see Chapter 2A for engine code details).*

Cylinder head
Maximum gasket face distortion:
- Petrol engines .... 0.05 mm
- 2.1 litre turbo diesel engines .... 0.05 mm
- 2.5 litre turbo diesel engines .... 0.03 mm
Swirl chamber protrusion - diesel engines only .... 0 to 0.03 mm

Valves

Valve head diameter:	Inlet	Exhaust
Petrol engines:		
2.0 litre engine	42.6 mm	34.5 mm
Diesel engines:		
2.1 litre turbo engine	33.9 mm	33.9 mm
2.5 litre engine	36.9 mm	36.9 mm
Valve stem diameter:		
Petrol engines:		
2.0 litre engines	7.984 +0 -0.015 mm	7.970 mm +0 -0.015 mm
Diesel engines:		
2.1 litre turbo engine	8.005 +0 -0.015 mm	7.975 +0 -0.015 mm
2.5 litre engine	6.990 +0 -0.015 mm	6.960 +0 -0.015 mm
Overall length:		
Petrol engines:		
2.0 litre engines	108.70 mm	108.25 mm
Diesel engines:		
2.1 litre turbo engine	122.30 mm	121.90 mm
2.5 litre engine	127.91 mm	127.51 mm

Cylinder block

Cylinder bore diameter:
  Petrol engines:
    2.0 litre engines:
      Standard . . . . 86.000 to 86.018 mm
      Oversize R1 . . . . 86.250 to 86.268 mm
      Oversize R2 . . . . 86.600 to 86.618 mm
  Diesel engines:
    2.1 litre turbo engines:
      Standard . . . . 85.000 to 85.018 mm
      Oversize A1 . . . . 85.030 to 85.048 mm
      Oversize R1 . . . . 85.250 to 85.268 mm
      Oversize R2 . . . . 85.600 to 85.618 mm
    2.5 litre engine:*
      Standard, cylinder No 1 . . . . 92.020 to 92.038 mm
      Oversize, cylinder No 1 . . . . 92.520 to 92.538 mm
      Standard, cylinder Nos 2 to 4 . . . . 92.010 to 92.028 mm
      Oversize, cylinder Nos 2 to 4 . . . . 92.510 to 92.528 mm

*__Note:__ *The offset design of the cylinder block on the 2.5 litre engine is such that the machining tolerance for cylinder No 1 is different to that of cylinders No 2, 3 and 4.*

Pistons

Piston diameter:
  Petrol engines:
    2.0 litre engines:
      Standard . . . . 85.967 to 85.976 mm
      Oversize R1 . . . . 86.217 to 86.226 mm
      Oversize R2 . . . . 86.567 to 86.576 mm
Piston diameter:
  Diesel engines:
    2.1 litre engine:
      Standard . . . . 84.920 to 84.929 mm
      1st oversize . . . . 84.950 to 84.959 mm
      2nd oversize . . . . 85.170 to 85.179 mm
      3rd oversize . . . . 85.520 to 85.529 mm
    2.5 litre engine:
      Standard . . . . 91.911 to 91.929 mm
      Oversize . . . . 92.411 to 92.429 mm

Crankshaft

Endfloat:
  2.5 litre turbo diesel engine . . . . 0.04 to 0.29 mm
  All other engines . . . . 0.07 to 0.32 mm
Main bearing journal diameter:
  Petrol engines:
    2.0 litre engines:
      Standard . . . . 60.0 +0 -0.019 mm
      Undersize . . . . 59.7 +0 -0.019 mm
  Diesel engines:
    2.1 litre turbo engine:
      Standard . . . . 60.0 +0 -0.019 mm
      Undersize . . . . 59.7 +0 -0.019 mm
    2.5 litre turbo engine:
      Standard . . . . 64.0 +0 -0.019 mm
      Undersize . . . . 63.7 +0 -0.019 mm
Big-end bearing journal diameter:
  Petrol engines:
    2.0 litre engines:
      Standard . . . . 50.0 +0 -0.016 mm
      Undersize . . . . 49.7 +0 -0.016 mm
  Diesel engines:
    2.1 litre turbo engine:
      Standard . . . . 50.0 +0 -0.016 mm
      Undersize . . . . 49.7 +0 -0.016 mm
    2.5 litre engine:
      Standard . . . . 54.0 +0 -0.019 mm
      Undersize . . . . 53.7 +0 -0.019 mm

Crankshaft (continued)

Maximum bearing journal out-of-round (all models) . . . . 0.007 mm
Main bearing running clearance:
   Petrol engines:
      2.0 litre engines:
         pre-1993 . . . . 0.045 to 0.109 mm
         1993 onwards . . . . 0.038 to 0.069 mm
   Diesel engines* . . . . 0.025 to 0.050 mm
Big-end bearing running clearance - all models* . . . . 0.025 to 0.050 mm
*These are suggested figures, typical for this type of engine - at the time of writing, no exact values were available from Citroën.

Piston rings

End gaps:
   Petrol engines:
      Top compression ring:
         2.0 litre engine . . . . 0.2 to 0.4 mm
      Second compression ring:
         2.0 litre engine . . . . 0.15 to 0.35 mm
      Oil control ring . . . . n/a
   Diesel engines:
      Top and second compression rings . . . . 0.30 to 0.50 mm
      Oil control ring . . . . 0.25 to 0.50 mm

Torque wrench settings

	Nm	lbf ft
2.0 litre petrol engine		
Big-end bearing cap nuts:		
Stage 1	20	15
Stage 2	Angle-tighten through 70°	
Crankshaft pulley retaining bolt	120	88
Cylinder head bolts:		
Stage 1	35	26
Stage 2	70	52
Stage 3	Angle-tighten through 160°	
Engine-to-transmission fixing bolts	45	33
Flywheel/driveplate retaining bolts	50	37
Front oil seal carrier bolts	16	12
Left-hand engine/transmission mounting:		
Mounting bracket-to-body	30	22
Rubber mounting-to-bracket bolts	30	22
Mounting stud-to-transmission	60	44
Mounting stud bracket-to-transmission	60	44
Centre nut	65	48
Lower engine movement limiter-to-driveshaft intermediate bearing housing	50	37
Lower engine movement limiter-to-subframe	85	62
Main bearing cap bolts	70	52
Oil pump retaining bolts	16	12
Piston oil jet spray tube bolt	10	7
Right-hand engine/transmission mounting:		
Mounting bracket-to-engine nuts/bolts	80	59
Mounting bracket-to-rubber mounting nut	45	33
Rubber mounting-to-body nut	40	29
Upper engine movement limiter bolts	50	37
Sump retaining bolts	16	12
Timing belt cover bolts	8	6
Timing belt tensioner pulley bolt	20	15
2.1 litre turbo diesel engines		
Big-end bearing cap nuts:		
Stage 1	20	15
Stage 2	Tighten through a further 70°	
Camshaft carrier bolts	25	18
Camshaft sprocket bolt	50	37
Crankshaft front oil seal housing bolts	16	12
Crankshaft pulley bolt:		
Stage 1	40	30
Stage 2	Tighten through a further 60°	
Cylinder head cover bolts	8	6

2C

Torque wrench settings	Nm	lbf ft
2.1 litre turbo diesel engines (continued)		
Cylinder head bolts (new bolts with guide bosses)*:		
Stage 1	20	15
Stage 2	60	44
Stage 3	Angle tighten through 180°	
Flywheel/driveplate bolts	50	37
Injection pump sprocket puller retaining screws	10	7
Injection pump sprocket nut	50	37
Left-hand engine/transmission mounting:		
Mounting bracket-to-body	30	22
Rubber mounting-to-bracket bolts	30	22
Mounting stud-to-transmission	60	44
Centre nut	65	48
Lower engine movement limiter-to-driveshaft intermediate bearing housing	50	37
Lower engine movement limiter-to-subframe	85	62
Main bearing cap bolts:		
Stage 1	15	11
Stage 2	Tighten through a further 60°	
Oil pump mounting bolts	13	10
Piston oil jet spray tube bolt	10	7
Right-hand engine/transmission mounting:		
Mounting bracket-to-engine nuts	45	33
Mounting bracket-to-rubber mounting nut	45	33
Rubber mounting-to-body nut	40	29
Upper engine movement limiter bolts	50	37
Sump bolts	16	12
Timing belt idler pulley	37	27
Timing belt tensioner nut/bolt	10	7
Transmission housing-to-engine bolts	55	40
2.5 litre turbo diesel engine		
Big-end bearing cap nuts:		
Stage 1	12	9
Stage 2	Angle tighten through 60°	
Crankshaft sprocket bolt:		
Stage 1	70	52
Stage 2	Angle tighten through 51°	
Cylinder head bolts:*		
M10 bolts:		
Stage 1	35	26
Stage 2	Angle tighten through 120°	
M12 bolts:		
Stage 1	50	37
Stage 2	Angle tighten through 120°	
Flywheel/driveplate bolts	50	37
Balance shaft housing bolts:		
High tensile bolts	34	25
All other bolts	25	18
Balance shaft sprocket nut:		
Rear balance shaft (LEFT HAND THREAD)	13	10
Front balance shaft	13	10
Balance shaft counter weight screws	12	9
Main bearing ladder inner bolts:		
Stage 1	20	15
Stage 2	Angle tighten through 60°	
Main bearing ladder outer bolts	10	7
Oil pump mounting bolts	9	7
Piston oil jet spray tube bolt	10	7
Sump bolts	8	6
Transmission housing-to-engine bolts	55	40

Note: Where new cylinder head bolts are fitted, there is no requirement to re-tighten the bolts after the engine has been started and warmed up for the first time; see Chapter 2B for details.

1 General information

1 Included in this Part of Chapter 2 are details of removing the engine/transmission from the car and general overhaul procedures for the cylinder head, cylinder block and all other engine internal components.

2 The information given ranges from advice concerning preparation for an overhaul and the purchase of replacement parts, to detailed step-by-step procedures covering removal, inspection, renovation and refitting of engine internal components.

3 After Section 7, all instructions are based on the assumption that the engine has been removed from the car. For information concerning in-car engine repair, as well as the removal and refitting of those external components necessary for full overhaul, refer to Part A or B of this Chapter (as applicable) and to Section 7. Ignore any preliminary dismantling operations described in Part A or B that are no longer relevant once the engine has been removed from the car.

4 Specifications relating to engine overhaul are at the beginning of this Part of Chapter 2.

2 Engine overhaul - general information

1 It is not always easy to determine when, or if, an engine should be completely overhauled, as a number of factors must be considered.

2 High mileage is not necessarily an indication that an overhaul is needed, while low mileage does not preclude the need for an overhaul. Frequency of servicing is probably the most important consideration. An engine which has had regular and frequent oil and filter changes, as well as other required maintenance, should give many thousands of miles of reliable service. Conversely, a neglected engine may require an overhaul very early in its life.

3 Excessive oil consumption is an indication that piston rings, valve seals and/or valve guides are in need of attention. Make sure that oil leaks are not responsible before deciding that the rings and/or guides are worn. Perform a compression test, as described in Part A (petrol engine) or B (Diesel engine) of this Chapter, to determine the likely cause of the problem.

4 Check the oil pressure with a gauge fitted in place of the oil pressure switch, and compare it with that specified. If it is extremely low, the main and big-end bearings, and/or the oil pump, are probably worn out.

5 Loss of power, rough running, knocking or metallic engine noises, excessive valve gear noise, and high fuel consumption may also point to the need for an overhaul, especially if they are all present at the same time. If a complete service does not remedy the situation, major mechanical work is the only solution.

6 An engine overhaul involves restoring all internal parts to the specification of a new engine. During an overhaul, the pistons and the piston rings are renewed. New main and big-end bearings are generally fitted; if necessary, the crankshaft may be renewed, to restore the journals. The valves are also serviced as well, since they are usually in less-than-perfect condition at this point. While the engine is being overhauled, other components, such as the starter and alternator, can be overhauled as well. The end result should be an as-new engine that will give many trouble-free miles. **Note:** *Critical cooling system components such as the hoses, thermostat and water pump should be renewed when an engine is overhauled. The radiator should be checked carefully, to ensure that it is not clogged or leaking. Also, it is a good idea to renew the oil pump whenever the engine is overhauled.*

7 Before beginning the engine overhaul, read through the entire procedure, to familiarise yourself with the scope and requirements of the job. Overhauling an engine is not difficult if you follow carefully all of the instructions, have the necessary tools and equipment, and pay close attention to all specifications. It can, however, be time-consuming. Plan on the car being off the road for a minimum of two weeks, especially if parts must be taken to an engineering works for repair or reconditioning. Check on the availability of parts and make sure that any necessary special tools and equipment are obtained in advance. Most work can be done with typical hand tools, although a number of precision measuring tools are required for inspecting parts to determine if they must be renewed. Often the engineering works will handle the inspection of parts and offer advice concerning reconditioning and renewal. **Note:** *Always wait until the engine has been completely dismantled, and until all components (especially the cylinder block and the crankshaft) have been inspected, before deciding what service and repair operations must be performed by an engineering works. The condition of these components will be the major factor to consider when determining whether to overhaul the original engine, or to buy a reconditioned unit. Do not, therefore, purchase parts or have overhaul work done on other components until they have been thoroughly inspected.* As a general rule, time is the primary cost of an overhaul, so it does not pay to fit worn or sub-standard parts.

8 As a final note, to ensure maximum life and minimum trouble from a reconditioned engine, everything must be assembled with care, in a spotlessly-clean environment.

3 Engine removal - methods and precautions

1 If you have decided that the engine must be removed for overhaul or major repair work, several preliminary steps should be taken.

2 Locating a suitable place to work is extremely important. Adequate work space, along with storage space for the car, will be needed. If a workshop or garage is not available, at the very least, a flat, level, clean work surface is required.

3 Cleaning the engine compartment and engine/transmission before beginning the removal procedure will help keep tools clean and organised.

4 An engine hoist or A-frame will also be necessary. Make sure the equipment is rated in excess of the combined weight of the engine and transmission. Safety is of primary importance, considering the potential hazards involved in lifting the engine/transmission out of the car.

5 If this is the first time you have removed an engine, an assistant should ideally be available. Advice and aid from someone more experienced would also be helpful. There are many instances when one person cannot simultaneously perform all of the operations required when lifting the engine out of the vehicle.

6 Plan the operation ahead of time. Before starting work, arrange for the hire of or obtain all of the tools and equipment you will need. Some of the equipment necessary to perform engine/transmission removal and installation safely and with relative ease (in addition to an engine hoist) is as follows: a heavy duty trolley jack, complete sets of spanners and sockets as described in the front of this manual, wooden blocks, and plenty of rags and cleaning solvent for mopping up spilled oil, coolant and fuel. If the hoist must be hired, make sure that you arrange for it in advance, and perform all of the operations possible without it beforehand. This will save you money and time.

7 Plan for the car to be out of use for quite a while. An engineering works will be required to perform some of the work which the do-it-yourselfer cannot accomplish without special equipment. These places often have a busy schedule, so it would be a good idea to consult them before removing the engine, in order to accurately estimate the amount of time required to rebuild or repair components that may need work.

8 Always be extremely careful when removing and refitting the engine/transmission. Serious injury can result from careless actions. Plan ahead and take your time, and a job of this nature, although major, can be accomplished successfully.

4 Petrol engine and manual transmission unit - removal, separation and refitting

Removal

1 Park the vehicle on firm, level ground. Chock the rear wheels, then firmly apply the parking brake. Jack up the front of the vehicle, and securely support it on axle stands. Remove both front roadwheels.

2C

2 Set the ride height control to its lowest position, then depressurise the hydraulic system - refer to Chapter 9 for details.

3 Set the bonnet in the upright position or, to improve access, remove it completely as described in Chapter 11.

4 Undo the retaining screws and remove the plastic undercover from beneath the engine/transmission unit. Also remove the plastic covers from the left- and right-hand wheelarches.

5 If the engine is to be dismantled, working as described in Chapter 1, first drain the oil and remove the oil filter. Clean and refit the drain plug, tightening it securely.

6 Disconnect both cables from the battery terminals. Remove the battery from its holder then unbolt the battery holder from the bodywork and remove it from the engine compartment.

7 Refer to Chapter 3 and drain the cooling system.

8 Drain the transmission oil as described in Chapter 7A.

9 Remove both driveshafts, as described in Chapter 8.

10 Refer to Chapter 2A and unbolt the rear engine mounting from the suspension subframe.

11 Remove the bolt and detach the metal clip(s) that secures the hydraulic system pipework to the front of the pressure regulator, at the front of the transmission casing.

12 Unbolt the exhaust downpipe from the exhaust manifold/turbocharger flange (as applicable), with reference to Chapter 4A/C.

13 Slacken the clips and detach the hoses from the coolant pump inlet and outlet ports. Similarly, disconnect the throttle housing coolant hose, and the radiator top and bottom hoses at their respective ports.

14 Refer to Chapter 7A and detach the three gear change/shift control rods from their ball joint pivots.

15 With reference to Chapters 9 and 11 as applicable, slacken the unions and disconnect the hydraulic system feed pipe, the power steering system supply pipe, the pressure regulator return pipe and the flow distributor return pipe from the regulator/distributor assembly.

16 Slacken the large worm drive clips and disconnect the air intake ducting from the throttle body. Where applicable, unplug the wiring at the connector and detach the section of ducting that incorporates the air flow meter. Store the air flow meter in a safe place. On turbo models, remove the securing screws, slacken the hose clips and disconnect the air cleaner to turbocharger, turbocharger to intercooler and intercooler to inlet manifold ducting.

17 Refer to Chapter 9 and detach the hose manifold from the hydraulic system fluid reservoir from its casing. Cover the open reservoir to prevent dirt ingress. Slacken the clip and detach the hydraulic pump feed pipe

from the hose manifold. Pad the exposed pipes on the underside of the manifold, to prevent damage and dirt ingress.

18 Working at the transmission, carry out the following:

a) *Refer to Chapter 7A and disconnect the speedometer drive cable or the transducer wiring from the transmission.*

b) *Unbolt the earth cable from the transmission casing.*

c) *Disconnect the clutch cable from the transmission with reference to Chapter 6.*

d) *Unplug the wiring from the reversing lamp switch at the connector.*

19 Disconnect the accelerator cable from throttle body as described in Chapter 4A.

20 Disconnect the main engine wiring harness at the connectors located on the top of the transmission casing.

21 Remove the electronic control unit(s) from the casing at the front right hand corner of the engine compartment. Slacken and withdraw the securing screws and detach the casing from the bodywork.

22 Slacken the hose clips and disconnect the fuel supply and return hoses from the fuel rail and pressure regulator - see Chapter 4A fro details.

23 Working at the engine compartment bulkhead, slacken the hose clips and detach the cabin heater supply and return hoses from their respective ports.

24 Support the engine securely on a hoist, or a lifting beam positioned across the engine compartment in line with the suspension mounting turrets.

25 Refer to Chapter 2A and carry out the following:

a) *Remove the centre nut and unbolt the transmission casing from the engine/transmission mounting.*

b) *Unbolt the engine/transmission mounting bracket from the chassis rail and remove it from the engine compartment.*

c) *Slacken and withdraw the centre bolt, then separate the two halves of the upper engine mounting.*

26 Make a final check that any components which would prevent the removal of the engine/transmission from the car have been removed or disconnected. Ensure that components such as the gearchange selector rod and driveshafts are secured so that they cannot be damaged on removal.

27 Lift the engine/transmission out of the car, ensuring that nothing is trapped or damaged. Enlist the help of an assistant during this procedure, as it will be necessary to tilt the assembly slightly to clear the body panels.

28 Once the engine is high enough, lift it out over the front of the body, and lower the unit to the ground.

Separation

29 With the engine/transmission assembly removed, support the assembly on suitable blocks of wood, on a workbench (or failing that, on a clean area of the workshop floor).

30 Undo the retaining bolts, and remove the flywheel lower cover plate from the transmission. On some models the plate has support struts attached to it, these will have to be unbolted from the side of the cylinder block.

31 Disconnect the wiring then undo the retaining bolts, and remove the starter motor from the transmission, noting the correct fitted position of the locating dowel (see Chapter 5A).

32 Ensure that both engine and transmission are adequately supported, then slacken and remove the remaining bolts securing the transmission housing to the engine. Note the correct fitted positions of each bolt (and the relevant brackets) as they are removed, to use as a reference on refitting.

33 With reference to Chapter 7A, carefully withdraw the transmission from the engine, ensuring that the weight of the transmission is not allowed to hang on the input shaft while it is engaged with the clutch friction disc.

34 If they are loose, remove the locating dowels from the engine or transmission (where applicable), and keep them in a safe place.

Refitting

35 If the engine and transmission have been separated, perform the operations described below in paragraphs 36 to 41. If not, proceed as described from paragraph 42 onwards.

36 Ensure the clutch plate and transmission input shaft splines are clean and dry. Do not apply grease to the splines as they have a special low-friction nickel coating.

37 Ensure the locating dowels are correctly positioned prior to installation and make sure the clutch release mechanism components are correctly fitted (see Chapter 6).

38 With reference to Chapter 7A, carefully offer the transmission to the engine, until the locating dowels are engaged. Ensure that the weight of the transmission is not allowed to hang on the input shaft as it is engaged with the clutch friction disc.

39 Refit the transmission housing-to-engine bolts, ensuring that all the necessary brackets are correctly positioned, and tighten them to the specified torque setting.

40 Refit the starter motor making sure its locating dowel is correctly positioned. Securely tighten its retaining bolts and reconnect the wiring (see Chapter 5A).

41 Refit the flywheel lower cover plate to the transmission, and tighten its retaining bolts to the specified torque.

42 Reconnect the hoist and lifting tackle to the engine lifting brackets. With the aid of an assistant, lift the assembly over the engine compartment.

43 The assembly should be tilted as necessary to clear the surrounding components, as during removal; lower the assembly into position in the engine compartment, manipulating the hoist and lifting tackle as necessary.

44 With the engine/transmission in position, refit the engine/transmission mounting bracket to the chassis rail and tighten the securing bolts to the specified torque. Pass the stud on the transmission casing through the engine/transmission mounting, then fit the nut to the stud and tighten it lightly by hand

45 Refit the right-hand engine mounting bolts and tighten them lightly by hand.

46 Rock the engine to settle it on its mountings. Centralise the right-hand mounting bracket in relation to the rubber mounting lug then tighten its retaining nut and bolts to their specified torque settings. Go around and tighten all the remaining mounting nuts and bolts to their specified torque settings and detach the hoist from the engine.

47 The remainder of the refitting procedure is a direct reversal of the removal sequence, noting the following points:

a) *Ensure that the wiring loom is correctly routed and retained by all the relevant retaining clips; all connectors should be correctly and securely reconnected.*

b) *Refit the cap and hose manifold to the hydraulic system fluid reservoir, then bleed the system as described in Chapter 9.*

c) *Prior to refitting the driveshafts to the transmission, the driveshaft oil seals should be renewed.*

d) *Ensure that all disturbed hoses are correctly reconnected, and securely retained by their retaining clips.*

e) *Adjust the clutch cable as described in Chapter 6.*

f) *Adjust the accelerator cable as described in Chapter 4A.*

g) *Refill the engine and transmission with correct quantity and type of oil, as described in Chapters 1A and 7A.*

h) *Refill the cooling system as described in Chapters 1A and 3.*

i) *Top up the fluid level in the hydraulic system as described in "Weekly checks".*

5 Petrol engine and automatic transmission unit - removal, separation and refitting

Removal

Note: *The engine can be removed from the car only as a complete unit with the transmission; the two are then separated for overhaul.*

1 The removal procedure is essentially the same as that described for the removal of the engine and manual transmission assembly, but note the following points:

a) *Where applicable, release the retaining clips and disconnect the coolant hoses from the transmission fluid cooler.*

b) *On models with electronic transmission control, release the retaining clips and disconnect the wiring connectors from the transmission electronic control unit (ECU).*

c) *Disconnect the selector cable from the transmission and position it clear of the unit as described in Chapter 7B.*

d) *Where applicable, disconnect the transmission kick-down cable at the throttle body.*

e) *Disconnect the earth cable from the stud on the transmission.*

f) *Remove the wiring harness bracket and the hose support bracket from the transmission.*

g) *Disconnect the wiring from the speedometer transducer (speedometer drive) and RPM sensor, then remove the RPM sensor from the bellhousing.*

h) *Remove the starter motor.*

i) *Label and disconnect any remaining wiring connectors and support brackets connected to the transmission.*

Separation

2 With the engine/transmission assembly removed, support the assembly on suitable blocks of wood, on a workbench (or failing that, on a clean area of the workshop floor).

3 Locate the access hole at the lower rear of the cylinder block, then turn the crankshaft, by means of a socket on the crankshaft pulley bolt, until one of the torque converter retaining bolts is accessible through the access hole.

4 Undo the accessible torque converter bolt then turn the crankshaft as necessary and undo the remaining two bolts.

5 Slacken and remove the bolts securing the transmission housing to the engine. Note the correct fitted positions of each bolt, and the necessary brackets, as they are removed, to use as a reference on refitting. Make a final check that all components have been disconnected, and are positioned clear of the transmission so that they will not hinder the removal procedure.

6 With the bolts removed, pull the transmission off the engine, to free it from its locating dowels. Once the transmission is free, and sufficient clearance exists, insert a bolt with a suitable washer, through the RPM sensor hole in the transmission bellhousing, to retain the torque converter on the transmission.

Preparation for reconnection

7 Prior to reconnection it is necessary to make a simple tool to align the torque converter with the driveplate as the transmission is refitted. To make the tool, obtain a bolt of the same size as the torque converter retaining bolts, but long enough to extend through the access hole in the cylinder block when the transmission is refitted.

8 Cut the head off the bolt and cut a slot (to enable it to be unscrewed) in the plain end. Check that the tool will slide easily through the torque converter retaining bolt hole in the driveplate.

9 Turn the engine crankshaft so that one of the torque converter retaining bolt holes in the driveplate, is aligned with the access hole in the cylinder block. Screw the alignment tool (finger tight only) into one of the retaining bolt holes in the torque converter. Turn the torque converter so that the alignment tool is in approximately the correct position, relative to the cylinder block access hole. As the transmission is refitted, the alignment tool will pass through the retaining bolt hole in the driveplate and through the access hole. It can then be unscrewed with a screwdriver and the first torque converter retaining bolt fitted in its place.

10 Check that the torque converter support bush fitted to the centre of the crankshaft is in good condition, and in place.

11 Ensure that the engine/transmission locating dowels are correctly positioned prior to installation.

Reconnection

12 The transmission is reconnected by a reversal of the removal procedure, bearing in mind the following points:

a) *Guide the transmission into position ensuring that the alignment tool passes through the driveplate and access hole.*

b) *Remove the bolt used to retain the torque converter in place, just before the transmission engages with the engine.*

c) *Once the transmission is bolted to the engine, remove the alignment tool and fit the first torque converter retaining bolt. Turn the crankshaft as necessary and fit the other two bolts.*

Refitting

13 Refit the starter motor, and securely tighten its retaining bolts.

14 Refit the engine unit to the vehicle as described in the relevant refitting paragraphs of Section 4.

15 The remainder of the refitting procedure is a reversal of the removal sequence, noting the following points:

a) *Ensure that the wiring loom is correctly routed, and retained by all the relevant retaining clips; all connectors should be correctly and securely reconnected.*

b) *Prior to refitting the driveshafts to the transmission, renew the driveshaft oil seals as described in Chapter 7B.*

c) *Ensure that all coolant hoses are correctly reconnected, and securely retained by their retaining clips.*

d) *Adjust the accelerator cable as described in Chapter 4A.*

e) *Refill the engine and transmission with correct quantity and type of lubricant, as described in Chapter 1A and 7B.*

f) *Refill the cooling system (see Weekly Checks).*

g) *Top up the hydraulic system with the specified grade of fluid, as described in "Weekly checks"*

2C

6.11a At the engine compartment front crossmember, remove the upper . . .

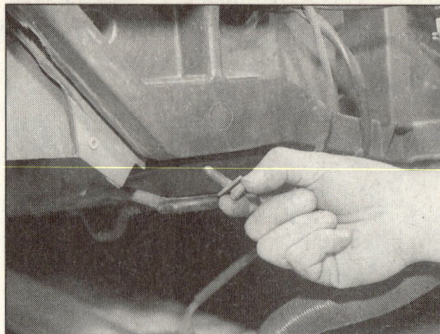

6.11b . . . and lower securing bolts . . .

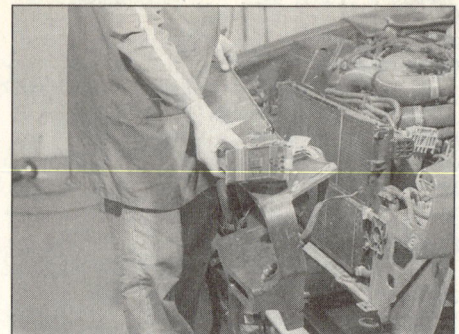

6.11c . . . then lift the crossmember away from the front of the car

6 Diesel engine and manual transmission unit - removal, separation and refitting

Removal

1 Park the vehicle on firm, level ground. Chock the rear wheels, then firmly apply the parking brake. Jack up the front of the vehicle, and securely support it on axle stands. Remove both front roadwheels.

2 Set the ride height control to its lowest position, then depressurise the hydraulic system - refer to Chapter 9 for details.

3 Set the bonnet in the upright position or, to improve access, remove it completely as described in Chapter 12.

4 Undo the retaining screws and remove the plastic undertray from beneath the engine/transmission unit. Also remove the plastic covers from the left- and right-hand wheelarches.

5 If the engine is to be dismantled, working as described in Chapter 1B, first drain the engine oil and remove the oil filter. Clean the sump drain plug (fitting a new sealing washer where applicable), then fit and tighten it securely.

6 Disconnect both cables from the battery terminals. Slacken the clamp bar and remove the battery from its holder (see Chapter 5A).

7 Refer to Chapter 3 and drain the cooling system.

8 Drain the transmission oil as described in Chapter 7A.

9 Refer to Chapter 4B and carry out the following:

a) Remove the air cleaner and its associated intake air ducting. On 2.1 litre models, slacken the clips and disconnect the inlet air ducting from the upper inlet manifold and intercooler.

b) Disconnect the accelerator cable from the fuel injection pump (excluding models with 'drive by wire' electronic fuel injection).

c) Slacken the clips and disconnect the fuel supply and return hoses.

d) Unplug the fuel injection pump wiring at the connectors.

10 Refer to Chapter 9 and detach the cap and hose manifold from the hydraulic system fluid reservoir casing. Cover the open reservoir to prevent dirt ingress. Slacken the clip and detach the hydraulic pump feed pipe from the hose manifold. Place the exposed pipes on the underside of the cap/manifold in a suitable container to prevent damage and dirt ingress, and tie them back away from the work area.

11 Remove the front bumper with reference to Chapter 12, then remove the headlights with reference to Chapter 13. Disconnect the bonnet release cable at the joint block, located underneath and behind the battery holder. Slacken and withdraw the securing bolts, then carefully lift off the engine compartment front crossmember, together with the cooling fans **(see illustrations)**. Unplug the horn and cooling fan wiring as the connectors become accessible.

12 Unbolt the fuel filter/priming pump, glow plug control unit and power supply unit from the battery tray. Label the connectors to aid refitting then unplug the wiring from the underside of the glow plug control and power supply units. Remove the securing screws and lift the battery tray out of the engine compartment.

6.13 Remove the electronic control units from their plastic casing, then remove the casing from the bodywork

13 Remove the electronic control unit(s) from their plastic casing, located at the front right hand corner of the engine compartment. Unplug each ECU from its respective wiring connector and store it in a safe place - avoid touching the connector pins. Slacken and withdraw the securing screws, then remove the ECU casing from the bodywork **(see illustration)**.

14 Refer to Chapter 3 and remove the radiator. Note that on 2.5 'litre models, the intercooler cooling radiator is secured to the front of the main radiator; both units can be removed as an assembly.

15 Refer to Chapter 8 and remove the right hand driveshaft and intermediate shaft. The left hand driveshaft must be withdrawn from the differential, but it does not need to be removed from the hub; (refer to Chapter 8 for details). Secure the shaft away from the transmission using a length of wire to ensure that it does impede the engine removal process. Ensure that the shaft is secured horizontally, so that the CV joints are not strained.

16 Unbolt the exhaust downpipe from the exhaust manifold/turbocharger flange (as applicable), with reference to Chapter 4C.

17 On 2.5 litre models, working underneath the car, unbolt the curved inlet air trunking from the turbocharger flange, and recover the O-ring seal **(see illustration)**. Slacken the

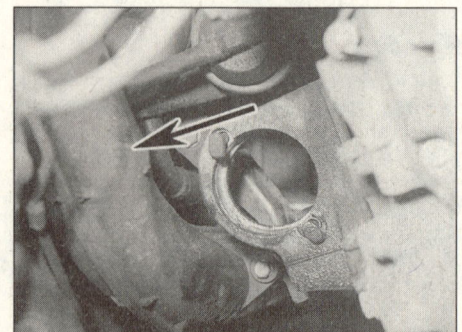

6.17 On 2.5 litre models, unbolt the inlet air trunking (arrowed) from the turbocharger flange (viewed through right hand wheel arch)

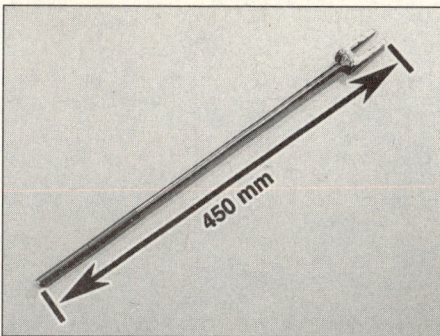

6.18a Improvised tool, used for separating the gear change cable ball joints on 2.5 litre models

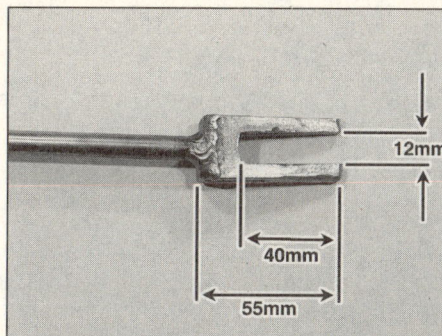

6.18b Fabricate the tapered, forked end to the dimensions shown

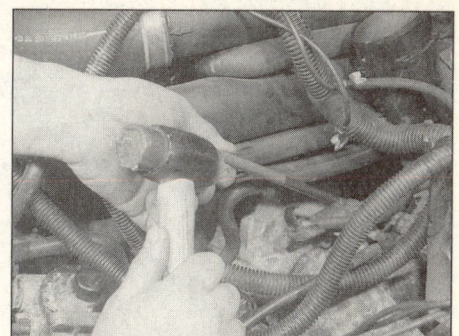

6.18c Insert the tool between the two halves of the balljoint and tap the end with a mallet to separate the joint

hose clip and detach the other end of the trunking from the rigid duct/coolant pump pulley cover assembly. Remove the trunking from the engine bay.

18 Refer to Chapter 7A and detach the three gear change/shift control rods from their ball joint pivots. On 2.5 litre models, disconnect the gear change and shift cables from the transmission as described in Chapter 7A.

Note: *Access to the gear change and shift cable ball joints at the transmission is extremely limited on the 2.5 litre model. The fabrication of a special tool for this purpose is recommended* **(see illustrations).**

19 Unplug the main engine wiring harness at the connectors, located above the transmission casing **(see illustration)**. Work around the engine and trace any remaining wiring that has not been isolated by the separation of the main wiring harness connectors. Unplug each section of wiring at its respective connector - label each connector carefully to aid reconnection later. Note that some (but not all) connectors are colour-coded.

20 Working at the transmission, carry out the following:

a) *Refer to Chapter 7A and disconnect the speedometer drive cable (or transducer wiring) and the reversing lamp switch wiring from the transmission.*

b) *Unbolt the earth cable from the transmission casing.*

c) *Disconnect the clutch cable (or slave cylinder, as applicable) from the transmission with reference to Chapter 6.*

d) *Remove the securing screw and withdraw the TDC sensor from the top of the transmission bellhousing.*

21 With reference to Chapters 9 and 11 as applicable, slacken the unions and disconnect the hydraulic system feed pipe, the power steering system supply pipe, the pressure regulator return pipe and the flow distributor return pipe from the regulator/distributor assembly (mounted on the front of the transmission assembly). It will be necessary to release the pipes from the metal clip(s) at the front of the pressure regulator.

22 Support the engine securely on a hoist. Attach the jib only to the lifting eyes provided, do not lift at any other point. Use all the available lifting eyes, to achieve even weight distribution and balance.

23 On 2.1 litre models, refer to Chapter 2B and carry out the following:

a) *Remove the auxiliary drivebelt.*

b) *Unbolt the auxiliary drivebelt tensioner assembly from the engine (where applicable).*

c) *Unbolt the auxiliary drivebelt pulley from the crankshaft timing/balance shaft sprocket.*

24 Ensure that the engine is securely supported on the hoist, then refer to Chapter

2B and unbolt and remove the engine mounting components.

25 Using the hoist, raise the engine slightly, then draw it away from the bulkhead slightly. Adjust the hoist so that the engine can be tilted towards the front of the car slightly.

26 Working at the engine compartment bulkhead, slacken the hose clips and detach the cabin heater supply and return hoses from their respective ports (see Chapter 3).

27 Make a final check that any components which would prevent the removal of the engine/transmission from the car have been removed or disconnected.

28 Ensure that components such as the gearchange selector rods/cables and hydraulic pipes are secured so that they cannot be damaged on removal.

29 Lift the engine/transmission upwards, until the bottom of the sump is just above the level of the suspension front crossmember. Check around the engine and ensure that nothing is trapped or damaged. Enlist the help of an assistant during this procedure, as it will be necessary to tilt the assembly slightly to clear the body panels.

30 Rotate the engine on the hoist, so that the timing belt end is furthest forward, allowing it to leave the engine compartment first. Manoeuvre the engine past the bodywork, taking great care to avoid the ABS hydraulic unit (where fitted). Lift the engine assembly out over the front crossmember using the hoist and lower the unit to a suitable work surface **(see illustration)**.

Separation

31 With the engine/transmission assembly removed, support the assembly on suitable blocks of wood, on a workbench (or failing that, on a clean area of the workshop floor).

32 Undo the retaining bolts, and remove the flywheel lower cover plate from the transmission. On some models the plate has support struts attached to it, these will have to be unbolted from the side of the cylinder block.

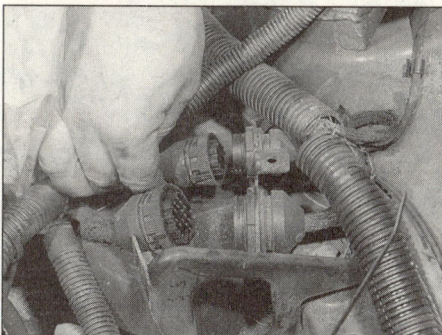

6.19 Unplug the main engine wiring harness at the connectors, located above the transmission casing

6.30 Lift the engine and transmission assembly out over the front crossmember

2C

6.33a On 2.5 litre models, unbolt the intercooler mounting bracket from the engine block

6.33b On 2.5 litre models, slacken the hose clips . . .

6.33c . . . remove the securing screws . . .

33 On 2.5 litre models, carry out the following **(see illustrations)**:
a) *Remove the intercooler, then unbolt the intercooler mounting bracket from the engine block (see Chapter 4B)*
b) *Slacken the hose clips, remove the securing screws, and lift off the combined coolant pump pulley cover/inlet air ducting assembly.*
c) *Remove the securing bolt and detach the fuel heater thermostat from the transmission bellhousing.*

34 Disconnect the wiring then undo the retaining bolts, and remove the starter motor

6.33d . . . and lift off the combined coolant pump pulley cover/inlet air ducting assembly

from the transmission, noting the correct fitted position of the locating dowel (see Chapter 5A).

35 Ensure that both engine and transmission are adequately supported, then slacken and remove the remaining bolts securing the transmission housing to the engine. Note the correct fitted positions of each bolt (and the relevant brackets) as they are removed, to use as a reference on refitting.

36 Carefully withdraw the transmission from the engine, ensuring that the weight of transmission is not allowed to hang on the input shaft while it is disengaged from the clutch friction disc **(see illustration)**. Note that on 2.5 litre models, some resistance may be encountered as the release bearing disengages from the release shaft forks.

37 If they are loose, remove the locating dowels from the engine or transmission (where applicable), and keep them in a safe place.

Refitting

38 If the engine and transmission have been separated, perform the operations described below in paragraphs 39 to 44. If not, proceed as described from paragraph 45 onwards.

39 Ensure the clutch plate and transmission input shaft splines are clean and dry. Do not apply grease to the splines as they have a special low-friction nickel coating.

40 Ensure the locating dowels are correctly positioned prior to installation and make sure the clutch release mechanism components are correctly fitted (see Chapter 6). Note that on 2.5 litre models, the clutch release bearing must be separated from the pressure plate and fitted to the clutch release fork - see Chapter 6 for details.

41 With reference to Chapter 7A, carefully offer the transmission to the engine, until the locating dowels are engaged. Ensure that the weight of the transmission is not allowed to hang on the input shaft as it is engaged with the clutch friction disc. On 2.5 models with a hydraulic clutch, refer to the information in Chapter 6 which describes how to reconnect the thrust bearing to the pressure plate.

42 Refit the transmission housing-to-engine bolts, ensuring that all the necessary brackets are correctly positioned, and tighten them to the specified torque setting **(see illustration)**.

43 Refit the starter motor making sure its locating dowel is correctly positioned. Securely tighten its retaining bolts and reconnect the wiring (see Chapter 5A).

44 On 2.5 litre engines, carry out the following:
a) *Refit the intercooler mounting bracket to the engine block, then refit the intercooler.*

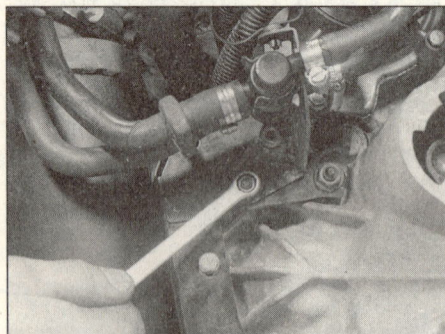

6.33e On 2.5 litre models, unbolt the fuel heater thermostat from the transmission bellhousing

6.36 Carefully withdraw the transmission from the engine, ensuring that the weight of the transmission is not allowed to hang on the input shaft

6.42 Refit the transmission bellhousing-to-engine bolts and tighten them to the specified torque setting

b) *Refit the combined coolant pump pulley cover/inlet air ducting assembly, and tighten the screws securely.*

c) *Refit the fuel heater thermostat.*

45 Refit the flywheel lower cover plate to the transmission, and tighten its retaining bolts to the specified torque.

46 Reconnect the hoist and lifting tackle to the engine lifting brackets. With the aid of an assistant, lift the assembly over the engine compartment.

47 The assembly should be tilted as necessary to clear the surrounding components, as during removal; lower the assembly into position in the engine compartment, manipulating the hoist and lifting tackle as necessary.

48 With the engine/transmission in position, refit the engine/transmission mountings as described in Chapter 2B. Do not fully tighten the mounting bolts at this stage.

49 Rock the engine to settle it on its mountings. Centralise the right-hand mounting bracket in relation to the rubber mounting lug then tighten its retaining nut and bolts to their specified torque settings. Go around and tighten all the remaining mounting nuts and bolts to their specified torque settings and detach the hoist from the engine. Where applicable, adjust the torque control mechanisms as described in Chapter 2B to ensure minimal engine movement.

50 The remainder of the refitting procedure is a direct reversal of the removal sequence, noting the following points:

a) *Ensure that the wiring loom is correctly routed and retained by all the relevant retaining clips; all connectors should be correctly and securely reconnected. Pay particular attention to body earth points - ensure that the connections surfaces are clean and free from corrosion.*

b) *Refit the cap and hose manifold to the hydraulic system fluid reservoir, then bleed the system as described in Chapter 9.*

c) *Prior to refitting the driveshafts to the transmission, the driveshaft oil seals should be renewed.*

d) *Ensure that all disturbed hoses are*

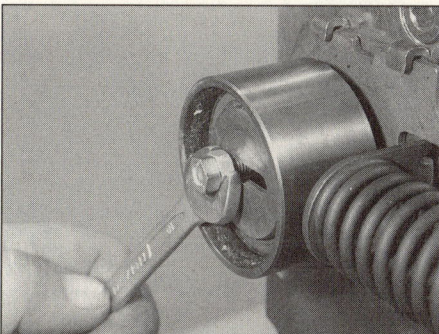

8.4a On 2.1 litre diesel models, remove the timing belt tensioner centre retaining stud by screwing on a second nut and locking the two nuts together . . .

correctly reconnected, and securely retained by their retaining clips.

e) *Adjust the clutch cable as described in Chapter 6.*

f) *Adjust the accelerator cable as described in Chapter 4B (where applicable).*

g) *Refill the engine and transmission with correct quantity and type of oil, as described in Chapters 1 and 7.*

h) *Refill the cooling system as described in Chapters 1B and 3.*

i) *Top up the fluid level and bleed the hydraulic system as described in Chapter 9.*

7 Engine overhaul - dismantling sequence

1 It is much easier to dismantle and work on the engine if it is mounted on a portable engine stand. These stands can often be hired from a tool hire shop. Before the engine is mounted on a stand, the flywheel/driveplate should be removed, so that the stand bolts can be tightened into the end of the cylinder block/crankcase.

2 If a stand is not available, it is possible to dismantle the engine with it blocked up on a sturdy workbench, or on the floor. Be extra-careful not to tip or drop the engine when working without a stand.

3 If you are going to obtain a reconditioned engine, all ancillary equipment and external components must be removed first, to be transferred to the replacement engine (just as they will if you are doing a complete engine overhaul yourself).

Note: *When removing the external components from the engine, pay close attention to details that may be helpful or important during refitting. Note the fitted position of gaskets, seals, spacers, pins, washers, bolts, and other small items.*

4 If you are obtaining a 'short' engine (which consists of the engine cylinder block/crankcase, crankshaft, pistons and connecting rods all assembled), then the cylinder head, sump, oil pump, and timing belt will have to be removed also.

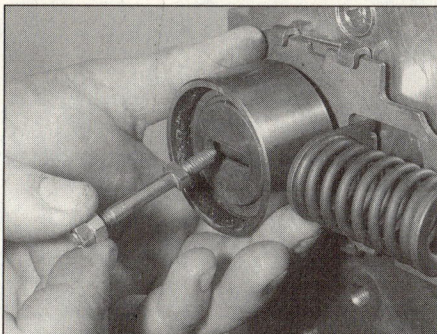

8.4b . . . then unscrew and remove the stud

5 If you are planning a complete overhaul, the engine can be dismantled, and the internal components removed, in the order given below, referring to Part A or B of this Chapter unless otherwise stated.

a) *Inlet and exhaust manifolds (Chapter 4).*

b) *Timing belt, sprockets and tensioner(s).*

c) *Cylinder head.*

d) *Flywheel/driveplate.*

e) *Sump.*

f) *Oil pump.*

g) *Balance shafts and housings (where applicable).*

h) *Pistons/connecting rods.*

i) *Crankshaft.*

6 Before beginning the dismantling and overhaul procedures, make sure that you have all of the correct tools necessary. See *Tools and working facilities* for further information.

8 Cylinder head - dismantling

Note: *New and reconditioned cylinder heads are available from the manufacturer, and from engine overhaul specialists. Be aware that some specialist tools are required for the dismantling and inspection procedures, and new components may not be readily available. It may therefore be more practical and economical for the home mechanic to purchase a reconditioned head, rather than dismantle, inspect and recondition the original head.*

1 Remove the cylinder head as described in Part A or B of this Chapter, or in this Part (as applicable).

2 If not already done, remove the inlet and exhaust manifolds with reference to the relevant Part of Chapter 4. Remove any remaining brackets or housings as required.

3 Remove the camshaft, followers and shims (as applicable) as described in Part A or B of this Chapter.

4 On diesel models, remove the glow plugs as described in Chapter 5C and the injectors as described in Chapter 4B. On 2.1 litre diesel models, remove the timing belt tensioner centre retaining stud by screwing on a second nut and locking the two nuts together. Unscrew the stud by means of the locked nuts. Undo the retracting cam retaining bolt and remove the tensioner assembly **(see illustrations)**.

5 On 2.5 litre diesel models, unbolt the camshaft timing belt and coolant pump drivebelt tensioner assemblies from either end of the cylinder head.

6 On all models, using a valve spring compressor, compress each valve spring in turn until the split collets can be removed. Release the compressor, and lift off the spring retainer, spring and spring seat. Using a pair of pliers, carefully extract the valve stem oil

2C

8.6a Compress each valve spring using a valve spring compressor and remove the split collets

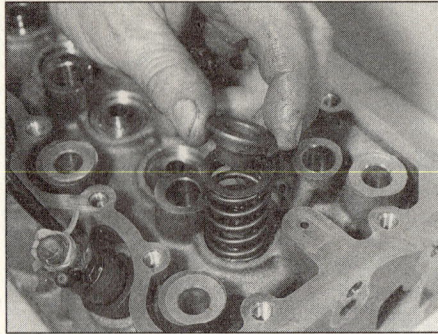

8.6b Lift off the spring retainer . . .

8.6c . . . followed by the spring . . .

8.6d . . . and the spring seat

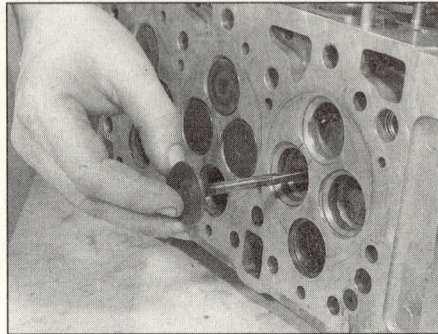

8.8 Withdraw the valve through the combustion chamber

8.9 Keep each valve and its associated components together in a labelled bag

seal from the top of the guide **(see illustrations)**.

7 If, when the valve spring compressor is screwed down, the spring retainer refuses to free and expose the split collets, gently tap the top of the tool, directly over the retainer, with a light hammer. This should free the retainer.

8 Withdraw the valve through the combustion chamber **(see illustration)**

9 It is essential that each valve is stored together with its collets, retainer, spring, and spring seat. The valves should also be kept in their correct sequence, unless they are so badly worn that they are to be renewed. If they are going to be kept and used again, place each valve assembly in a labelled polythene bag or similar small container **(see illustration)**. Note that No 1 valve is nearest to the transmission (flywheel/driveplate) end of the engine.

| 9 | **Cylinder head and valves -** cleaning and inspection |

1 Thorough cleaning of the cylinder head and valve components, followed by a detailed inspection, will enable you to decide how much valve service work must be carried out during the engine overhaul. **Note:** *If the engine has been severely overheated, it is best*

to assume that the cylinder head is warped - check carefully for signs of this.

Cleaning

2 Scrape away all traces of old gasket material from the cylinder head.

3 Scrape away the carbon from the combustion chambers and ports, then wash the cylinder head thoroughly with paraffin or a suitable solvent. Similarly, scrape off any heavy carbon deposits that may have formed on the valves, then use a power-operated wire brush to remove deposits from the valve heads and stems.

4 Where applicable on 2.5 litre diesel engines, unscrew the plug together with it

9.4a Remove the plug and sealing washer . . .

sealing washer and remove the oil filter cartridge from the cylinder head casting **(see illustrations)**.

Inspection

Note: *Be sure to perform all the following inspection procedures before concluding that the services of a machine shop or engine overhaul specialist are required. Make a list of all items that require attention.*

Cylinder head

5 Inspect the head very carefully for cracks, evidence of coolant leakage, and other damage. If cracks are found, a new cylinder head should be obtained.

9.4b . . . then withdraw the oil filter cartridge from the cylinder head casting - 2.5 litre diesel models

6 Use a straight-edge and feeler blade to check that the cylinder head gasket surface is not distorted **(see illustration)**. If it is, it may be possible to have it machined, provided that the cylinder head is not reduced to less than the specified height. **Note:** *On diesel engines, it will be necessary to recut the combustion chambers and valve seats if more than 0.1 mm has been machined off the cylinder head. This is necessary in order to maintain the correct dimensions between the valve heads, valve guides and cylinder head gasket face.*

7 Examine the valve seats in each of the combustion chambers. If they are severely pitted, cracked, or burned, they will need to be renewed or re-cut by an engine overhaul specialist. If they are only slightly pitted, this can be removed by grinding-in the valve heads and seats with fine valve-grinding compound, as described below.

8 Check the valve guides for wear by inserting the relevant valve, and checking for side-to-side motion of the valve. A very small amount of movement is acceptable. If the movement seems excessive, remove the valve. Measure the valve stem diameter (see below), and renew the valve if it is worn. If the valve stem is not worn, the wear must be in the valve guide, and the guide must be renewed. The renewal of valve guides is best carried out by a Citroën dealer or engine overhaul specialist, who will have the necessary tools available. Where no valve stem diameter is specified, seek the advice of a Citroën dealer on the best course of action.

9 If renewing the valve guides, the valve seats should be re-cut or re-ground only *after* the guides have been fitted.

10 On diesel models, inspect the swirl chambers for burning or damage such as cracking. Small cracks in the chambers are acceptable; renewal of the chambers will only be required if chamber tracts are badly burned and disfigured, or if they are no longer a tight fit in the cylinder head. If there is any doubt as to the swirl chamber condition, seek the advice of a Citroën dealer or a suitable repairer who specialises in diesel engines. Swirl chamber renewal should be entrusted to

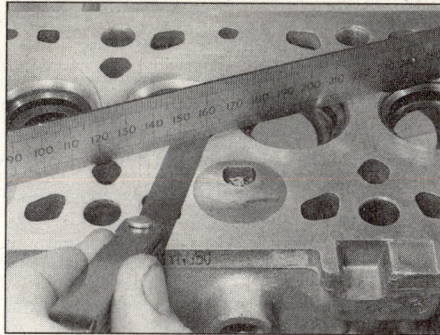

9.6 Use a straight-edge and feeler blade to check that the cylinder head gasket surface is not distorted

9.10 Using a dial test indicator, check that the swirl chamber protrusion is within the limits given in the Specifications

a specialist. Using a dial test indicator, check that the swirl chamber protrusion is within the limits given in the Specifications **(see illustration)**. Zero the dial test indicator on the gasket surface of the cylinder head, then measure the protrusion of the swirl chamber. If the protrusion is not within the specified limits, the advice of a Citroën dealer or suitable repairer who specialises in diesel engines should be sought.

Valves

11 Examine the head of each valve for pitting, burning, cracks, and general wear. Check the valve stem for scoring and wear ridges. Rotate the valve, and check for any obvious indication that it is bent. Look for pits or excessive wear on the tip of each valve stem. Renew any valve that shows any such signs of wear or damage.

12 If the valve appears satisfactory at this stage, measure the valve stem diameter at several points using a micrometer **(see illustration)**. Any significant difference in the readings obtained indicates wear of the valve stem. Should any of these conditions be apparent, the valve(s) must be renewed.

13 If the valves are in satisfactory condition, they should be ground (lapped) into their respective seats, to ensure a smooth, gas-tight seal. If the seat is only lightly pitted, or if it has been re-cut, fine grinding compound *only* should be used to produce the required

finish. Coarse valve-grinding compound should *not* be used, unless a seat is badly burned or deeply pitted. If this is the case, the cylinder head and valves should be inspected by an expert, to decide whether seat re-cutting, or even the renewal of the valve or seat insert (where possible) is required.

14 Valve grinding is carried out as follows. Place the cylinder head upside-down on a bench.

15 Smear a trace of (the appropriate grade of) valve-grinding compound on the seat face, and press a suction grinding tool onto the valve head **(see illustration)**. With a semi-rotary action, grind the valve head to its seat, lifting the valve occasionally to redistribute the grinding compound. A light spring placed under the valve head will greatly ease this operation.

16 If coarse grinding compound is being used, work only until a dull, matt even surface is produced on both the valve seat and the valve, then wipe off the used compound, and repeat the process with fine compound. When a smooth unbroken ring of light grey matt finish is produced on both the valve and seat, the grinding operation is complete. *Do not* grind-in the valves any further than absolutely necessary, or the seat will be prematurely sunk into the cylinder head.

17 When all the valves have been ground-in, carefully wash off *all* traces of grinding compound using paraffin or a suitable solvent, before reassembling the cylinder head.

Valve components

18 Examine the valve springs for signs of damage and discoloration. No minimum free length is specified by Citroën, so the only way of judging valve spring wear is by comparison with a new component.

19 Stand each spring on a flat surface, and check it for squareness. If any of the springs are damaged, distorted or have lost their tension, obtain a complete new set of springs. It is normal to renew the valve springs as a matter of course if a major overhaul is being carried out.

20 Renew the valve stem oil seals regardless of their apparent condition.

9.12 Measure the valve stem diameter at several points using a micrometer

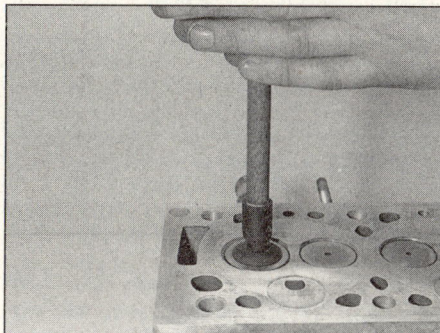

9.15 Grinding-in a valve

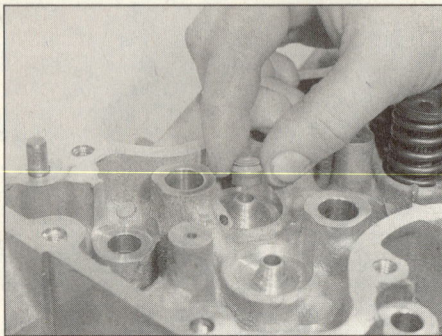

10.2a Carefully locate the new valve stem oil seal over the valve and onto the guide

10.2b Use a suitable socket or metal tube to press the seal firmly onto the guide

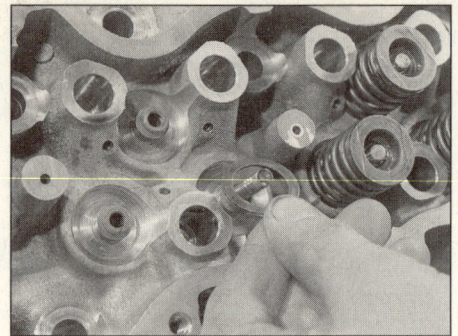

10.3a Fit the spring seat . . .

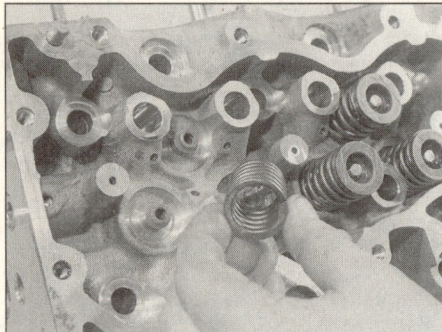

10.3b . . . locate the valve spring on top of its seat . . .

10.3c . . . then refit the spring retainer

10 Cylinder head - reassembly

1 Lubricate the stems of the valves, and insert the valves into their original locations. If new valves are being fitted, insert them into the locations to which they have been ground.
2 Refit the spring seat then, working on the first valve, dip the new valve stem seal in fresh engine oil. Carefully locate it over the valve and onto the guide. Take care not to damage the seal as it is passed over the valve stem. Use a suitable socket or metal tube to press the seal firmly onto the guide **(see illustrations)**.
3 Fit the spring seat, locate the valve spring on top of its seat, then refit the spring retainer **(see illustrations)**

10.4 Valve spring compressor in use

4 Compress the valve spring, and locate the split collets in the recess in the valve stem. Release the compressor, then repeat the procedure on the remaining valves **(see illustration)**.

Use a little dab of grease to hold the collets in position on the valve stem while the spring compressor is released

5 With all the valves installed, support the cylinder head and, using a hammer and interposed block of wood, tap the end of each valve stem to settle the components.
6 Refit the camshaft, followers and shims (as applicable) as described in Part A or B of this Chapter.
7 Refit any remaining components using the reverse of the removal sequence and with new seals or gaskets as necessary. On 2.1 litre diesel models, refit the timing belt tensioner using thread locking compound on the centre stud. Tighten the stud using the locked nuts, then remove the second nut from the end of the stud.
8 The cylinder head can then be refitted as described in Part A or B of this Chapter (as applicable).

11 Balance shaft housings (2.5 litre Diesel engine)- removal and refitting

Removal

1 Brace the balance shaft timing belt sprocket to prevent rotation **(see Tool Tip in Part B of this chapter)**, then slacken and remove the sprocket securing nut. Note that on the rear balance shaft (ie the one below the exhaust manifold), the nut has LEFT HAND THREAD. Slide the sprocket off the end of the shaft and recover the woodruff key.
2 Progressively slacken and withdraw the balance shaft housing bolts. Note that the bolts are of different sizes and tensile strengths - make a note of their order of fitment, to aid refitting.
3 Lift the housing away from the cylinder block and recover the gasket.
4 Pull the balance shaft oil seal from the housing using a hooked instrument. Alternatively, drill a small hole in the oil seal and use a self-tapping screw and a pair of pliers to remove it. Take great care to avoid drilling into the balance shaft or the oil seal housing sealing surfaces.
5 Clean the oil seal housing and the balance shaft sealing surface.
6 Slacken the securing screws, then lift the balance weights and plastic shields from the shaft. Make a note of their correct fitted position, to aid refitting later.
7 Withdraw the balance shaft from the housing. Inspect the bearing and oil seal mating surfaces for signs of wear.
8 Clean the balance shaft housing-to-crankcase sealing surface thoroughly, removing any traces of the old gasket.

Refitting

9 Lightly oil the bearing surfaces then slide the balance shaft into the housing. Refit the

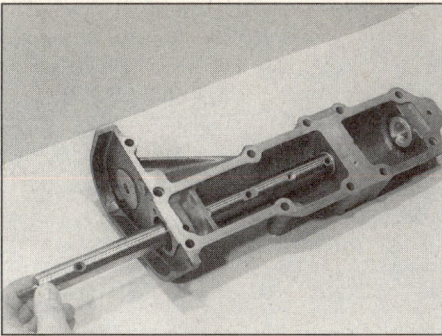

11.9a Slide the balance shaft into the housing . . .

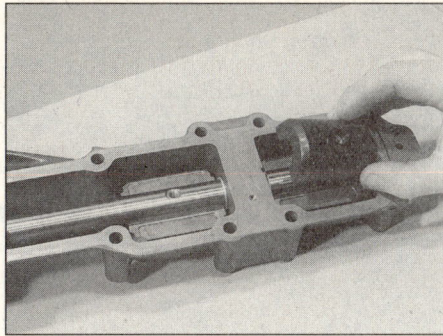

11.9b Refit the plastic covers . . .

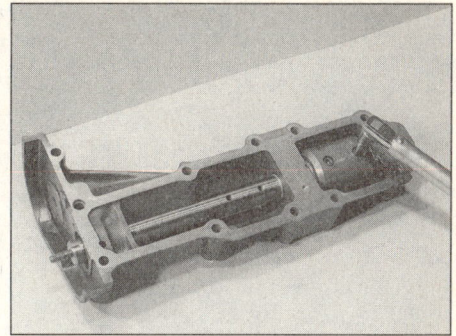

11.9c . . . and balance weights and tighten the securing screws to the specified torque

balance weights and plastic covers, according to the notes made during removal. Tighten the securing screws to the specified torque **(see illustrations)**.

10 Fit a new balance shaft housing-to-crankcase gasket, then offer up the balance shaft housing to the cylinder block. Refit the balance shaft housing securing bolts, ensuring that they are refitted in the correct order, according to their length and size, then tighten them to the specified torque **(see illustrations)**.

11 Fit a new oil seal over the end of the balance shaft, open end first **(see illustration)**. A piece of thin plastic or tape wound around the front of the balance shaft is useful to prevent damage to the oil seal as it is fitted.

12 Press the seal into the housing until it is flush with the end face of the seal housing.

13 Fit the Woodruff key and engage the balance shaft sprocket with the end of the balance shaft. Brace the sprocket to prevent rotation, then tighten the sprocket securing nut to the specified torque **(see illustrations)**.

12 Piston/connecting rod assembly - removal

1 Remove the cylinder head, sump and oil pump as described in Part A or B of this Chapter (as applicable).

2 If there is a pronounced wear ridge at the top of any bore, it may be necessary to remove it with a scraper or ridge reamer, to avoid piston damage during removal. Such a ridge indicates excessive wear of the cylinder bore.

2C

11.10a Fit a new balance shaft housing-to-crankcase gasket

11.10b . . . offer up the balance shaft housing to the cylinder block . . .

11.10c . . . and refit the balance shaft housing securing bolts and tighten them to the specified torque

11.11 Fit a new oil seal over the end of the balance shaft

11.13a Fit the Woodruff key and engage the sprocket with the end of the balance shaft

11.13b Tighten the sprocket securing nut to the specified torque

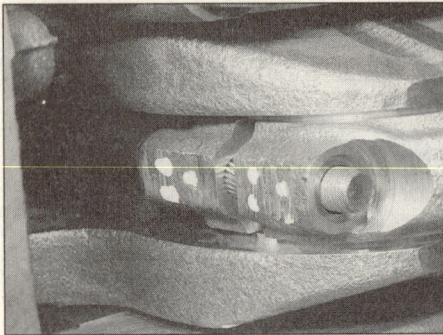

12.3 Connecting rod and big end bearing cap marked for identification

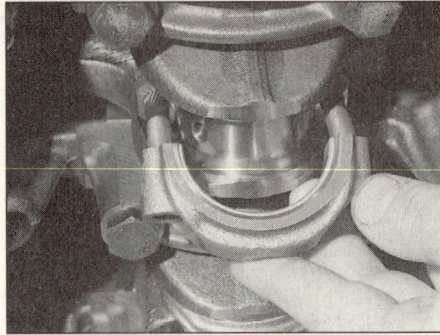

12.5 Removing a big-end bearing cap and shell

12.6 To prevent the possibility of damage to the crankshaft bearing journals, tape over the connecting rod stud threads

3 Using a hammer and centre-punch, paint or similar, mark each connecting rod big-end bearing cap with its respective cylinder number on the flat machined surface provided; if the engine has been dismantled before, note carefully any identifying marks made previously **(see illustration)**. Note that No 1 cylinder is at the transmission (flywheel) end of the engine.
4 Turn the crankshaft to bring pistons 1 and 4 to BDC (bottom dead centre).
5 Unscrew the nuts from No 1 piston big-end bearing cap. Take off the cap, and recover the bottom half bearing shell **(see illustration)**. If the bearing shells are to be re-used, tape the cap and the shell together.
6 To prevent the possibility of damage to the

crankshaft bearing journals, tape over the connecting rod stud threads **(see illustration)**.
7 Using a hammer handle, push the piston up through the bore, and remove it from the top of the cylinder block. Recover the bearing shell, and tape it to the connecting rod for safe-keeping.
8 Loosely refit the big-end cap to the connecting rod, and secure with the nuts - this will help to keep the components in their correct order.
9 Remove No 4 piston assembly in the same way.
10 Turn the crankshaft through 180° to bring pistons 2 and 3 to BDC (bottom dead centre), and remove them in the same way.

13 Crankshaft - removal

1 Remove the crankshaft sprocket and the oil pump as described in Part A or B of this Chapter (as applicable).
2 Remove the pistons and connecting rods, as described in Section 12. If no work is to be done on the pistons and connecting rods, there is no need to remove the cylinder head, or to push the pistons out of the cylinder bores. The pistons should just be pushed far enough up the bores so that they are positioned clear of the crankshaft journals.
3 Check the crankshaft endfloat as described in Section 16, then proceed as follows.
4 Where applicable on XU and XUD engines, slacken and remove the retaining bolts, and remove the oil seal carrier from the front (timing belt) end of the cylinder block, along with its gasket **(see illustration)**.
5 Remove the oil pump drive chain, and slide the drive sprocket and spacer (where fitted) off the end of the crankshaft. Remove the Woodruff key, and store it with the sprocket for safe-keeping **(see illustrations)**.
6 The main bearing caps should be numbered 1 to 5, starting from the transmission (flywheel/driveplate) end of the engine **(see illustration)**. If not, mark them accordingly using a centre-punch. Also note

13.4 Remove the oil seal carrier from the front (timing belt) end of the cylinder block - XU and XUD engines

13.5a Remove the oil pump drive chain . . .

13.5b . . . and slide the drive sprocket and spacer (where fitted) off the end of the crankshaft

13.5c Remove the Woodruff key, and store it with the sprocket for safe-keeping

13.6 Main bearing cap identification marks (arrowed)

13.8 Removing No 2 main bearing cap; note the thrustwasher (arrowed) - XU and XUD engines

13.9 Lifting out the crankshaft - XU and XUD engines

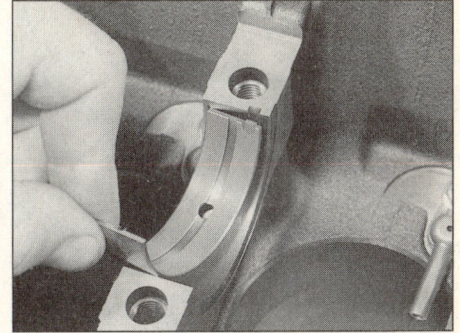

13.10 Recover the upper bearing shells from the cylinder block

the correct fitted depth of the rear crankshaft oil seal in the bearing cap.

7 On 2.5 litre diesel engines, progressively slacken the bearing ladder retaining bolts, then lift the ladder away from the crankshaft. Hold the lower bearing shells as you do this, to prevent them falling out. Tape them to their respective housings in the bearing ladder for safe-keeping. Note that the lower thrustwashers are integral with the No 2 main bearing half shell.

8 On all engines (excluding the 2.5 litre diesel) slacken and remove the main bearing cap retaining bolts/nuts, and lift off each bearing cap. Recover the lower bearing shells, and tape them to their respective caps for safe-keeping. Also recover the lower thrustwasher halves from the side of No 2 main bearing cap **(see illustration)**. Remove the rubber sealing strips from the sides of No 1 main bearing cap, and discard them.

9 Lift out the crankshaft, and discard the oil seal(s) **(see illustration)**. Place it on a clean work surface and chock it with blocks of wood to prevent it rolling.

10 Recover the upper bearing shells from the cylinder block, and tape them to their respective caps/ladder housings for safe-keeping **(see illustration)**. Remove the upper thrustwasher halves from the side of No 2 main bearing, and store them with the lower halves. Note that on 2.5 litre engines, the thrustwashers are integral with the No 2 upper main bearing shell.

the retaining bolt, then remove the piston oil jets tube from inside the cylinder block.

4 Scrape all traces of gasket/sealant from the cylinder block/crankcase, and from the main bearing ladder (where fitted), taking care not to damage the gasket/sealing surfaces.

5 Remove all oil gallery plugs (where fitted). The plugs are usually very tight - they may have to be drilled out, and the holes re-tapped. Use new plugs when reassembling.

6 If any of the castings are extremely dirty, all should be steam-cleaned.

7 After the castings are returned, clean all oil holes and oil galleries one more time. Flush all internal passages with warm water until the water runs clear. Dry thoroughly, and apply a light film of oil to all mating surfaces, to prevent rusting. On cast-iron block engines, also oil the cylinder bores. If you have access to compressed air, use it to speed up the drying process, and to blow out all the oil holes and galleries.

> ⚠ **Warning: Wear eye protection when using compressed air!**

8 If the castings are not very dirty, you can do an adequate cleaning job with hot, soapy water and a stiff brush. Take plenty of time, and do a thorough job. Regardless of the cleaning method used, be sure to clean all oil holes and galleries very thoroughly, and to dry all components well. On cast-iron block engines, protect the cylinder bores as described above, to prevent rusting.

9 All threaded holes must be clean, to ensure

accurate torque readings during reassembly. To clean the threads, run the correct-size tap into each of the holes to remove rust, corrosion, thread sealant or sludge, and to restore damaged threads. If possible, use compressed air to clear the holes of debris produced by this operation.

> **HAYNES HiNT** *A good alternative is to inject water-dispersant lubricant into each hole, using the long spout usually supplied. However, if the hole is blind, ensure that the excess fluid is removed before reassembly.*

> ⚠ **Warning: Wear eye protection when cleaning out these holes in this way!**

10 Apply suitable sealant to the new oil gallery plugs, and insert them into the holes in the block. Tighten them securely.

11 Where applicable, clean the threads of the piston oil jet retaining bolt, and apply a drop of thread-locking compound to the bolt threads. Refit the piston oil jet spray tube to the cylinder block, and tighten its retaining bolt to the specified torque setting **(see illustrations)**.

12 If the engine is not going to be reassembled right away, cover it with a large plastic bag to keep it clean; protect all mating surfaces and the cylinder bores as described above, to prevent rusting.

2C

---

14 Cylinder block/crankcase - cleaning and inspection 🔧

Cleaning

1 Remove all external components and electrical switches/sensors from the block.

2 For complete cleaning, the core plugs should ideally be removed. Drill a small hole in the plugs, then insert a self-tapping screw into the hole. Pull out the plugs by pulling on the screw with a pair of grips, or by using a slide hammer.

3 Where applicable, slacken and withdraw

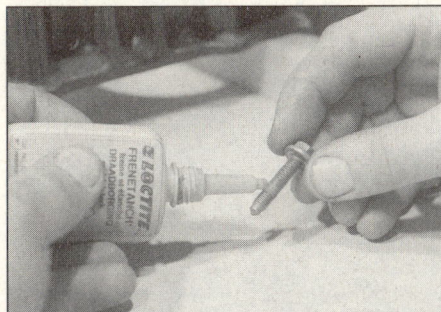

14.11a Apply a drop of thread-locking compound to the threads of the piston oil jet retaining bolt

14.11b Refit the piston jets . . .

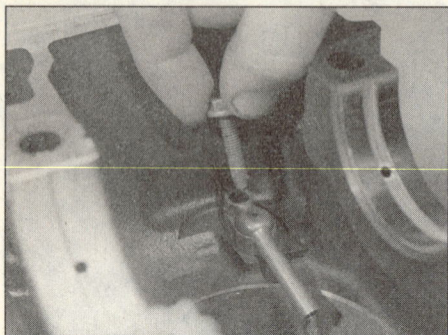

14.11c . . . then insert the retaining bolts and tighten them to the specified torque

Inspection

13 Visually check the castings for cracks and corrosion. Look for stripped threads in the threaded holes. If there has been any history of internal water leakage, it may be worthwhile having an engine overhaul specialist check the cylinder block/crankcase with special equipment. If defects are found, have them repaired if possible, or renew the assembly.

14 Check each cylinder bore for scuffing and scoring. Check for signs of a wear ridge at the top of the cylinder, indicating that the bore is excessively worn.

15 If the necessary measuring equipment is available, measure the bore diameter of each cylinder liner at the top (just under the wear ridge), centre, and bottom of the cylinder bore, parallel to the crankshaft axis.

16 Next, measure the bore diameter at the same three locations, at right-angles to the crankshaft axis. Compare the results with the figures given in the Specifications. Where no figures are stated by Citroën, if there is any doubt about the condition of the cylinder bores seek the advice of a Citroën dealer or suitable engine reconditioning specialist.

17 At the time of writing, it was not clear whether oversize pistons were available for all models. Consult your Citroën dealer for the latest information on piston availability. If oversize pistons are available, then it may be possible to have the cylinder bores rebored and fit the oversize pistons. If oversize pistons are not available, and the bores are worn, a new block seems to be the only option.

15 Piston/connecting rod assembly - inspection

1 Before the inspection process can begin, the piston/connecting rod assemblies must be cleaned, and the original piston rings removed from the pistons.

2 Carefully expand the old rings over the top of the pistons. The use of two or three old feeler blades will be helpful in preventing the rings dropping into empty grooves (see illustration). Be careful not to scratch the piston with the ends of the ring. The rings are brittle, and will snap if they are spread too far. They are also very sharp - protect your hands and fingers. Note that the third ring incorporates an expander. Always remove the rings from the top of the piston. Keep each set of rings with its piston if the old rings are to be re-used.

3 Scrape away all traces of carbon from the top of the piston. A hand-held wire brush (or a piece of fine emery cloth) can be used, once the majority of the deposits have been scraped away.

4 Remove the carbon from the ring grooves in the piston, using an old ring. Break the ring in half to do this (be careful not to cut your fingers - piston rings are sharp). Be careful to remove only the carbon deposits - do not remove any metal, and do not nick or scratch the sides of the ring grooves.

5 Once the deposits have been removed, clean the piston/connecting rod assembly with paraffin or a suitable solvent, and dry thoroughly. Make sure that the oil return holes in the ring grooves are clear.

6 If the pistons and cylinder bores are not damaged or worn excessively, and if the cylinder block does not need to be rebored, the original pistons can be refitted. Normal piston wear shows up as even vertical wear on the piston thrust surfaces, and slight looseness of the top ring in its groove. New piston rings should always be used when the engine is reassembled.

7 Carefully inspect each piston for cracks around the skirt, around the gudgeon pin holes, and at the piston ring 'lands' (between the ring grooves).

8 Look for scoring and scuffing on the piston skirt, holes in the piston crown, and burned areas at the edge of the crown. If the skirt is scored or scuffed, the engine may have been suffering from overheating, and/or abnormal combustion which caused excessively high operating temperatures. The cooling and lubrication systems should be checked thoroughly. Scorch marks on the sides of the pistons show that blow-by has occurred. A hole in the piston crown, or burned areas at the edge of the piston crown, indicates that abnormal combustion (pre-ignition, knocking, or detonation) has been occurring.

9 If any of the above problems exist, the causes must be investigated and corrected, or the damage will occur again. The causes may include incorrect ignition/injection pump timing, or a faulty injector (as applicable).

10 Corrosion of the piston, in the form of pitting, indicates that coolant has been leaking into the combustion chamber and/or the crankcase. Again, the cause must be corrected, or the problem may persist in the rebuilt engine.

11 Examine each connecting rod carefully for signs of damage, such as cracks around the big-end and small-end bearings. Check that the rod is not bent or distorted. Damage is highly unlikely, unless the engine has been seized or badly overheated. Detailed checking of the connecting rod assembly can only be carried out by a Citroën dealer or engine repair specialist with the necessary equipment.

12 On all engines, due to the tightening procedure for the connecting rod big-end cap retaining nuts, it is highly recommended that the big-end cap nuts and bolts are renewed as a complete set prior to refitting.

13 On all petrol engines, the gudgeon pins are an interference fit in the connecting rod small-end bearing. Therefore, piston and/or connecting rod renewal should be entrusted to a Citroën dealer or engine repair specialist, who will have the necessary tooling to remove and install the gudgeon pins.

14 On diesel engines, the gudgeon pins are of the floating type, secured in position by two circlips. On these engines, the pistons and connecting rods can be separated as follows.

15 Using a small flat-bladed screwdriver, prise out the circlips, and push out the gudgeon pin (see illustrations). Hand

15.2 Using an old feeler blade to aid the removal of the piston rings

15.15a On diesel engines, prise out the circlips . . .

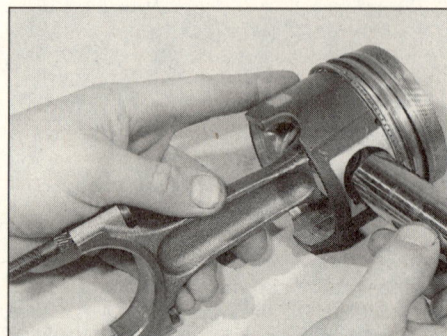

15.15b . . . and push out the gudgeon pin - 2.1 litre model shown

15.19a On diesel engines, the piston to connecting rod and bearing cap orientation is given by the inlet valve cutouts on the piston crowns

15.19b Slide gudgeon pin into the piston, through the connecting rod small-end . . .

15.19c . . . then fit two new circlips

pressure should be sufficient to remove the pin. Identify the piston and rod to ensure correct reassembly. Discard the circlips - new ones *must* be used on refitting.

16 Examine the gudgeon pin and connecting rod small-end bearing for signs of wear or damage. Wear can be cured by renewing both the pin and bush. Bush renewal, however, is a specialist job - press facilities are required, and the new bush must be reamed accurately.

17 The connecting rods themselves should not be in need of renewal, unless seizure or some other major mechanical failure has occurred. Check the alignment of the connecting rods visually, and if the rods are not straight, take them to an engine overhaul specialist for a more detailed check.

18 Examine all components, and obtain any new parts from your Citroën dealer. If new pistons are purchased, they will be supplied complete with gudgeon pins and circlips. Circlips can also be purchased individually.

19 Position the piston so that the arrow on the piston crown is positioned as shown, in relation to the connecting rod big-end bearing shell cutouts. On 2.5 litre diesel engines, the orientation is given by the inlet valve cut-outs on the piston crowns. Apply a smear of clean engine oil to the gudgeon pin. Slide it into the piston and through the connecting rod small-end. Check that the piston pivots freely on the rod, then secure the gudgeon pin in position with two new circlips. Ensure that each circlip is correctly located in its groove in the piston **(see illustrations)**.

16 Crankshaft - inspection

Checking crankshaft endfloat

1 If the crankshaft endfloat is to be checked, this must be done when the crankshaft is still installed in the cylinder block/crankcase, but is free to move.

2 Check the endfloat using a dial gauge in contact with the end of the crankshaft. Push the crankshaft fully one way, and then zero

the gauge. Push the crankshaft fully the other way, and check the endfloat. The result can be compared with the specified amount, and will give an indication as to whether new thrustwashers are required **(see illustration)**.

3 If a dial gauge is not available, feeler gauges can be used. First push the crankshaft fully towards the flywheel end of the engine, then use feeler gauges to measure the gap between the web of No 2 crankpin and the thrustwasher.

Inspection

4 Clean the crankshaft using paraffin or a suitable solvent, and dry it, preferably with compressed air if available.

⚠ *Warning: Wear eye protection when using compressed air! Be sure to clean the oil holes with a pipe cleaner or similar probe, to ensure that they are not obstructed.*

5 Check the main and big-end bearing journals for evidence of uneven wear, scoring, pitting and cracking.

6 Big-end bearing wear is accompanied by distinct metallic knocking when the engine is running (particularly noticeable when the engine is pulling from low speed) and by some loss of oil pressure.

7 Main bearing wear is accompanied by severe engine vibration and rumble - getting progressively worse as engine speed increases - and again by loss of oil pressure.

8 Check the bearing journal for roughness by

running a finger lightly over the bearing surface. Any roughness (which will be accompanied by obvious bearing wear) indicates that the crankshaft requires regrinding (where possible) or renewal.

9 If the crankshaft has been reground, check for burrs around the crankshaft oil holes (the holes are usually chamfered, so burrs should not be a problem unless regrinding has been carried out carelessly). Remove any burrs with a fine file or scraper, and thoroughly clean the oil holes as described previously.

10 Using a micrometer, measure the diameter of the main and big-end bearing journals, and compare the results with the Specifications **(see illustration)**. By measuring the diameter at a number of points around each journal's circumference, you will be able to determine whether or not the journal is out-of-round. Take the measurement at each end of the journal, near the webs, to determine if the journal is tapered. Compare the results obtained with those given in the Specifications. Where no specified journal diameters are quoted, seek the advice of a Citroën dealer.

11 Check the oil seal contact surfaces at each end of the crankshaft for wear and damage. If the seal has worn a deep groove in the surface of the crankshaft, consult an engine overhaul specialist; repair may be possible, but otherwise a new crankshaft will be required.

12 At the time of writing, it was not clear whether Citroën produce oversize bearing

2C

16.2 Checking the crankshaft endfloat

16.10 Checking the diameter of a big-end journal

shells for all of the engines covered in this manual. On some engines, if the crankshaft journals have not already been reground, it may be possible to have the crankshaft reconditioned, and to fit oversize shells (see Section 20). If no oversize shells are available and the crankshaft has worn beyond the specified limits, it will have to be renewed. Consult your Citroën dealer or engine specialist for further information on parts availability.

17 Main and big-end bearings - inspection

1 Even though the main and big-end bearings should be renewed during the engine overhaul, the old bearings should be retained for close examination, as they may reveal valuable information about the condition of the engine. The bearing shells are graded by thickness, the grade of each shell being indicated by the colour code marked on it.
2 Bearing failure can occur due to lack of lubrication, the presence of dirt or other foreign particles, overloading the engine, or corrosion **(see illustration)**. Regardless of the cause of bearing failure, the cause must be corrected (where applicable) before the engine is reassembled, to prevent it from happening again.
3 When examining the bearing shells, remove them from the cylinder block/crankcase, the main bearing ladder/caps (as appropriate), the connecting rods and the connecting rod big-end bearing caps. Lay them out on a clean surface in the same general position as their location in the engine. This will enable you to match any bearing problems with the corresponding crankshaft journal. *Do not* touch any shell's bearing surface with your fingers while checking it, or the delicate surface may be scratched.

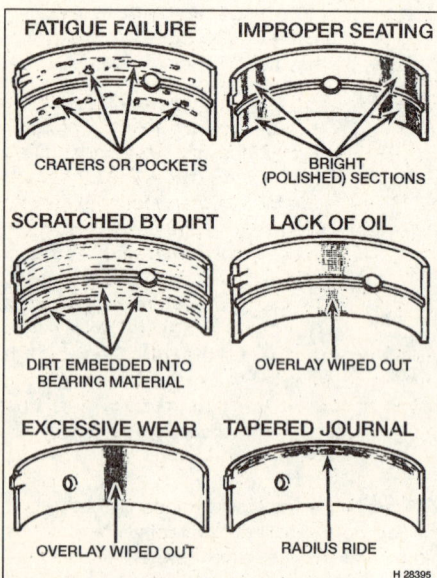

FATIGUE FAILURE — CRATERS OR POCKETS

IMPROPER SEATING — BRIGHT (POLISHED) SECTIONS

SCRATCHED BY DIRT — DIRT EMBEDDED INTO BEARING MATERIAL

LACK OF OIL — OVERLAY WIPED OUT

EXCESSIVE WEAR — OVERLAY WIPED OUT

TAPERED JOURNAL — RADIUS RIDE

H 28395

17.2 Typical bearing failures

4 Dirt and other foreign matter gets into the engine in a variety of ways. It may be left in the engine during assembly, or it may pass through filters or the crankcase ventilation system. It may get into the oil, and from there into the bearings. Metal chips from machining operations and normal engine wear are often present. Abrasives are sometimes left in engine components after reconditioning, especially when parts are not thoroughly cleaned using the proper cleaning methods. Whatever the source, these foreign objects often end up embedded in the soft bearing material, and are easily recognised. Large particles will not embed in the bearing, and will score or gouge the bearing and journal. The best prevention for this cause of bearing failure is to clean all parts thoroughly, and keep everything spotlessly-clean during engine assembly. Frequent and regular engine oil and filter changes are also recommended.
5 Lack of lubrication (or lubrication breakdown) has a number of interrelated causes. Excessive heat (which thins the oil), overloading (which squeezes the oil from the bearing face) and oil leakage (from excessive bearing clearances, worn oil pump or high engine speeds) all contribute to lubrication breakdown. Blocked oil passages, which usually are the result of misaligned oil holes in a bearing shell, will also oil-starve a bearing, and destroy it. When lack of lubrication is the cause of bearing failure, the bearing material is wiped or extruded from the steel backing of the bearing. Temperatures may increase to the point where the steel backing turns blue from overheating.
6 Driving habits can have a definite effect on bearing life. Full-throttle, low-speed operation (labouring the engine) puts very high loads on bearings, tending to squeeze out the oil film. These loads cause the bearings to flex, which produces fine cracks in the bearing face (fatigue failure). Eventually, the bearing material will loosen in pieces, and tear away from the steel backing.
7 Short-distance driving leads to corrosion of bearings, because insufficient engine heat is produced to drive off the condensed water and corrosive gases. These products collect in the engine oil, forming acid and sludge. As the oil is carried to the engine bearings, the acid attacks and corrodes the bearing material.
8 Incorrect bearing installation during engine assembly will lead to bearing failure as well. Tight-fitting bearings leave insufficient bearing running clearance, and will result in oil starvation. Dirt or foreign particles trapped behind a bearing shell result in high spots on the bearing, which lead to failure.
9 *Do not* touch any shell's bearing surface with your fingers during reassembly; there is a risk of scratching the delicate surface, or of depositing particles of dirt on it.
10 As mentioned at the beginning of this Section, the bearing shells should be renewed as a matter of course during engine overhaul;

to do otherwise is false economy. Refer to Sections 20 and 21 for details of bearing shell selection.

18 Engine overhaul - reassembly sequence

1 Before reassembly begins, ensure that all new parts have been obtained, and that all necessary tools are available. Read through the entire procedure to familiarise yourself with the work involved, and to ensure that all items necessary for reassembly of the engine are at hand. In addition to all normal tools and materials, thread-locking compound will be needed. A suitable tube of liquid sealant will also be required for the joint faces that are fitted without gaskets. It is recommended that Citroën's own product(s) are used, which are specially formulated for this purpose; the relevant product names are quoted in the text of each Section where they are required.
2 In order to save time and avoid problems, engine reassembly can be carried out in the following order:
a) *Crankshaft (Section 20).*
b) *Piston/connecting rod assemblies (Section 21).*
c) *Oil pump (See Chapters 2A or 2B - as applicable).*
d) *Sump (See Chapters 2A or 2B - as applicable).*
e) *Flywheel (See Chapters 2A or 2B - as applicable).*
f) *Cylinder head (See Chapters 2A or 2B, or this Part - as applicable).*
g) *Timing belt tensioner and sprockets, and timing belt (See Chapters 2A or 2B - as applicable).*
h) *Engine external components.*
3 At this stage, all engine components should be absolutely clean and dry, with all faults repaired. The components should be laid out (or in individual containers) on a completely clean work surface.

19 Piston rings - refitting

1 Before fitting new piston rings, the ring end gaps must be checked as follows.
2 Lay out the piston/connecting rod assemblies and the new piston ring sets, so that the ring sets will be matched with the same piston and cylinder during the end gap measurement and subsequent engine reassembly.
3 Insert the top ring into the first cylinder, and push it down the bore using the top of the piston. This will ensure that the ring remains square with the cylinder walls. Position the ring near the bottom of the cylinder bore, at the lower limit of ring travel. Note that the top and second compression rings are different. The second ring is easily identified by the step

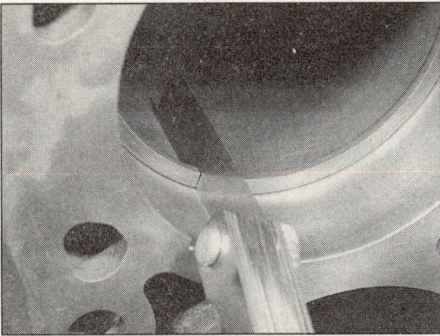

19.4 Measuring a piston ring end gap using a feeler blade

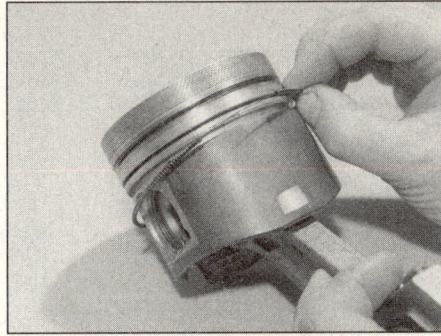

19.10a Fit the expander ring . . .

19.10b . . . then the oil control ring

on its lower surface, and by the fact that its outer face is tapered.

4 Measure the end gap using feeler gauges **(see illustration)**.

5 Repeat the procedure with the ring at the top of the cylinder bore, at the upper limit of its travel, and compare the measurements with the figures given in the Specifications. Where no figures are given, seek the advice of a Citroën dealer or engine reconditioning specialist.

6 If the gap is too small (unlikely if genuine Citroën parts are used), it must be enlarged, or the ring ends may contact each other during engine operation, causing serious damage. Ideally, new piston rings providing the correct end gap should be fitted. As a last resort, the end gap can be increased by filing the ring ends very carefully with a fine file. Mount the file in a vice equipped with soft jaws, slip the ring over the file with the ends contacting the file face, and slowly move the ring to remove material from the ends. Take care, as piston rings are sharp, and are easily broken.

7 With new piston rings, it is unlikely that the end gap will be too large. If the gaps are too large, check that you have the correct rings for your engine and for the particular cylinder bore size.

8 Repeat the checking procedure for each ring in the first cylinder, and then for the rings

in the remaining cylinders. Remember to keep rings, pistons and cylinders matched up.

9 Once the ring end gaps have been checked and if necessary corrected, the rings can be fitted to the pistons.

10 Fit the piston rings using the same technique as for removal. Fit the bottom (oil control) ring first, and work up. When fitting the oil control ring, first insert the expander (where fitted), then fit the ring with its gap positioned 180° from the expander gap. Ensure that the second compression ring is fitted the correct way up, with its identification mark (either a dot of paint or the word 'TOP' stamped on the ring surface) at the top, and the stepped surface at the bottom **(see illustrations)**. Arrange the gaps of the top and second compression rings 120° either side of the oil control ring gap. **Note:** *Always follow any instructions supplied with the new piston ring sets - different manufacturers may specify different procedures. Do not mix up the top and second compression rings, as they have different cross-sections.*

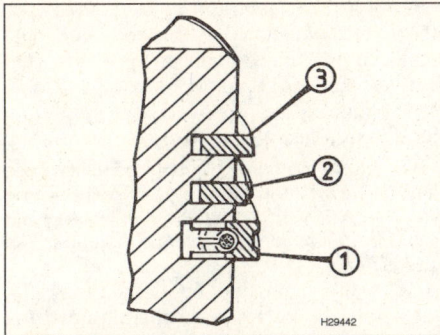

19.10c Sectional view of piston rings fitted to piston

1 Oil control ring and expander ring
2 Second compression ring
3 Top compression ring

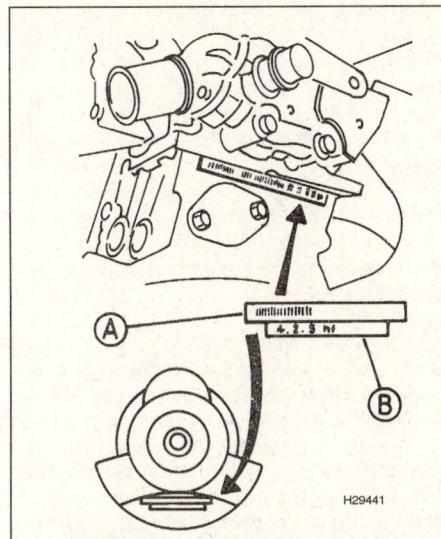

20.4 Cylinder block and crankshaft main bearing reference marking locations - XU petrol engines

A Bar code (production use only)
B Reference markings

20 Crankshaft - refitting and main bearing running clearance check

Selection of new bearing shells

XU series (petrol) engine

1 On some early engines, both the upper and lower bearing shells were of the same thickness.

2 However, on later engines the main bearing running clearance was significantly reduced. To enable this to be done, four different grades of bearing shell were introduced. The grades are indicated by a colour-coding marked on the edge of each shell, which denotes the shell's thickness, as listed in the following table. The upper shell on all bearings is of the same size, and the running clearance is controlled by fitting a lower bearing shell of the required thickness.

Bearing colour code	Thickness (mm)	
	Standard	Undersize
Upper bearing		
Black	1.847	N/A
Lower bearing		
Blue (Class A)	1.844	N/A
Black (Class B)	1.857	N/A
Green (Class C)	1.866	N/A
Red (Class D)	1.877	N/A

Note: *On all engines, upper shells are easily distinguished from lower shells, by their grooved bearing surface and oilway drilling; the lower shells have a plain surface. It was not clear at the time of writing whether undersize bearing shells were available for 1998 cc engine. Refer to your Citroën dealer for further information.*

3 On most later engines, new bearing shells can be selected using the reference marks on the cylinder block/crankcase. The cylinder block marks identify the diameter of the bearing bores and the crankshaft marks, the diameter of the crankshaft journals. Where no marks are present, the bearing shells can only be selected by checking the running clearance (see below).

4 The cylinder block reference marks are on the left-hand (flywheel/driveplate) end of the block, and the crankshaft reference marks are on the end web of the crankshaft **(see illustration)**. These marks can be used to

2C

20.6 Main bearing shell selection chart for use with XU series petrol engines (see text for details)

The chart is labelled with crankshaft references across the top (0 1 2 3 4 5 6 7 8 9 A B C D E F G H J L N P R T U Y) and cylinder block references down the side (0 1 2 3 4 5 6 7 8 9 A B C D E F G H J L), with regions marked BE, NR, VE and RG, and grade letters A, B, C, D indicating the lower bearing shell class.

H2944U

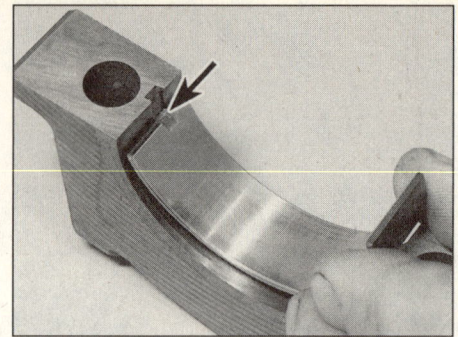

20.12 Press the bearing shells into their locations, ensuring that the tab (arrowed) on each shell engages in the notch in the bearing cap

select bearing shells of the required thickness as follows.

5 On both the crankshaft and block there are two lines of identification: a bar code, which is used by Citroën during production, and a row of five letters. The first letter in the sequence refers to the size of No 1 bearing (at the flywheel/driveplate end). The last letter in the sequence (which is followed by an arrow) refers to the size of No 5 main bearing. These marks can be used to select the required bearing shell grade as follows.

6 Obtain the identification number/letter of both the relevant crankshaft journal and the cylinder block bearing bore. Noting that the crankshaft references are listed across the top of the chart, and the cylinder block references down the side, trace a vertical line down from the relevant crankshaft reference, and a horizontal line across from the relevant cylinder block reference, and find the point at which both lines cross. This crossover point will indicate the grade of lower bearing shell required to give the correct main bearing running clearance. For example, the illustration shows crankshaft reference 6, and cylinder block reference H, crossing at a point within the RED area, indicating that a Red-coded (Class D) lower bearing shell is required to give the correct main bearing running clearance **(see illustration)**.

7 Repeat this procedure so that the required bearing shell grade is obtained for each of the five main bearing journals.

8 Seek the advice of your Citroën dealer on parts availability, and on the best course of action when ordering new bearing shells. **Note:** *On early models, at overhaul it is recommended that the later bearing shell arrangement is fitted. This, however, should only be done if the lubrication system components are upgraded (necessitating replacement of the oil pump relief valve piston and spring as well as the pump sprocket and drive chain) at the same time. If the new bearing arrangement is to be used without uprating the lubrication system, Blue (Class A) lower bearing shells should be fitted. Refer to your Citroën dealer for further information.*

Diesel engines

9 On 2.1 litre diesel engines both the upper and lower bearing shells are of the same thickness. Citroën produce both a standard set of shells and an undersize set of shells. On 2.5 litre engines, the upper and lower shells are of different thickness. Citroën produce both standard and undersize sets of shells. The upper shells are easily distinguished from lower shells, by their grooved bearing surface and oilway drilling; the lower shells have a plain surface.

Main bearing running clearance check

XU series (petrol) engine

10 On early engines, if the later bearing shells are to be fitted, obtain a set of new upper bearing shells, and new blue (as applicable) lower bearing shells (see paragraph 2). On later engines where the modified bearing shells are already fitted, the running clearance check can be carried out using the original bearing shells. However, it is preferable to use a new set, since the results obtained will be more conclusive.

11 Clean the backs of the bearing shells, and the bearing locations in both the crankcase and the main bearing caps/ladder.

12 Press the bearing shells into their locations, ensuring that the tab on each shell engages in the notch in the crankcase or bearing cap/ladder **(see illustration)**. Take care not to touch any shell's bearing surface with your fingers. Note that the upper bearing shells all have a grooved bearing surface, whereas the lower shells have a plain bearing surface. If the original bearing shells are being used for the check, ensure that they are refitted in their original locations.

13 The clearance can be checked in either of two ways.

14 One method (which will be difficult to achieve without a range of internal micrometers or internal/external expanding calipers) is to refit the main bearing caps to the cylinder block/crankcase, with bearing shells in place. With the cap retaining bolts tightened to the specified torque, measure the internal diameter of each assembled pair of bearing shells. If the diameter of each corresponding crankshaft journal is measured and then subtracted from the bearing internal diameter, the result will be the main bearing running clearance.

15 The second, and more accurate, method is to use Plastigauge. This consists of a fine thread of perfectly-round plastic, which is compressed between the bearing shell and the journal. When the shell is removed, the plastic is deformed, and can be measured with a special card gauge supplied with the kit. The running clearance is determined from this gauge. Plastigauge should be available from your Citroën dealer; otherwise, enquiries at one of the larger specialist motor factors should produce the name of a stockist in your area. The procedure for using Plastigauge is as follows.

16 With the main bearing upper shells in place, carefully lay the crankshaft in position. Do not use any lubricant; the crankshaft journals and bearing shells must be perfectly clean and dry.

17 Cut several lengths of the appropriate-size Plastigauge (they should be slightly shorter than the width of the main bearings),

20.17 Plastigauge in place on a crankshaft main bearing journal

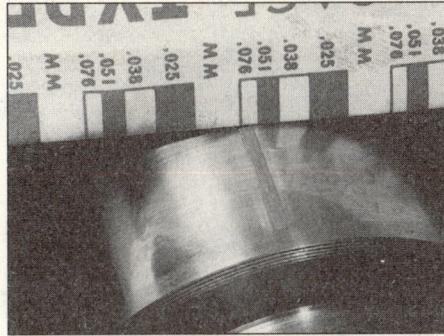

20.20 Measure the width of the deformed Plastigauge using the scale on the card provided

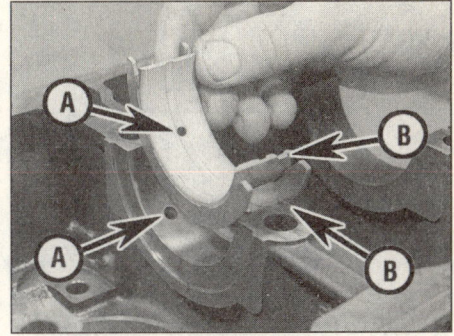

20.27a Place the bearing shells in their locations, ensuring that the location lugs in the shells and crankcase are correctly engaged, and that the oilways are aligned

A Oilways   B Location lugs

and place one length on each crankshaft journal axis **(see illustration)**.

18 With the main bearing lower shells in position, refit the main bearing caps and tighten them as described later in this Section. Take care not to disturb the Plastigauge, and *do not* rotate the crankshaft at any time during this operation.

19 Remove the main bearing caps, again taking great care not to disturb the Plastigauge or rotate the crankshaft.

20 Compare the width of the crushed Plastigauge on each journal to the scale printed on the Plastigauge envelope, to obtain the main bearing running clearance **(see illustration)**. Compare the clearance measured with that given in the Specifications at the start of this Chapter.

21 If the clearance is significantly different from that expected, the bearing shells may be the wrong size (or excessively worn, if the original shells are being re-used). Before deciding that different-size shells are required, make sure that no dirt or oil was trapped between the bearing shells and the caps or block when the clearance was measured. If the Plastigauge was wider at one end than at the other, the crankshaft journal may be tapered.

22 If the clearance is not as specified, use the reading obtained, along with the shell thicknesses quoted above, to calculate the necessary grade of bearing shells required. When calculating the bearing clearance

required, bear in mind that it is always better to have the running clearance towards the lower end of the specified range, to allow for wear in use.

23 Where necessary, obtain the required grades of bearing shell, and repeat the running clearance checking procedure as described above.

24 On completion, carefully scrape away all traces of the Plastigauge material from the crankshaft and bearing shells. Use your fingernail, or a wooden or plastic scraper which is unlikely to score the bearing surfaces.

Diesel engines

25 The running clearance check can be carried out using the original bearing shells. However, it is preferable to use a new set, since the results obtained will be more conclusive. Perform the check using the information given in the preceding paragraphs.

Final crankshaft refitting

26 Carefully lift the crankshaft out of the cylinder block once more.

2.5 litre diesel engines

27 Place the bearing shells in their locations, ensuring that the location lugs in the shells and crankcase are correctly engaged, and that the oilways are aligned. Note that the upper bearing shells all have a grooved

bearing surface, whereas the lower shells have a plain bearing surface. Note that the thrustwashers are integral with the No 2 main bearing half shell. If the original bearing shells are being used for the check, ensure that they are refitted in their original locations. If new shells are being fitted, ensure that all traces of protective grease are cleaned off using paraffin. Wipe dry the shells and connecting rods with a lint-free cloth. Liberally lubricate each bearing shell in the cylinder block/crankcase with clean engine oil **(see illustrations)**.

28 Lower the crankshaft into position so that Nos 2 and 3 cylinder crankpins are at TDC; Nos 1 and 4 cylinder crankpins will be at BDC, ready for fitting No 1 piston **(see illustration)**. Check the crankshaft endfloat, referring to Section 15. Ensure that the oil pump drive chain is not trapped between the crankshaft and the crankcase.

29 Place the lower bearing shells in the bearing ladder and lubricate them liberally with clean engine oil. Make sure that the locating lugs on the shells engage with the corresponding recesses in the ladder journals. Note that the thrustwashers are integral with the No 2 main bearing half shell **(see illustration)**.

20.27b Liberally lubricate each bearing shell in the cylinder block/ crankcase with clean engine oil

20.28 Lower the crankshaft into position so that Nos 2 and 3 cylinder crankpins are at TDC - 2.5 litre engine shown

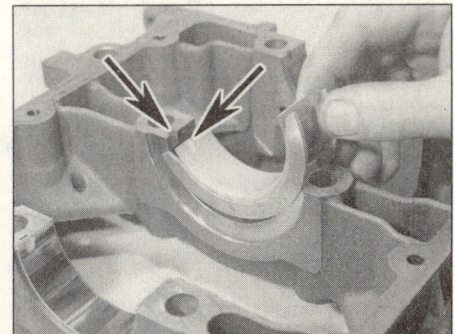

20.29 Place the lower bearing shells in the bearing ladder. Make sure that the locating lugs on the shells engage with the corresponding recesses in the bearing ladder (arrowed)

2C

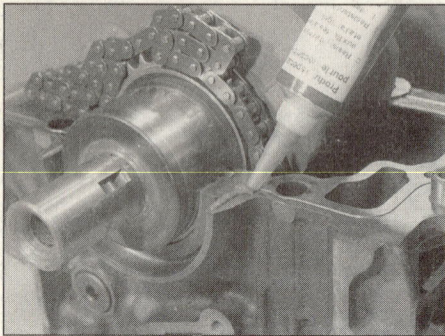
20.30a Apply a bead of suitable sealant to the recess in the bearing ladder/crankcase mating surface

20.30b Lower the bearing ladder into position over the crankshaft

20.30c Insert and progressively tighten the bearing ladder inner . . .

30 Apply a bead of suitable sealant to the recess in the bearing ladder/crankcase mating surface. Lower the bearing ladder into position over the crankshaft, ensuring that the lower bearing shells stay in position. Progressively tighten the bearing ladder retaining bolts in the specified sequence as shown, and to the specified torque wrench settings **(see illustrations)**.

31 Fit new crankshaft front and rear oil seals as described in Part B of this Chapter.

32 Refit the piston/connecting rod assemblies to the crankshaft as described in Section 21.

33 Ensuring that the drive chain is correctly located on the sprocket, refit the oil pump and sump as described in Part A or B of this Chapter.

Petrol and 2.1 litre diesel engines

34 Using a little grease, stick the upper thrustwashers to each side of the No 2 main bearing upper location. Ensure that the oilway grooves on each thrustwasher face outwards (away from the cylinder block) **(see illustration)**.

35 Place the bearing shells in their locations, ensuring that the location lugs in the shells and crankcase are correctly aligned, as described earlier. If new shells are being fitted, ensure that all traces of protective grease are cleaned off using paraffin. Note that the upper bearing shells all have a grooved bearing surface, whereas the lower shells have a plain bearing surface. If the original bearing shells are being used for the check, ensure that they are refitted in their original locations. Wipe dry the

shells and connecting rods with a lint-free cloth. Liberally lubricate each bearing shell in the cylinder block/crankcase and cap with clean engine oil.

36 Lower the crankshaft into position so that Nos 2 and 3 cylinder crankpins are at TDC; Nos 1 and 4 cylinder crankpins will be at BDC, ready for fitting No 1 piston. Check the crankshaft endfloat, referring to Section 15.

37 Lubricate the lower bearing shells in the main bearing caps with clean engine oil. Make sure that the locating lugs on the shells engage with the corresponding recesses in the caps.

38 Fit main bearing caps Nos 2 to 5 to their correct locations, ensuring that they are fitted the correct way round (the bearing shell tab recesses in the block and caps must be on the same side). Insert the bolts/nuts, tightening them only loosely at this stage.

39 Apply a small amount of sealant to the No 1 main bearing cap mating face on the cylinder block, around the sealing strip holes **(see illustration)**.

40 Locate the tab of each sealing strip over the pins on the base of No 1 bearing cap, and press the strips into the bearing cap grooves. It is now necessary to obtain two thin metal strips, of 0.25 mm thickness or less, in order to prevent the strips moving when the cap is being fitted. Citroën garages use the tool shown, which acts as a clamp. Metal strips (such as old feeler blades) can be used, provided all burrs which may damage the sealing strips are first removed **(see illustration)**.

20.30d . . . and outer retaining bolts in the specified sequence, and to the specified torque wrench settings

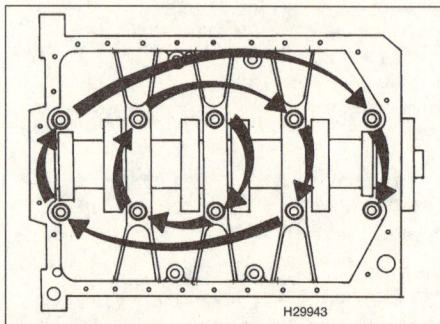
20.30e Tightening sequence of the bearing ladder retaining bolts

H29943

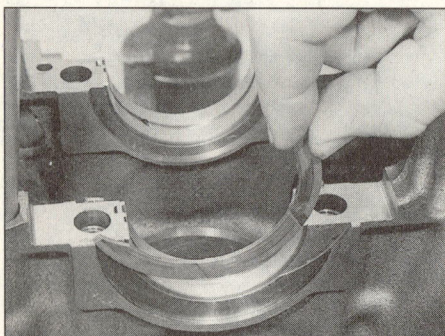
20.34 Fitting a thrustwasher to No 2 main bearing upper location

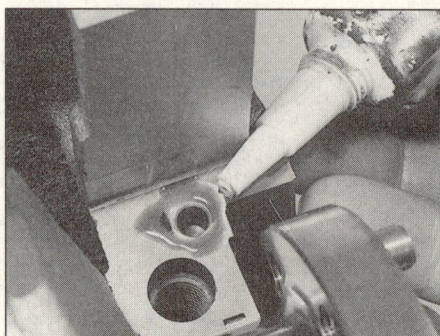
20.39 Apply a small amount of sealant to the No 1 main bearing cap mating face on the cylinder block, around the sealing strip holes

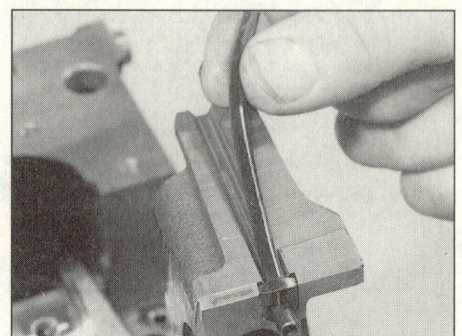
20.40 Fitting a new sealing strip to No 1 main bearing cap

20.41a Fitting No 1 main bearing cap, using metal strips to retain the side seals

20.41b Removing the metal strips from No 1 main bearing cap

20.42a Tighten all the main bearing cap bolts/nuts evenly to the specified torque . . .

41 Where applicable, oil both sides of the metal strips, and hold them on the sealing strips. Fit the No 1 main bearing cap, insert the bolts loosely, then carefully pull out the metal strips in a horizontal direction, using a pair of pliers **(see illustrations)**.

42 Tighten all the main bearing cap bolts/nuts evenly to the specified torque. Using a sharp knife, trim off the ends of the No 1 bearing cap sealing strips (where applicable), so that they protrude above the cylinder block/crankcase mating surface by approximately 1 mm **(see illustrations)**.

43 Fit a new crankshaft rear oil seal as described in Part A or B of this Chapter (as applicable).

44 Refit the piston/connecting rod assemblies to the crankshaft as described in Section 21.

45 Refit the Woodruff key, then slide on the oil pump drive sprocket and spacer (where fitted), and locate the drive chain on the sprocket.

46 Where applicable, ensure that the mating surfaces of the front oil seal carrier and cylinder block are clean and dry. Note the correct fitted depth of the oil seal then, using a large flat-bladed screwdriver, lever the old seal out of the housing.

47 Where applicable, apply a smear of suitable sealant to the oil seal carrier mating

surface. Ensure that the locating dowels are in position, then slide the carrier over the end of the crankshaft and into position on the cylinder block. Tighten the carrier retaining bolts to the specified torque.

48 Fit a new crankshaft front oil seal as described in Part A or B of this Chapter.

49 Ensuring that the drive chain is correctly located on the sprocket, refit the oil pump and sump as described in Part A or B of this Chapter.

50 Where removed, refit the cylinder head as described in Part A or B of this Chapter.

21 Pistons/connecting rods - refitting and big-end bearing running clearance check

Selection of bearing shells

1 On most engines, there are two sizes of big-end bearing shell produced by Citroën; a standard size for use with the standard crankshaft, and an oversize for use once the crankshaft journals have been reground.

2 Consult your Citroën dealer for the latest information on parts availability. To be safe, always quote the diameter of the crankshaft big-end crankpins when ordering bearing shells.

3 Prior to refitting the piston/connecting rod assemblies, it is recommended that the big-end bearing running clearance is checked as follows.

Big-end bearing running clearance check

4 Clean the backs of the bearing shells, and the bearing locations in both the connecting rod and bearing cap.

5 Press the bearing shells into their locations, ensuring that the tab on each shell engages in the notch in the connecting rod and cap. Take care not to touch any shell's bearing surface with your fingers **(see illustrations)**. If the original bearing shells are being used for the check, ensure they are refitted in their original locations. The clearance can be checked in either of two ways.

6 One method is to refit the big-end bearing cap to the connecting rod, ensuring they are fitted the correct way around (see paragraph 20), with the bearing shells in place. With the cap retaining nuts correctly tightened, use an internal micrometer or vernier caliper to measure the internal diameter of each assembled pair of bearing shells. If the diameter of each corresponding crankshaft journal is measured and then subtracted from the bearing internal diameter, the result will be the big-end bearing running clearance.

2C

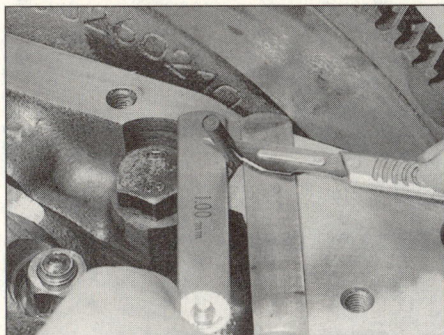

20.42b . . . Using a sharp knife, trim off the ends of the No 1 bearing cap sealing strips, so that they protrude above the cylinder block/crankcase mating surface by approximately 1 mm

21.5a Press the bearing shells into the connecting rod . . .

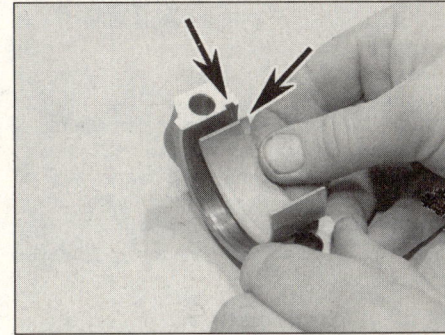

21.5b . . . and big end bearing cap, ensuring that the tabs (arrowed) on each shell engages in the notches in the connecting rod and cap

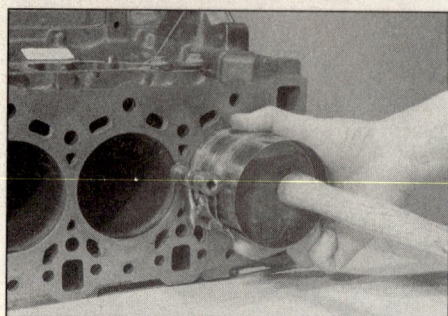
21.19 Using a block of wood or hammer handle against the piston crown, tap the assembly into the cylinder

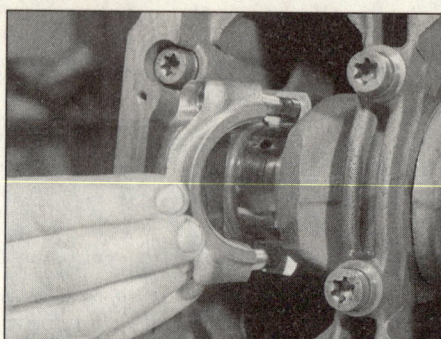
21.21a Refit the big-end bearing cap . . .

21.21b . . . and retaining nuts

7 The second, and more accurate method is to use Plastigauge as follows.

8 Ensure that the bearing shells are correctly fitted. Place a strand of Plastigauge on each (cleaned) crankpin journal.

9 Refit the (clean) piston/connecting rod assemblies to the crankshaft, and refit the big-end bearing caps, using the marks made or noted on removal to ensure they are fitted the correct way around.

10 Tighten the bearing cap nuts as described below in paragraph 21 or 22 (as applicable). Take care not to disturb the Plastigauge or rotate the connecting rod during the tightening sequence.

11 Dismantle the assemblies without rotating the connecting rods. Use the scale printed on the Plastigauge envelope to obtain the big-end bearing running clearance.

12 If the clearance is significantly different from that expected, the bearing shells may be the wrong size (or excessively worn, if the original shells are being re-used). Make sure that no dirt or oil was trapped between the bearing shells and the caps or block when the clearance was measured. If the Plastigauge was wider at one end than at the other, the crankshaft journal may be tapered.

13 Note that Citroën do not specify a recommended big-end bearing running clearance. The figure given in the Specifications is a guide figure, which is typical for this type of engine. Before condemning the components concerned, refer to your Citroën dealer or engine reconditioning specialist for further information on the specified running clearance. Their advice on the best course of action to be taken can then also be obtained.

14 On completion, carefully scrape away all traces of the Plastigauge material from the crankshaft and bearing shells. Use your fingernail, or some other object which is unlikely to score the bearing surfaces.

Final piston/connecting rod refitting

15 Note that the following procedure assumes that the crankshaft and main bearing caps (or ladder) are in place (see Section 20).

16 Ensure that the bearing shells are correctly fitted to the connecting rods and big end bearing caps, as described earlier. If new shells are being fitted, ensure that all traces of the protective grease are cleaned off using paraffin. Wipe dry the shells and connecting rods with a lint-free cloth.

17 Lubricate the cylinder bores, the pistons, and piston rings, then lay out each piston/connecting rod assembly in its respective position.

18 Start with assembly No 1. Make sure that the piston rings are still spaced as described in Section 18, then clamp them in position with a piston ring compressor.

19 Insert the piston/connecting rod assembly into the top of cylinder/liner No 1. On petrol engines, ensure that the arrow on the piston crown is pointing towards the timing belt end of the engine and on diesel engines, ensure that the cloverleaf-shaped cut-out on the piston crown is towards the front (oil filter side) of the cylinder block. Using a block of wood or hammer handle against the piston crown, tap the assembly into the cylinder/liner until the piston crown is flush with the top of the cylinder/liner **(see illustration)**.

20 Ensure that the bearing shell is still correctly installed. Liberally lubricate the crankpin and both bearing shells. Taking care not to mark the cylinder/liner bores, pull the piston/connecting rod assembly down the bore and onto the crankpin.

21 Refit the big-end bearing cap, tightening its retaining nuts finger-tight at first. Note that the faces with the identification marks must match (which means that the bearing shell locating tabs abut each other) **(see illustrations)**.

22 Tighten the bearing cap retaining nuts evenly and progressively to the stage 1 torque setting. Once both nuts have been tightened to the stage 1 setting, angle-tighten them through the specified stage 2 angle, using a socket and extension bar **(see illustrations)**.

23 On all engines, once the bearing cap retaining nuts have been correctly tightened, rotate the crankshaft. Check that it turns freely; some stiffness is to be expected if new components have been fitted, but there should be no signs of binding or tight spots.

24 Refit the remaining three piston/connecting rod assemblies in the same way. Note that on 2.5 litre diesel engines, the cut-out in the lower edge of No 4 piston skirt must be oriented towards the oil filter housing flange, on the side of the cylinder block.

25 Refit the cylinder head and oil pump as described in Part A or B of this Chapter, or this Part (as applicable).

22 Engine - initial start-up after overhaul

1 With the engine refitted in the vehicle, double-check the engine oil and coolant levels. Make a final check that everything has

21.22a Tighten the bearing cap retaining nuts evenly and progressively to the stage 1 torque setting . . .

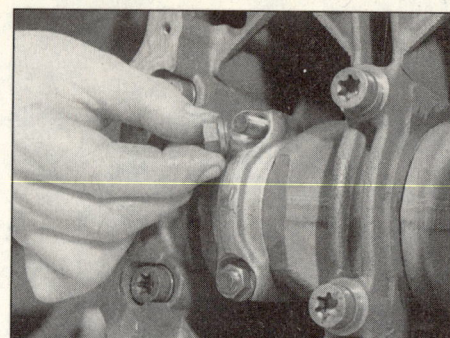
21.22b . . . then angle-tighten them through the specified stage 2 angle, using a socket and extension bar

been reconnected, and that there are no tools or rags left in the engine compartment.

Petrol engine models

2 Remove the spark plugs and disable the fuel system by disconnecting the wiring connectors from the fuel injectors, referring to Chapter 4A for further information. Disable the ignition system (see Chapter 2A, section 2).
3 Turn the engine on the starter until the oil pressure warning light goes out. Refit the spark plugs, and reconnect the ignition system components (see Chapter 2A, section 2)

Diesel engine models

4 To prevent the engine starting, disconnect the wiring from the stop solenoid on the injection pump or remove the fuel injection system relay (Chapter 4B), then turn the

engine on the starter motor until the oil pressure warning light goes out. Reconnect the wire to the stop solenoid.
5 Prime the fuel system (refer to Chapter 4B).
6 Fully depress the accelerator pedal, turn the ignition key to position 'M', and wait for the preheating warning light to go out.

All models

7 Start the engine, noting that this may take a little longer than usual, due to the fuel system components having been disturbed.
8 While the engine is idling, check for fuel, water and oil leaks. Don't be alarmed if there are some odd smells and smoke from parts getting hot and burning off oil deposits.
9 Assuming all is well, keep the engine idling until hot water is felt circulating through the top hose, then switch off the engine.

10 Check the ignition timing (petrol engines) or injection pump timing (diesel engines), and the idle speed settings (as appropriate), then switch the engine off.
11 After a few minutes, recheck the oil and coolant levels as described in *Weekly Checks*, and top-up as necessary.
12 If the cylinder head bolts were tightened as described, there is no need to re-tighten them once the engine has first run after reassembly.
13 If new pistons, rings or crankshaft bearings have been fitted, the engine must be treated as new, and run-in for the first 500 miles (800 km). *Do not* operate the engine at full-throttle, or allow it to labour at low engine speeds in any gear. It is recommended that the oil and filter be changed at the end of this period.

2C

Notes

Chapter 3
Cooling, heating and air conditioning systems

Contents

Degrees of difficulty

Easy, suitable for novice with little experience	**Fairly easy,** suitable for beginner with some experience	**Fairly difficult,** suitable for competent DIY mechanic	**Difficult,** suitable for experienced DIY mechanic	**Very difficult,** suitable for expert DIY or professional

Specifications

General
Maximum system pressure .... 1.4 bars

Thermostat
Opening temperatures:
  All except 2.5 litre diesel engines:
    Starts to open .... 85°C
    Fully open .... 100°C
  2.5 litre diesel engines:
    Main thermostat:
      Starts to open .... 85°C
      Fully open .... 100°C
    Secondary thermostat:
      Starts to open .... 84°C
      Fully open .... 88°C

1 General information and precautions

General information

The cooling system is of pressurised type, comprising a pump driven by the timing belt (all except 2.5 litre diesel engines), or the auxiliary drivebelt (2.5 litre diesel engines), an aluminium crossflow radiator, electric cooling fan, and a thermostat. The system functions as follows. Cold coolant from the radiator passes through the hose to the coolant pump, where it is pumped around the cylinder block and head passages. After cooling the cylinder bores, combustion surfaces and valve seats, the coolant reaches the underside of the thermostat, which is initially closed. The coolant passes through the heater, and is returned via the cylinder block to the coolant pump.

When the engine is cold, the coolant circulates only through the cylinder block, cylinder head and heater. When the coolant reaches a predetermined temperature, the thermostat opens and the coolant passes through to the radiator. As the coolant circulates through the radiator, it is cooled by the inrush of air when the car is in forward motion. Airflow is supplemented by the action of the electric cooling fan when necessary. Once the coolant has passed through the radiator, and has cooled, the cycle is repeated.

The electric cooling fan, mounted on the rear of the radiator, is controlled by a thermostatic switch. At a predetermined coolant temperature, the switch actuates the fan. On models with air conditioning, the cooling fans are controlled via the air conditioning and engine management ECUs.

An expansion tank is fitted to allow for the expansion of the coolant when hot. The expansion tank is connected to the top of the radiator.

Refer to Section 10 for information on the air conditioning system.

Precautions

⚠️ **Warning: Do not attempt to remove the expansion tank filler cap, or disturb any part of the cooling system, while the engine is hot; there is a high risk of scalding. If the expansion tank filler cap must be removed before the engine and radiator have fully cooled (even though this is not recommended) the pressure in the cooling system must first be relieved. Cover the cap with a thick layer of cloth, to avoid scalding, and slowly unscrew the filler cap until a hissing sound can be heard. When the hissing has stopped, indicating that the pressure has reduced, slowly unscrew the filler cap until it can be removed; if more hissing sounds are heard, wait until they have stopped before unscrewing the cap completely. At all times, keep well away from the filler cap opening.**

⚠️ **Warning: Do not allow antifreeze to come into contact with skin, or with the painted surfaces of the vehicle. Rinse off spills immediately, with plenty of water. Never leave antifreeze lying around in an open container, or in a puddle on the driveway or garage floor. Children and pets are attracted by its sweet smell, but antifreeze can be fatal if ingested.**

⚠️ **Warning: If the engine is hot, the electric cooling fan may start rotating even if the engine is not running; be careful to keep hands, hair and loose clothing well clear when working in the engine compartment.**

⚠️ **Warning: Refer to Section 10 for precautions to be observed when working on models equipped with air conditioning.**

2 Cooling system hoses – disconnection and renewal

Note: *Refer to the warnings given in Section 1 of this Chapter before proceeding. Do not*

2.3 Disconnecting the radiator top hose

attempt to disconnect any hose while the system is still hot.

1 If the checks described in Chapter 1 reveal a faulty hose, it must be renewed as follows.
2 First drain the cooling system (see Chapter 1). If the coolant is not due for renewal, it may be re-used if it is collected in a clean container.
3 Before disconnecting a hose, first note its routing in the engine compartment, and whether it is secured by any clips or ties. Use a screwdriver to slacken the clips, then move the clips along the hose, clear of the relevant inlet/outlet union. Carefully work the hose free **(see illustration)**.
4 Note that the radiator inlet and outlet unions are fragile; do not use excessive force when attempting to remove the hoses. If a hose proves to be difficult to remove, try to release it by rotating the hose ends before attempting to free it.

> **HAYNES HINT** *If all else fails, cut the coolant hose with a sharp knife, then slit it so that it can be peeled off in two pieces. Although this may prove expensive if the hose is otherwise undamaged, it is preferable to buying a new radiator.*

5 When fitting a hose, first slide the clips onto the hose, then work the hose into position. If clamp-type clips were originally fitted, it is a good idea to replace them with screw-type

clips when refitting the hose. If the hose is stiff, use a little soapy water (washing-up liquid is ideal) as a lubricant, or soften the hose by soaking it in hot water.
6 Work the hose into position, checking that it is correctly routed and secured. Slide each clip along the hose until it passes over the flared end of the relevant inlet/outlet union, before tightening the clips securely.
7 Refill the cooling system with reference to Chapter 1.
8 Check thoroughly for leaks as soon as possible after disturbing any part of the cooling system.

3 Radiator – removal, inspection and refitting

Removal

1 Using a screwdriver, release the securing clip, and remove the cover from the battery, then disconnect the battery negative lead.
2 Drain the cooling system as described in Chapter 1. Where applicable, disconnect the fan switch wiring.
3 The radiator is removed by lifting it out of the engine compartment but, on some models, there is insufficient clearance for the radiator to pass between the front body panel and the engine, in which case, proceed as follows **(see illustrations)**.
a) Remove the sidelight units as described in Chapter 13.
b) Unscrew the front grille panel securing bolts (one on each side) from the sidelight apertures.
c) Unscrew the four upper front body panel securing bolts.
d) Working under the front wheel arches, prise the covers from the wheel arch liners for access to the front bumper side securing bolts (one on each side of the vehicle), then unscrew the securing bolts.
4 Slacken the hose clips, and disconnect the upper and lower radiator hoses. Where applicable, release the hose clips, and disconnect the remaining coolant hoses from the top of the radiator **(see illustrations)**.

3.3a Unscrew the front grille panel securing bolts (arrowed) from the sidelight apertures . . .

3.3b . . . and unscrew the upper front body panel securing bolts

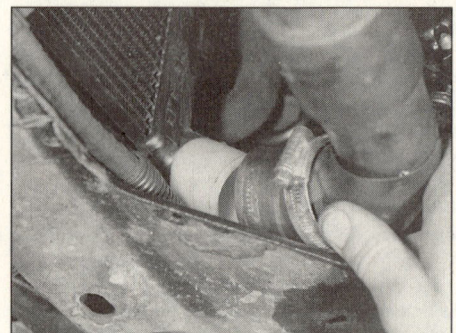

3.4a Disconnecting the lower radiator hose

3.4b Disconnecting the smaller hose from the top of the radiator – 2.5 litre diesel engine model

3.5 Depressing a radiator securing clip – 2.5 litre diesel engine model

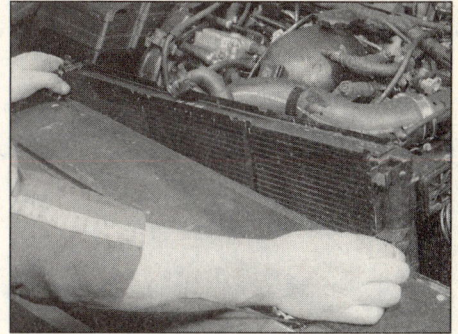

3.8 Lifting out the radiator

5 Working at the top corners of the radiator, depress the two radiator securing clips then, where applicable, tilt the radiator back to allow access to the two coolant hoses connected to the lower right-hand corner of the radiator **(see illustration)**.

6 On the 2.5 litre diesel engine, unbolt the fuel filter housing from the battery tray, and move it to one side, leaving the hoses connected.

7 If there is insufficient clearance for the radiator to pass between the front body panel and the engine, carefully pull the body panel forwards (the securing bolts should have been removed – see paragraph 3) to give sufficient clearance to remove the radiator. Take care not to damage the body panel or surrounding components, as not all the securing bolts have been removed.

8 Carefully lift the radiator from the engine compartment, taking care not to damage the radiator fins on surrounding components **(see illustration)**.

Inspection

9 If the radiator has been removed due to suspected blockage, reverse-flush it as described in Chapter 1. Clean dirt and debris from the radiator fins, using an air line (in which case, wear eye protection) or a soft brush. Be careful, as the fins are sharp, and easily damaged.

10 If necessary, a radiator specialist can perform a 'flow test' on the radiator, to establish whether an internal blockage exists.

11 A leaking radiator must be referred to a specialist for permanent repair. Do not attempt to weld or solder a leaking radiator, as damage to the plastic components may result.

12 In an emergency, minor leaks from the radiator can be cured by using a suitable radiator sealant, in accordance with its manufacturer's instructions, with the radiator *in situ*.

13 If the radiator is to be sent for repair or renewed, remove all hoses.

14 Inspect the condition of the radiator mounting rubbers, and renew them if necessary.

Refitting

15 Refitting is a reversal of removal, bearing in mind the following points.
a) *If there is limited clearance for the radiator to pass between the body front panel and the engine, it may prove helpful to hold the radiator securing clips in the released position, using cable-ties or string, until the radiator has been refitted.*
b) *Ensure that the lower radiator locating pins engage correctly with the mounting rubbers.*
c) *On completion, refill the cooling system as described in Chapter 1.*

---

4 Thermostat – removal, testing and refitting

Petrol engines

Note: *A new thermostat sealing ring should be used on refitting.*

Removal

1 On petrol engines, the thermostat is located at the left-hand end of the cylinder head.

2 Using a screwdriver, release the securing clip, and remove the cover from the battery, then disconnect the battery negative lead.

4.6 Unscrew the thermostat cover bolts . . .

3 Drain the cooling system as described in Chapter 1A.

4 Where necessary, release any relevant hoses and wiring from the retaining clips and/or brackets, and move them clear of the thermostat housing to improve access.

5 If necessary, slacken the clip(s) and disconnect the coolant hose(s) from the thermostat cover. Note that this is not necessary on all models, as the thermostat cover can usually be moved sufficiently to remove the thermostat without disconnecting the hose(s).

6 Unscrew the securing nuts or bolts (as applicable), and carefully lift off the thermostat cover to expose the thermostat **(see illustration)**.

7 Lift the thermostat from the housing, and recover the sealing ring **(see illustration)**.

Testing

8 A rough test of the thermostat's operation may be made by suspending it with a piece of string in a container full of water. Heat the water to bring it to the boil - the thermostat must open by the time the water boils. If not, renew it.

9 The opening temperature is usually marked on the thermostat. If a thermometer is available, the precise opening temperature of the thermostat may be determined, and compared with the value marked on the thermostat.

10 A thermostat which fails to close as the water cools must also be renewed.

4.7 . . . and lift out the thermostat and sealing ring – petrol engine

3

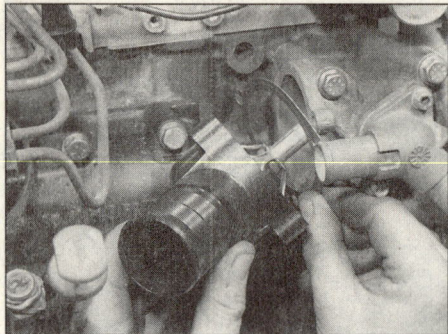

4.24 Removing the main thermostat cover/thermostat and gasket – 2.5 litre diesel engine (viewed with engine removed for clarity)

4.27 Tightening the main thermostat cover/thermostat securing bolts – 2.5 litre diesel engine (viewed with engine removed for clarity)

4.32 Removing the auxiliary thermostat cover/thermostat assembly – 2.5 litre diesel engine (viewed with engine removed for clarity)

Refitting

11 Refitting is a reversal of removal, bearing in mind the following points.
a) *Always fit a new sealing ring, and ensure that the thermostat is fitted the correct way round, with the spring(s) facing into the housing.*
b) *On completion, refill the cooling system as described in Chapter 1A.*

2.1 litre diesel engine

Note: *A new thermostat sealing ring should be used on refitting.*

Removal

12 The thermostat is located at the rear right-hand corner of the cylinder block.
13 Using a screwdriver, release the securing clip, and remove the cover from the battery, then disconnect the battery negative lead.
14 If desired, to improve access, apply the parking brake, then jack up the front of the vehicle and support securely on axle stands (see *Jacking and Vehicle Support*).
15 Drain the cooling system as described in Chapter 1B.
16 Proceed as described in paragraphs 4 to 7.

Testing

17 Proceed as described in paragraphs 8 to 10.

Refitting

18 Refitting is a reversal of removal, bearing in mind the following points.
a) *Always fit a new sealing ring, and ensure that the thermostat is fitted the correct way round, with the spring(s) facing into the housing on non-turbo models, or into the cover on turbo models.*
b) *On completion, refill the cooling system as described in Chapter 1B.*

2.5 litre diesel engine – main thermostat

Note: *A new thermostat cover gasket will be required on refitting.*

Removal

19 The thermostat is located in the rear of the coolant pump housing, at the left-hand end of the cylinder block. The thermostat is integral with the cover, and the two components are not available separately.
20 To enable access to the thermostat, unbolt the coolant reservoir, and move it to one side, leaving the hoses connected.
21 Drain the cooling system as described in Chapter 1B.
22 Slacken the clip, and disconnect the coolant hose from the thermostat cover.
23 Unscrew the securing bolts, and remove the thermostat cover/thermostat assembly.
24 Recover the gasket (see illustration).

Testing

25 Proceed as described in paragraphs 8 to 10.

Refitting

26 Thoroughly clean the mating faces of the thermostat cover and the housing.
27 Refit the thermostat cover/thermostat assembly using a new gasket, and tighten the securing bolts (see illustration).
28 Further refitting is a reversal of removal, but on completion, refill the cooling system as described in Chapter 1B.

2.5 litre diesel engine – auxiliary thermostat

Removal

29 The auxiliary thermostat is located at the top of the coolant housing, at the front left-hand corner of the cylinder head. The thermostat is integral with the cover, and the two components are not available separately.
30 Drain the cooling system as described in Chapter 1B.
31 Slacken the clips, and disconnect the coolant hoses from the thermostat cover.
32 Unscrew the securing bolts, and remove the thermostat cover/thermostat assembly (see illustration).
33 Recover the gasket.

Testing

34 Proceed as described in paragraphs 8 to 10.

Refitting

35 Proceed as described in paragraphs 26 to 28.

5 Electric cooling fan – testing, removal and refitting

Testing

1 Current supply to the cooling fan(s) is via the ignition switch (see Chapter 5A) and a fuse (see Chapter 13). The circuit is completed by the cooling fan thermostatic switch. On models with air conditioning, the cooling fans are controlled via the air conditioning and engine management ECUs, and testing should be entrusted to a Citroën dealer or specialist.
2 If a fan does not appear to work, run the engine until normal operating temperature is reached, then allow it to idle. The fan should cut in within a few minutes (before the temperature gauge needle enters the red section, or before the coolant temperature warning light comes on). If not, switch off the ignition and disconnect the wiring plug from the cooling fan switch. Bridge the two contacts in the wiring plug using a length of spare wire, and switch on the ignition. If the fan now operates, the switch is probably faulty, and should be renewed.
3 If the fan still fails to operate, check that battery voltage is available at the feed wire to the switch; if not, then there is a fault in the feed wire (possibly due to a fault in the fan motor, or a blown fuse). If there is no problem with the feed, check that there is continuity between the switch earth terminal and a good earth point on the body; if not, then the earth connection is faulty, and must be re-made.
4 If the switch and the wiring are in good condition, the fault must lie in the motor itself. The motor can be checked by disconnecting it from the wiring loom, and connecting a 12-volt supply directly to it.

Removal

5 Using a screwdriver, release the securing clip, and remove the cover from the battery, then disconnect the battery negative lead.
6 Remove the front grille panel as described in Chapter 12.

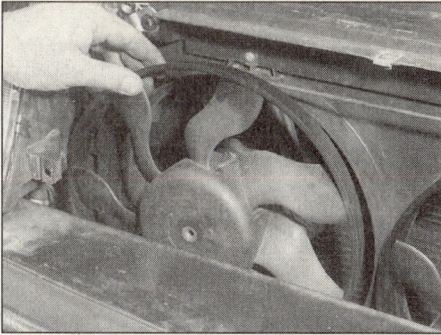

5.7 Removing the cooling fan blades

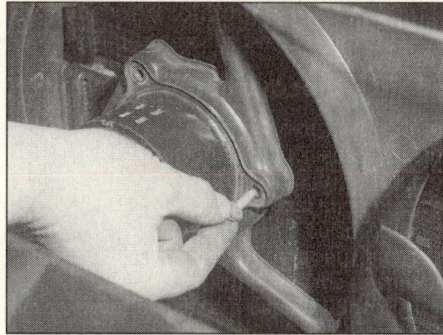

5.8a Unscrew the securing bolts . . .

5.8b . . . and withdraw the fan motor

7 Remove the securing screw and withdraw the cooling fan blades (see illustration).
8 Unscrew the three securing bolts, and withdraw the fan motor from the housing (see illustrations).

Refitting

9 Refitting is a reversal of removal, bearing in mind the following points.
a) Ensure that the motor is fitted correct way round so that wiring plug engages with the connector.
b) Make sure fan blades engage with flat on motor shaft.

6 Cooling system electrical switches and sensors – testing, removal and refitting

Cooling fan switch – petrol engine

1 The switch is located at the right-hand end of the radiator. Note that on models with air conditioning, the cooling fans are controlled via the air conditioning and engine management ECUs.

Testing

2 Testing of the switch is described in Section 5 as part of the electric cooling fan test procedure.

Removal

3 Disconnect the battery negative lead.
4 Partially drain the cooling system to just below the level of the switch (see Chapter 1). Alternatively, have ready a suitable bung to plug the switch aperture in the radiator when the switch is removed. If this method is used, take great care not to damage the threads in the radiator, and do not use anything which will allow foreign matter to enter the cooling system.
5 Disconnect the wiring plug from the switch.
6 Carefully unscrew the switch from the radiator. If the system has not been drained, plug the switch aperture to prevent further coolant loss.

Refitting

7 If the original switch is being refitted, and the switch was originally fitted using sealing compound, clean the switch threads thoroughly.
8 Where applicable, coat the switch threads with fresh sealing compound.
9 Refitting is a reversal of removal. Tighten the switch, and refill (or top-up) the cooling system as described in Chapter 1 or Weekly checks, as applicable.
10 On completion, start the engine and run it until it reaches normal operating temperature. Continue to run the engine, and check that the cooling fan cuts in and out correctly.

Cooling fan switch – 2.1 litre diesel engine

11 The switch is located in the coolant housing at the front left-hand corner of the cylinder head. Note that on models with air conditioning, the cooling fans are controlled via the air conditioning and engine management ECUs.

Testing

12 Testing of the switch is described in Section 5 as part of the electric cooling fan test procedure.

Removal and refitting

13 Proceed as described in paragraphs 3 to 10.

Cooling fan switch – 2.5 litre diesel engine

14 The sensor is located in the auxiliary thermostat housing, at the left-hand end of the engine (see illustration). Note that on models with air conditioning, the cooling fans

6.14 Cooling fan switch location (arrowed) – 2.5 litre diesel engine model

are controlled via the air conditioning and engine management ECUs.

Testing

15 Testing of the switch is described in Section 5 as part of the electric cooling fan test procedure.

Removal and refitting

16 Proceed as described in paragraphs 3 to 10.

Coolant temperature sensor – petrol engine

17 The sensor is located in the thermostat housing at the left-hand end of the cylinder head.

Testing

18 The sensor contains a thermistor – an electronic component whose electrical resistance decreases at a predetermined rate as its temperature rises. The engine management ECU supplies the sensor with a set voltage and then, by measuring the current flowing in the sensor circuit, it determines the engine temperature. This information is then used, in conjunction with other inputs, to control the injector opening time (pulse width). On some models, the idle speed and/or the ignition timing settings are also temperature-dependent.
19 If the sensor circuit should fail to provide adequate information, the ECU back-up facility will override the sensor signal. In this event, the ECU assumes a predetermined setting which will allow the engine management system to run, albeit at reduced efficiency. When this occurs, the warning light on the instrument panel will come on, and the advice of a Citroën dealer should be sought. The sensor circuit can be tested using suitable diagnostic equipment. Do not attempt to test the circuit using any other equipment, as there is a risk of damaging the ECU.

Removal and refitting

20 The procedure is similar to that described previously in this Section for the cooling fan switch.

Coolant temperature sensor – 2.1 litre diesel engine

Testing

21 The sensor is located in the coolant housing at the front left-hand corner of the

3

6.23 Coolant temperature sensor location (arrowed) – 2.5 litre diesel engine

6.35 Temperature gauge sender location (arrowed) – 2.5 litre diesel engine

cylinder head. The sensor contains a thermistor (see paragraph 18), which provides a signal to control the exhaust gas recirculation system.

Removal and refitting

22 The procedure is similar to that described previously in paragraphs 3 to 10 for the cooling fan switch.

Coolant temperature sensor – 2.5 litre diesel engine

Testing

23 The sensor is located in the thermostat housing, at the left-hand end of the engine **(see illustration)**. The sensor contains a thermistor (see paragraph 18), which provides a signal to the engine management ECU.

Removal and refitting

24 The procedure is similar to that described previously in paragraphs 3 to 10 for the cooling fan switch.

Temperature gauge sender – petrol engine

Testing

25 The sensor is located in the thermostat housing at the left-hand end of the cylinder head.

26 The temperature gauge is fed with a stabilised voltage from the instrument panel feed. The gauge earth is controlled by the sender. The sender contains a thermistor (see paragraph 18). When the coolant is cold, the sender resistance is high, current flow through the gauge is reduced, and the gauge needle points towards the blue (cold) end of the scale. As the coolant temperature rises, and the sender resistance falls, current flow increases, and the gauge needle moves towards the upper end of the scale.

27 On models with a temperature warning light, the light is fed with a voltage from the instrument panel. The light earth is controlled by the sender. The sender is effectively a switch, which operates at a predetermined temperature to earth the light and complete the circuit. The senders for the light and temperature gauge are incorporated in a single unit with two wires, one each for gauge and light earths.

28 If the gauge develops a fault, first check the other instruments. If they do not work at all, check the instrument panel electrical feed. If the readings are erratic, there may be a fault in the voltage stabiliser, which will necessitate renewal of the stabiliser (the stabiliser is integral with the instrument panel printed circuit board). If the fault lies in the temperature gauge alone, check it as follows.

29 If the gauge needle remains at the 'cold' end of the scale when the engine is hot, disconnect the sender wiring plug, and earth the relevant wire to the cylinder head. If the needle then deflects when the ignition is switched on, then sender unit is proved faulty, and should be renewed. If the needle still does not move, remove the instrument panel (see Chapter 13) and check the continuity of the wire between the sender and the gauge. Also check the feed to the gauge unit. If continuity is shown, and the fault still exists, the gauge is faulty, and should be renewed.

30 If the gauge needle remains at the 'hot' end of the scale when the engine is cold, disconnect the sender wire. If the needle then returns to the 'cold' end of the scale when the ignition is switched on, the sender unit proved faulty, and should be renewed. If the needle still does not move, check the remainder of the circuit as described previously.

31 The same basic principles apply to testing the warning light. The light should illuminate when the relevant sender wire is earthed.

Removal and refitting

32 The procedure is similar to that described previously in paragraphs 3 to 10 for the cooling fan switch.

Temperature gauge sender – 2.1 litre diesel engine

Testing

33 The sensor is located in the coolant housing at the front left-hand corner of the cylinder head. The testing procedure is as described for petrol engine models in paragraphs 25 to 31.

Removal and refitting

34 The procedure is similar to that described previously in paragraphs 3 to 10 for the cooling fan switch.

Temperature gauge sender – 2.5 litre diesel engine

Testing

35 The sensor is located in the thermostat housing, at the left-hand end of the engine **(see illustration)**. The testing procedure is as described for petrol engine models in paragraphs 25 to 31.

Removal and refitting

36 The procedure is similar to that described previously in paragraphs 3 to 10 for the cooling fan switch.

7 Coolant pump – removal and refitting

All except 2.5 litre diesel engine
Removal

1 The coolant pump is located at the right-hand end of the engine, and is driven by the timing belt.

2 Drain the cooling system as described in Chapter 1.

3 Remove the timing belt as described in Chapter 2, Part A.

4 Unscrew the securing bolts, and lift the coolant pump from the engine **(see illustrations)**.

5 Recover the gasket.

7.4a Unscrew the securing bolts (arrowed) . . .

7.4b . . . and remove the coolant pump – petrol engine model

7.12 Removing the coolant pump – 2.5 litre diesel engine (viewed with engine removed for clarity)

7.14 Coat the mating face of the coolant pump housing with sealing compound – 2.5 litre diesel engine (viewed with engine removed for clarity)

9.15 Prise out the alarm bulb trim plate to expose the switch/heater control/display trim panel securing screw (arrowed)

Refitting

6 Thoroughly clean the mating faces of the coolant pump and cylinder block.

7 Fit the pump using a new gasket, and tighten the securing bolts securely.

8 Refit the timing belt as described in Chapter 2, Part A.

9 Refill the cooling system as described in Chapter 1.

2.5 litre diesel engine

Removal

10 The coolant pump is located at the left-hand end of the engine, and is driven via an auxiliary drivebelt from a pulley on the camshaft.

11 Remove the coolant pump drivebelt as described in Chapter 1B.

12 Unscrew the securing bolts, and lift the coolant pump from the engine **(see illustration)**.

Refitting

13 Commence refitting by thoroughly cleaning the mating faces of the coolant pump and its housing.

14 Coat the mating face of the coolant pump housing with a thin layer of sealing compound **(see illustration)**.

15 Refit the pump, and tighten the securing bolts securely.

16 Refit and tension the pump drivebelt as described in Chapter 1B.

---

8 Heating and ventilation system – general information

1 The heater/ventilation system consists of a blower motor (housed behind the facia), face-level vents in the centre and at each end of the facia, and air ducts to the front footwells, and the rear passenger compartment (located in the centre pillars).

2 The control unit is located in the facia, and the controls operate flap valves and a coolant valve, to deflect and mix the air flowing through the various parts of the heater/ventilation system. The flap valves are

contained in the air distribution housing, which acts as a central distribution unit, passing air to the various ducts and vents.

3 Cold air enters the system through the grille at the rear of the engine compartment.

4 The air (boosted by the blower fan if required) then flows through the various ducts, according to the settings of the controls. Stale air is expelled through ducts at the rear of the vehicle. If warm air is required, the cold air is passed through the heater matrix, which is heated by the engine coolant.

5 A recirculation switch enables the outside air supply to be closed off, while the air inside the vehicle is recirculated. This can be useful to prevent unpleasant odours entering from outside the vehicle, but should only be used briefly, as the recirculated air inside the vehicle will soon deteriorate.

6 A manual or automatic control unit may be fitted, depending on model. The automatic control unit provides electronic control of the heater/ventilation system and the air conditioning system (where applicable), and maintains the temperature inside the car at the desired pre-set value.

---

9 Heater/ventilation system components – removal and refitting

Heater/ventilation control unit - models up to 1994

Removal

1 Using a screwdriver, release the securing clip, and remove the cover from the battery, then disconnect the battery negative lead.

2 Working at each side of the instrument panel in turn, prise out the cover plate, and unscrew the instrument panel surround securing screw.

3 Carefully prise up the front edge of the instrument panel surround, and withdraw it forwards, taking care to release the three retaining lugs at the rear of the panel.

4 Carefully pull the knobs from the heater/ventilation controls.

5 Unscrew the two securing screws (located in the blower motor and air temperature control apertures), then depress the two clips in the centre air duct and remove the heater/ventilation control unit trim panel.

6 Unscrew the four screws securing the heater/ventilation control unit to the facia.

7 Move the adjuster to fully lower the steering column.

8 Working underneath the facia, on either side of the steering column, unscrew the two lower switch/heater control/display trim panel securing screws.

9 Working at the front of the instrument panel aperture, unscrew the two upper switch/heater control/display trim panel securing screws.

10 Pull the top of the switch/heater control/display trim panel forwards from the facia to release the securing clips then, working at the rear of the panel, disconnect the wiring plugs from the panel-mounted components, noting the locations of the plugs, and withdraw the panel.

11 Pull the heater/ventilation control panel forwards from the facia, and disconnect the wiring plug(s) and, where applicable, the control cables from the rear of the unit, noting their locations.

12 Withdraw the heater/ventilation control panel.

Refitting

13 Refitting is a reversal of removal, but make sure that the control cables and wiring plug(s), as applicable, are correctly reconnected as noted before removal.

Heater/ventilation control unit – models from 1995

Removal

14 Remove the steering column shrouds as described in Chapter 12.

15 Carefully prise out the alarm bulb trim plate (or the blanking plate, as applicable) from the passenger's side end of the panel, to expose the upper switch/heater control/display trim panel securing screw **(see illustration)**. Remove the screw.

16 Working on either side of the steering

3

9.16a Prise out the covers . . .

9.16b . . . then remove the remaining two upper switch/heater control/display trim panel securing screws (arrowed)

9.17a Prise out the heater control surround panel . . .

column, prise out the screw covers, then remove the remaining two upper switch/heater control/display trim panel securing screws **(see illustrations)**.

17 Using a small flat-bladed screwdriver, carefully prise the heater control surround panel from the switch/heater control/display trim panel, then unscrew the four screws securing the heater control panel to the trim panel **(see illustrations)**.

18 Unscrew the five lower securing screws from the bottom of the switch/heater control/display trim panel.

19 Lever the driver's side of the panel to release the securing clip, then withdraw the panel forward from the facia **(see illustration)**. Working at the rear of the panel, disconnect the wiring plugs from the panel-mounted

components, noting the locations of the plugs, and withdraw the panel.

20 Proceed as described in paragraphs 11 and 12.

Refitting

21 Refitting is a reversal of removal, but make sure that the control cables and wiring plug(s), as applicable, are correctly reconnected as noted before removal, and refit the steering column shrouds with reference to Chapter 12.

Heater blower motor

Removal

22 Using a screwdriver, release the securing clip, and remove the cover from the battery, then disconnect the battery negative lead.

23 Working on one side of the centre console, remove the side trim panel as follows.
a) Unscrew the two screws securing the front air vent, and pull out the air vent.
b) Unscrew the screw securing the rear air vent, and withdraw the vent.
c) Working through the slit in the carpet trim, unscrew the side trim panel rear securing nut.
d) Withdraw the trim panel.

24 Remove the securing screws and clips, and withdraw the passenger's side lower facia panel **(see illustration)**.

25 Unscrew the three securing screws, then lower the motor assembly from the heater assembly, and disconnect the two wiring connectors **(see illustrations)**.

9.17b . . . then unscrew the four heater control panel securing screws (arrowed)

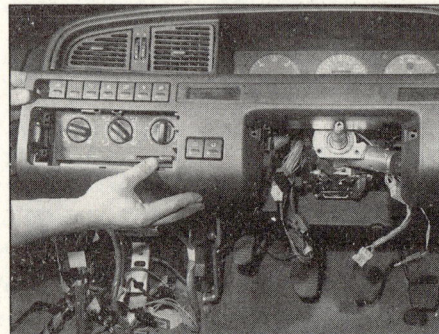

9.19 Withdraw the switch/heater control/display trim panel

9.24 Removing the passenger's side lower facia panel

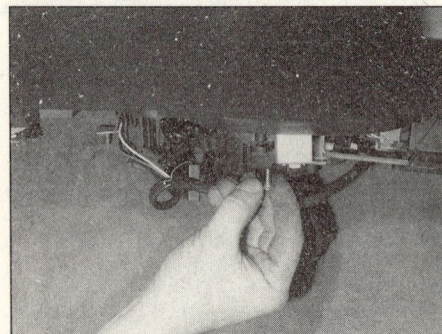

9.25a Remove the securing screws . . .

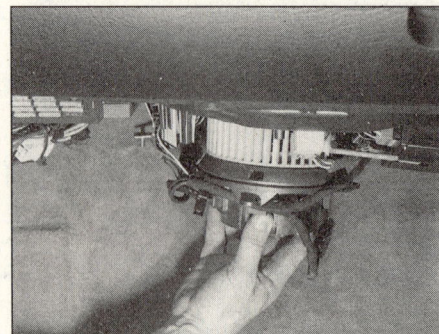

9.25b . . . and lower the heater blower motor assembly

9.26a Pull the motor wiring plug out of the housing . . .

9.26b . . . then depress the motor retaining rubbers . . .

9.26c . . . and pull the motor and fan blades assembly from the housing

9.30 Withdrawing the heater blower motor control unit

26 To remove the motor from its housing, proceed as follows (see illustrations).
a) *Pull the motor wiring plug out of the housing.*
b) *Using a screwdriver inserted through the slots in the housing, depress the three motor retaining rubbers.*
c) *Pull the motor and fan blades assembly from the housing.*

Refitting

27 Refitting is a reversal of removal but, if the motor has been removed from the housing, ensure that the retaining rubbers are securely engaged.

Heater blower motor control unit

Removal

28 Remove the heater blower motor, then remove the blower motor from its housing, as described previously in this Section.
29 Disconnect the wiring connectors from the rear of the control unit.
30 Unscrew the securing screws, then withdraw the control unit from the motor housing (see illustration).

Refitting

31 Refitting is a reversal of removal.

Heater matrix – right-hand-drive models

Removal

Note: *During this procedure, the heater matrix will almost certainly be irreparably damaged –*

do not remove the matrix unless it is to be renewed. New heater matrix pipe O-rings will be required on refitting, and suitable adhesive or adhesive tape will be required to repair the matrix housing – see text.
32 Drain the cooling system as described in Chapter 1.
33 Remove the steering column, as described in Chapter 11.
34 Remove the securing screws and withdraw the carpet trim panel from under the driver's side of the facia.
35 Unscrew the securing screws and lower the driver's side lower facia panel. Unclip the instrument panel dimmer switch and the engine diagnostic connector from the panel, and withdraw the panel.
36 Place a suitable container beneath the heater matrix pipe connections, to catch the coolant which will escape as the pipes are disconnected.

> **HAYNES HiNT**
>
> *Chock the front wheels, and release the parking brake before disconnecting the heater matrix pipes – this will move the parking brake pedal clear of the flow of coolant.*

37 Remove the screw securing the heater matrix pipe clamping bracket, then slide the bracket back along the pipes (see illustration).
38 Pull the pipes from the matrix, and allow

the coolant to drain into the container. Recover the O-rings if they are loose (see illustration).
39 Unscrew the heater matrix securing screw(s) (see illustration).
40 Simultaneously release the three heater matrix securing clips, and pull the matrix from the housing. The matrix will almost certainly be irreparably damaged as it is removed, as the fins are pushed against the surrounding panels (see illustration).

Refitting

41 In order to avoid damaging the new heater matrix during refitting, it is necessary to cut away part of the housing – if care is taken, it will be possible to glue the removed portion back in position once the matrix has been refitted.
42 Using a suitable tool, cut away a section

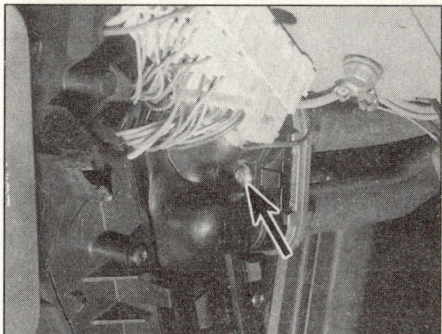

9.37 Remove the screw (arrowed) securing the heater matrix pipe clamping bracket . . .

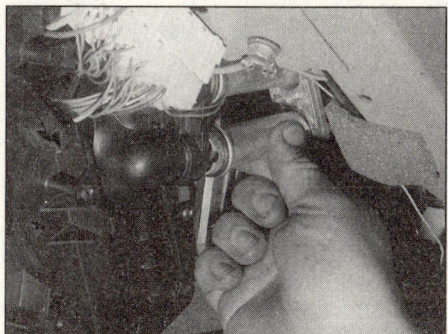

9.38 . . . then pull the pipes from the matrix

9.39 Unscrew the heater matrix securing screws

9.40 Removing the heater matrix

3

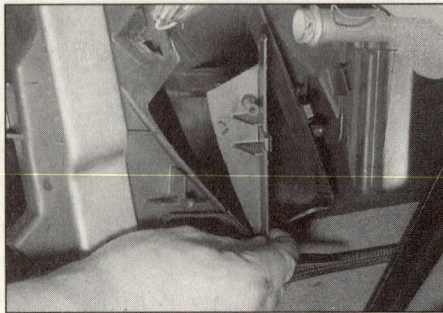

9.42a Cut away the front corner of the heater matrix housing . . .

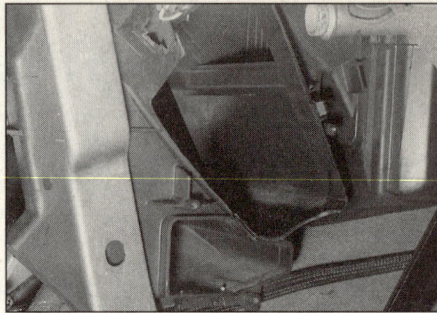

9.42b . . . to allow clearance for the new matrix to be fitted without damage

from the front edge of the heater matrix housing, as shown **(see illustrations)**.

43 Fit new O-rings to the heater matrix pipes, ensuring that they are correctly located **(see illustration)**.

44 Carefully slide the new matrix into position in the housing, taking great care not to damage the fins **(see illustration)**. Push the matrix fully home in the housing until the securing clips engage, then refit the matrix securing screw(s).

45 Using suitable adhesive, glue the removed section back into position at the front edge of the heater matrix housing. Alternatively, high-strength adhesive tape can be used to secure the removed section.

46 Reconnect the pipes to the matrix, ensuring that the O-rings locate correctly, then slide the clamping bracket into position, and refit and tighten the securing screw.

47 Further refitting is a reversal of removal, bearing in mind the following points.
a) *Refit the steering column as described in Chapter 11.*
b) *On completion, refill the cooling system as described in Chapter 1.*

Heater matrix – left-hand-drive models

Removal

48 New heater matrix pipe O-rings will be required on refitting.

49 Remove the securing screws and withdraw the carpet trim panel from under the passenger's side of the facia.

50 If necessary, remove the heater blower motor, as described previously in this Section,

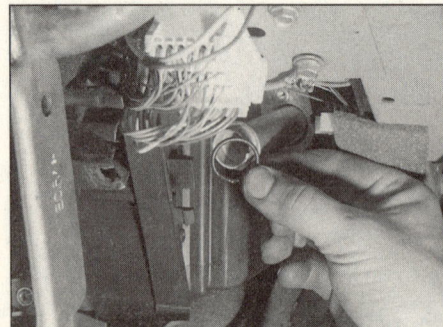

9.43 Fit new O-rings to the heater matrix pipes . . .

to provide sufficient clearance to withdraw the heater matrix.

51 If not already done, remove the securing screws and clips, and withdraw the passenger's side lower facia panel.

52 Place a suitable container beneath the heater matrix pipe connections, to catch the coolant which will escape as the pipes are disconnected.

53 Proceed as described in paragraphs 37 to 40.

Refitting

54 Refitting is a reversal of removal, bearing in mind the following points.
a) *Fit new O-rings to the heater matrix pipes, ensuring that they are correctly located.*
b) *Take care not to damage the heater matrix fins during refitting.*
c) *Make sure that the O-rings locate correctly when reconnecting the pipes to the matrix.*
d) *On completion, refill the cooling system as described in Chapter 1.*

10 Air conditioning system – general information and precautions

General information

1 Air conditioning is available on certain models. It enables the temperature of incoming air to be lowered, and also dehumidifies the air, which makes for rapid demisting and increased comfort.

2 The cooling side of the system works in the same way as a domestic refrigerator. Refrigerant

9.44 . . . then slide the new matrix into the housing

gas is drawn into a belt-driven compressor, and passes into a condenser mounted in front of the radiator, where it loses heat and becomes liquid. The liquid passes through an expansion valve to an evaporator, where it changes from liquid under high pressure to gas under low pressure. This change is accompanied by a drop in temperature, which cools the evaporator. The refrigerant returns to the compressor, and the cycle begins again.

3 Air blown through the evaporator passes to the heater assembly, where it is mixed with hot air blown through the heater matrix, to achieve the desired temperature in the passenger compartment.

4 The heating side of the system works in the same way as on models without air conditioning (see Section 8).

5 The operation of the system is controlled electronically. Any problems with the system should be referred to a Citroën dealer or an air conditioning specialist.

Precautions

6 It is necessary to observe special precautions whenever dealing with any part of the system, its associated components, and any items which necessitate disconnection of the system.

⚠ *Warning: The refrigeration circuit contains a liquid refrigerant (Freon). This refrigerant is potentially dangerous, and should only be handled by qualified persons. If it is splashed onto the skin, it can cause frostbite. It is not itself poisonous, but in the presence of a naked flame it forms a poisonous gas; inhalation of the vapour through a lighted cigarette could prove fatal. Uncontrolled discharging of the refrigerant is dangerous, and potentially damaging to the environment. It is therefore dangerous to disconnect any part of the system without specialised knowledge and equipment. If for any reason the system must be disconnected, entrust this task an authorised dealer or an air conditioning specialist.*

7 Do not operate the air conditioning system if it is known to be short of refrigerant, as this may damage the compressor.

11 Air conditioning system components – removal and refitting

⚠ *Warning: Do not attempt to open the refrigerant circuit. Refer to the precautions in Section 10.*

The only operation which can be carried out easily without discharging the refrigerant is the renewal of the compressor drivebelt (see Chapter 1). All other operations must be referred to a Citroën dealer or air conditioning specialist.

If necessary, the compressor can be unbolted and moved to one side, without disconnecting the flexible hoses, after removing the drivebelt.

Chapter 4 Part A:
Fuel system - petrol injection models

Contents

Degrees of difficulty

Easy, suitable for novice with little experience	Fairly easy, suitable for beginner with some experience	Fairly difficult, suitable for competent DIY mechanic	Difficult, suitable for experienced DIY mechanic	Very difficult, suitable for expert DIY or professional

Specifications

System type

2.0 injection models (up to 1994)	Bosch LE2 Jetronic or Magneti Marelli
2.0 injection models (from 1994)	Bosch Motronic MP 5.1
2.0 turbo models	Bosch Motronic MP 3.2

Fuel system data

Fuel pump type	Electric, immersed in tank
Fuel pump regulated constant pressure (at specified idle speed):	
Bosch LE2 Jetronic/Magneti Marelli system	2.5 to 3.0 bar
Bosch MP3.2/MP5.1 system	3.0 bar
Specified idle speed:	
2.0 litre injection models	
Manual transmission	850 ± 50 rpm
Automatic transmission	750 ± 50 rpm
2.0 litre 16v models:	
Manual transmission	800 ± 50 rpm
Automatic transmission	750 ± 50 rpm
Air conditioning	880 ± 50 rpm
2.0 litre turbo models:	
Manual transmission	800 ± 50 rpm
Automatic transmission	850 ± 50 rpm
Air conditioning	900 ± 50 rpm
Idle mixture CO content:	
Non-catalyst models	1.5 - 2.0 %
Catalyst models	Less than 0.5 %

Recommended fuel

Minimum octane rating	95 RON unleaded (UK unleaded premium). Leaded fuel **not** to be used

Torque wrench setting

	Nm	lbf ft
Inlet manifold nuts	35	26
Knock sensor	20	15

4A

1 General information and precautions

The fuel supply system consists of a fuel tank (which is mounted under the rear of the car, with an electric fuel pump immersed in it), a fuel filter, fuel feed and return lines. The fuel pump supplies fuel to the fuel rail, which acts as a reservoir for the four fuel injectors which inject fuel into the inlet tracts. The fuel filter incorporated in the feed line from the pump to the fuel rail ensures that the fuel supplied to the injectors is clean.

Refer to Section 6 for further information on the operation of the individual fuel injection systems. Throughout this Section, it is also occasionally necessary to identify vehicles by their engine codes rather than by engine capacity alone. Refer to the relevant Part of Chapter 2 for information on engine code identification.

⚠️ **Warning: Many procedures in this Chapter require the removal of fuel lines and connections, which may result in fuel spillage. Before carrying out any operation on the fuel system, refer to the precautions given in Safety first! at the beginning of this** manual, and follow them implicitly. Petrol is a highly dangerous and volatile liquid, and the precautions necessary when handling it cannot be overstressed.

Note: *Residual pressure will remain in the fuel lines long after the vehicle was last used. When disconnecting any fuel line, first depressurise the fuel system as described in Section 7.*

2 Air cleaner assembly and inlet ducts - removal and refitting

Removal

1 Slacken the retaining clips and disconnect the intake duct from the throttle housing and air cleaner housing lid. On LE2 Jetronic models, slacken the clips and detach the two hose sections from the air flow meter **(see illustration)**.
2 Undo the screws securing the lid to the air cleaner casing. Lift off the lid and take out the filter element (see Chapter 1A).
3 Slacken and remove the screws securing the air cleaner casing to the bodywork. Lift the air cleaner casing upwards to disengage it from the lower locating lugs. On some models, as the housing is lifted up, it will be necessary to disengage a small plastic retaining tag at the front securing the housing to the cold air intake duct underneath.
4 On turbo models, the intake ducting is complex arrangement consisting of flexible hoses and rigid ducts connecting the air cleaner assembly, intercooler and turbocharger. The intake ducts pass over the top and rear of the engine and also along the top of the transmission bellhousing. To remove the underside ducts it will be necessary to jack up the front of the car and support it on axle stands, then remove the engine splash guard.
5 To remove a section of intake ducting, slacken the retaining clips at each end and undo the bolts securing the relevant duct to its mounting bracket or support. The air inlet and outlet ducts are secured to the turbocharger flanges by a number of screws - these are accessible from the underside of the engine compartment.

Refitting

6 Refitting is a reversal of the removal procedure, ensuring that all hoses are properly reconnected, and that all ducts are correctly seated and securely held by their retaining clips.

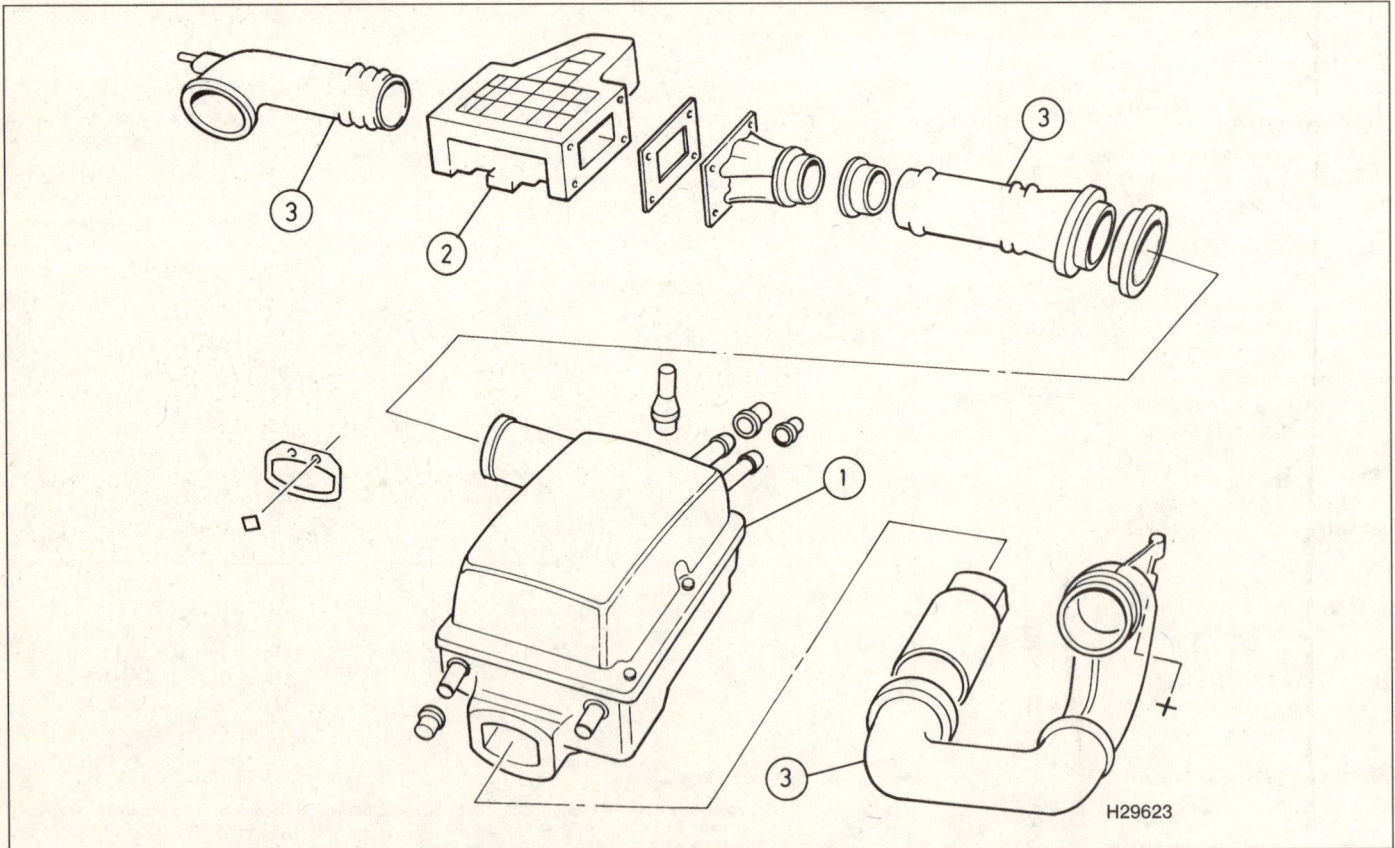

2.1 Air cleaner and intake ducting arrangement - models with LE2 Jetronic fuel injection
1 Air cleaner housing   2 Air flow meter   3 Intake ducts

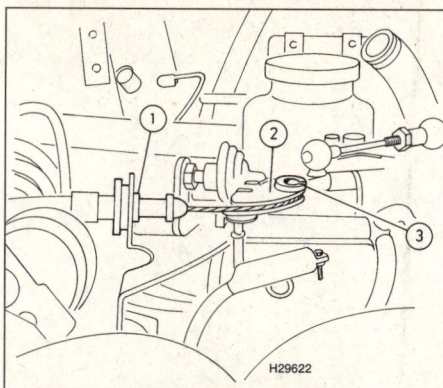

3.1 Disconnect the accelerator cable from the throttle housing

1 Retaining clip
2 Throttle lever
3 Accelerator inner cable end stop

3 Accelerator cable - removal, refitting and adjustment

Removal

1 Working in the engine compartment, free the accelerator inner cable end stop from the throttle lever, then pull the outer cable out from its mounting bracket rubber grommet. Slide the flat washer off the end of the cable, and remove the spring clip **(see illustration)**.
2 Working back along the length of the cable, free it from any retaining clips or ties, noting its correct routing.
3 Release the retaining clips, and remove the panel from underneath the driver's side of the facia panel.
4 Release the retaining clip, and detach the inner cable from the top of the accelerator pedal.
5 Release the outer cable from its retainer on the pedal mounting bracket, then tie a length of string to the end of the cable.
6 Return to the engine compartment, release the cable grommet from the bulkhead and withdraw the cable. When the end of the cable appears, untie the string and leave it in position - it can then be used to draw the cable back into position on refitting.

Refitting

7 Tie the string to the end of the cable, then use the string to draw the cable into position through the bulkhead. Once the cable end is visible, untie the string, then clip the outer cable into its pedal bracket retainer, and clip the inner cable into position in the pedal end.
8 Check that the cable is securely retained, then refit the cover panel to the facia.

9 From within the engine compartment, ensure that the outer cable is correctly seated in the bulkhead grommet, then work along the cable, securing it in position with the retaining clips and ties, and ensuring that the cable is correctly routed.
10 Slide the flat washer onto the cable end, and refit the spring clip.
11 Pass the outer cable through the mounting bracket grommet on the throttle housing, and reconnect the inner cable to the throttle cam. Adjust the cable as described below.

Adjustment

12 Remove the spring clip from the accelerator outer cable. Ensuring that the throttle cam is fully against its stop, gently pull the cable through its grommet until all free play is removed from the inner cable.
13 With the cable held in this position, refit the spring clip to the last exposed outer cable groove in front of the rubber grommet and washer. When the clip is refitted and the outer cable is released, there should be only a small amount of free play in the inner cable.
14 Have an assistant depress the accelerator pedal, and check that the throttle cam opens fully and returns smoothly to its stop.
15 On models with automatic transmission, once the accelerator cable is correctly adjusted, check the kick-down cable adjustment as described in Chapter 7B.

4 Accelerator pedal - removal and refitting

Removal

1 Disconnect the accelerator cable from the pedal as described in Section 3.
2 Slacken and withdraw the two screws (or screw and nut on left hand drive models), then lift the bearing away from the pedal pivot shaft **(see illustration)**.
3 Lift the pedal from the mounting bracket.
4 Examine the pivot shaft for signs of wear or damage and, if necessary, renew the pedal assembly.

Refitting

5 Refitting is a reversal of the removal procedure, applying a little multi-purpose grease to the pedal pivot point. On completion, refit and adjust the accelerator cable as described in Section 3.

5 Unleaded petrol - general information and usage

Note: *The information given in this Chapter is correct at the time of writing. If the latest information is required, check with a Citroën dealer. If travelling abroad, consult one of the*

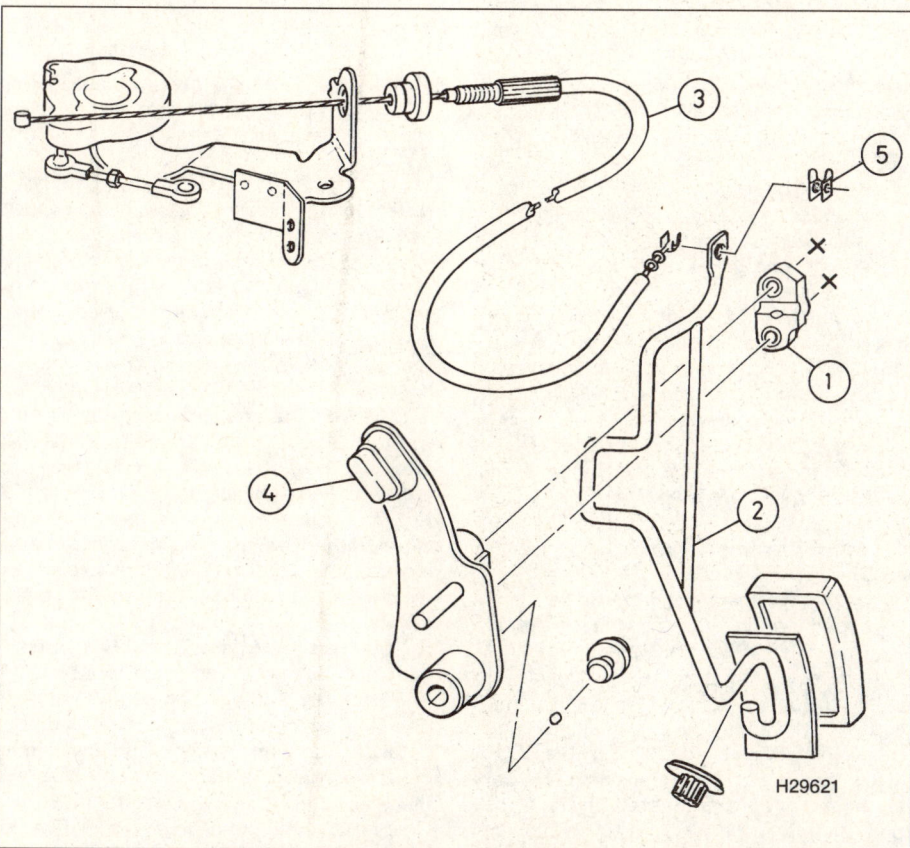

4.2 Accelerator pedal and cable mounting details - right hand drive version shown

1 Bearing   2 Pedal   3 Accelerator cable   4 Pedal mounting bracket   5 Clip

4A

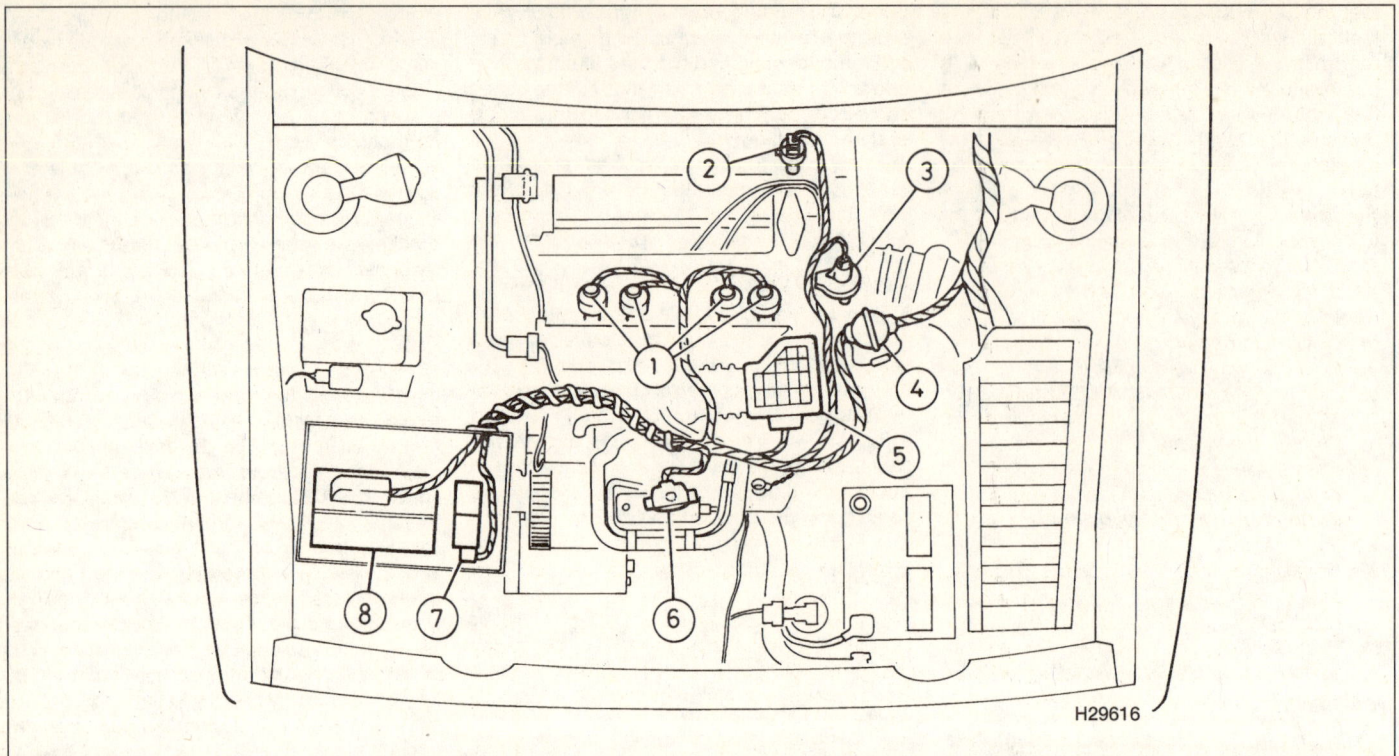

6.1 Component locations - models with Bosch LE2 Jetronic injection

1 Fuel injectors	3 Auxiliary air valve	5 Air flow meter	7 Fuel injection system relay
2 Coolant temperature sensor	4 Wiring harness connector	6 Throttle position switch	8 ECU

motoring organisations (or a similar authority) for advice on the fuel available.

1 The fuel recommended by Citroën is given in the Specifications, followed by the equivalent petrol currently on sale in the UK.
2 Some earlier models are not fitted with a catalytic converter, but note that *all* catalytic converter models must be run on unleaded fuel **only**. Under no circumstances should leaded fuel (UK '4-star') be used, as this may damage the converter.

6 Fuel injection systems - general information

Bosch LE2 Jetronic system

1 The Bosch LE2 Jetronic system is fitted to earlier, non-catalyst 2.0 litre injection models. Unlike the Bosch Motronic systems, which are self-contained engine management systems, controlling both the fuel injection and ignition, the LE2 injection system controls only the fuel injection **(see illustration)**.
2 Fuel is supplied to the engine via four fuel injectors which are mounted on the inlet manifold. The injectors are electro-magnetically operated pintle valves that open and close under the control of an Electronic Control Unit (ECU). The ECU calculates the injection timing and duration according to electronic signals received from sensors mounted on and around the engine. These include:

a) A flap-type air flow meter. This device supplies an analogue voltage to the ECU, which varies in proportion to the air flowing through it (and hence into the engine).
b) An intake air temperature sensor.
c) A coolant temperature sensor.
d) A throttle position switch - informs the ECU when the engine is at full load, part load or idling/overrunning.
e) The ignition coil - supplies the ECU with a voltage signal that alternates in proportion to the speed of the engine.
f) The ignition key - informs the ECU when engine cranking is taking place.
g) The battery - the ECU monitors the battery voltage and compensates accordingly when measuring the signals from all the other fuel system sensors.

3 To compensate for the extra frictional load when the engine is cold, the engine is supplied with additional air, via an auxiliary air valve. This device is controlled by the ECU .
4 Idle speed and mixture are both adjustable by means of screws located on the throttle body and air flow meter respectively - refer to the relevant Section for details.
Note: *The Bosch LE2 Jetronic fuel injection system was fitted concurrently with the*

Magneti Marelli engine management system; either system may be fitted to models of this age and specification

Bosch Motronic MP5.1 system

Note: *The fuel injection ECU is of the adaptive type, meaning that as it operates, it also monitors and stores the settings which give optimum engine performance under all conditions. When the battery is disconnected, these settings are lost and the ECU reverts to the base settings programmed into its memory at the factory. On restarting, this may lead to the engine running/idling roughly for a short while, until the ECU has re-learned the optimum settings. This process is best accomplished by taking the vehicle on a road test (for approximately 15 minutes), covering all engine speeds and loads, concentrating mainly in the 2,500 to 3,500 rpm region. Allow the engine to idle for several minutes on your return.*

5 The Bosch Motronic MP5.1 engine management (fuel injection/ignition) system is fitted to later 2.0 injection catalyst models. The system incorporates a closed-loop catalytic converter and an evaporative emission control system, and complies with the latest emission control standards. Refer to Chapter 5B for information on the ignition side of the system; the fuel side of the system operates as follows.
6 The fuel pump (which is immersed in the

fuel tank) supplies fuel from the tank to the fuel rail, via a filter mounted underneath the rear of the vehicle. Fuel supply pressure is controlled by the pressure regulator in the fuel rail. When the optimum operating pressure of the fuel system is exceeded, the regulator allows excess fuel to return to the tank.

7 The electrical control system consists of the ECU, along with the following sensors:

a) *Throttle potentiometer - informs the ECU of the throttle position, and the rate of throttle opening/closing.*

b) *Coolant temperature sensor - informs the ECU of engine temperature.*

c) *Inlet air temperature sensor - informs the ECU of the temperature of the air passing through the throttle housing.*

d) *Lambda sensor - informs the ECU of the oxygen content of the exhaust gases (explained in greater detail in Part C of this Chapter).*

e) *Crankshaft sensor - informs the ECU of the crankshaft position and speed of rotation.*

f) *Manifold Absolute Pressure (MAP) sensor - informs the ECU of the load on the engine (expressed in terms of inlet manifold vacuum).*

g) *Vehicle speed sensor - informs the ECU of the vehicle speed.*

8 All the above signals are analysed by the ECU which selects the fuelling response appropriate to those values. The ECU controls the fuel injectors (varying the pulse width - the length of time the injectors are held open - to provide a richer or weaker mixture, as appropriate). The mixture is constantly varied by the ECU, to provide the best setting for cranking, starting (with either a hot or cold engine), warm-up, idle, cruising and acceleration.

9 The ECU also has full control over the engine idle speed, via a stepper motor which acts directly on the throttle valve. The ECU controls rotation of the stepper, which in turn alters the opening of the throttle valve. This regulates the amount of air entering the manifold, and so controls the idle speed.

10 The ECU also controls the exhaust and evaporative emission control systems, which are described in detail in Part C of this Chapter.

11 An electric heating element is fitted to the throttle housing; the heater is supplied with current by the ECU, and warms the throttle housing on cold starts to prevent possible icing of the throttle valve.

12 If there is an abnormality in any of the readings obtained from either the coolant temperature sensor or the lambda sensor, the ECU enters its back-up mode. In this event, it ignores the abnormal sensor signal and assumes a pre-programmed value which will allow the engine to continue running (albeit at reduced efficiency). If the ECU enters this back-up mode, the warning light on the instrument panel will come on, and the relevant fault code will be stored in the ECU memory.

13 If the warning light comes on, the vehicle should be taken to a Citroën dealer at the earliest opportunity. A complete test of the engine management system can then be carried out, using a special electronic diagnostic test unit which is simply plugged into the system's diagnostic connector located near the fusebox on the facia.

Bosch Motronic MP 3.2 system

Note: *The fuel injection ECU is of the adaptive type, meaning that as it operates, it also monitors and stores the settings which give optimum engine performance under all conditions. When the battery is disconnected, these settings are lost and the ECU reverts to the base settings programmed into its memory at the factory. On restarting, this may lead to the engine running/idling roughly for a short while, until the ECU has re-learned the optimum settings. This process is best accomplished by taking the vehicle on a road test (for approximately 15 minutes), covering all engine speeds and loads, concentrating mainly in the 2,500 to 3,500 rpm region. Allow the engine to idle for several minutes on your return.*

14 The Motronic MP3.2 is fitted to all turbo-charged 2.0 litre petrol models. Refer to Chapter 5B for information on the ignition side of the system.

15 The MP 3.2 system is very similar in operation to the MP5.1 system described above. The main difference is that the system is able to dynamically control the boost pressure supplied by the turbocharger. The ECU has an on board manifold pressure/depression sensor and a vacuum modulator valve output to support this function.

Magneti Marelli system

16 The Magneti Marelli engine management (fuel injection/ignition) system is fitted to certain variants of early 2.0 litre injection non-catalyst models.

17 The system is very similar in operation to the Bosch MP5.1 system described above, apart from the idle speed control system.

18 On the Magneti Marelli system, the idle speed is controlled by the ECU via a stepper motor fitted to the throttle housing. The motor has a pushrod controlling the opening of an air passage which bypasses the throttle valve. When the throttle valve is closed, the ECU controls the movement of the motor pushrod, which regulates the amount of air which flows through the throttle housing passage, so controlling the idle speed. The bypass passage is also used as an additional air supply during cold starting.

Note: *The Magneti Marelli engine management system was fitted concurrently with the Bosch LE2 Jetronic fuel injection system; either system may be fitted to models of this age and specification.*

7 Fuel injection system - depressurisation

Note: *Refer to the warning note in Section 1 before proceeding.*

1 The fuel system referred to in this Section is defined as the tank-mounted fuel pump, the fuel filter, the fuel injectors, the fuel rail and the pressure regulator, and the metal pipes and flexible hoses of the fuel lines between these components. All these contain fuel which will be under pressure while the engine is running, and/or while the ignition is switched on. The pressure will remain for some time after the ignition has been switched off, and must be relieved in a controlled fashion when any of these components are disturbed for servicing work.

2 On non-catalyst models only, locate the fuel injection pump relay (where applicable) and disconnect it from its base. Crank the engine on the starter motor. If the engine fires, allow it to run for a few seconds until it stalls through fuel starvation. This will relieve the line fuel pressure via the fuel injectors. DO NOT attempt this procedure on models equipped with a catalytic converter - there is a danger that partially burnt fuel will be supplied to the catalyst and this could cause irreparable damage.

3 Disconnect the battery negative terminal.

4 Place a small container beneath the connection/union to be disconnected, and have a large rag ready to soak up any escaping fuel not being caught by the container.

5 Slowly loosen the connection or union nut to avoid a sudden release of pressure, and position the rag around the connection, to catch any fuel spray which may be expelled. Once the pressure is released, disconnect the fuel line. Plug the pipe ends, to minimise fuel loss and prevent the entry of dirt into the fuel system.

⚠️ **Warning: This procedure will merely relieve the pressure in the fuel system - remember that fuel will still be present in the system components and take precautions accordingly before disconnecting them.**

8 Fuel pump - removal and refitting

Note: *Refer to the warning note in Section 1 before proceeding.*

Removal

1 Disconnect the battery negative lead.

2 For access to the fuel pump, tilt the rear seat forwards and pull back the carpet from the right-hand side of the rear passenger compartment.

4A

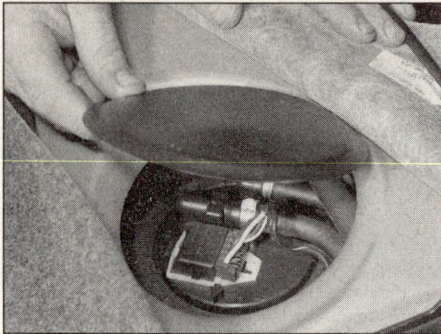
8.3a Remove the plastic cover . . .

8.3b . . . for access to the fuel pump

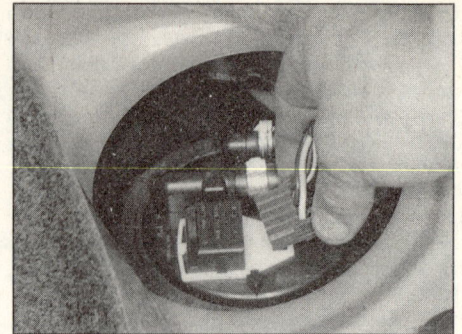
8.4 Disconnecting the wiring from the fuel pump

3 Using a screwdriver, carefully prise the plastic access cover from the floor to expose the fuel pump **(see illustrations)**.

4 Disconnect the wiring connector from the fuel pump, and tape the connector to the vehicle body, to prevent it from disappearing behind the tank **(see illustration)**.

5 Mark the hoses for identification purposes, then slacken the feed and return hose retaining clips. Where the crimped-type Citroën hose clips are fitted, cut the clips and discard them; use standard worm-drive hose clips on refitting. Disconnect both hoses from the top of the pump, and plug the hose ends.

6 Noting the alignment marks on the tank, pump cover and the locking ring, unscrew the ring and remove it from the tank **(see illustration)**. This is best accomplished by using a screwdriver on the raised ribs of the locking ring. Carefully tap the screwdriver to turn the ring anti-clockwise until it can be unscrewed by hand.

7 Displace the pump cover, then reach into the tank and unclip the pump from the tank base. Lift the fuel pump assembly out of the fuel tank, taking great care not to damage the filter, or to spill fuel onto the interior of the vehicle. Recover the rubber sealing ring and discard it - a new one must be used on refitting.

8 Note that the fuel pump is only available as a complete assembly - no components are available separately.

Refitting

9 Ensure that the fuel pump pick-up filter is clean and free of debris. Fit the new sealing ring to the top of the fuel tank.

10 Carefully manoeuvre the pump assembly into the fuel tank, and clip it into position in the base of the tank.

11 Align the mark on the fuel pump cover with the centre of the three alignment marks on the fuel tank, then refit the locking ring. Securely tighten the locking ring, then check that the locking ring, pump cover and tank marks are all correctly aligned.

12 Reconnect the feed and return hoses to the top of the fuel pump, using the marks made on removal to ensure that they are correctly reconnected, and securely tighten their retaining clips.

13 Reconnect the pump wiring connector.

14 Reconnect the battery negative terminal and start the engine. Check the fuel pump feed and return hoses unions for signs of leakage. If all is well, refit the plastic access cover. Tilt or refit the rear seat as described in Chapter 12 (as applicable).

9 Fuel gauge sender unit - removal and refitting

The fuel gauge sender unit is incorporated in the fuel pump - refer to Section 8 for details.

10 Fuel tank - removal and refitting

Removal

1 Before removing the fuel tank, all fuel must be drained out. Since a fuel tank drain plug is not provided, it is therefore preferable to carry out the removal operation when the tank is nearly empty. Before proceeding, disconnect the battery negative lead and syphon or hand-pump the remaining fuel from the tank.

2 Jack up the rear of the car and support on axle stands (see *Jacking and Vehicle Support*). Remove the right-hand rear wheel, then extract the fixings and remove the wheel arch liner.

3 With reference to Chapter 4C, unbolt the rear section of the exhaust pipe from the

8.6 Note the alignment marks on the fuel pump and locking ring

intermediate section. Unhook the silencers from their rubber mountings and lower the exhaust pipe away from the car.

4 Unhook the front silencer unit from its rubber mountings, and lower it away from the floorpan. Unbolt and remove the heat shield above the silencer.

5 Remove the metal clip and disconnect front end of the rear suspension height manual control linkage from the side of the control lever mechanism (see Chapter 11). Remove the bolt and disconnect the rear end of the control linkage from the rear corrector. Slide the linkage out of its mounting bracket and remove it from the car.

6 Remove the covers from the hydraulic pipes located on the underbody, leaving the pipes attached to the covers.

7 Disconnect the rigid fuel supply and return lines at the rubber hoses connections - be prepared for some fuel spillage.

8 Release the rear ABS wiring from the clips and cable supports on the floor pan.

9 Where necessary disconnect the wiring for the hydractive suspension from the clips on the underside of the vehicle.

10 Loosen the clips and disconnect the filler and vent hoses from the fuel tank. Make a careful note of their order of connection.

11 Unclip the heat shield, then release the fuel filter from its mounting and move it to one side.

12 Remove the screws and lower the retaining bars from the underside of the fuel tank **(see illustration)**.

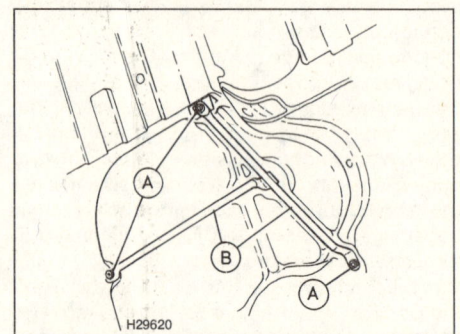
10.12 Remove the screws (A) and lower the retaining bars (B) from the underside of the fuel tank

13 Place a trolley jack with an interposed block of wood beneath the tank, then raise the jack carefully until it is supporting the weight of the tank.

14 Unscrew the fuel tank mounting bolts, then using the jack, lower the fuel tank slightly. Release the wiring and fuel hoses from the clips and straps on fuel tank, as they become accessible.

15 Disconnect the wiring from the fuel pump by pressing on the plastic tab.

16 Loosen the clips and disconnect the hoses from the fuel pump.

17 Lower the fuel tank slowly to the ground, checking that nothing else remains connected. Remove the tank from under the car.

18 If the tank is contaminated with sediment or water, remove the fuel pump and pick-up unit, and swill the tank out with clean fuel. The tank is injection-moulded from a synthetic material - if seriously damaged, it should be renewed. However, in certain cases, it may be possible to have small leaks or minor damage repaired. Seek the advice of a specialist before attempting to repair the fuel tank.

Refitting

19 Refitting is the reverse of the removal procedure, noting the following points:

a) *When lifting the tank back into position, take care to ensure that none of the hoses become trapped between the tank and vehicle body.*

b) *Ensure that all pipes and hoses are correctly routed, and securely held in position with their retaining clips.*

c) *Before reconnecting the height operation linkage, set the height control to LOW.*

d) *On completion, refill the tank with a small amount of fuel, and check for leakage prior to taking the vehicle out on the road.*

11 Fuel injection system - testing and adjustment

Testing

1 If a fault appears in the fuel injection system, first ensure that all the system wiring connectors are securely connected and free of corrosion. Ensure that the fault is not due to poor maintenance; ie, check that the air cleaner filter element is clean, the spark plugs are in good condition and correctly gapped, the cylinder compression pressures are correct, the ignition timing is correct, and that the engine breather hoses are clear and undamaged, referring to Chapters 1, 2 and 5 for further information.

2 If these checks fail to reveal the cause of the problem, the vehicle should be taken to a suitably-equipped Citroën dealer or fuel injection specialist for testing. A wiring block connector is incorporated in the engine management circuit, into which a special

electronic diagnostic tester can be plugged; the connector is located near the fusebox on the facia. The tester will locate the fault quickly and simply, alleviating the need to test all the system components individually, which is a time-consuming operation that carries a risk of damaging the ECU.

Adjustment

Bosch LE2 Jetronic models

Note: *The following adjustments should be carried out when the engine has just reached normal operating temperature after starting from cold, rather than after a long drive. Ideally, the adjustments should be completed before the engine temperature exceeds 100°C, or before the auxiliary cooling fan cuts in.*

3 When carrying out the adjustment, the air cleaner should be in position, with all vacuum and breather hoses connected. The instrument pack tachometer is sufficiently accurate to allow idle speed adjustment.

4 Connect an exhaust gas analyser to the exhaust tailpipe/sample pipe, in accordance with the manufacturers instructions.

5 Start the engine and allow it to idle. On vehicles with automatic transmission, select 'PARK'. Ensure that all ancillaries are switched off, including the air conditioning (where fitted).

6 The idle speed adjustment screw is located at the side of the throttle body **(see illustration)**.

7 Turn the idle speed adjustment screw until the idle speed is within the limits given in the Specifications. Accelerate the engine briefly, then allow it to idle again.

8 Re-check the idle speed and adjust if necessary. Switch off the engine on completion.

9 To adjust the exhaust gas CO content, prise the tamperproof plug from the CO adjustment screw hole, which is located on the underside of the air flow meter housing **(see illustration)**.

10 Start the engine and accelerate it to roughly 2000 rpm. Maintain this speed for about 10 seconds, then allow it to return to idle.

11 Check the CO reading on the exhaust gas analyser. If it is not as specified, insert an Allen key into the CO adjustment screw hole and engage it with the end of the screw.

12 Turn the adjustment screw as required to correct the exhaust gas CO content; clockwise enrichens the mixture and anticlockwise weakens it.

13 Re-check the exhaust gas CO content after accelerating the engine, as described from paragraph 10 onwards, and continue adjusting the mixture as required. Only turn the adjustment screw through half a turn at a time, between measurements, to avoid overshooting.

14 On completion, disconnect all test instruments and fit a new tamperproof plug to the mixture adjustment screw tube.

All other models

15 Experienced home mechanics with a considerable amount of skill and equipment (including a tachometer and an accurately calibrated exhaust gas analyser) may be able to *check* the exhaust CO level and the idle speed, as described in the previous subsection. However, if these are found to be in need of *adjustment*, the car *must* be taken to a Citroën dealer or fuel injection specialist for further testing. The mixture (exhaust gas CO content) and idle speed are controlled dynamically by the engine management system are not adjustable without the aid of dedicated test equipment.

12 Throttle housing - removal and refitting

Removal

1 Disconnect the negative cable from the battery and position it away from the terminal.

Bosch LE2 Jetronic and Magneti-Marelli models

2 Remove the air cleaner/air flow meter-to-throttle housing duct, with reference to Section 2.

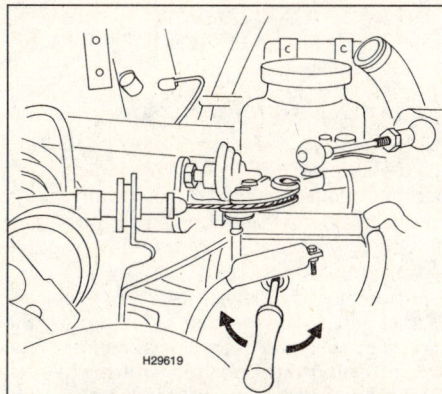

11.6 Idle speed adjustment screw location - models with LE2 Jetronic fuel injection

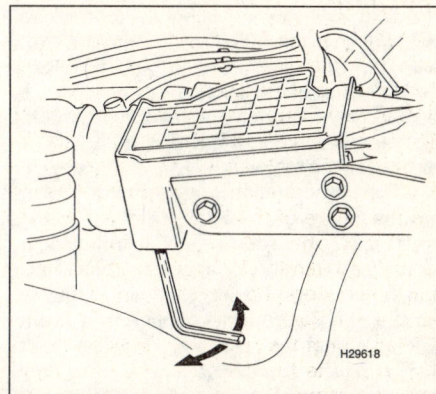

11.9 Exhaust CO/mixture adjustment screw location - models with LE2 Jetronic fuel injection

4A

3 Disconnect the accelerator inner cable from the throttle cam then withdraw the outer cable from the mounting bracket along with its flat washer and spring clip. Where necessary, also disconnect the kick-down cable as described in Chapter 7B.

4 Depress the retaining clips, and disconnect the wiring connectors from the throttle potentiometer, the electric heating element, the air temperature sensor and idle control stepper motor (as applicable).

5 Release the retaining clips (where fitted), and disconnect all the relevant vacuum and breather hoses from the throttle housing. Make identification marks on the hoses, to ensure that they are connected correctly on refitting.

6 Slacken and remove the retaining screws, and remove the throttle housing from the inlet manifold.

7 Remove the gasket from the manifold, and discard it - a new one must be used on refitting.

Bosch MP5.1 models

8 Slacken the retaining clip then disconnect the inlet duct from the throttle housing and recover the sealing ring.

9 Disconnect the accelerator inner cable from the throttle cam, then withdraw the outer cable from the mounting bracket, along with its flat washer and spring clip.

10 Depress the retaining clip and disconnect the wiring connector(s) from the throttle potentiometer, and, where necessary, from the electric heating element, and the air temperature sensor.

11 Slacken and remove the three retaining screws and remove the throttle housing from the inlet manifold. Recover the O-ring from manifold and discard it; a new one must be used on refitting.

Bosch MP3.1 models

12 Slacken the retaining clip, disconnect the intake duct from the end of the throttle housing, and recover the rubber sealing ring (where fitted).

13 Disconnect the accelerator inner cable from the throttle cam. Slacken and remove the bolt and nut securing the outer cable mounting bracket to the manifold, then withdraw the bracket. Remove the flat washer from the end of the cable for safekeeping. On models with automatic transmission, free the kick-down cable from the throttle cam.

14 Depress the retaining clip, and disconnect the wiring connector from the throttle potentiometer.

15 Relieve any pressure in the cooling system by unscrewing the filler cap. Slacken the retaining clips, and disconnect the two coolant hoses from the base of the throttle housing. Plug the hose ends, working quickly to minimise coolant loss.

16 Release the retaining clips (where fitted), and disconnect all the relevant vacuum and breather hose(s) from the throttle housing. Make identification marks on the hoses,

ensure they are correctly reconnected on refitting.

17 Undo the securing screws and remove the throttle housing from the manifold. Remove the O-ring from the manifold, and discard it - a new one must be used on refitting.

Bosch MP3.2 models

18 Slacken the retaining clip then disconnect the inlet duct from the throttle housing and recover the sealing ring.

19 Slacken the clip and detach the auxiliary air valve hose from the port at the front of the of throttle body.

20 Disconnect the accelerator inner cable from the throttle cam, then withdraw the outer cable from the mounting bracket, along with its flat washer and spring clip.

21 Depress the retaining clip and disconnect the wiring connector(s) from the throttle potentiometer, and, where necessary, from the electric heating element, and the air temperature sensor.

22 Slacken and remove the three retaining screws and remove the throttle housing from the inlet manifold. Recover the gasket from manifold and discard it; a new one must be used on refitting.

Refitting

23 Refitting is a reversal of the removal procedure, noting the following points:
a) Fit a new gasket to the manifold, then refit the throttle housing and securely tighten its retaining nuts or screws (as applicable).
b) Ensure that all hoses are correctly reconnected and, where necessary, are securely held in position by the retaining clips.
c) Ensure that all wiring is correctly routed, and that the connectors are securely reconnected.
d) On completion, adjust the accelerator cable as described in Section 3 and, where necessary, the kick-down cable as described in Chapter 7B.

13 Bosch Motronic MP5.1 system components - removal and refitting

Fuel rail and injectors

Note: *Refer to the warning note in Section 1 before proceeding.*
Note: *If a faulty injector is suspected, before condemning the injector, it is worth trying the effect of one of the proprietary injector-cleaning treatments.*

1 Disconnect the battery negative terminal.
2 Disconnect the vacuum pipe from the fuel pressure regulator then slacken and remove the retaining nut and bolt and release the wiring/hose retaining clip from the end of the fuel rail.
3 Bearing in mind the information given in Section 7, slacken the retaining clips and

disconnect the fuel feed and return hoses from the fuel rail. Where the original crimped-type Citroën hose clips are still fitted, cut them and discard; replace them with standard worm-type hose clips on refitting.

4 Depress the retaining tangs and disconnect the wiring connectors from the four fuel injectors.

5 Slacken and remove the fuel rail retaining bolts and nuts then carefully ease the fuel rail and injector assembly out from the inlet manifold and remove it from the vehicle. Remove the O-rings from the end of each injector and discard them; they must be renewed whenever they are disturbed.

6 Slide out the retaining clip(s) and remove the relevant injector(s) from the fuel rail. Remove the upper O-ring from each disturbed injector and discard; all disturbed O-rings must be renewed.

7 Refitting is a reversal of the removal procedure, noting the following points.
a) Fit new O-rings to all disturbed injector unions.
b) Apply a smear of engine oil to the O-rings to aid installation then ease the injectors and fuel rail into position ensuring that none of the O-rings are displaced.
c) On completion start the engine and check for fuel leaks.

Fuel pressure regulator

Note: *Refer to the warning note in Section 1 before proceeding.*

8 Disconnect the vacuum pipe from the regulator. Note that access to the regulator is poor with the fuel rail in position, if necessary, remove the fuel rail as described earlier then remove the regulator.

9 Place a wad of rag over the regulator, to catch any fuel spray which may be released, then remove the retaining clip and ease the regulator out from the fuel rail.

10 Refitting is a reversal of the removal procedure. Examine the regulator seal for signs of damage or deterioration and renew if necessary.

Throttle potentiometer

11 Disconnect the battery negative terminal.
12 Depress the retaining clip and disconnect the wiring connector from the throttle potentiometer.
13 Slacken and remove the two retaining screws then disengage the potentiometer from the throttle valve spindle and remove it from the vehicle.

14 Refitting is a reverse of the removal procedure ensuring that the potentiometer is correctly engaged with the throttle valve spindle.

Electronic Control Unit (ECU)

15 The ECU is located in a plastic box which is mounted on the right-hand front wheel arch.
16 To remove the ECU first disconnect the battery.
17 Unclip the cover from the box then lift the

retaining clip and disconnect the wiring connector from the ECU.
18 Lift the ECU from its mounting box.
19 If necessary undo the retaining screws and remove the mounting box.
20 Refitting is a reverse of the removal procedure ensuring that the wiring connector is securely reconnected.

Throttle stepper motor

21 The throttle stepper motor is mounted on the throttle body.
22 To remove it first disconnect the battery negative terminal.
23 Depress the retaining clip and disconnect the wiring connector from the motor.
24 Slacken and withdraw the retaining screws then remove the stepper motor from the throttle body.
25 Recover the O-ring seal and examine it carefully - renew it if it shows signs of wear or deterioration.
26 Refitting is a reversal of the removal procedure. Examine the mounting rubber for signs of deterioration and renew it if necessary.

Manifold absolute pressure (MAP) sensor

27 The MAP sensor is situated on the underside of the inlet manifold casting.
28 Unplug the wiring connector from the sensor. On early models, unscrew the sensor body from the inlet manifold. On later models, undo the screws and withdraw the MAP sensor from the manifold.
29 Recover the sealing gasket. Note that a new item must be used on refitting.
30 Refitting is the reverse of the removal procedure.

Coolant temperature sensor

31 Refer to Chapter 3.

Inlet air temperature sensor

32 The inlet air temperature sensor is screwed into the top of the air cleaner housing. To remove the sensor first disconnect the battery negative terminal.
33 Disconnect the wiring connector then unscrew the sensor and remove it from the vehicle.
34 Refitting is the reverse of removal.

Crankshaft sensor

35 The crankshaft sensor is situated on the front face of the transmission clutch housing.
36 To remove the sensor first disconnect the battery negative terminal.
37 Trace the wiring back from the sensor to the wiring connector and disconnect it from the main harness.
38 Prise out the rubber grommet then undo the retaining bolt and withdraw the sensor from the transmission.
39 Refitting is reverse of the removal procedure ensuring that the sensor retaining bolt is securely tightened and the grommet is correctly seated in the transmission housing.

Fuel injection system relay unit

40 The relay unit is mounted in the ECU casing, in the front right corner of the engine compartment.
41 To remove the relay unit, first remove the ECU as described earlier in this Section.
42 Disconnect the wiring connector and remove the relay unit from the mounting plate.
43 Refitting is the reverse of removal, ensuring that the relay unit is securely clipped in position.

14 Magneti Marelli system components - removal and refitting

Fuel rail and injectors

Note: *Refer to the warning note in Section 1 before proceeding. If a faulty injector is suspected, before condemning the injector, it is worth trying the effect of one of the proprietary injector-cleaning treatments.*
1 Disconnect the battery negative terminal.
2 Remove the air cleaner-to-throttle housing duct, referring to Section 2.
3 Disconnect the vacuum pipe from the fuel pressure regulator.
4 Release the retaining clip, and free the various hoses from the top of the fuel rail.
5 Bearing in mind the information given in Section 7, slacken the retaining clip, and disconnect the fuel feed and return hoses from the ends of the fuel rail. Where the original crimped-type Citroën hose clips are still fitted, cut them off and discard them; use standard worm-drive hose clips on refitting.
6 Depress the retaining clips, and unplug the wiring connectors from the four injectors.
7 Slacken and remove the three fuel rail retaining bolts, then carefully ease the fuel rail and injector assembly out from the inlet manifold, and remove it from the vehicle. Remove the O-rings from the end of each injector, and discard them; these must be renewed whenever they are disturbed.
8 Slide out the retaining clip(s), and remove the relevant injector(s) from the fuel rail. Remove the upper O-ring from each injector as it is removed, and discard it; all O-rings must be renewed once they have been disturbed.
9 Refitting is a reversal of the removal procedure, noting the following points:
a) *Fit new O-rings to all disturbed injectors.*
b) *Apply a smear of engine oil to the O-rings to aid installation, then ease the injectors and fuel rail into position, ensuring that none of the O-rings are displaced.*
c) *On completion, start the engine and check for fuel leaks.*

Fuel pressure regulator

10 Refer to Section 13.

Throttle potentiometer

11 Remove the throttle housing (Section 12).
12 Disconnect the wiring then undo the two retaining screws, and remove the potentiometer from the throttle housing.
13 On refitting, ensure that the potentiometer is correctly engaged with the throttle valve spindle, and securely tighten its screws.
14 Refit the throttle housing (see Section 12).

Electronic control unit (ECU)

15 The ECU is located in a plastic box which is mounted on the right-hand front wheel arch. To remove the ECU, first disconnect the battery negative terminal.
16 Unclip the lid from the plastic box, and disconnect the wiring connector from the ECU.
17 Slide the ECU out of the box and, if necessary, undo the retaining nuts and separate it from its mounting plate.
18 Refitting is the reverse of removal, ensuring that the wiring connector is securely reconnected.

Idle speed control stepper motor

19 The idle speed control stepper motor is located on the front of the throttle housing assembly. To remove the motor, first disconnect the battery negative terminal.
20 Release the retaining clip, and disconnect the wiring connector from the motor.
21 Slacken and remove the two retaining screws, and withdraw the motor from the throttle housing.
22 Refitting is a reversal of removal.

Manifold absolute pressure (MAP) sensor

23 The MAP sensor is situated on the right-hand front wheel arch. To remove the sensor, first disconnect the battery negative terminal.
24 Undo the three nuts, and free the sensor from the underside of the mounting bracket.
25 Depress the retaining clip, disconnect the wiring connector and vacuum hose from the sensor, and remove the sensor from the engine compartment.
26 Refitting is a reversal of removal.

Coolant temperature sensor

27 Refer to Chapter 3.

Inlet air temperature sensor

28 The inlet air temperature sensor is located in the base of the throttle housing.
29 To remove the sensor, first remove the throttle housing as described in Section 12, then undo the two retaining screws and remove the throttle potentiometer from the base of the housing.
30 Trace the wiring back from the sensor to its wiring connector, and remove the screw securing the connector to the throttle housing.
31 Carefully ease the sensor out of position, and remove it from the throttle housing. Examine the sensor O-ring for damage or deterioration, and renew if necessary.
32 Refitting is a reversal of removal, using a new O-ring where necessary.

4A

Crankshaft sensor

33 Refer to Section 13.

Fuel injection system relay unit

34 Refer to Section 13.

---

15 Bosch Motronic MP 3.2 system components - removal and refitting

Note: *Check parts availability with a Citroën dealer prior to removing individual components. At the time of writing, certain components are only available as part of a larger assembly - eg the throttle potentiometer is only available as part of the throttle housing assembly.*

Fuel rail and injectors

Note: *Refer to the warning note in Section 1 before proceeding. If a faulty injector is suspected, before condemning the injector, it is worth trying the effect of one of the proprietary injector-cleaning treatments.*

1 Disconnect the battery negative terminal.
2 Disconnect the vacuum pipe from the fuel pressure regulator.

3 Bearing in mind the information given in Section 7, slacken the retaining clips, and disconnect the fuel feed and return hoses from the either end of the fuel rail. Where the original crimped-type Citroën hose clips are still fitted, cut them off and discard them; use standard worm-drive hose clips on refitting.
4 Depress the retaining tangs, and disconnect the wiring connectors from the four injectors.
5 Slacken and remove the two fuel rail retaining bolts, then carefully ease the fuel rail and injector assembly out from the inlet manifold and remove it from the vehicle **(see illustration)**. Remove the O-rings from the end of each injector, and discard them; these must be renewed whenever they are disturbed.
6 Slide out the retaining clip(s), and remove the relevant injector(s) from the fuel rail. Remove the upper O-ring from each injector as it is removed, and discard it; all O-rings must be renewed once they have been disturbed.
7 Refitting is a reversal of the removal procedure, noting the following points:
a) *Fit new O-rings to all disturbed injectors.*
b) *Apply a smear of engine oil to the O-rings*

to aid installation, then ease the injectors and fuel rail into position, ensuring that none of the O-rings are displaced.
c) *On completion, start the engine and check for fuel leaks.*

Fuel pressure regulator

Note: *Refer to the warning note in Section 1 before proceeding.*

8 Disconnect the vacuum pipe from the regulator.
9 Place a wad of clean rag over the regulator, to catch any fuel spray which may be released. Remove the retaining clip, and slide the regulator out of the end of the fuel rail.
10 Refitting is a reversal of the removal procedure. Examine the regulator seal for signs of damage or deterioration, and renew if necessary.

Throttle potentiometer

11 Disconnect the battery negative terminal.
12 Depress the retaining clip, and disconnect the wiring connector from the throttle potentiometer.
13 Slacken and remove the two retaining screws, then disengage the potentiometer from the throttle valve spindle and remove it from the vehicle.
14 Refitting is a reverse of the removal procedure, ensuring that the potentiometer is correctly engaged with the throttle valve spindle.

Electronic Control Unit (ECU)

15 The ECU is located in the rear left-hand corner of the engine compartment. To remove the ECU, first disconnect the battery negative terminal.
16 Depress the retaining clip, and disconnect the wiring connector from the idle mixture adjustment potentiometer.
17 Slacken and remove the upper bolt securing the ECU mounting bracket, then loosen the lower bolt. Withdraw the bracket and ECU assembly from the engine compartment, disconnecting the wiring connector and vacuum pipe from the ECU as they become accessible.
18 With the assembly on the bench, undo the bolts securing the ECU to the bracket, and separate the two components.
19 Refitting is a reversal of the removal procedure, ensuring that the wiring connector and vacuum pipe are securely reconnected.

Knock sensor

20 The knock sensor is bolted to the front of the engine block.
21 Unplug the wiring from the sensor flying lead, then unbolt the sensor from the engine block. Recover the washer (where fitted).
22 Refitting is a reversal of removal, but observe the correct torque when refitting the sensor securing bolt, as this has an effect on the operation of the sensor.

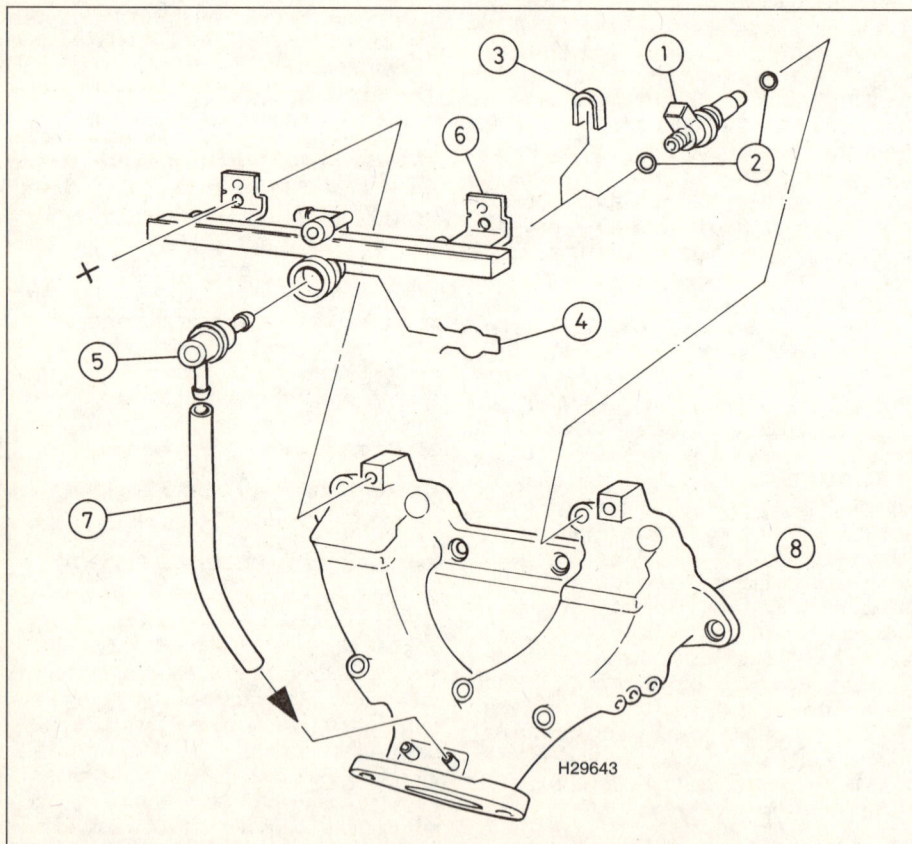

15.5 Fuel rail, pressure regulator and injector assembly - turbo models with Bosch Motronic MP 3.2 fuel injection

1 *Injector*
2 *O-ring seals*
3 *Injector retaining clips*
4 *Pressure regulator retaining clip*
5 *Fuel pressure regulator*
6 *Fuel rail*
7 *Vacuum hose*
8 *Inlet manifold*

Auxiliary air valve

23 The auxiliary air valve is mounted on the left-hand end of the engine, adjacent to the battery holder.

24 Depress the retaining clip, and disconnect the wiring connector from the air valve.

25 Slacken the retaining clips, and disconnect the hoses from either end of the auxiliary air valve.

26 Undo the two retaining bolts, and remove the auxiliary air valve from the engine compartment.

27 Refitting is a reversal of the removal procedure.

Manifold absolute pressure (MAP) sensor

28 The MAP sensor is an integral part of the electronic control unit (ECU).

Coolant temperature sensor

29 Refer to Chapter 3.

Intake air temperature sensor

30 The intake air temperature sensor is located in the throttle housing. To remove the sensor, first disconnect the battery negative terminal.

31 Disconnect the wiring connector, then unscrew the sensor from the throttle housing and remove it from the vehicle.

32 Refitting is the reverse of removal.

Crankshaft sensor

33 The crankshaft sensor is mounted on top of the transmission housing, next to the left-hand end of the cylinder block. To remove the sensor, first disconnect the battery negative terminal.

34 Access to the sensor is poor, and it will be necessary to remove the battery and battery tray and/or the intake duct assembly to improve access (depending on model and specification). On some models, it will also be necessary to remove the metal plate from the top of the transmission housing; the plate is retained by one of the engine-to-transmission bolts, and by a second bolt securing the plate to the top of the transmission.

35 Trace the wiring back from the sensor to its wiring connector, and disconnect it from the main wiring harness. Undo the retaining bolt, and remove the sensor from the top of the transmission housing.

36 Refitting is a reversal of the removal procedure, ensuring that the sensor wiring is correctly routed.

Fuel injection system relay unit

37 The fuel injection system relay unit is mounted on the front of the engine compartment junction box. To remove the relay unit, first disconnect the battery negative terminal.

38 Open up the junction box lid, then slacken and remove the relay mounting nut and washer. Release the retaining clip, then disconnect the wiring connector and remove the relay unit from the engine compartment.

39 Refitting is the reverse of removal.

16 Bosch LE2 Jetronic system components - removal and refitting

Fuel pressure regulator

Note: *Refer to the warning note in Section 1 before proceeding.*

1 Disconnect the vacuum pipe from the regulator.

2 Place a wad of clean rag over the regulator, to catch any fuel spray which may be released. Slacken and remove the hose clips, then remove the fuel return hose from the regulator. Have a small container ready to catch any fuel spills.

3 Slacken and withdraw the screws, then pull the regulator from the fuel rail.

4 Refitting is a reversal of the removal procedure. Renew the regulator-to-fuel rail O-ring seal as a matter of course.

Throttle position switch

5 Disconnect the battery negative terminal.

6 Depress the retaining clip, and disconnect the wiring connector from the throttle potentiometer. Mark the relationship between the switch body and its mounting bracket.

7 Slacken and remove the two retaining screws, then disengage the switch from the throttle valve spindle and remove it from the vehicle **(see illustration)**.

8 Refitting is a reverse of the removal procedure, ensuring that the potentiometer is correctly engaged with the throttle valve spindle. Tighten the switch mounting screws when the switch body is aligned with the markings made during removal.

Electronic Control Unit (ECU)

9 The ECU is located in the front right-hand corner of the engine compartment in plastic case. To remove the ECU, first disconnect the battery negative terminal.

10 Remove the cover from the ECU casing.

11 Withdraw the ECU assembly from the casing, disconnecting the wiring connector and vacuum pipe (where applicable) from the ECU as they become accessible.

12 Refitting is a reversal of the removal procedure, ensuring that the wiring connector and vacuum pipe are securely reconnected.

Air flow meter

13 Disconnect the battery negative cable from the terminal.

14 Unplug the wiring connector from the air flow meter at the connector.

15 Slacken the large hose clips and detach the intake ducting from either side of the air flow meter housing.

16 Remove the bolts and lift the air flow meter away from its mounting bracket.

17 Refitting is a reversal of the removal procedure, ensuring that the wiring connector and intake duct clips are securely reconnected.

Auxiliary air valve

18 The auxiliary air valve is mounted on the left-hand end of the engine, adjacent to the battery holder.

19 Depress the retaining clip, and disconnect the wiring connector from the air valve.

20 Slacken the retaining clips, and disconnect the air hoses from either end of the auxiliary air valve.

21 Slacken the retaining bolts, and slide the auxiliary air valve from its mounting bracket and remove it from the engine compartment.

22 Refitting is a reversal of the removal procedure.

Fuel rail and injectors

23 Refer to Section 7 and depressurise the fuel system.

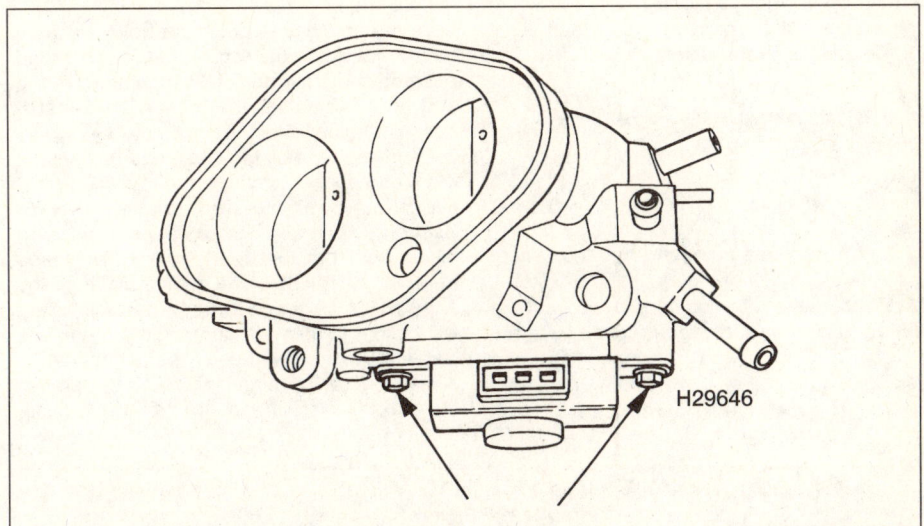

16.7 Remove the throttle position switch mounting screws (arrowed) - models with Bosch LE2 Jetronic fuel injection

4A

24 Disconnect the negative battery cable and position it away from the terminal.

25 Remove the fuel pressure regulator as described earlier in this section.

26 Unplug the wiring from each of the fuel injectors and the connectors.

27 Slacken the clip and disconnect the fuel supply hose from the fuel rail. Be prepared for some fuel spillage.

28 Remove the screws and carefully lift the fuel rail away from the manifold. The injectors will remain connected to the rail - recover the lower injector O-ring seals from the manifold **(see illustration)**.

29 To remove an injector, slide the metal securing clip to one side and pull the injector from the fuel rail. Recover the Upper O-ring seal.

30 Refitting is a reversal of the removal procedure, noting the following points:

a) *Fit new O-rings to all disturbed injectors.*

b) *Apply a smear of engine oil to the O-rings to aid installation, then ease the injectors and fuel rail into position, ensuring that none of the O-rings are displaced.*

c) *On completion, start the engine and check for fuel leaks.*

Coolant temperature sensor

31 Refer to Chapter 3.

Inlet air temperature sensor

32 The intake air temperature sensor is an integral part of the air flow meter assembly and cannot be renewed separately.

Fuel pump relay

33 The relay is mounted at the front right hand corner of the engine compartment, inside the ECU casing.

34 To remove the relay, ensure that the battery negative cable is disconnected, the remove the relay securing screw (where applicable). Pull the relay from its base.

35 Refitting is a reversal of removal.

Manifold Absolute Pressure (MAP) sensor

36 The MAP sensor is situated on the right-hand front wheel arch. To remove the sensor, first disconnect the battery negative terminal.

37 Undo the three nuts, and free the sensor from the underside of the mounting bracket.

38 Depress the retaining clip, disconnect the wiring connector and vacuum hose from the sensor, and remove the sensor from the engine compartment.

39 Refitting is a reversal of removal.

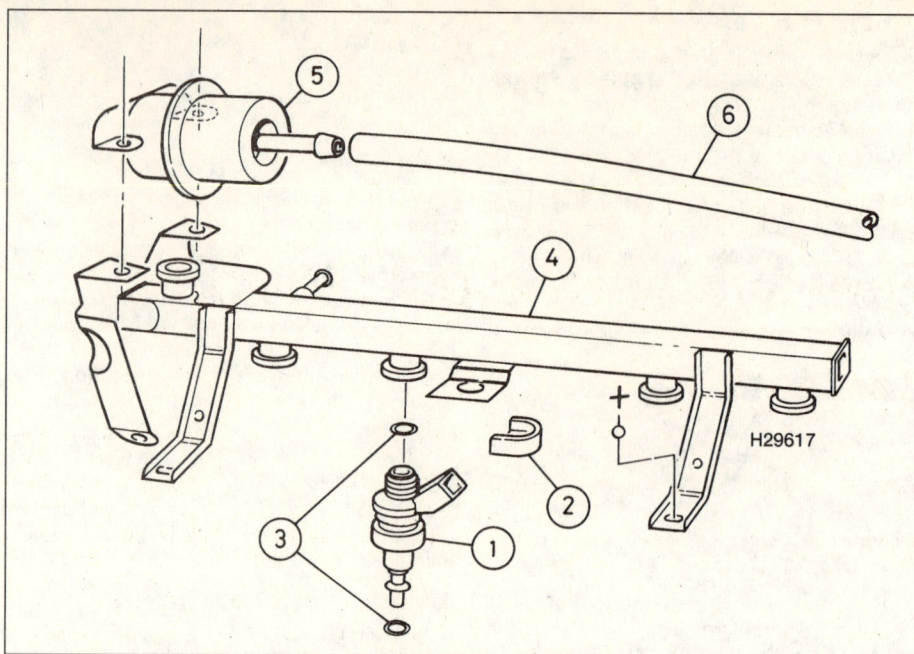

16.28 Fuel rail, pressure regulator and injector assembly - models with Bosch LE2 Jetronic fuel injection

1 *Injector*
2 *Injector retaining clip*
3 *O-ring seals*
4 *Fuel rail*
5 *Fuel pressure regulator*
6 *Vacuum hose*

17 Inlet manifold - removal and refitting

Removal

1 Disconnect the battery negative terminal and proceed as described under the relevant sub-heading.

2 Remove the air cleaner-to-throttle housing duct as described in Section 2.

3 Remove the throttle housing as described in Section 12.

4 Where applicable, undo the bolts securing the wiring tray to the top of the manifold, and position the tray, and its associated wiring and hoses, clear of the manifold so that it does not hinder removal.

5 Depress the retaining clips, and disconnect the wiring connectors from the four injectors.

6 Bearing in mind the information given in Section 7, slacken the retaining clips, and disconnect the fuel feed and return hoses from the either side of the manifold. Where the original crimped-type Citroën hose clips are still fitted, cut them off and discard them; use standard worm-drive hose clips on refitting.

7 Slacken the retaining clip(s), and disconnect the braking system vacuum servo unit hose, and all the relevant vacuum/breather hoses, from the top of the manifold. Where necessary, make identification marks on the hoses, to ensure that they are correctly reconnected on refitting.

8 Where applicable, slacken and remove the bolt securing the dipstick tube to the side of the manifold.

9 Undo the manifold retaining nuts, and withdraw the manifold from the engine compartment. Recover the manifold seal(s), and discard them - new ones must be used on refitting.

Refitting

10 Refitting is a reverse of the relevant removal procedure, noting the following points:

a) *Ensure that the manifold and cylinder head mating surfaces are clean and dry, then locate the new seals in their recesses in the manifold. Refit the manifold and tighten its retaining nuts and bolts to the specified torque.*

b) *Ensure that all relevant hoses are reconnected to their original positions and are securely held (where necessary) by the retaining clips.*

c) *Adjust the accelerator cable as described in Section 3 then, where necessary, adjust the kick-down cable (Chapter 7B).*

Chapter 4 Part B:
Fuel system - diesel models

Contents

Degrees of difficulty

Easy, suitable for novice with little experience	**Fairly easy,** suitable for beginner with some experience	**Fairly difficult,** suitable for competent DIY mechanic	**Difficult,** suitable for experienced DIY mechanic	**Very difficult,** suitable for expert DIY or professional

4B

Specifications

General

Type . . . .   Rear mounted fuel tank, distributor type fuel injection pump with integral lift pump, naturally aspirated and turbocharged/intercooled variants, indirect injection (mechanically or electronically controlled, depending on specification).

Application:
   2.1 litre non-turbo models (XUD11A engine) . . . .   Lucas Roto Diesel DPC 061 or Bosch VE 533
   2.1 litre turbo models:
     Early models (XUD11ATE engine) . . . .   Bosch VE 532, VE 531 or Lucas Roto-Diesel 062
     Later models (XUD11BTE engine) . . . .   Lucas/PSA EPIC electronic fuel injection
   2.5 litre turbo models (DK5ATE engine) . . . .   Bosch VP36 electronic fuel injection
Firing order . . . .   1-3-4-2 (No 1 at flywheel/driveplate end)

Injection pump

Static timing:
   Bosch VE531 . . . .   0.88 mm ATDC on No 4 cylinder
   Bosch VE533 . . . .   0.76 mm ATDC on No 4 cylinder
   Bosch VE532 . . . .   0.84 mm ATDC on No 4 cylinder
   Lucas Roto-Diesel 062 . . . .   (Stamped on pump operating lever)
   Lucas Roto-Diesel 061 . . . .   (Stamped on pump operating lever)
   Lucas EPIC . . . .   Static and dynamic timing managed by ECU*
   Bosch VP36 . . . .   Static and dynamic timing managed by ECU*
*Not adjustable

Injectors

Type	Single stage pintle
Opening pressure:	
With Bosch VE531 pump	175 bar
With Bosch VE533 pump	140 bar
With Bosch VE532 pump	150 bar
With Lucas Roto-Diesel 062 pump	150 bar
With Lucas Roto-Diesel 061 pump	130 bar
With Lucas EPIC electronic fuel injection	163.5 ± 3.5 bar
Bosch VP36 electronic fuel injection	172.5 ± 2.5 bar

Torque wrench settings

	Nm	lbf ft
Stop solenoid	20	15
Fuel pipe union nuts	25	18
Injection pump timing hole blanking plug:		
Lucas pump	6	4
Bosch pump	15	11
Injection pump mounting nuts/bolts	20	15
Injection pump sprocket nut	50	37
Injectors to cylinder head		
Bosch VP36	55	41
All other models	90	66

1 General information and precautions

General information

Up to 1994, two diesel models were available, both using variants of the XUD11 engine. The naturally aspirated 2138 cc engine and the turbocharged 2088 cc engines all employed mechanical fuel injection, using a distributor-type fuel injection pump, manufactured either by Bosch or Lucas. From mid-1995 onwards, the turbocharged variant was fitted with an electronic fuel injection control system (Lucas EPIC). Additionally, in 1994, the range was extended by the introduction of a turbocharged 2445 cc engine, which used Bosch VP36 electronic diesel engine management, incorporating a VE-type distributor fuel injection pump.

The fuel system itself consists of a rear-mounted fuel tank, a fuel filter with integral water separator, an electronically regulated fuel injection pump, injectors and associated components. Before entering the filter, the fuel passes through a heater core, mounted on the left hand end of the cylinder head and is heated by coolant flowing through the cylinder head. An air-to-air intercooler is fitted to 2.1 litre turbocharged models, and an air-to-water intercooler is used on the 2.5 litre models.

Fuel is drawn from the fuel tank to the fuel injection pump by a lift pump incorporated in the fuel injection pump. Before reaching the pump, the fuel passes through a fuel filter, where foreign matter and water are removed. Excess fuel lubricates the moving components of the pump, and is then returned to the tank.

The fuel injection pump is driven at half-crankshaft speed by the timing belt. The high pressure required to inject the fuel into the compressed air in the swirl chambers is achieved by a cam and piston arrangement in the pump. The fuel passes through a central rotor with a single outlet drilling which aligns with ports leading to the injector pipes.

Fuel metering on mechanical injection pumps is controlled by a centrifugal governor, which reacts to accelerator pedal position and engine speed. The governor is linked to a metering valve, which increases or decreases the amount of fuel delivered at each pumping stroke. A separate device also increases fuel delivery with increasing boost pressure.

Basic injection timing is determined when the pump is fitted. When the engine is running, it is varied automatically by an internal mechanism within the pump, acting under the control of the system ECU.

The four fuel injectors produce a homogeneous spray of fuel into the swirl chambers located in the cylinder head. The injectors are calibrated to open and close at critical pressures to provide efficient and even combustion. Each injector needle is lubricated by fuel, which accumulates in the spring chamber and is channelled to the injection pump return hose by leak-off pipes.

The Lucas EPIC system provides programmed electronic control of the fuel injection pump, and electronic control of the exhaust gas recirculation system, via the EPIC electronic control module (ECU). For the ECU to assess fuel system requirements under all operating conditions, sensors are provided to monitor accelerator pedal position, manifold absolute pressure, crankshaft position/speed, engine coolant temperature and intake air temperature. Operation of the EGR valve is also controlled by the ECU in conjunction with a solenoid valve.

The primary functions of the Bosch VP36 system are:

a) *Electronic control of the fuel injection timing and duration*

b) *Engine idle speed control*
c) *Exhaust gas recirculation (EGR) system control*
d) *Preheater glow plug control.*

Secondary functions include cruise control system operation and anti-theft system control. The system is managed by an electronic control unit (ECU) which receives inputs from sensors which primarily monitor engine load, engine speed and crankshaft position, injector operation, engine coolant temperature, intake air temperature, brake and clutch pedal position and atmospheric pressure. Further information on the various sensors and actuators are given in the relevant Sections later in this Chapter.

A unique feature of the Lucas EPIC and Bosch VP36 systems is the 'drive-by-wire' throttle control. Instead of the accelerator cable being connected to the fuel injection pump, as it is in the normal mechanical system, the cable is connected to a pedal position sensor. This sensor sends pedal position signals to the ECU, which in turn controls the fuel injection pump electronically.

Cold starting is assisted by preheater or 'glow' plugs fitted to each swirl chamber, and a coolant-filled, thermostatically controlled fuel heater, mounted on the left hand end of the cylinder head. On the Lucas EPIC and Bosch VP36 systems, the fast idle is automatically regulated by the injection pump. The fast idle function is also regulated by the system ECU. A stop solenoid cuts the fuel supply to the injection pump rotor when the ignition is switched off.

Provided that the specified maintenance is carried out, the fuel injection equipment will give long and trouble-free service. The injection pump itself may well outlast the engine. The main potential cause of damage to the injection pump and injectors is dirt or water in the fuel.

2.2a Hand operated fuel system priming bulb - 2.1 litre models with EPIC electronic fuel injection

2.2b Hand operated fuel system priming pump - 2.5 litre models and 2.1 litre models with mechanical fuel injection

2.3 Loosen the fuel system bleed screw, at the side of the filter housing

Servicing of the injection pump and injectors is very limited for the home mechanic, and any dismantling or adjustment other than that described in this Chapter must be entrusted to a Citroën dealer or fuel injection specialist.

Precautions

⚠️ **Warning: It is necessary to take certain precautions when working on the fuel system components, particularly the fuel injectors. Before carrying out any operations on the fuel system, refer to the precautions given in 'Safety first!' at the beginning of this manual, and to any additional warning notes at the start of the relevant Sections.**

When working on the Lucas EPIC or Bosch VP36 system, the following additional precautions should be observed.

a) *Always disconnect the battery negative lead before removing any of the electronic control system's electrical connectors.*
b) *When installing a battery, be particularly careful to avoid reversing the positive and negative battery leads.*
c) *Do not subject any components of the system (especially the ECU) to severe impact during removal or installation.*
d) *Never attempt to work on the ECU, to test it (with any kind of test equipment), or to open its cover.*
e) *If you are inspecting electronic control system components during rainy weather, make sure that water does not enter any part. When washing the engine compartment, do not spray these parts or their electrical connectors with water.*

2 Fuel system - priming and bleeding

1 After disconnecting part of the fuel supply system it will necessary to prime the system and bleed off any air which may have entered the system components. This procedure should also be followed after running out of fuel.

2 All models are fitted with a hand-operated priming pump. On models with Lucas EPIC electronic fuel injection, this is the form of a rubber bulb, mounted at the right hand side of the engine compartment. On all other models, the priming pump is located on the top of the fuel filter housing, at the front left hand side of the engine compartment, next to the battery **(see illustrations)**.
3 To prime the system, loosen the bleed screw in the fuel supply union, at the side of the filter housing **(see illustration)**.
4 Pump the priming plunger (or squeeze the rubber priming bulb, as applicable) until fuel free from air bubbles emerges from the bleed screw. Retighten the bleed screw mid-stroke, to ensure that air is not drawn back in.
5 Switch on the ignition to activate the injection pump stop solenoid - this will allow any air bubbles present to be forced through the injection pump. Continue operating the priming pump until firm resistance is felt, then pump a few more times.
6 Crank the engine on the starter motor, with the accelerator pedal pressed down through three quarters of its travel, until the engine starts. If the engine does not fire within fifteen seconds or so, or will not idle smoothly, wait fifteen seconds then continue cranking in ten second bursts. If the engine still does not run, repeat the steps in paragraphs 3 to 5 inclusive.
7 If a large volume of air has entered the pump, place a wad of rag around the fuel return union on the pump (to absorb spilt fuel), then slacken the union. Operate the priming pump (with the ignition switched on to activate the stop solenoid), or crank the engine on the starter motor in 10 second bursts, until fuel free from air bubbles emerges from the fuel union. Tighten the union and mop up spilt fuel.

⚠️ **Warning: Be prepared to stop the engine if it should fire, to avoid excessive fuel spray and spillage.**

8 If air has entered the injector pipes, place plenty of rag wadding around the injector pipe unions at the injectors (to absorb spilt fuel),

then slacken the unions. Crank the engine on the starter motor until fuel emerges from the unions, then stop cranking the engine and retighten the unions. Mop up spilt fuel. Refer to the warning given in the previous paragraph - the fuel in the injector lines is under extremely high pressure.
9 Start the engine with the accelerator pedal fully depressed. Additional cranking may be necessary to finally bleed the system before the engine starts.

3 Air cleaner assembly and intake ducts - removal and refitting

Removal
Air cleaner

1 Refer to Chapter 1B and remove the air filter element.
2 Slacken the retaining clips and disconnect the intake duct from the air cleaner housing lid.
3 Slacken and withdraw the air cleaner housing securing screws. Lift the housing body upward to disengage it from the lower locating lugs. On some models, as the housing is lifted up, it will be necessary to disengage a small plastic retaining tag at the front securing the housing to the cold air intake duct underneath **(see illustrations)**.

3.3a Removing the air cleaner housing

4B

3.3b On some models, as the housing is lifted up, it will be necessary to disengage a small plastic retaining tag at the front securing the housing to the cold air intake duct underneath

Intake ducts

4 On all engine types the intake ducting is complex arrangement consisting of flexible hoses and rigid ducts connecting the air cleaner assembly and, where applicable, intercooler and turbo-charger. The intake ducts pass over the top and rear of the engine and, on 2.1 litre models, across the top of the transmission.

5 To remove a section of intake ducting, slacken the retaining clips at each end and undo the bolts securing the relevant duct to its mounting bracket or support. On 2.1 litre models, when removing the front duct over the engine, it will be necessary to disconnect the inlet air temperature sensor wiring connector.

6 Release the ends of the duct then work it from its location.

Refitting

7 Refitting is the reverse of the relevant removal procedure.

4 Accelerator cable - removal, refitting and adjustment

Removal

1 On all 2.1 litre non-turbo and early 2.1 litre turbo models, operate the pump control lever on the fuel injection pump, and release the inner cable from the accelerator lever. Pull out the clip and withdraw the outer cable from the grommet in the fuel injection pump bracket **(see illustration)**.

2 On later 2.1 litre turbo and all 2.5 litre turbo models, release the inner cable from the lever on the accelerator pedal position sensor, located on the right-hand side of the engine compartment, behind the ECU casing (right hand drive models) or at the rear of the engine compartment, next to the hydraulic fluid reservoir (left hand drive models). Remove the clip and pull the outer cable from the

4.1 Pulling the accelerator cable outer from the fuel injection pump bracket

grommet in the pedal position sensor bracket.

3 On all models, release the cable from the remaining clips and brackets in the engine compartment, noting its routing.

4 Working from inside the vehicle, release the retaining clips, and remove the panel from underneath the driver's side of the facia panel.

5 Remove the clip (where applicable) and unhook the inner cable from the top of the accelerator pedal.

6 Release the outer cable from its retainer on the pedal mounting bracket, then tie a length of string to the end of the cable.

7 Return to the engine compartment, release the cable grommet from the bulkhead and withdraw the cable. When the end of the cable appears, untie the string and leave it in position - it can then be used to draw the cable back into position on refitting.

Refitting

8 Tie the string to the end of the cable, then use the string to draw the cable into position through the bulkhead. Once the cable end is visible, untie the string, then clip the outer cable into its pedal bracket retainer, and clip the inner cable into position in the pedal end.

9 Check that the cable is securely retained, then refit the lower panel to the facia.

10 Within the engine compartment, ensure the outer cable is correctly seated in the bulkhead grommet, then work along the cable, securing it in position with the retaining clips and ties, and ensuring that the cable is correctly routed.

11 Slide the flat washer onto the cable end, and refit the spring clip.

12 Pass the outer cable through the grommet on the injection pump bracket or pedal position sensor bracket and reconnect the inner cable to the lever. Adjust the cable as described below.

Adjustment

13 Remove the spring clip from the accelerator outer cable. Ensuring the control lever or pedal position sensor lever is against its stop, gently pull the cable out of its grommet until all free play is removed from the inner cable.

14 With the cable held in this position, refit the spring clip to the last exposed outer cable

groove in front of the rubber grommet and washer. When the clip is refitted and the outer cable is released, there should be only a small amount of free play in the inner cable.

15 Have an assistant depress the accelerator pedal, and check that the control lever or sensor lever opens fully and returns smoothly to its stop.

16 On models with automatic transmission, once the accelerator cable is correctly adjusted, check the kick-down cable adjustment as described in Chapter 7B.

5 Accelerator pedal - removal and refitting

Refer to the information given in Chapter 4A

6 Fuel gauge sender and pick-up unit - removal and refitting

The fuel gauge sender and pick-up unit is in the same position as the fuel pump on petrol models, and removal/refitting is similar. Refer to the information given in Chapter 4A.

7 Fuel tank - removal and refitting

Refer to the information given in Chapter 4A.

8 Fuel injection pump - adjustments

Note: *The following procedure only applies to the Bosch and Roto-Diesel fuel injection pumps fitted to 2.1 litre non-turbo and early 2.1 litre turbo models (see Specifications for details). Idle, fast idle and anti-stall speed settings on the Lucas EPIC and Bosch VP36 systems are controlled by the ECU.*

1 The usual type of tachometer (rev counter), which works from ignition system pulses, cannot be used on diesel engines. A diagnostic socket is provided for the use of Citroën test equipment, but this will not normally be available to the home mechanic. If it is felt that adjusting the idle speed by means of the vehicle's dashboard tachometer will not be satisfactory, it will be necessary to purchase or hire an appropriate tachometer, or else leave the task to a Citroën dealer or other suitably equipped specialist.

2 Check the accelerator cable adjustment as

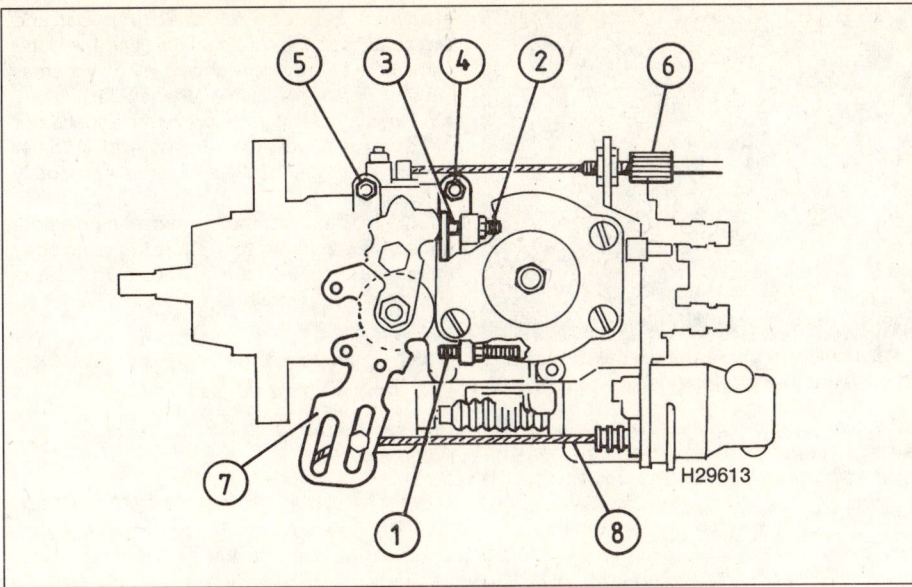

8.2a Bosch fuel injection pump - adjustment points

1 Maximum speed adjustment screw
2 Anti-stall adjustment screw
3 Shim for anti-stall adjustment
4 Idle speed adjustment screw
5 Fast idle speed adjustment screw
6 Fast idle cable adjustment nut
7 Accelerator lever
8 Accelerator cable

follows, but before making any adjustments to the fuel injection pump **(see illustrations)**:

a) Have an assistant depress the accelerator pedal to the floor and hold it there.

b) Check that the control lever on the injection pump is in contact with the maximum speed adjustment screw **(refer to illustration 8.2)**. If this is not the case, remove the spring clip from the accelerator outer cable and reposition the

cable as necessary. Refit the spring clip to the last exposed outer cable groove.

c) Release the accelerator cable and check that the control lever is in contact with the anti-stall adjustment screw.

Fast idle speed adjustment

Note: *The engine must be cold before you carry out this adjustment.*

3 With the engine stopped, check that the fast

idle lever is in contact with the fast idle adjustment screw. If necessary, slacken the locknut and turn the fast idle cable knurled adjustment nut, located on the fast idle control diaphragm bracket **(refer to illustration 8.2)**. Tighten the locknut after adjustment.

4 Locate the thermostatic sensor, which is mounted in the coolant outlet elbow, and note the position of the control cable leading from it. Start the engine and allow it to warm up. As the coolant temperature rises, the wax capsule inside the sensor should expand, allowing the control cable to extend. The total change in the length of the control cable should be greater than 6 mm.

Note: *Before proceeding with the following adjustments, warm up the engine to normal operating temperature, ensuring that the radiator cooling fan has operated at least twice.*

Idle speed adjustment

5 Loosen the locknut, and unscrew the anti-stall adjustment screw until it is clear of the pump control lever.

6 Loosen the locknut and turn the idle speed adjustment screw as required, then retighten the locknut **(refer to illustration 8.2)**. On completion, check the anti-stall speed adjustment.

Anti-stall speed adjustment

7 Insert a 4.0 mm shim or feeler blade between the pump control lever and the anti-stall adjustment screw **(refer to illustration 8.2)**.

8 Start the engine and allow it to idle. The engine speed should be as specified for the anti-stall speed.

9 If adjustment is necessary, loosen the locknut and turn the anti-stall adjustment screw as required. Retighten the locknut.

10 Remove the shim or feeler blade and allow the engine to return to idle.

4B

9 Stop solenoid - description, removal and refitting

Caution: *Be careful not to allow dirt into the injection pump during this procedure.*
Note: *The following procedure does not apply to models with Bosch VP36 or LUCAS EPIC fuel injection (refer to Specifications for application details).*

Description

1 The stop solenoid is located on the end of the fuel injection pump. Its purpose is to cut the fuel supply when the ignition is switched off. If an open-circuit occurs in the solenoid or supply wiring, it will be impossible to start the engine, as the fuel will not reach the injectors. The same applies if the solenoid plunger jams in the 'stop' position. If the solenoid jams in the 'run' position, the engine will not stop when the ignition is switched off.

2 If the solenoid has failed and the engine will not run, a temporary repair may be made by removing the solenoid as described in the

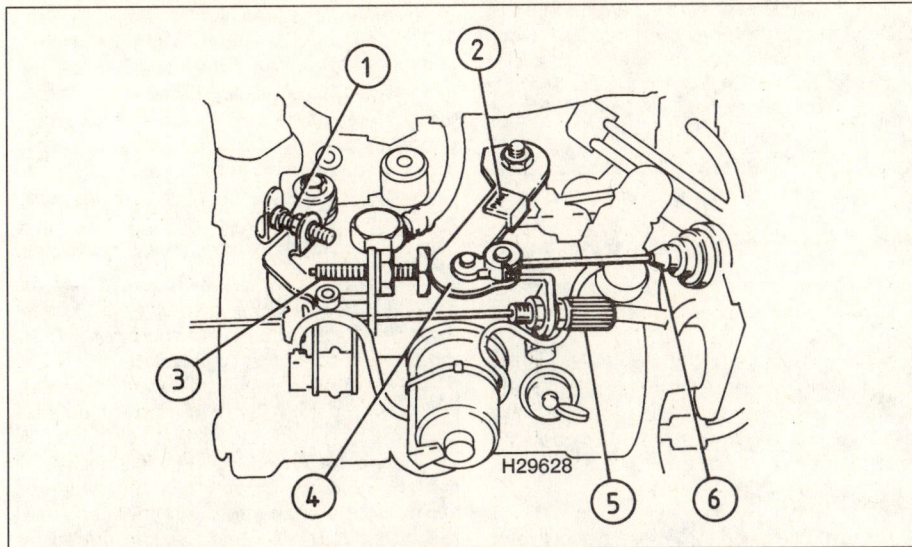

8.2b Lucas fuel injection pump - adjustment points (shown with accelerator damper removed for clarity)

1 Idle speed adjustment screw
2 Static injection timing stamped here
3 Anti-stall adjustment screw
4 Accelerator lever
5 Fast idle adjustment screw
6 Accelerator cable

9.4 Removing the stop solenoid wiring cover

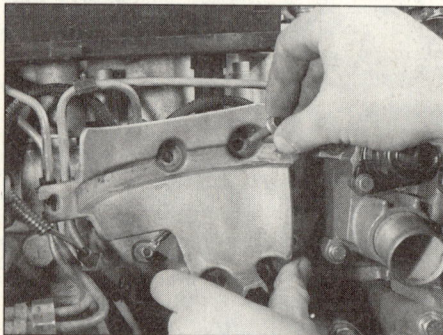

10.4a At the coolant expansion tank, slacken the lock screw . . .

10.4b . . . release the collar and lift off the tank

following paragraphs. Refit the solenoid body without the plunger and spring. Tape up the wire so that it cannot touch earth. The engine can now be started as usual, but it will be necessary to use the manual stop lever on the side of the fuel injection pump (or to stall the engine in gear) to stop it.

Removal

3 Disconnect the battery negative terminal.
4 Withdraw the rubber boot (if fitted), then unscrew the terminal nut and detach the wire from the top of the solenoid (see illustration).
5 Carefully clean around the solenoid, then unscrew and withdraw the solenoid, and recover the sealing washer or O-ring (as applicable). Recover the solenoid plunger and

spring if they remain in the pump. Operate the hand-priming pump (see Section 2) as the solenoid is removed, to flush away any dirt.

Refitting

6 Refitting is a reversal of removal, using a new sealing washer or O-ring and tightening the solenoid to the specified torque setting.

10 Bosch VP36 fuel injection pump - removal and refitting

Caution: Be careful not to allow dirt into the injection pump or injector pipes during this procedure. New sealing rings should

be used on the fuel pipe banjo unions when refitting.

Removal

1 Disconnect the battery negative terminal.
2 Raise the front of the car and support it securely on axle stands, so that the roadwheels are just clear of the ground.
3 Slacken and withdraw the fixings and remove the cover panel from the top of the engine.
4 With reference to Chapter 3, slacken the lock screw, release the collar and lift the coolant expansion tank from its mountings. Move it to one side, without disconnecting the coolant pipes. Unbolt the support bracket from the cylinder head (see illustrations).
5 Release the hose clips and disconnect the fuel supply and return hoses from the fuel injection pump ports. Cover the open end of the hose and the injection pump ports to keep dirt out (see illustration).
6 Remove the screws and lift the wiring harness/hydraulic hose support rail away from the cylinder head (see illustration).
7 Disconnect all wiring from the pump, including the wiring for the needle lift sensor, integral with No 3 injector. Label each cable to aid correct refitting. The main pump wiring connector is located on the right hand side of the engine compartment. To separate its two halves, first remove the plastic cover, then release the locking bar by sliding it to one side using the tip of a screwdriver (see illustrations).

10.4c Unbolt the cylinder head support bracket from the cylinder head

10.5 Release the hose clips and disconnect the fuel supply and return hoses from the fuel injection pump ports

10.6 Remove the wiring harness/hydraulic hose support rail from the cylinder head

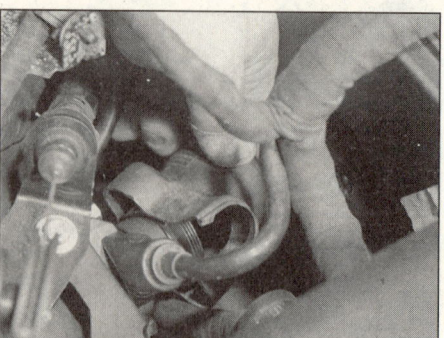

10.7a At the main pump wiring connector, first remove the plastic cover . . .

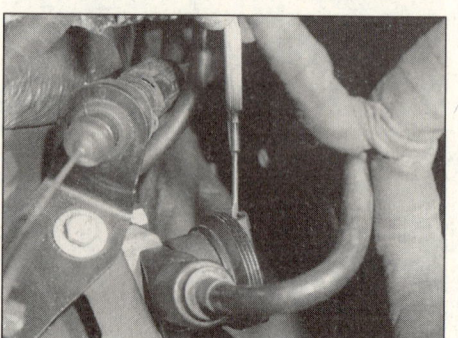

10.7b . . . then release the locking bar by sliding it to one side using the tip of a screwdriver . . .

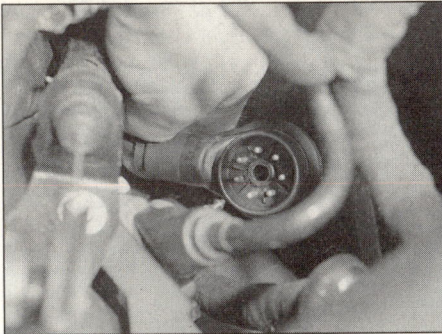

10.7c . . . then separate the two halves

10.8 Slacken the unions and remove the injector pipes from the rear of the injection pump

10.13a Unscrew the three fuel injection pump front mounting screws . . .

8 Unscrew the union nuts securing the injector pipes to the fuel injection pump and injectors. Counterhold the unions on the pump, while unscrewing the pipe-to-pump union nuts. Remove the pipes as a set. Cover open unions to keep dirt out (see illustration).
9 Remove the upper timing belt cover, with reference to Chapter 2B.
10 Still working from Chapter 2B, section 12, set the engine to TDC on cylinder No 4 and lock it in position as described, using the timing belt camshaft sprocket and flywheel locking pins.
11 Remove the timing belt from the fuel injection pump sprocket, as described in Chapter 2B.
12 Unbolt the injection pump sprocket from the pump shaft, as described in Chapter 2B.
13 Unscrew the three fuel injection pump front mounting screws. Unscrew and remove the rear mounting nut and bolt, then carefully withdraw the pump from its mounting bracket (see illustrations).

Refitting

14 Offer up the injection pump to the mounting bracket, and refit the pump mounting screws, nuts and bolts tightening them to the specified torque.
15 Refit the injection pump sprocket to the pump shaft, as described in Chapter 2B. Insert and lightly tighten the sprocket securing bolts.

16 Still working from Chapter 2B, refit the timing belt, then check the timing belt tension and the injection pump/valve timing. Note that injection timing is set by the fuel system ECU; the position of the injection pump in relation to the engine cannot be adjusted to alter the injection timing.
17 The remainder of the refitting procedure is a reversal of removal, noting the following points:
a) Use new worm drive hose clips when reconnecting the fuel supply and return hoses, if the original clips were of the crimped type (see illustration).
b) Reconnect all electrical wiring according to the notes made during removal.
c) On completion, prime and bleed the fuel system as described in Section 2.

11 Lucas EPIC fuel injection pump - removal and refitting

Caution: Be careful not to allow dirt into the injection pump or injector pipes during this procedure. New sealing rings should be used on the fuel pipe banjo unions when refitting.

Removal

1 Disconnect the negative cable from the battery terminal.

2 Jack up the front of the vehicle and support it securely on axle stands. Remove the right hand road wheel.
3 Extract the fixings and remove both the engine compartment undertray and right hand wheel arch liner.
4 Remove the fuel injection system electronic control unit (ECU) and the ECU module box as described in Section 17.
5 Remove the intake ducting as described in Section 3 and the upper section of the inlet manifold as described in Section 18.
6 Remove the auxiliary drivebelt, with reference to Chapter 1B.
7 Remove the two nuts and unbolt the auxiliary drive belt guide roller from the side of the engine. Undo the screws and remove the belt casing.
8 Undo the bolts securing the wiring harness carrier to the engine and move the carrier clear of the injection pump. According to equipment fitted, it may be necessary to disconnect specific wiring connectors to enable the harness carrier to be moved sufficiently. Label all disconnected wiring to aid refitting.
9 Remove the timing belt cover over the injection pump sprocket with reference to Chapter 2B.
10 Undo the banjo unions, and disconnect the fuel supply and return hoses from the pump. Recover the sealing washers from the banjo unions. Cover the open end of the

4B

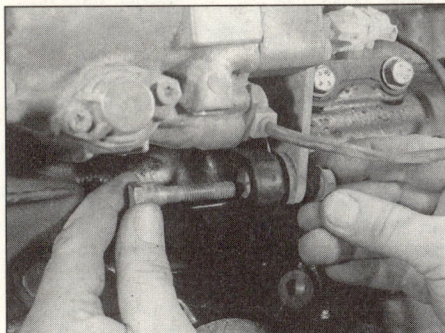

10.13b . . . then remove the rear mounting nut and bolt . . .

10.13c . . . and carefully withdraw the pump from its mounting bracket

10.17 Use new hose clips when reconnecting the fuel supply and return hoses to the fuel injection pump

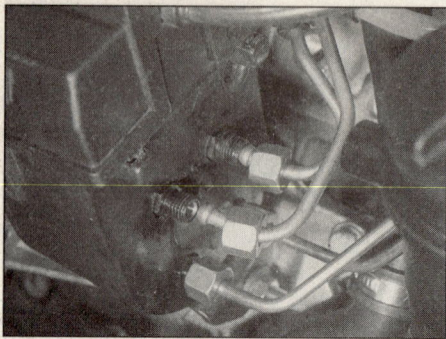

11.12a Unscrew the union nuts securing the injector pipes to the fuel injection pump . . .

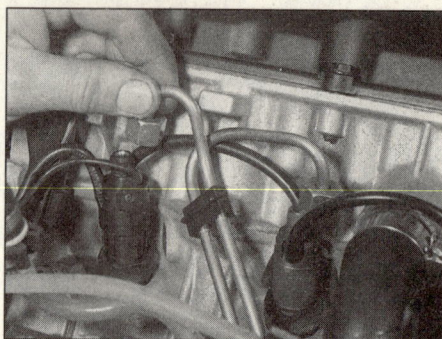

11.12b . . . and to the fuel injectors

hoses, and refit and cover the banjo bolt to keep dirt out.

11 Disconnect all remaining wiring and hose clips and brackets from the pump. Release the main multiway wiring connector by depressing the locking tang using a small flat bladed screwdriver, and rotating the connector body anticlockwise.

12 Unscrew the union nuts securing the injector pipes to the fuel injection pump and injectors (see illustrations). Counterhold the unions on the pump, while unscrewing the pipe-to-pump union nuts. Remove the pipes as a set. Cover open unions to keep dirt out, using the fingers cut from an old rubber glove secured with elastic bands.

13 Using a socket on the crankshaft pulley, turn the crankshaft in the normal direction of rotation until the two bolt holes in the fuel injection pump sprocket are aligned with the corresponding holes in the engine front plate. It will be easier to turn the engine if the glow plugs are removed (see Chapter 5C).

14 Insert two M8 bolts through the holes, and hand-tighten them. Note that the bolts must retain the sprocket while the fuel injection pump is removed, thereby making it unnecessary to remove the timing belt.

15 Mark the fuel injection pump in relation to the mounting bracket, using a scriber or felt tip pen. This will ensure that the correct pump timing is retained when refitting.

16 Unscrew the three front mounting nuts, and recover the washers. Unscrew and remove the rear mounting nut and bolt, noting

11.16a Unscrew the three fuel injection pump front mounting nuts (one arrowed)

the locations of the washers, and support the injection pump on a block of wood (see illustrations).

17 Release the injection pump sprocket from the pump shaft, as described in Chapter 2B, section 15. Note that the sprocket can be left engaged with the timing belt as the pump is withdrawn from its mounting bracket. Refit the M8 bolts to retain the sprocket in position while the pump is removed.

18 Carefully withdraw the pump. Recover the Woodruff key from the end of the pump shaft if it is loose, and similarly recover the bush from the rear of the mounting bracket (where fitted).

Refitting

19 Commence refitting the injection pump by fitting the Woodruff key to the shaft groove (if removed).

20 Offer the pump to the mounting bracket, and support on a block of wood, as during removal.

21 Engage the pump shaft with the sprocket, and refit the sprocket as described in Chapter 2B, section 15. Ensure that the Woodruff key does not fall out of the shaft as the sprocket is engaged.

22 Align the marks made on the pump and mounting bracket before removal. If a new pump is being fitted, transfer the mark from the old pump to give an approximate setting.

23 Refit and lightly tighten the pump mounting nuts and bolt.

24 Set up the injection timing, as described in Sections 13 and 15 .

11.16b Unscrew and remove the rear mounting nut and bolt (arrowed)

25 Refit and reconnect the injector fuel pipes.

26 Reconnect all relevant wiring to the pump. Secure the main multiway wiring connector by pressing the male and female halves of the connector together and then rotating the connector body clockwise.

27 The remainder of the refitting procedure is a reversal of removal, noting the following points:

a) *Use new sealing washers when reconnecting the fuel supply and return hoses banjo unions (where applicable).*

b) *Refit the auxiliary drivebelt as described in Chapter 1B.*

c) *On completion, prime and bleed the fuel system as described in Section 2.*

12 Lucas Roto-diesel/Bosch fuel injection pump - removal and refitting

Note: *This Section applies to 2.1 litre non-turbo and early 2.1 litre turbo models, fitted with Roto-Diesel or Bosch fuel injection pumps. It does not apply to models with Lucas EPIC or Bosch VP36 electronic fuel injection systems.*

Removal

1 Disconnect the negative cable from the battery terminal.

2 Jack up the front of the car, support it securely on axle stands and remove the right hand front road wheel.

3 Extract the fixings and remove the plastic liner from the right hand front wheel arch.

4 On models with ABS and/or Hydractive suspension, remove the ECU and casing from the front right hand side of the engine compartment, with reference to Chapter 2C, section 6 (as applicable).

5 Remove the inlet air trunking from the front of the upper section of the inlet manifold.

6 Refer to Section 18 and remove the upper section of the inlet manifold.

7 Relieve the tension on the auxiliary drivebelt and release it from the hydraulic pump pulley - refer to Chapter 1B for details.

8 Unbolt the auxiliary drivebelt guide roller from the side of the engine.

9 Refer to Chapter 2B and remove the upper timing belt cover, to expose the fuel injection pump sprocket.

10 Disconnect the accelerator cable from the fuel injection pump, with reference to Section 4 .

11 Disconnect the fast idle cable from the fuel injection pump, by unhooking the nipple from control lever and unscrewing the locknut from the support bracket.

12 Undo the banjo unions, and disconnect the fuel supply and return hoses from the pump. Recover the sealing washers from the banjo unions. Cover the open end of the hoses, and refit and cover the banjo bolt to keep dirt out.

13 Disconnect the injector overflow return

pipe from the port at the top of the fuel injection pump.

14 Unplug all the electrical wiring from the fuel injection pump. Label each connection carefully to aid refitting later.

15 Unscrew the union nuts securing the injector pipes to the fuel injection pump and injectors. Counterhold the unions on the pump, while unscrewing the pipe-to-pump union nuts. Remove the pipes as a set. Cover open unions to keep dirt out.

16 Set the engine to TDC on cylinder No 4 using flywheel and timing belt sprocket locking tools - refer to Chapter 2B for details of this procedure; note that it will be easier to turn the engine manually if the glow plugs are first removed.

17 Still working from Chapter 2B, release the timing belt sprocket from the fuel injection pump shaft. Note that it will not be necessary to remove the timing belt from the pump sprocket - the TDC locking tools will hold the sprocket in place after the pump has been removed.

18 Mark the relationship between the fuel injection pump body and its mounting bracket. This will help to give an approximate basic timing setting, when the pump is refitted.

19 Slacken and remove the fuel injection pump-to-mounting bracket bolts (three at the front of the pump, one at the rear).

20 Carefully withdraw the pump from its mounting bracket and recover the Woodruff key from the machined recess in the pump shaft. Cover the open unions and ports to prevent the ingress of dirt.

Refitting

21 Commence refitting the injection pump by fitting the Woodruff key to the recess the in the pump shaft.

22 Offer the pump to the mounting bracket, and support it on a block of wood.

23 Engage the pump shaft with the sprocket, and refit the sprocket as described in Chapter 2B. Ensure that the Woodruff key does not fall out of the shaft as the sprocket is engaged.

24 Align the marks made on the pump and mounting bracket before removal. If a new pump is being fitted, transfer the mark from the old pump to give an approximate setting.

25 Refit and lightly tighten the pump mounting nuts and bolt.

26 Set up the injection timing, as described in Sections 13 and 14 or 15 (as applicable).

27 Refit and reconnect the injector fuel pipes.

28 The remainder of the refitting procedure is a reversal of removal, noting the following points:

a) *Use new sealing washers when reconnecting the fuel supply and return hoses banjo unions.*

b) *Reconnect all relevant wiring to the pump, using the labels made during removal.*

c) *Refit the hydraulic system pump as described in Chapter 9.*

d) *Refit the auxiliary drivebelt as described in Chapter 1B.*

e) *On completion, prime and bleed the fuel system as described in Section 2.*

13 Injection timing - checking methods and adjustment

General information

1 Checking the fuel injection pump timing is **not** a routine operation. It is only necessary after the fuel injection pump has been disturbed. The information given in Sections 14 and 15 applies to 2.1 litre turbo and non-turbo models, fitted with Lucas and Bosch injection pumps, including those with Lucas EPIC electronic fuel injection. It does not apply to 2.5 litre models with Bosch VP36 electronic fuel injection - on these models the injection timing is controlled electronically and the basic static setting is fixed.

2 Dynamic timing equipment does exist, but it is unlikely to be available to the home mechanic. The equipment works by converting pressure pulses in an injector pipe into electrical signals. If such equipment is available, use it in accordance with its maker's instructions.

3 Static timing as described in this Chapter gives good results if carried out carefully. If working on the Bosch fuel injection pump, a dial test indicator will be needed, with suitable probes and adapters. When working on models with Lucas (mechanical) Roto-Diesel fuel injection pumps, a Citroën special tool will be needed. On models using the Lucas EPIC injection system, a home-made alternative to the required special tool can be used. Read through the procedures before starting work, to find out what is involved.

14 Injection timing (Bosch fuel injection pump) - checking and adjustment

Caution: Some of the injection pump settings and access plugs may be sealed by the manufacturers at the factory, using paint or locking wire and lead seals. Do not disturb the seals if the vehicle is still within the warranty period, otherwise the warranty will be invalidated. Also do not attempt the timing procedure unless accurate instrumentation is available.

Note: A dial test indicator (DTI) gauge and an adapter will be required for this procedure. If the Citroën adapter cannot be obtained, suitable alternatives to fit a range of Bosch fuel injection pumps can be purchased from most motor factors or diesel injection specialists.

1 If the injection timing is being checked with the pump in position on the engine, rather than as part of the pump refitting procedure,

disconnect the battery negative terminal and cover the alternator with a clean cloth or plastic bag to prevent the possibility of fuel being spilt onto it.

2 Remove the injector pipes as described in Sections 12 and 16.

3 Referring to Chapter 2B, section 12, set the engine to TDC on No 4 cylinder by aligning the engine assembly/valve timing holes, then turn the crankshaft **backwards** (anti-clockwise) approximately a quarter of a turn (but no further).

4 Unscrew the access plug, situated in the centre of the four injector pipe unions, from the rear of the injection pump. As the plug is removed, position a suitable container beneath the pump to catch any escaping fuel. Mop up any spilt fuel with a clean cloth.

5 Screw the adapter into the rear of the pump and mount the dial gauge in the adapter **(see illustration)**. Position the dial gauge so that its plunger is at the mid-point of its travel and securely tighten the adapter locknut.

6 Slowly rotate the crankshaft back and forth whilst observing the dial gauge, to determine when the injection pump piston is at the bottom of its travel (BDC). When the piston is correctly positioned, zero the dial gauge.

7 Rotate the crankshaft slowly in the correct direction until the crankshaft locking tool can be re-inserted.

8 The reading obtained on the dial gauge should be equal to the specified pump timing measurement given in the Specifications at the start of this Chapter. If adjustment is necessary, slacken the front and rear pump mounting nuts and bolts and slowly rotate the pump body until the point is found where the specified reading is obtained. When the pump is correctly positioned, tighten both its front and rear mounting nuts and bolts securely.

9 Rotate the crankshaft through one and three quarter rotations in the normal direction of rotation. Find the injection pump piston BDC as described in earlier and zero the dial gauge.

10 Rotate the crankshaft slowly in the correct direction of rotation until the crankshaft

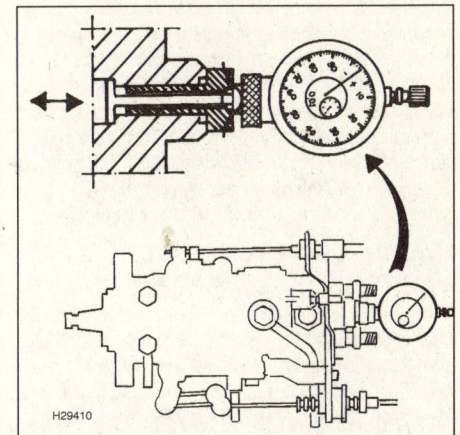

14.5 DTI gauge and adapter mounted on the Bosch fuel injection pump

4B

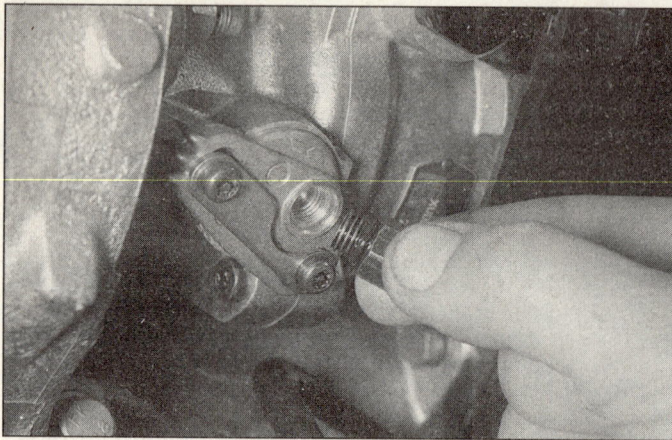

15.1 Location of the injection pump timing hole - models with Lucas EPIC fuel injection

15.5 Insert the home-made setting tool into the timing hole on the side of the pump

locking tool can be re-inserted (bringing the engine back to TDC). Recheck the timing measurement.

11 If adjustment is necessary, slacken the pump mounting nuts and bolts and repeat the operations in paragraphs 8 to 10.

12 When the pump timing is correctly set, unscrew the adapter and remove the dial gauge.

13 Refit the screw and sealing washer to the pump and tighten it securely.

14 If the procedure is being carried out as part of the pump refitting sequence, proceed as described in Section 12.

15 If the procedure is being carried out with the pump fitted to the engine, refit the injector pipes tightening their union nuts to the specified torque setting. Reconnect the battery then bleed the fuel system as described in Section 2. Start the engine and carry out the adjustments in Section 8.

15 Injection timing (Lucas fuel injection pump) - checking and adjustment

Caution: Some of the injection pump settings and access plugs may be sealed by the manufacturers at the factory, using paint or locking wire and lead seals. Do not disturb the seals if the vehicle is still within the warranty period, otherwise the warranty will be invalidated. Also do not attempt the timing procedure unless accurate instrumentation is available.

Models with Lucas EPIC electronic fuel injection

Note: *A pump timing setting rod (available as a Citroën special tool) will be required for the following procedure. Alternatively, a short length of approximately 1.5 mm diameter rod (ie welding rod) shaped as described in the text, can be used.*

1 Locate the injection pump timing hole, at the side of the pump body **(see illustration)**.

Slacken (but do not remove) the fuel injector pipe unions at the rear of the injection pump body. Be prepared for some fuel spillage.

2 Referring to Chapter 2B, section 12, set the engine to TDC on No 4 cylinder, so that the engine assembly/valve timing holes are aligned, then turn the crankshaft **backwards** (anti-clockwise) approximately a quarter of a turn (but no further).

3 Unscrew the cap over the timing hole on the side of the injection pump. As the cap is removed, position a suitable container beneath the pump to catch any escaping fuel. Mop up any spilt fuel with a clean cloth.

4 If the Citroën setting rod is not available, obtain a short length of approximately 1.5 mm diameter rod (welding rod will do) and taper one end to form a point.

5 Insert the Citroën setting rod or the home-made alternative into the timing hole on the side of the pump **(see illustration)**. While keeping pressure on the tool, slowly turn the crankshaft in the correct direction of rotation until the setting rod moves in further slightly to engage with a slot in the internal mechanism of the pump. This is very much a trial-and-error operation (especially if the home-made tool is being used) and it is not always immediately obvious if the setting rod has engaged internally or not.

6 With the setting rod engaged, the crankshaft should be at TDC and it should be possible to insert the locking tool into the engine assembly/valve timing hole in the crankshaft.

7 The engagement of the setting tool when the engine is at TDC (on cylinder No 4) indicates that the basic timing setting is correct. If adjustment is necessary, remove the pump setting rod and turn the crankshaft to the TDC position. Insert the locking tool into the engine assembly/valve timing hole in the flywheel.

8 Slacken the front and rear pump mounting nuts and bolts and rotate the pump away from the engine. Insert the setting rod into the timing hole on the side of the pump. While

keeping pressure on the tool as before, slowly rotate the pump toward the engine until the setting rod engages internally. Tighten the front pump mounting bolts followed by the rear mounting bolt, to the specified torque.

9 Remove the setting rod and refit the cap over the timing hole.

10 Remove the locking tool from the crankshaft.

11 Where applicable, refit all the components removed for access as described in Section 12.

Models with mechanical fuel injection

Note: *Access to Citroën special tool No. 4093-T (Lucas/Roto-diesel fuel injection pump setting tool) is essential to the completion of this operation.*

12 Referring to Section 12 disconnect the fuel supply lines and electrical wiring as necessary to gain access to the timing hole on the top of the injection pump **(see illustration)**. Unscrew the timing hole plug.

13 Referring to Chapter 2B, section 12, set the engine to TDC on No 4 cylinder, so that the engine assembly/valve timing holes are aligned, then turn the crankshaft **backwards** (anti-clockwise) approximately a quarter of a turn (but no further).

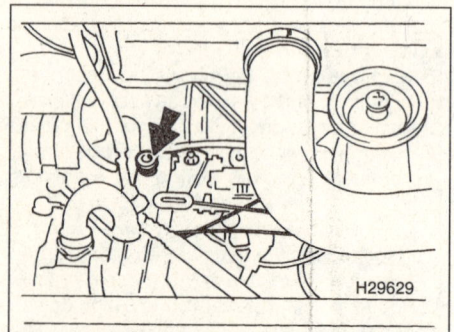

15.12 Fuel injection pump timing hole plug (arrowed) - models with mechanical fuel injection

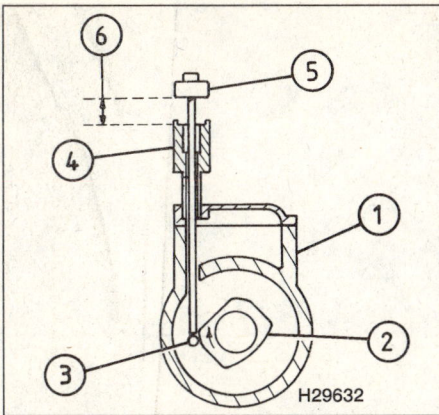

15.15 Sectional view of Lucas fuel injection pump with setting tool engaged

1 *Injection pump body*
2 *Injection pump rotor*
3 *Rotor alignment slot*
4 *Setting tool guide tube*
5 *Setting tool central rod*
6 *Measurement point*

14 Manufacturing tolerances mean that the injection timing varies from pump to pump; this figure is stamped on a plastic disc attached to the pump accelerator lever.
15 Insert the Citroën setting rod into the timing hole on the side of the pump. While keeping pressure on the tool, slowly turn the crankshaft in the correct direction of rotation until the setting rod moves in further slightly to engage with a slot in the internal rotor mechanism of the pump. This is very much a trial-and-error operation (especially if the home-made tool is being used) and it is not always immediately obvious if the setting rod has engaged internally or not **(see illustration)**.
16 With the setting rod engaged, the crankshaft should be at TDC and it should be possible to insert the locking tool into the engine assembly/valve timing hole in the crankshaft (see Chapter 2B, section 12).
17 The engagement of the setting tool when the engine is at TDC (on cylinder No 4) indicates that the basic timing setting can now be measured. Measure the distance between the top of the special tool's guide tube and the ferrule on the centre rod **(refer to**

illustration 15.15). Check that this measurement accords with that stamped on the pump accelerator lever.
18 If adjustment is necessary, slacken the front and rear pump mounting nuts and bolts and rotate the pump towards from the engine, until the correct dimension is obtained.
19 On completion, remove the setting rod and refit the cap over the timing hole.
20 Remove the locking tool from the crankshaft.
21 Where applicable, refit all the components removed for access as described in Section 12. If the fuel lines were disturbed, prime and bleed the fuel system as described in Section 2.

16 Fuel injectors - testing, removal and refitting

⚠️ *Warning: Exercise extreme caution when working on the fuel injectors. Never expose the hands or any part of the body to injector spray, as the high working pressure can cause the fuel to penetrate the skin, with possibly fatal results. You are strongly advised to have any work which involves testing the injectors under pressure carried out by a dealer or fuel injection specialist.*

Testing

1 Injectors do deteriorate with prolonged use, and it is reasonable to expect them to need reconditioning or renewal after 60 000 miles (100 000 km) or so. Accurate testing, overhaul and calibration of the injectors must be left to a specialist. A defective injector which is causing knocking or smoking can be located without dismantling as follows.
2 Run the engine at a fast idle. Slacken each injector union in turn, placing rag around the union to catch spilt fuel, and being careful not to expose the skin to any spray. When the union on the defective injector is slackened, the knocking or smoking will stop.

Removal

Note: *To remove the injector fitted with the needle lift sensor, a suitable slotted socket will*

16.6 Unscrew the union nuts and disconnect the pipes from the injectors

be required to allow clearance for the sensor wire. These sockets are available from Citroën dealers (as a special tool) or from diesel injection specialists.
3 On 2.1 litre models remove the air intake ducting as described in Section 3 and the inlet manifold upper part (see Section 18).
4 On 2.5 litre models, remove the plastic cover panel, then unbolt the coolant expansion tank and move it to one side.
5 Carefully clean around the injectors and injector pipe union nuts.
6 Unscrew the union nuts and disconnect the pipes from the injectors **(see illustration)**. If necessary, the injector pipes may be completely removed. Note carefully the locations of the pipe clamps, for use when refitting. Cover the ends of the injectors, to prevent dirt ingress.
7 Pull the leak-off pipes from the injectors.
8 Unscrew the union nuts securing the injector pipes to the fuel injection pump. Counterhold the unions on the pump when unscrewing the nuts. Cover open unions to keep dirt out, using small plastic bags, or fingers cut from discarded (but clean!) rubber gloves.
9 Bearing in mind the information given in the Note at the start of this Section, unscrew the injectors using a deep socket or box spanner and remove them from the cylinder head. Note that on 2.5 litre models, injector No 3 has an integral needle lift sensor and a flying lead connection. A slotted socket will be required to remove/refit the injector **(see illustrations)**.

4B

16.9a Unscrew the injectors and remove them from the cylinder head

16.9b On 2.5 litre models, injector No 3 has an integral needle lift sensor and a flying lead connection (arrowed) . . .

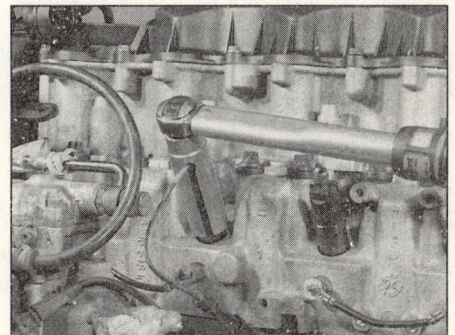

16.9c . . . a slotted socket will be required to remove/refit the injector

16.10a Recover the copper washers . . .

16.10b . . . and on 2.1 litre models, the fire seal washers . . .

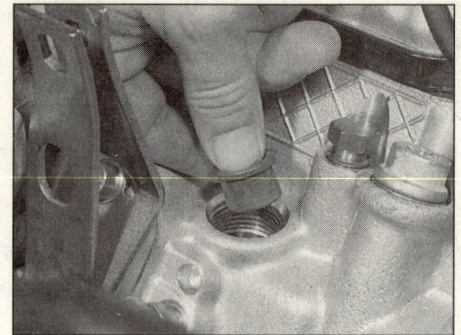

16.10c . . . and sleeves

10 Recover the copper washers (where applicable) from the cylinder head (see illustrations).

Refitting

11 Obtain new copper washers and fire seal washers. Also renew the sleeves, if they are damaged.

12 Take care not to drop the injectors, or allow the needles at their tips to become damaged. The injectors are precision-made to fine limits, and must not be handled roughly. In particular, never mount them in a bench vice.

13 Commence refitting by inserting the sleeves and then the fire seal washers (where applicable) convex face uppermost, and copper washers.

14 Insert the injectors and tighten them to the specified torque.

15 Refit the injector pipes and tighten the union nuts. Make sure the pipe clamps are in their previously-noted positions. If the clamps are wrongly positioned or missing, problems may be experienced with pipes breaking or splitting.

16 Reconnect the leak-off pipes (see illustration).

17 Refit the air intake ducting (Section 3) and, where applicable, the inlet manifold upper part (Section 18).

18 Start the engine. If difficulty is experienced, bleed the fuel system as described in Section 2.

16.16 Reconnect the injector leak-off pipes

17 Injection system sensors and actuators (Lucas EPIC & Bosch VP36) - removal and refitting

General information

ECU (Electronic Control Unit)

1 This component is the heart of the entire engine management system, controlling the fuel injection, and emission control systems. The ECU receives signals from various sensors, which monitor changing engine operating conditions such as inlet air temperature, coolant temperature, engine speed, accelerator pedal position, etc. These signals are used by the ECU to determine the correct fuel metering by the injection pump.

Crankshaft (RPM) sensor

2 This is an inductive pulse generator bolted to the transmission bellhousing, to scan the ridges between holes machined in the inboard face of the flywheel. As each ridge passes the sensor tip, a signal is generated, which is used by the ECU to determine engine speed.

3 The ridge between one of the holes is missing - this step in the incoming signals is used by the ECU to determine crankshaft (ie, piston) position.

Coolant temperature sensor

4 This component is an NTC (Negative Temperature Coefficient) thermistor - that is, a semi-conductor whose electrical resistance decreases as its temperature increases. It provides the ECU with a constantly-varying (analogue) voltage signal, corresponding to the temperature of the engine coolant. This is used to refine the calculations made by the ECU, when determining fuel metering.

Inlet air temperature sensor

5 This component is also an NTC thermistor - see the previous paragraph - providing the ECU with a signal corresponding to the temperature of air passing into the engine. This is also used to refine fuel metering calculations.

Accelerator pedal position sensor

6 The 'drive-by-wire' throttle control information is provided by this sensor. The accelerator cable is connected to the pedal position sensor which converts accelerator pedal movement into an electrical signal. After processing this signal (and refining it using information received from the other sensors) the ECU controls the fuel injection pump electronically so that the correct fuel metering is achieved to obtain the desired road speed.

Manifold absolute pressure sensor/Barometric pressure sensor

7 The manifold absolute pressure sensor measures inlet manifold pressure and supplies this information to the ECU for calculation of engine load at any given throttle position.

Injector needle lift sensor

8 The needle lift sensor is an integral part of one of the fuel injectors and sends a signal to the ECU whenever the injector opens. This provides feedback allowing the ECU to accurately sense the start of injection and so measure and correct the injection advance.

Preheating system control unit

9 This unit is a relay, controlled by the ECU, to operate the preheating system glow plugs during cold starting conditions.

Vehicle speed sensor

10 The vehicle speed sensor consists of a transducer incorporated into the speedometer drive unit. The ECU uses inputs from the sensor to modify fuel metering in accordance with vehicle speed.

EGR valve

11 Introduction of part of the exhaust gas back into the inlet manifold is controlled by the ECU in conjunction with an EGR solenoid valve and EGR valve. Vacuum from the engine vacuum pump is directed to the EGR valve, via the solenoid valve according to engine speed, load and altitude.

Testing

12 If a fault appears in the system, first ensure that all the system wiring connectors are securely connected and free of corrosion. Ensure that the fault is not due to poor maintenance; ie, check that the air cleaner

17.23 The crankshaft sensor is situated on the top face of the transmission clutch housing

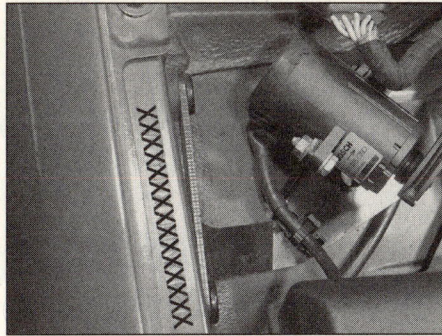

17.30 Accelerator pedal position sensor - right hand drive 2.5 litre model shown

17.33 MAP sensor - right hand drive 2.5 litre model shown

filter element is clean, the cylinder compression pressures are correct, and that the engine breather hoses are clear and undamaged.

13 If these checks fail to reveal the cause of the problem, the vehicle should be taken to a suitably-equipped Citroën dealer for testing. A wiring block connector is incorporated in the engine management circuit, into which a special electronic diagnostic tester can be plugged. The tester will locate the fault quickly and simply, alleviating the need to test all the system components individually, which is a time-consuming operation that also carries a risk of damaging the ECU.

Removal and refitting

General

14 Before disconnecting any of these components, always disconnect the battery negative lead first.

ECU

Note: *The ECU is fragile. Take care not to drop it or subject it to any other kind of impact, and do not subject it to extremes of temperature, or allow it to get wet. Once disconnected from its wiring harness, do not touch the exposed ECU connector pins, as stray static electricity can easily damage the internal components.*

15 The ECU is located in a plastic casing, which is mounted in the right-hand front corner of the engine compartment.
16 Lift off the ECU module box lid.
17 Release the wiring connector by lifting the locking lever on top of the connector upwards. Lift the connector at the rear, disengage the tag at the front and carefully withdraw the connector from the ECU pins.
18 Lift the ECU upwards and remove it from its location.
19 To remove the ECU module box, turn the injection pump wiring connector on the top of the box clockwise, disengage the retaining lug using a screwdriver, then turn the connector anti-clockwise and lift off.

20 Undo the two screws securing the wiring connector base to the module box and lift off the connector base.
21 Undo the internal and external retaining bolts and remove the module box.
22 Refitting is a reversal of removal.

Crankshaft (TDC) sensor

23 The crankshaft sensor is situated on the top face of the transmission clutch housing **(see illustration)**.
24 Trace the wiring back from the sensor to the wiring connector and disconnect it from the main harness.
25 Prise out the rubber grommet then undo the retaining bolt and withdraw the sensor from the transmission.
26 Refitting is reversal of removal ensuring that the sensor retaining bolt is securely tightened and the grommet is correctly seated in the transmission housing.

Coolant temperature sensor

27 Refer to Chapter 3.

Inlet air temperature sensor

28 Disconnect the wiring connector then unscrew the sensor from the intake ducting over the front of the engine.
29 Refitting is a reversal of removal.

Accelerator pedal position sensor

30 Release the accelerator inner cable from the lever on the pedal position sensor located on the right-hand side of the engine compartment (or the rear of the engine

18.1 Disconnect the air intake duct from the manifold upper part

compartment, on left-hand drive models) **(see illustration)**. Pull the outer cable from the grommet in the pedal position sensor bracket.
31 Disconnect the wiring connector, undo the mountings and withdraw the sensor complete with mounting bracket. If it is necessary to remove the sensor from the mounting bracket, mark the position of the sensor in relation to the bracket, then undo the two screws and remove the unit. When refitting, align the marks made on removal. If a new unit is to be fitted, align it centrally within the mounting holes initially, then have the sensor adjusted by a Citroën dealer.
32 The remainder of refitting is a reversal of removal.

Manifold absolute pressure sensor

33 The sensor may be located beneath the air cleaner assembly or in various locations at the front left-hand side of the engine compartment **(see illustration)**.
34 Disconnect the vacuum hose and wiring multiplug then undo the two sensor mounting bolts.
35 Withdraw the sensor from its location.
36 Refitting is a reversal of removal.

Injector needle lift sensor

37 The needle lift sensor is an integral part of the fuel injectors. Fuel injector removal and refitting procedures are contained in Section 16.

Preheating system control unit

38 Refer to Chapter 5C.

EGR valve

39 Refer to Part C of this Chapter.

18 Inlet manifold - removal and refitting

2.1 litre models

Removal - upper part

1 Slacken the retaining clip and disconnect the air intake duct from the manifold upper part **(see illustration)**.

4B

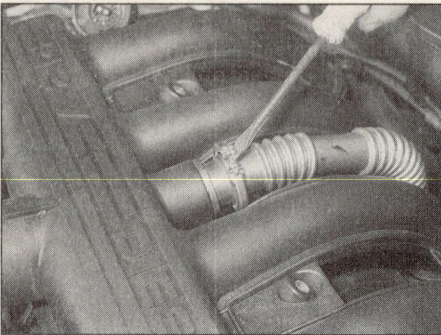

18.2 Remove the clip securing the flexible portion of the EGR pipe to the manifold

18.4a Undo the four retaining bolts (arrowed) . . .

18.4b . . . and lift off the manifold upper part - 2.1 litre model

2 Remove the clip securing the flexible portion of the EGR pipe to the manifold. If the original crimped clip is still in place, cut it off; new clips are supplied by Citroën parts stockists with a screw clamp fixing (see illustration). If a screw clamp type clip is fitted, undo the screw and manipulate the clip off the pipe.

3 Undo the four retaining bolts and lift off the manifold upper part.

4 Recover the four rubber connecting tubes from the lower part (see illustrations).

Refitting - upper part

5 Refitting is a reversal of removal, bearing in mind the following points.

a) Renew the four rubber connecting tubes as a set if any one shows signs of deterioration.

b) Tighten all fixings securely.

c) Secure the EGR pipe with a new screw clamp type clip, if a crimped type was initially fitted (see illustration).

Removal - lower part

6 Remove the engine/transmission as described in Chapter 2C.

7 Remove the manifold upper part as described previously.

18.4c Recover the four rubber connecting tubes - 2.1 litre model

8 Undo the manifold retaining bolts, noting the location of the pipe support bracket at the right-hand end of the manifold (see illustration).

9 Withdraw the manifold from the cylinder head, ease the EGR pipe aside, and manipulate the manifold out from between the EGR pipe and head (see illustration).

10 Remove the manifold gasket.

Refitting - lower part

11 Refitting is a reversal of removal, bearing in mind the following points.

a) Renew the gasket when refitting the manifold.

18.5 Secure the EGR pipe with a new screw clamp type clip (arrowed) when refitting

b) Tighten all fixings securely.

c) Refit the manifold upper part as described previously.

d) Refit the engine/transmission as described in Chapter 2C.

2.5 litre models

Removal - upper part

12 Slacken the retaining clip and disconnect the air intake duct from the port at the side of manifold upper part (see illustration).

13 Remove the securing screws and detach the EGR solenoid valve and bracket, as an

18.8 Note the location of the pipe support bracket (arrowed) when removing the lower section of the inlet manifold - 2.1 litre model

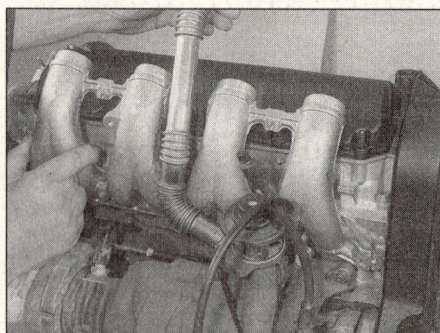

18.9 Withdraw the lower section of the inlet manifold from the cylinder head

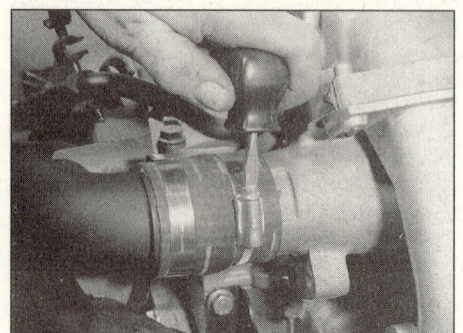

18.12 Slacken the retaining clip and disconnect the air intake duct from the port at the side of manifold upper part - 2.5 litre engine

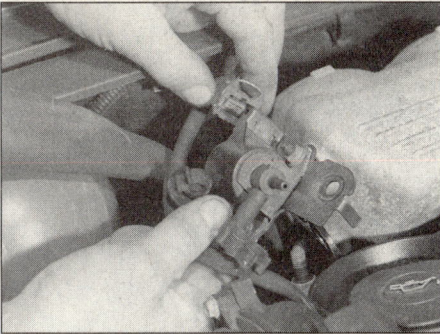
18.13 Removing the EGR solenoid valve and bracket assembly

18.14a Undo the four retaining screws . . .

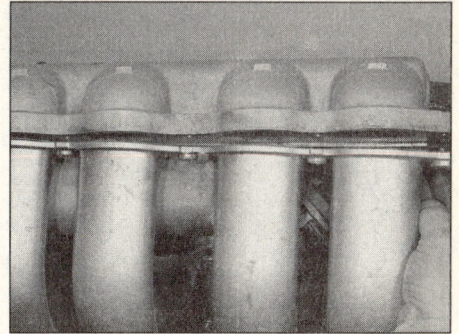
18.14b . . . lift off the manifold upper part . . .

18.14c . . . and recover the gasket

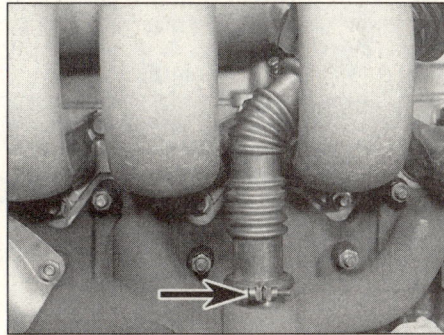
18.15 Slacken the clip (arrowed) and disconnect the EGR pipe from the exhaust manifold

assembly, from the side of the inlet manifold (see illustration).
14 Undo the four retaining screws and lift off the manifold upper part. Recover the gasket (see illustrations).
15 Slacken the clip and disconnect the EGR pipe from the exhaust manifold (see illustration).

Refitting - upper part
16 Refitting is a reversal of removal. Use a new upper-to-lower manifold gasket, and tighten all fixings securely.

Removal - lower part
17 Remove the manifold upper part as described previously. Undo the manifold retaining bolts, then withdraw the manifold from the cylinder head and recover the gasket (see illustrations).

Refitting - lower part
18 Refitting is a reversal of removal, bearing in mind the following points.
a) Renew the gaskets when refitting the manifold.
b) Tighten all fixings securely.
c) Refit the manifold upper part as described previously.
d) Refit the engine/transmission as described in Chapter 2C.

19 Intercooler - removal and refitting

2.1 litre models (air-to-air intercooler)
Removal
1 The intercooler is located behind the radiator. To remove it first remove the radiator as described in Chapter 3.
2 Disconnect the air hoses from each end of the intercooler then remove the unit from the front cross panel.
Refitting
3 Refitting is a reversal of removal; refer to Chapter 3 when refitting the radiator.

2.5 litre models (air-to-water inercooler)
Removal
4 Slacken the large worm-drive clips, then detach the air inlet and outlet hoses from the top of the intercooler (see illustration).
5 Partially drain the cooling system, with reference to Chapter 3.
6 Slacken the clips and detach the two

18.17a Remove the inlet manifold lower section from the cylinder head . . .

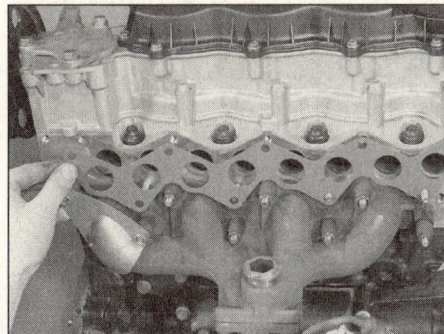
18.17b . . . and recover the gasket

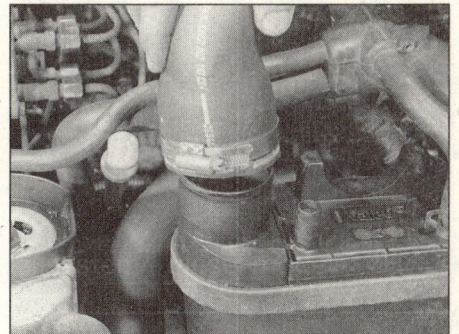
19.4 Detach the air inlet and outlet hoses from the top of the intercooler - 2.5 litre models

4B

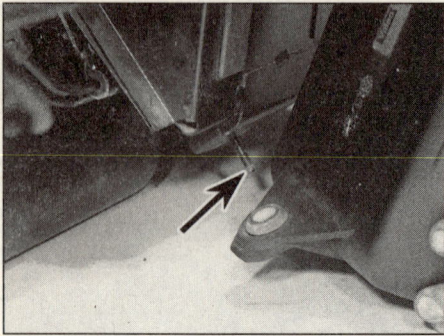

19.7 Slacken and remove the nut from the stud (arrowed) at the base of the intercooler

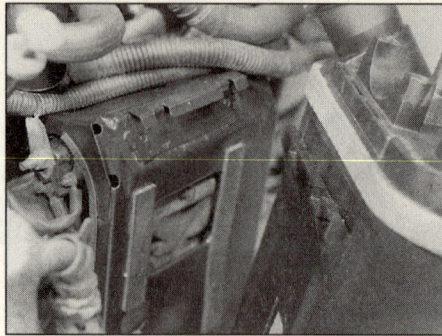

19.8 Carefully slide the intercooler upwards off its mounting bracket and remove it from the engine compartment

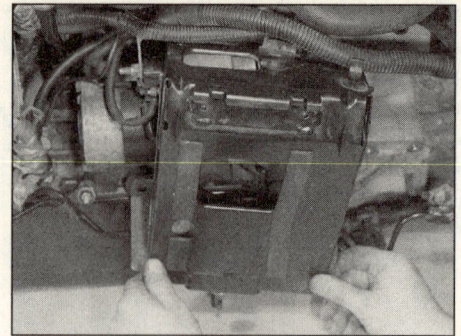

19.9 Removing the intercooler mounting bracket from the engine block

coolant hoses from the ports at the top of the intercooler. Be prepared for some coolant loss as you do this.

7 Slacken and remove the nut from the stud at the base of the intercooler **(see illustration)**.

8 Carefully slide the intercooler upwards off its mounting bracket and remove it from the engine compartment **(see illustration)**. Some coolant will remain in the heat exchanger matrix, so store the unit upright, unless it is to drained.

9 If required, the intercooler mounting bracket can be unbolted from the engine block, to allow access to the starter motor **(see illustration)**.

Refitting

10 Refitting is a reversal of removal. Refill and bleed the cooling system, with reference to Chapters 3 and 1B, as applicable.

Chapter 4 Part C:
Exhaust and emission control systems

Contents

Degrees of difficulty

Easy, suitable for novice with little experience	**Fairly easy,** suitable for beginner with some experience	**Fairly difficult,** suitable for competent DIY mechanic	**Difficult,** suitable for experienced DIY mechanic	**Very difficult,** suitable for expert DIY or professional

4C

Specifications

Turbocharger

Manufacturer/type:
    2.0 litre C.T. petrol models . . . .   Garret T025
    2.1 litre turbo diesel models . . . .   Mitsubishi TD 0411B
    2.5 litre turbo diesel models . . . .   Garret T2
Boost pressure (at 3000 rpm on full load):
    2.0 litre C.T. petrol . . . .   0.65 bar
    2.1 litre turbo diesel . . . .   0.9 bar
    2.5 litre turbo diesel . . . .   0.9 bar

Torque wrench settings	**Nm**	**lbf ft**
Turbo oil return union screws to cylinder block	20	15
Exhaust manifold nuts	30	22
Turbocharger support bracket	20	15
Oil supply union to cylinder block	30	22
Oil supply union to turbo:		
2.1 litre engine	20	15
2.5 litre engine:		
Stage 1	10	7
Stage 2	Angle tighten through 45°	
Turbocharger mounting nuts:		
2.1 litre engine	60	44
2.5 litre engine	25	18

1 General information

Emission control systems

1 All petrol engined models covered in this manual are controlled by fuel injection or engine management systems that are 'tuned' to give the best compromise between driveability, fuel consumption and exhaust emission production. In addition, a number of systems are fitted that help to minimise other harmful emissions: a crankcase emission-control system that reduces the release of pollutants from the engines lubrication system is fitted to all models, catalytic converters that reduce exhaust gas pollutants are fitted to most models and an evaporative loss emission control system that reduces the release of gaseous hydrocarbons from the fuel tank is fitted to some models.

2 All diesel engined models are also equipped with a crankcase emission control system. In addition, all models are fitted with an Exhaust Gas Recirculation (EGR) system to reduce exhaust emissions.

Crankcase emission control

3 To reduce the emission of unburned hydrocarbons from the crankcase into the atmosphere, the engine is sealed and the blow-by gases and oil vapour are drawn from inside the crankcase, through a wire mesh oil separator, into the inlet tract to be burned by the engine during normal combustion.

4 Under conditions of high manifold depression (idling, deceleration) the gases will be sucked positively out of the crankcase. Under conditions of low manifold depression (acceleration, full-throttle running) the gases are forced out of the crankcase by the (relatively) higher crankcase pressure; if the engine is worn, the raised crankcase pressure (due to increased blow-by) will cause some of the flow to return under all manifold conditions. On certain engines, a pressure regulating valve (mounted on the camshaft cover) controls the flow of gases from the crankcase.

Exhaust emission control - petrol models

5 To minimise the amount of pollutants which escape into the atmosphere, most models are fitted with a catalytic converter in the exhaust system. On all models where a catalytic converter is fitted, the fuelling system is of the closed-loop type, in which a lambda sensor in the exhaust system provides the engine management system ECU with constant feedback, enabling the ECU to adjust the air/fuel mixture to optimise combustion.

6 The lambda sensor has a heating element built-in that is controlled by the ECU through the lambda sensor relay to quickly bring the sensor's tip to its optimum operating temperature. The sensor's tip is sensitive to oxygen and relays a voltage signal to the ECU that varies according on the amount of oxygen in the exhaust gas. If the intake air/fuel mixture is too rich, the exhaust gases are low in oxygen so the sensor sends a low-voltage signal, the voltage rising as the mixture weakens and the amount of oxygen rises in the exhaust gases. Peak conversion efficiency of all major pollutants occurs if the intake air/fuel mixture is maintained at the chemically-correct ratio for the complete combustion of petrol of 14.7 parts (by weight) of air to 1 part of fuel (the 'stoichiometric' ratio). The sensor output voltage alters in a large step at this point, the ECU using the signal change as a reference point and correcting the intake air/fuel mixture accordingly by altering the fuel injector pulse width.

Exhaust emission control - diesel models

7 An Exhaust Gas Recirculation (EGR) system is fitted to all diesel engined models. This reduces the level of nitrogen oxides produced during combustion by introducing a proportion of the exhaust gas back into the inlet manifold, under certain engine operating conditions, via a plunger valve.

Evaporative emission control - petrol models

8 To minimise the escape of unburned hydrocarbons into the atmosphere, an evaporative loss emission control system is fitted to certain petrol models. The fuel tank filler cap is sealed and a charcoal canister is mounted underneath the right-hand wing to collect the petrol vapours released from the fuel contained in the fuel tank. It stores them until they can be drawn from the canister (under the control of the fuel-injection/ignition system ECU) via the purge valve(s) into the inlet tract, where they are then burned by the engine during normal combustion.

9 To ensure that the engine runs correctly when it is cold and/or idling and to protect the catalytic converter from the effects of an over-rich mixture, the purge control valve(s) are not opened by the ECU until the engine has warmed up, and the engine is under load; the valve solenoid is then modulated on and off to allow the stored vapour to pass into the inlet tract.

Exhaust systems

10 The exhaust system comprises the exhaust manifold, a number of silencer units (depending on model and specification), a catalytic converter (where fitted), a number of mounting brackets and a series of connecting pipes.

11 Refer to Section 4 for details of the turbocharger fitted to 2.0 litre turbo petrol, 2.1 and 2.5 litre turbo diesel models.

2 Exhaust manifold (petrol models) - removal and refitting

Removal

1 Chock the rear wheels, then jack up the front of the vehicle and support it on axle stands (see *Jacking and Vehicle Support*).

2 On turbo models, refer to Section 5 and remove the turbocharger from the exhaust manifold.

3 On non-turbo models, undo the nuts securing the front pipe to the manifold. Recover the springs and spring cups, and withdraw the bolts then disconnect the front pipe from the manifold, and recover the gasket.

4 Undo the nuts securing the manifold to the cylinder head. Manoeuvre the manifold out of the engine compartment, complete with gasket. Space is very limited and hence care should be taken.

5 Undo the two retaining bolts and separate the manifold and gasket, noting the spacers which are fitted between the gasket and manifold.

Refitting

6 Refitting is the reverse of the removal procedure, noting the following points:

a) *Examine all the exhaust manifold studs for signs of damage and corrosion; remove all traces of corrosion, and repair or renew any damaged studs.*

b) *Ensure that the manifold and cylinder head sealing faces are clean and flat, and fit the new manifold gasket(s). Tighten the manifold nuts to the specified torque.*

c) *Reconnect the front pipe to the manifold or refit the turbocharger with reference to Section 5.*

3 Exhaust manifold (diesel models) - removal and refitting

Removal

Note: *Access to the exhaust manifold with the engine in situ is extremely limited, particularly on 2.1 litre models.*

1 Remove the upper and lower sections of the inlet manifold, as described in Chapter 4B, to improve access.

2 Where applicable, remove the clip securing the flexible portion of the EGR pipe to the manifold. If the original crimped clip is still in place, cut it off; new clips are supplied by Citroën parts stockists with a screw clamp fixing. If a screw clamp type clip is fitted, undo the screw and manipulate the clip off the pipe.

3 On turbo models, unscrew the union nut and disconnect the oil feed pipe from the top of the turbocharger.

4 Undo the two bolts and separate the oil return pipe flange from the base of the turbocharger (turbo models only). Recover the gasket.

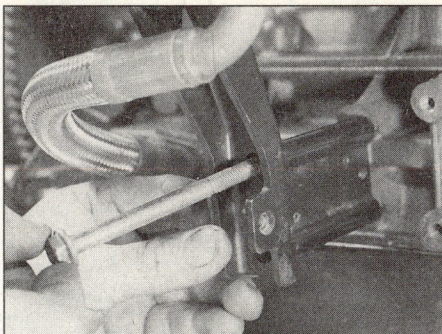
3.5 Removing the turbocharger steady bracket bolts

3.6a Note the position of the various support brackets and heat shields . . .

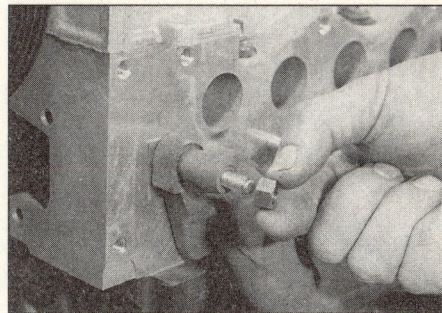
3.6b . . . then undo the nuts securing the exhaust manifold to the cylinder head studs . . .

5 Undo the turbocharger steady bracket bolt (where applicable) (see illustration).
6 Undo the nuts securing the manifold to the cylinder head studs, noting the position of the various support brackets and heat shields. Recover the spacers (see illustrations).
7 Withdraw the manifold, complete with turbocharger, from the cylinder head and recover the gasket (see illustrations).

Refitting

8 Refitting is a reversal of removal, bearing in mind the following points.
a) Renew all gaskets and manifold securing nuts when refitting.
b) Tighten all fixings to the specified torque, where given.
c) On turbo models, secure the EGR pipe with a new screw clamp type clip, if a crimped type was initially fitted.
d) Refit the inlet manifold as described in Chapter 4B.

4 Turbocharger - description and precautions

Description

A turbocharger is fitted to certain petrol and diesel engines covered by this manual. It operates by raising the pressure in the inlet manifold above atmospheric pressure.

Instead of the inlet air being sucked into the combustion chambers, it is forced in under pressure. This leads to a greater charge pressure increase during combustion and improved fuel burning, which raises the thermal efficiency of the engine. Under these conditions, additional fuel is supplied by the fuel injection system, in proportion to the increased air flow.

Energy for the operation of the turbocharger comes from the exhaust gas. The gas flows through a specially-shaped housing (the turbine housing) and in so doing, spins the turbine wheel. The turbine wheel is attached to a shaft, at the end of which is another vaned wheel known as the compressor wheel. The compressor wheel spins in its own housing, and compresses the inlet air on the way to the inlet manifold.

Between the turbocharger and the inlet manifold, the compressed air passes through an intercooler. This is an air-to-air heat exchanger, mounted in front of the radiator and supplied with cooling air from the front grille and electric cooling fans. The temperature of the inlet air rises due to the compression action of the turbocharger - the purpose of the intercooler is to cool the inlet air again, before it enters the engine. Because cool air is denser than hot air, this allows a greater mass of air (occupying the same volume) to be forced into the combustion chambers, resulting in a further increase the engine's thermal efficiency.

Boost pressure (the pressure in the inlet manifold) is limited by a wastegate, which diverts the exhaust gas away from the turbine wheel in response to a pressure-sensitive actuator. On petrol models, the wastegate valve is controlled by the engine management system ECU, via an electronic vacuum modulator valve. The ECU uses the modulator valve to open the wastegate valve in a series of rapid pulses - the duty ratio of the pulses depends primarily on engine speed and load. In this manner, the ECU keeps the turbocharger boost pressure constant throughout the engine speed range, resulting in an almost flat torque curve.

The turbo shaft is pressure-lubricated by an oil feed pipe from the main oil gallery. The shaft 'floats' on a cushion of oil. A drain pipe returns the oil to the sump. On petrol models the turbo charger is water cooled and has a dedicated system of coolant supply and return pipes.

Precautions

The turbocharger operates at extremely high speeds and temperatures. Certain precautions must be observed, to avoid premature failure of the turbo, or injury to the operator.

Do not operate the turbo with any of its parts exposed, or with any of its hoses removed. Foreign objects falling onto the rotating vanes could cause excessive damage, and (if ejected) personal injury. Suction can build up very quickly and without warning at the turbocharger air intake when the engine speed is raised.

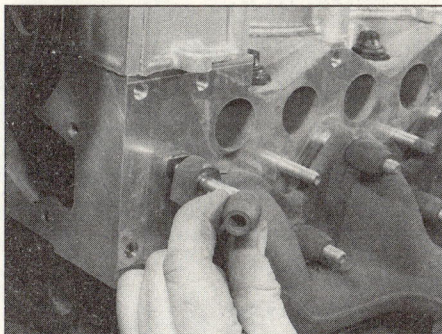
3.6c . . . and recover the exhaust manifold stud spacers

3.7a Withdraw the manifold, complete with turbocharger, from the cylinder head . . .

3.7b . . . and recover the gasket - 2.5 litre turbo diesel model shown

4C

Do not race the engine immediately after start-up, especially if it is cold. Give the oil a few seconds to circulate.

Always allow the engine to return to idle speed before switching it off - do not blip the throttle and switch off, as this will leave the turbo spinning without lubrication.

Allow the engine to idle for several minutes before switching off after a high-speed run; this will allow the oil to reduce the temperature of the turbocharger.

Observe the recommended intervals for oil and filter changing, and use a reputable oil of the specified quality. Neglect of oil changing, or use of inferior oil, can cause carbon formation on the turbo shaft, leading to subsequent failure.

5 Turbocharger - removal and refitting

Note: *On 2.1 and 2.5 litre diesel models, access to the turbocharger is extremely limited.*

2.1 litre diesel models

Removal

1 Disconnect the battery negative cable and position it away from the terminal.

2 Slacken and remove the screws that secure the air cleaner-to- turbocharger ducting to its mountings.

3 Apply the parking brake, then jack up the front of the vehicle and support securely on axle stands (see *Jacking and Vehicle Support*).

4 Remove the fixings and lower the undertray away from the engine bay.

5 Refer to Chapter 7 and disconnect the gear change/selector link rods from their respective ball joints at the transmission.

6 Unbolt the exhaust front pipe from the flange at the end of the turbocharger elbow.

7 Unbolt and remove the support bracket(s) from the base of the turbocharger and the engine block.

8 Slacken and withdraw the turbocharger-to-manifold securing bolts. Access is extremely limited - you will need to use a socket extension bar with a flexible or universal joint.

9 At the rear of the engine, below the intermediate driveshaft bearing support bracket, remove the bolt that secures the bracket to the engine torque rod. This will allow the engine to tilt slightly, giving greater clearance when the turbocharger is removed.

10 Tilt the engine forward slightly and wedge a piece of wood (approx. 50mm thick) between the transmission casing and the suspension subframe to keep it in this position.

11 Slacken the clip and disconnect the air ducting from the turbocharger outlet port. Remove the bolt and disconnect the air ducting from inlet port at the top of the turbocharger. Cover the turbocharger openings with clean rag.

5.23a Disconnect the air ducting from the ports on the turbocharger . . .

12 Unscrew the union nut, and disconnect the oil feed pipe from the top of the turbocharger.

13 Undo the union and separate the oil return pipe from the base of the turbocharger. Recover the gasket.

14 Slacken and withdraw the turbocharger-to-manifold securing bolts.

15 Rotate the turbocharger and elbow assembly through half a turn and carefully withdraw it from the engine.

16 Lower the turbocharger and remove it from under the car. If it is to be refitted, store the turbocharger carefully, and plug its openings to prevent dirt ingress.

Refitting

17 Refitting is a reversal of removal, bearing in mind the following points:

a) Renew all gaskets and oil supply and return union seals.

b) Tighten all fixings to the correct torque (where specified).

c) If a new turbo is being fitted, change the engine oil and filter. Also renew the filter (where fitted) in the oil feed pipe.

d) Do not fully tighten the oil feed pipe unions until both ends of the pipe are in place. When tightening the oil return pipe union, position it so that the return hose is not strained.

e) Before starting the engine, prime the turbo lubrication circuit by disconnecting the stop solenoid lead at the fuel pump, and cranking the engine on the starter for three ten-second bursts.

5.24 Unbolting the turbocharger support bracket from the base of the turbocharger

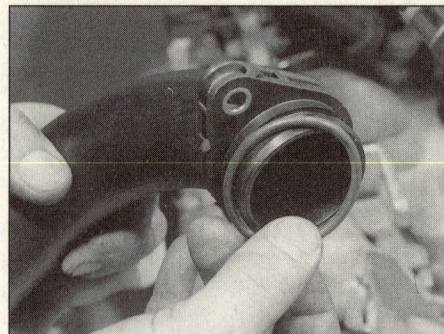

5.23b . . . and recover the O-ring seals - 2.5 litre model shown

2.5 litre diesel models

Removal

18 Disconnect both battery cables from their respective terminals. Lift out the battery then unbolt the battery tray and remove it from the engine bay.

19 Raise the front of the car and support it securely on axle stands. Remove the right hand front wheel.

20 Unscrew the fixings and remove the engine bay undertray. Similarly, remove the rear section of the right hand wheel arch plastic liner.

21 Refer to Chapter 8 and remove the right hand intermediate shaft/driveshaft assembly from the vehicle.

22 Slacken the clamps and release the heat shield sleeving from the air ducting, at top and side of the turbocharger (where applicable).

23 Remove the securing screws and disconnect the air inlet and outlet ducting from the ports on the turbocharger. Recover the O-ring seals from the ends of the ducting **(see illustrations)**.

24 Unbolt the turbocharger support bracket from the base of the turbocharger and the engine block **(see illustration)**.

25 Slacken the clamp and disconnect the exhaust front pipe from the turbocharger exhaust outlet elbow. Unbolt the exhaust outlet elbow from the side of the turbocharger.

26 Remove the two bolts and disconnect the oil return union at the cylinder block **(see illustration)**

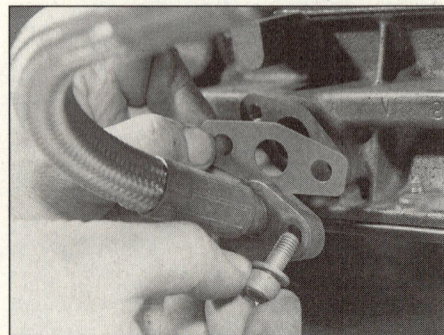

5.26 Remove the two bolts and disconnect the oil return union at the cylinder block

27 Undo the union and disconnect the oil supply union from the top of the turbocharger **(see illustration)**.
28 Slacken and remove the turbocharger-to-manifold nuts **(see illustration)**, then lower the turbocharger away from the engine and remove it from the vehicle.

Refitting

29 Refitting is a reversal of removal, bearing in mind the following points **(see illustrations)**:
a) *Renew all gaskets and oil supply and return union seals*
b) *Tighten all fixings to the correct torque (where specified).*
c) *If a new turbo is being fitted, change the engine oil and filter. Also renew the filter (where fitted) in the oil feed pipe.*
d) *Do not fully tighten the oil feed pipe unions until both ends of the pipe are in place. When tightening the oil return pipe union, position it so that the return hose is not strained.*
e) *Renew the air duct O-ring seals.*
f) *Refer to Chapter 8 for details of the driveshaft refitting procedure.*
g) *Before starting the engine, prime the turbo lubrication circuit by disconnecting the stop solenoid lead at the fuel pump, and cranking the engine on the starter for three ten-second bursts.*

2.0 litre petrol models

Removal

30 Disconnect the battery negative cable and position it away from the terminal.
31 Slacken and remove the screws that secure the air cleaner-to-turbocharger ducting to its mountings.
32 Apply the parking brake, then jack up the front of the vehicle and support securely on axle stands (see *Jacking and Vehicle Support*).
33 Remove the fixings and lower the undertray away from the engine bay.
34 Unbolt the exhaust front pipe from the turbocharger exhaust flange.
35 Unbolt and remove the support bracket(s) from the base of the turbocharger and the engine block.
36 Clamp the hoses to minimise coolant loss,

5.27 Undo the union and disconnect the oil supply union from the top of the turbocharger

then slacken the clips and disconnect the coolant hoses from the turbocharger housing.
37 Slacken and withdraw the turbocharger-to-manifold securing bolts. Access is extremely limited - you will need to use a socket extension bar with a flexible or universal joint.
38 At the rear of the engine, below the intermediate driveshaft bearing support bracket, remove the bolt that secures the bracket to the engine torque rod. This will allow the engine to tilt slightly, giving greater clearance when the turbocharger is removed.
39 Tilt the engine forward slightly and wedge a piece of wood (approx. 50 mm thick) between the transmission casing and the suspension subframe to keep it in this position.
40 Slacken the clip and disconnect the air ducting from the turbocharger outlet port.
41 Remove the bolt and disconnect the air ducting from inlet port at the top of the turbocharger. Cover the turbocharger openings with clean rag.
42 Unscrew the union nut, and disconnect the oil feed pipe from the top of the turbocharger.
43 Undo the union and separate the oil return pipe from the base of the turbocharger. Recover the gasket.
44 Slacken and withdraw the turbocharger-to-manifold securing bolts.
45 Rotate the turbocharger and elbow assembly through half a turn and carefully withdraw it from the engine.

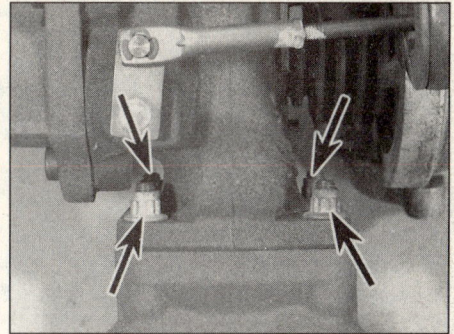

5.28 Slacken and remove the turbocharger-to-manifold nuts (arrowed) - 2.5 litre model shown

46 Lower the turbocharger and remove it from under the car. If it is to be refitted, store the turbocharger carefully, and plug its openings to prevent dirt ingress.

Refitting

47 Refitting is a reversal of removal, bearing in mind the following points:
a) *Renew all gaskets and oil supply and return union seals.*
b) *Tighten all fixings to the correct torque (where specified).*
c) *If a new turbo is being fitted, change the engine oil and filter. Also renew the filter (where fitted) in the oil feed pipe.*
d) *Do not fully tighten the oil feed pipe unions until both ends of the pipe are in place. When tightening the oil return pipe union, position it so that the return hose is not strained.*
e) *Before starting the engine, prime the turbo lubrication circuit by removing the fuel pump relay and cranking the engine on the starter for three ten-second bursts.*

6 Turbocharger - examination and renovation

1 With the turbocharger removed, inspect the housing for cracks or other visible damage.
2 Spin the turbine or the compressor wheel, to verify that the shaft is intact and to feel for excessive shake or roughness. Some play is

5.29a Renew all gaskets and oil supply and return union seals

5.29b Unscrew the oil supply line union at the cylinder block . . .

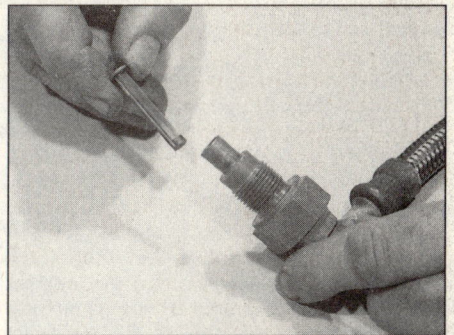

5.29c . . . and renew the oil filter cartridge

normal, since in use, the shaft is suspended on a film of oil. Check that the wheel vanes are undamaged.

3 If the exhaust or induction passages are oil-contaminated, the turbo shaft oil seals have probably failed. (On the induction side, this will also have contaminated the intercooler, which should if necessary be flushed with a suitable solvent.)

4 On petrol models, check that the coolant passages are clear and free from excessive corrosion.

5 No DIY repair of the turbo is possible. A new unit may be available on an exchange basis.

7 Exhaust system - general information, removal and refitting

General information

1 On all models, the exhaust system is made up of three sections (excluding the exhaust manifolds): the down pipe which incorporates the front silencer (or catalytic converter where fitted), the intermediate silencer, and the tail section which contains the two rear silencers.

2 The exhaust system is suspended along its entire length by rubber mountings, which are secured to the underside of the vehicle by metal brackets. The downpipe is secured to the transmission by means of a bracket and U-bolts.

3 The connection between the exhaust manifold/turbocharger and the downpipe is a gasketed flange joint, secured with bolts. The connection between the downpipe and the intermediate silencer is also a flange joint and is fitted with a sealing olive. A clamping ring, secured with a bolt is used to connect the intermediate silencer to the tail section.

Removal

4 Each exhaust section can be removed individually or, alternatively, the complete system can be removed as a unit.

5 Before removing any part of the system, first jack up the front or rear of the car, as applicable, and support it on axle stands. Alternatively position the car over an inspection pit or on car ramps.

Downpipe

6 Place blocks of wood under the catalytic converter, or the lowest point of the downpipe, to act as a support. Where applicable, unbolt and remove the lambda sensor from the exhaust pipe. This may entail the use of a special slotted socket which can either be fabricated from an old long reach socket or borrowed/hired from a Citroën dealer.

7 Slacken and remove the nuts securing the down pipe to the intermediate silencer (as applicable). Remove the bolts and recover the sealing olive from the joint.

8 Slacken and withdraw the bolts that secure the downpipe bracket to the base of the transmission casing (where applicable).

9 Undo the nuts and separate the downpipe from the exhaust manifold(s)/turbocharger Recover the gasket then withdraw the downpipe from underneath the vehicle.

Intermediate silencer

10 Slacken and remove the nuts securing the down pipe to the catalytic converter or the intermediate silencer. Remove the bolts and recover the sealing olive from the joint .

11 Slacken the catalytic converter to intermediate pipe clamping ring nut and bolt.

12 Free the catalytic converter from the intermediate pipe then withdraw it from underneath the vehicle.

Tailpipe

13 Slacken the clamping ring bolts and disengage the tailpipe at the joint.

14 Unhook the tailpipe from its mounting rubbers and remove it from the vehicle. Note: *Where applicable, silencers in the tail section can be separated from the exhaust system, to allow individual renewal, by slackening the clamp rings and pulling the silencers apart. Refer to a Citroën dealer or an exhaust specialist for further advice.*

Complete system

15 Disconnect the front pipe from the manifold(s) or turbocharger as described above. Where applicable, unplug the lambda sensor wiring at the connector - refer to the *Downpipe* sub-section for details.

16 With the aid of an assistant, free the system from all its mounting rubbers and manoeuvre it out from underneath the vehicle.

Refitting

17 Each section is refitted by a reverse of the removal sequence, noting the following points.

a) *Ensure that all traces of corrosion have been removed from the flanges and renew all necessary gaskets.*

b) *Inspect the rubber mountings for signs of damage or deterioration and renew as necessary.*

c) *Renew the sealing olive in the intermediate silencer-to-downpipe joint.*

d) *On joints which are secured by clamping rings, apply a smear of exhaust system jointing paste to the joint mating surfaces to ensure an air-tight seal. Tighten the clamping ring nuts securely.*

e) *Prior to tightening the exhaust system fasteners, ensure all rubber mountings are correctly located and that there is adequate clearance between the exhaust system and vehicle underbody.*

8 Catalytic converters - general information and precautions

1 The catalytic converter is a reliable and simple device, with no moving parts and as such requires no maintenance. There are, however, some facts of which an owner should be aware if the converter is to function properly for its full service life.

Petrol models

a) *DO NOT use leaded petrol in a car equipped with a catalytic converter - the lead will coat the precious metals reagents, reducing their converting efficiency and will eventually destroy the converter.*

b) *Always keep the ignition and fuel systems well-maintained in accordance with the manufacturer's schedule.*

c) *If the engine develops a misfire, do not drive the car at all (or at least as little as possible) until the fault is cured.*

d) *DO NOT push- or tow-start the car - this will soak the catalytic converter in unburned fuel, causing it to overheat when the engine does start.*

e) *DO NOT switch off the ignition at high engine speeds.*

f) *In some cases a sulphurous smell (like that of rotten eggs) may be noticed from the exhaust. This is common to many catalytic converter-equipped cars. Low quality fuel with a high sulphur content will exacerbate this effect.*

g) *The catalytic converter, used on a well-maintained and well-driven car, should last for between 50 000 and 100 000 miles - if the converter is no longer effective it must be renewed.*

Petrol and diesel models

h) *DO NOT use fuel or engine oil additives - these may contain substances harmful to the catalytic converter.*

i) *DO NOT continue to use the car if the engine burns oil to the extent of leaving a visible trail of blue smoke.*

j) *Remember that the catalytic converter operates at very high temperatures and its external casing can take a while to cool down. DO NOT, therefore, park the car in dry undergrowth, over long grass, or piles of dead leaves after a long run.*

k) *Remember that the catalytic converter is FRAGILE - do not strike it with tools during servicing work.*

9 Crankcase emission system - general information

1 The crankcase emission control system consists of a series of hoses that connect the

crankcase vent to the camshaft cover vent and the air intake, a pressure regulating valve (where applicable) and an oil separator unit.

2 The components of this system require no attention, other than to check at regular intervals that the hose(s) are free of blockages and undamaged.

10 Exhaust Gas Recirculation (EGR) system - general information and component renewal

General information

1 Several different versions of the EGR control system have been fitted to the XM. The type of system fitted and its mode of operation will depend on the age of the vehicle, the engine type and the market for which the vehicle is specified.

2 In depth coverage of each system type is beyond the scope of this manual, therefore the procedures described in this section are limited to the renewal of the major components that are common to each system.

Vacuum supply pump

Removal

3 The pump is located inside the right hand wheel arch, behind the plastic liner. Access is improved if the right hand road wheel is removed.

4 Ensure that the ignition is switched off, then unplug the wiring from the pump at the connector.

5 Make a careful note of their order of connection, then pull the vacuum hoses from the ports on the vacuum pump.

6 Undo the screws and remove the pump from the wheel arch.

Refitting

7 Refitting is a reversal of removal, but ensure that the vacuum hoses are reconnected to the correct ports on the pump.

EGR valve

Removal

8 Ensure that the engine has cooled completely before starting work. Jack up the front of the car and rest it securely on axle stands.

9 Unplug the vacuum hose from the port on the top of the EGR valve.

10 Remove the inlet manifold as described in Chapter 4B.

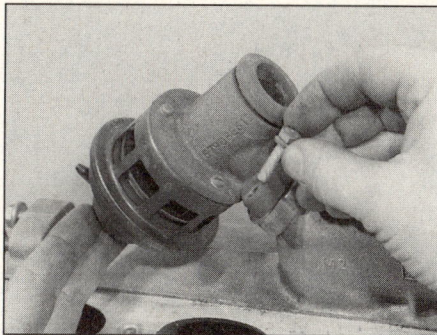

10.11 Remove the screws and lift the EGR valve from the manifold - 2.5 litre model shown

11 Slacken and remove the screws and lift the EGR valve from the manifold (see illustration).

Refitting

12 Refitting is a reversal of removal. Use new gasket and ensure that the securing nuts are tightened to the correct torque.

Vacuum solenoid valve

13 The solenoid valve is mounted on the right hand end of the inlet manifold (see Chapter 4B).

14 Disconnect the hoses from the valve, noting their order of connection to avoid confusion during refitting.

15 Remove the screws, lift the valve from its mountings and remove it from the engine bay (see illustration).

Refitting

16 Refitting is a reversal of removal.

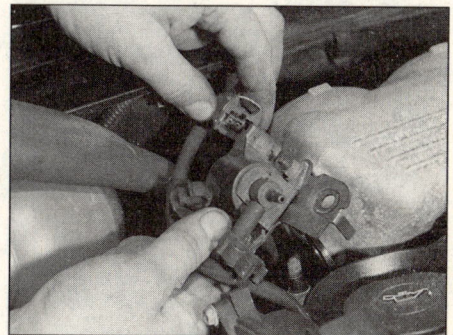

10.15 Removing the EGR vacuum solenoid valve

4C

Notes

Chapter 5 Part A:
Starting and charging systems

Contents

Degrees of difficulty

Easy, suitable for novice with little experience	**Fairly easy,** suitable for beginner with some experience	**Fairly difficult,** suitable for competent DIY mechanic	**Difficult,** suitable for experienced DIY mechanic	**Very difficult,** suitable for expert DIY or professional

5A

Specifications

System type . . . . 12-volt, negative earth

Battery

Type . . . . Low maintenance or 'maintenance-free' sealed for life
Charge condition:
  Poor . . . . 12.5 volts
  Normal . . . . 12.6 volts
  Good . . . . 12.7 volts

Alternator

Type . . . . Valeo or Mitsubishi (depending on model)

Starter motor

Type . . . . Valeo or Bosch (depending on model)

Torque wrench settings

At the time of writing, no torque wrench settings were available.

1 General information and precautions

General information

The engine electrical system consists mainly of the charging and starting systems. Because of their engine-related functions, these components are covered separately from the body electrical devices such as the lights, instruments, etc (which are covered in Chapter 13). On petrol engine models refer to Part B for information on the ignition system, and on diesel models refer to Part C for information on the preheating system.

The electrical system is of the 12-volt negative earth type.

The battery is of the low maintenance or 'maintenance-free' (sealed for life) type and is charged by the alternator, which is belt-driven from the crankshaft pulley.

The starter motor is of the pre-engaged type incorporating an integral solenoid. On starting, the solenoid moves the drive pinion into engagement with the flywheel ring gear before the starter motor is energised. Once the engine has started, a one-way clutch prevents the motor armature being driven by the engine until the pinion disengages from the flywheel.

Precautions

Further details of the various systems are given in the relevant Sections of this Chapter. While some repair procedures are given, the usual course of action is to renew the component concerned. The owner whose interest extends beyond mere component renewal should obtain a copy of the *Automobile Electrical & Electronic Systems Manual*, available from the publishers of this manual.

It is necessary to take extra care when working on the electrical system to avoid damage to semi-conductor devices (diodes and transistors), and to avoid the risk of personal injury. In addition to the precautions given in *Safety first!* at the beginning of this manual, observe the following when working on the system:

Always remove rings, watches, etc before working on the electrical system. Even with the battery disconnected, capacitive discharge could occur if a component's live terminal is earthed through a metal object. This could cause a shock or nasty burn.

Do not reverse the battery connections. Components such as the alternator, electronic control units, or any other components having semi-conductor circuitry could be irreparably damaged.

If the engine is being started using jump leads and a slave battery, connect the batteries positive-to-positive and negative-to-engine earth (see Booster battery (jump) starting).

Never disconnect the battery terminals, the alternator, any electrical wiring or any test instruments when the engine is running.

Do not allow the engine to turn the alternator when the alternator is not connected.

Never test for alternator output by flashing the output lead to earth.

Never use an ohmmeter of the type incorporating a hand-cranked generator for circuit or continuity testing.

Always ensure that the battery negative lead is disconnected when working on the electrical system.

Before using electric-arc welding equipment on the car, disconnect the battery, alternator and components such as the fuel injection/ignition electronic control unit to protect them from the risk of damage.

The radio/cassette unit fitted as standard equipment by Citroën is equipped with a built-in security code to deter thieves. If the power source to the unit is cut, the anti-theft system will activate. Even if the power source is immediately reconnected, the radio/cassette unit will not function until the correct security code has been entered. Therefore, if you do not know the correct security code for the radio/cassette unit do not disconnect the battery negative terminal of the battery or remove the radio/cassette unit from the vehicle. Refer to the Radio/cassette unit anti-theft system precaution Section for further information.

2 Electrical fault finding - general information

Refer to the information given in Chapter 13.

3 Battery - testing and charging

Standard and low maintenance battery - testing

1 If the vehicle covers a small annual mileage, it is worthwhile checking the specific gravity of the electrolyte every three months to determine the state of charge of the battery. Use a hydrometer to make the check and compare the results with the following table.

	Above 25°C	Below 25°C
Fully-charged	1.210 to 1.230	1.270 to 1.290
70% charged	1.170 to 1.190	1.230 to 1.250
Discharged	1.050 to 1.070	1.110 to 1.130

2 If the battery condition is suspect, first check the specific gravity of electrolyte in each cell. A variation of 0.040 or more between any cells indicates loss of electrolyte or deterioration of the internal plates.

3 If the specific gravity variation is 0.040 or more, the battery should be renewed. If the cell variation is satisfactory but the battery is discharged, it should be charged as described later in this Section.

Maintenance-free battery - testing

4 In cases where a 'sealed for life' maintenance-free battery is fitted, topping-up and testing of the electrolyte in each cell is not possible. The condition of the battery can therefore only be tested using a battery condition indicator or a voltmeter.

5 Certain models may be fitted with a maintenance-free battery, with a built-in charge condition indicator. The indicator is located in the top of the battery casing, and indicates the condition of the battery from its colour. If the indicator shows green, then the battery is in a good state of charge. If the indicator shows black, then the battery requires charging, as described later in this Section. If the indicator shows blue, then the electrolyte level in the battery is too low to allow further use, and the battery should be renewed. Do not attempt to charge, load or jump start a battery when the indicator shows blue.

6 If testing the battery using a voltmeter, connect the voltmeter across the battery and compare the result with those given in the Specifications under 'charge condition'. The test is only accurate if the battery has not been subjected to any kind of charge for the previous six hours. If this is not the case, switch on the headlights for 30 seconds, then wait four to five minutes before testing the battery after switching off the headlights. All other electrical circuits must be switched off, so check that the doors, boot and/or tailgate are fully shut when making the test.

7 If the voltage reading is less than 12.2 volts, then the battery is discharged, whilst a reading of 12.2 to 12.4 volts indicates a partially discharged condition.

8 If the battery is to be charged, remove it from the vehicle (Section 4) and charge it as described later in this Section.

Standard and low maintenance battery - charging

Note: *The following is intended as a guide only. Always refer to the manufacturer's recommendations (often printed on a label attached to the battery) before charging a battery.*

9 Charge the battery at a rate of 3.5 to 4 amps and continue to charge the battery at this rate until no further rise in specific gravity is noted over a four hour period.

10 Alternatively, a trickle charger charging at the rate of 1.5 amps can safely be used overnight.

11 Specially rapid 'boost' charges which are claimed to restore the power of the battery

in 1 to 2 hours are not recommended, as they can cause serious damage to the battery plates through overheating.

12 While charging the battery, note that the temperature of the electrolyte should never exceed 37.8°C (100°F).

Maintenance-free battery - charging

Note: *The following is intended as a guide only. Always refer to the manufacturer's recommendations (often printed on a label attached to the battery) before charging a battery.*

13 This battery type takes considerably longer to fully recharge than the standard type, the time taken being dependent on the extent of discharge.

14 A constant voltage type charger is required, to be set, when connected, to 13.9 to 14.9 volts with a charger current of 5.0 amps maximum. Using this method, the battery should be charged within twenty four hours.

4 Battery - removal and refitting

Note: *On models equipped with a Citroën anti-theft alarm system, disable the alarm before disconnecting the battery (see Chapter 13). If a Citroën radio/cassette unit is fitted, refer to Radio/cassette unit anti-theft system - precaution.*

Removal

1 The battery is located on the left-hand side of the engine compartment. Where applicable, remove the plastic cover from the battery holder.

2 Slacken the clamp bolts and disconnect the clamp from the battery negative (earth) terminal.

3 Remove the insulation cover (where fitted) and disconnect the positive terminal lead(s) in the same way.

4 Unscrew the bolt(s) and remove the battery retaining bar.

5 Lift the battery out of the engine compartment. If necessary, release all the relevant clips securing the wiring harness to the battery holder. Remove the screws and detach the fusebox(es) from the side of the battery tray. On diesel models, remove the screws and detach the fuel filter/priming pump and glow plug control unit from the side of the battery holder **(see illustration).**

6 Remove the securing bolts and lift the battery holder from the bodywork **(see illustration).** Where necessary, release the bonnet release cable from the battery holder, with reference to Chapter 12.

Refitting

7 Refitting is a reversal of removal, but smear petroleum jelly on the terminals when reconnecting the leads, and always reconnect the positive lead first, and the negative lead last.

5 Charging system - testing

Note: *Refer to Safety first! and Section 1 of this Chapter before starting work.*

1 If the ignition warning light fails to illuminate when the ignition is switched on, first check the alternator wiring connections for security. If satisfactory, check that the warning light bulb has not blown, and that the bulbholder is secure in its location in the instrument panel. If the light still fails to illuminate, check the continuity of the warning light feed wire from the alternator to the bulbholder. If all is satisfactory, the alternator is at fault and should be renewed or taken to an auto-electrician for testing and repair.

2 If the ignition warning light illuminates when the engine is running, stop the engine and check that the drivebelt is correctly tensioned (see Chapter 1A or 1B) and that the alternator connections are secure. If all is so far satisfactory, have the alternator checked by an auto-electrician for testing and repair.

3 If alternator output is suspect even though the warning light functions correctly, the regulated voltage may be checked as follows.

4 Connect a voltmeter across the battery terminals and start the engine.

5 Increase the engine speed until the reading stabilises; the reading should be around 12 to 13 volts, and no more than 14 volts.

6 Switch on several electrical accessories (eg, the headlights, heated rear window and heater blower), and check that the regulated voltage is maintained around 13 to 14 volts.

7 If the regulated voltage is not as stated, the fault may be due to worn brushes, weak brush springs, a faulty voltage regulator, a faulty diode, a severed phase winding or worn or damaged slip rings. The alternator should be renewed or taken to an auto-electrician for testing and repair.

6 Alternator drivebelt - removal, refitting and tensioning

Refer to the procedure given for the auxiliary drivebelt in Chapter 1A or 1B.

7 Alternator - removal and refitting

Removal

1 Disconnect the battery negative lead.

2 Slacken the auxiliary drivebelt (Chapter 1) and disengage it from the alternator pulley.

3 On 2.1 litre diesel models, unbolt the hydraulic system pump and move it to one side without disconnecting any hydraulic pipes or hoses (see Chapter 9).

4 Remove the rubber covers (where fitted) from the alternator terminals, then unscrew the nuts and disconnect the wiring from the rear of the alternator **(see illustrations).**

5 Unscrew the nut/bolt securing the alternator to the upper mounting bracket, then unscrew the lower mounting bolt. Note that, where a long through-bolt is used to secure

5A

4.5 On diesel models, detach the fuel filter/priming pump from the side of the battery holder

4.6 Lift the battery holder from the engine compartment

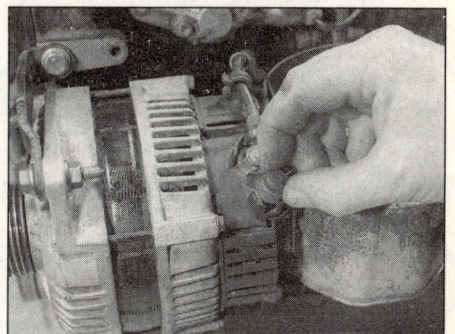

7.4a Remove the rubber covers from the alternator terminals . . .

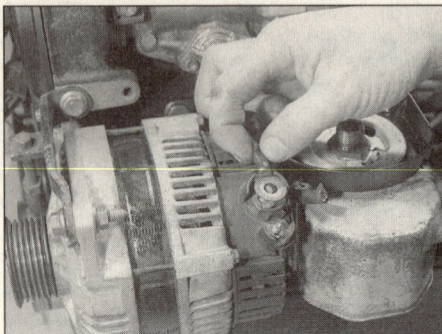

7.4b . . . then unscrew the nuts and disconnect the wiring from the rear of the alternator

7.5a Unscrew the nut and bolt securing the alternator to the upper mounting bracket . . .

7.5b . . . then unscrew the lower mounting bolt - 2.5 litre models shown

the alternator in position, the bolt does not need to be fully removed; the alternator can be disengaged from the bolt once it has been slackened sufficiently **(see illustrations)**. On some models, you may need to remove the drivebelt idler/tensioner pulley to access the alternator mounting nuts and bolts (depending on model). On diesel models, prise out the cover shield fastener to allow the cover to be lifted to improve access to the upper bolt.

6 Manoeuvre the alternator away from its brackets and out from the engine bay.

Refitting

7 Refitting is a reversal of removal. Tension the auxiliary drivebelt (Chapter 1A or 1B) on models with a manually adjusted tensioner, and ensure the alternator mountings are tightened.

8 Alternator -
testing and overhaul

If the alternator is thought to be suspect, it should be removed from the vehicle and taken to an auto-electrician for testing. Most auto-electricians will be able to supply and fit brushes at a reasonable cost. However, check on the cost of repairs before proceeding as it may prove more economical to obtain a new or exchange alternator.

9 Starting system - testing

Note: *Refer to the precautions given in Safety first! and in Section 1 of this Chapter before starting work.*

1 If the starter motor fails to operate when the ignition key is turned to the appropriate position, the following possible causes may be to blame.
a) *The battery is faulty.*
b) *The electrical connections between the switch, solenoid, battery and starter motor are somewhere failing to pass the necessary current from the battery through the starter to earth.*

c) *The solenoid is faulty.*
d) *The starter motor is mechanically or electrically defective.*

2 To check the battery, switch on the headlights. If they dim after a few seconds, this indicates that the battery is discharged - recharge (see Section 3) or renew the battery. If the headlights glow brightly, operate the ignition switch and observe the lights. If they dim, then this indicates that current is reaching the starter motor, therefore the fault must lie in the starter motor. If the lights continue to glow brightly (and no clicking sound can be heard from the starter motor solenoid), this indicates that there is a fault in the circuit or solenoid - see following paragraphs. If the starter motor turns slowly when operated, but the battery is in good condition, then this indicates that either the starter motor is faulty, or there is considerable resistance somewhere in the circuit.

3 If a fault in the circuit is suspected, disconnect the battery leads (including the earth connection to the body), the starter/solenoid wiring and the engine/transmission earth strap. Thoroughly clean the connections, and reconnect the leads and wiring, then use a voltmeter or test lamp to check that full battery voltage is available at the battery positive lead connection to the solenoid, and that the earth is sound. Smear petroleum jelly around the battery terminals to prevent corrosion - corroded connections are amongst the most frequent causes of electrical system faults.

4 If the battery and all connections are in good condition, check the circuit by disconnecting the wire from the solenoid blade terminal. Connect a voltmeter or test lamp between the wire end and a good earth (such as the battery negative terminal), and check that the wire is live when the ignition switch is turned to the 'start' position. If it is, then the circuit is sound - if not the circuit wiring can be checked (see Chapter 13).

5 The solenoid contacts can be checked by connecting a voltmeter or test lamp between the battery positive connection on the starter side of the solenoid, and earth. When the ignition switch is in the 'start' position, there should be a reading or lighted bulb, as

applicable. No reading nor lighted bulb means the solenoid is faulty and should be renewed.

6 If the circuit and solenoid are proved sound, the fault must lie in the starter motor. In this event, it may be possible to have the starter motor overhauled by a specialist, but check on the cost of spares before proceeding, as it may prove more economical to obtain a new or exchange motor.

10 Starter motor -
removal and refitting

Removal

1 Disconnect the battery negative lead.

2 So that access to the motor can be gained both from above and below, chock the rear wheels then jack up the front of the vehicle and support it on axle stands (see *Jacking and Vehicle Support*). Where applicable, to improve access to the motor remove the air cleaner housing as described in Chapter 4A or 4B.

3 On 2.5 litre turbo diesel models, carry out the following:
a) *Remove the intercooler and its mounting bracket as described in Chapter 4B.*
b) *Remove the cooling system expansion tank, with reference to Chapter 3.*

4 Slacken and remove the two retaining nuts and disconnect the wiring from the starter motor solenoid. Recover the washers under the nuts **(see illustrations)**.

10.4a Disconnect the solenoid wiring from starter motor . . .

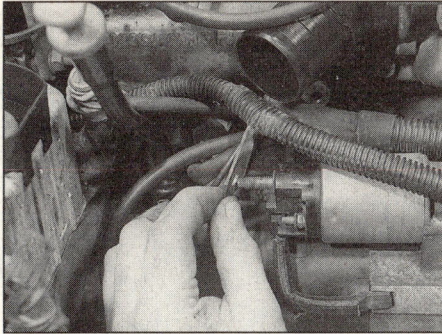

10.4b . . . and then the main supply wiring

10.5 Undo the three starter motor mounting bolts (arrowed) - 2.5 litre diesel model shown

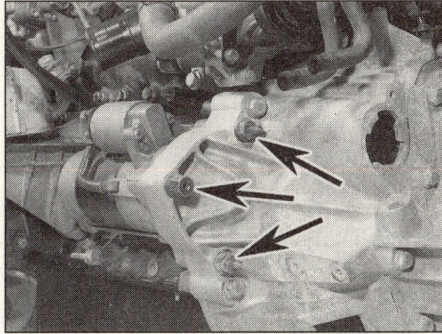

10.6 Removing the starter motor- 2.5 litre diesel model shown

5 Undo the three mounting bolts, supporting the motor as the bolts are withdrawn **(see illustration)**. Recover the washers from under the bolt heads and note the locations of any wiring or hose brackets secured by the bolts.

6 Manoeuvre the starter motor out from underneath the engine and recover the locating dowel(s) from the motor/transmission (as applicable) **(see illustration)**.

Refitting

7 Refitting is a reversal of removal, ensuring that the locating dowel(s) are correctly positioned (where applicable). Also make sure that any wiring or hose brackets are in place under the bolt heads as noted prior to removal.

11 Starter motor - testing and overhaul

If the starter motor is thought to be suspect, it should be removed from the vehicle and taken to an auto-electrician for testing. Most auto-electricians will be able to supply and fit brushes at a reasonable cost. However, check on the cost of repairs before proceeding as it may prove more economical to obtain a new or exchange motor.

12 Ignition switch - removal and refitting

The ignition switch is integral with the steering column lock, and can be removed as described in Chapter 11.

13 Oil pressure warning light switch - removal and refitting

Removal

1 The switch is located at the front of the cylinder block, above the oil filter mounting. Note that on some models access to the switch may be improved if the vehicle is jacked up and supported on axle stands so that the switch can be reached from below (see *Jacking and Vehicle Support*).

2 Disconnect the battery negative lead.

3 Remove the protective sleeve from the wiring plug (where applicable), then disconnect the wiring from the switch.

4 Unscrew the switch from the cylinder block, and recover the sealing washer. Be prepared for oil spillage, and if the switch is to be left removed from the engine for any length of time, plug the hole in the cylinder block.

Refitting

5 Examine the sealing washer for signs of damage or deterioration and if necessary renew.

6 Refit the switch, complete with washer, and tighten it securely. Reconnect the wiring connector.

7 Lower the vehicle to the ground then check and, if necessary, top-up the engine oil as described in *"Weekly checks"*.

14 Oil level sensor - removal and refitting

1 On 2.5 litre turbo diesel models, the sensor is located on the front of the cylinder block, just above the sump joint face. On all other models, the sensor is located on the rear facing side of the cylinder block at the flywheel end.

2 The removal and refitting procedure is as described for the oil pressure switch in Section 13. Access is most easily obtained from underneath the vehicle **(see illustrations)**. Withdraw the sensor from its mounting with care, as it has a long probe that can easily be damaged.

14.2a Removing the oil level sensor - 2.1 litre turbo diesel model

14.2b Removing the oil level sensor - 2.5 litre turbo diesel model

5A

Notes

Chapter 5 Part B:
Ignition system (petrol models)

Contents

Degrees of difficulty

Easy, suitable for novice with little experience	**Fairly easy,** suitable for beginner with some experience	**Fairly difficult,** suitable for competent DIY mechanic	**Difficult,** suitable for experienced DIY mechanic	**Very difficult,** suitable for expert DIY or professional

Specifications

System type
Engine code:
R6A	Breakerless distributor with transistor ignition amplifier
RFZ	Static, Bosch Motronic 5.1.
RGY, RGX	Static, Bosch Motronic 3.2.

Firing order . . . . 1-3-4-2 (No 1 cylinder at transmission end)

Ignition timing
Engine code:
R6A	5° BTDC at 850 rpm at normal operating temperature, with vacuum advance hose disconnected and plugged.
RFZ, RGY, RGX	Ignition timing controlled by engine management ECU (see text).

Ignition HT coil resistances
Engine code:
R6A :
Primary windings	0.8 Ω
Secondary windings	6.5 kΩ
RFZ, RGY, RGX	not available at time of writing

Miscellaneous
Distributor electromagnetic sensor, coil resistance (R6A engine) . . . . . 300 Ω

Torque wrench settings
At the time of writing, no torque figures were available for the ignition systems components.

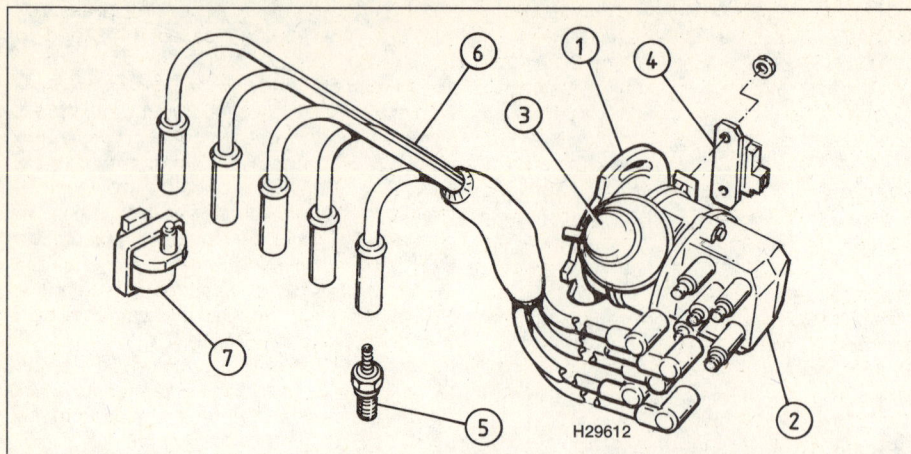

1.1a Ignition system HT components - models with a breakerless distributor ignition system

1 Distributor body	3 Vacuum advance	5 Spark plug
2 Distributor cap	diaphragm housing	6 HT leads
	4 Ignition amplifier module	7 Ignition coil

1 Ignition system - general information

Models with breakerless distributor ignition system

1 The ignition system consists of a breakerless distributor, mounted at the left hand end of the cylinder head and driven by the camshaft, a single ended HT coil, a transistorised ignition amplifier module, mounted on the side of the distributor body, and five HT leads (see illustrations).

2 The distributor contains an electronic trigger mechanism which is made up of a four pole trigger wheel mounted on the distributor shaft and an electromagnetic sensor. When the distributor shaft rotates, the trigger wheel poles pass under the sensor and alter the current flowing through it. This causes the sensor to send a trigger signal to the amplifier module, which then interrupts the LT supply to the coil primary windings. The electromagnetic field surrounding the primary windings collapses and this induces a large HT voltage in the secondary windings. The HT voltage is then directed to the appropriate spark plug, via the HT leads, by the distributor rotor arm passing under the distributor cap terminals.

3 The static ignition timing is set by altering the position of the distributor on its mountings, using a strobe timing lamp to illuminate timing marks on the flywheel and transmission bellhousing (refer to the relevant Section for greater detail). Dwell angle is set by the amplifier module and as such is not adjustable.

4 Dynamic ignition timing is controlled by mechanisms within the distributor. Engine speed advance correction is handled by a centrifugal advance mechanism. Engine load

1.1b The ignition amplifier module is mounted on the side of the distributor body

advance correction is carried out by a vacuum diaphragm module, mounted on the side of the distributor.

Models with Bosch Motronic engine management

5 The ignition system is integrated with the fuel injection system to form a combined engine management system under the control of one ECU (See the relevant Part of Chapter 4 for further information).

6 The ignition side of the system is of the static (distributorless) type, consisting only of a four output ignition coil (see illustration). The ignition coil actually consists of two separate, double-ended HT coils, which supply two cylinders each. Under the control of the ECU, the ignition coil operates on the 'wasted spark' principle, ie. each spark plug sparks twice for every cycle of the engine, once on the compression stroke and once on the exhaust stroke (hence the ' wasted spark'). The ECU uses its inputs from the various sensors to calculate the required ignition advance setting and coil charging time. The ignition timing is under the control of the engine management system ECU and is not manually adjustable without access to dedicated electronic test equipment. Static ignition timing is derived from a signal received from the crankshaft sensor. A basic ignition setting at idle speed is not quoted, because the ignition timing is constantly being altered to control engine idle speed (in conjunction with the idle actuator valve). The vehicle must be taken to a Citroën dealer or Bosch engine management system specialist if the ignition timing requires checking or adjustment.

7 A knock sensor is incorporated into the ignition system. The sensor is mounted onto the cylinder block and is sensitive to the high frequency vibrations which occur when the engine starts to 'pink' (pre-ignite), usually under load. Under these conditions, the knock sensor sends an electrical signal to the ECU, which in turn retards the ignition advance setting until the 'pinking' ceases. The ECU then advances the ignition again in small steps, to maintain optimum engine performance and economy. This process is

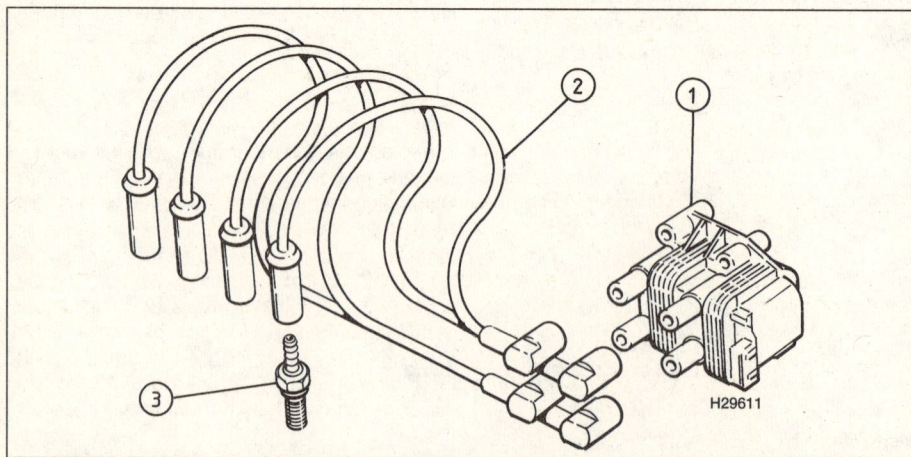

1.6 Ignition system HT components - models with Bosch Motronic engine management

1 Ignition coil   2 HT leads   3 Spark plugs

repeated whenever engine knock is detected.

8 It should be noted that comprehensive fault diagnosis of all the engine management systems described in this Chapter is only possible with dedicated electronic test equipment. Problems with the systems operation that cannot be pinpointed by following the basic guidelines in Section 2 should therefore be referred to a Citroën dealer for assessment. Once the fault has been identified, the removal/refitting sequences detailed in the following Sections will then allow the appropriate component(s) to renewed as required.

2 Ignition system - testing

⚠ *Warning: Voltages produced by an electronic ignition system are considerably higher than those produced by conventional ignition systems. Extreme care must be taken when working on the system with the ignition switched on. Persons with surgically-implanted cardiac pacemaker devices should keep well clear of the ignition circuits, components and test equipment.*

Models with breakerless distributor ignition systems

General

1 Most ignition system faults are likely to be due to loose or dirty connections or to 'tracking' (unintentional earthing) of HT voltage due to dirt, dampness or damaged insulation, rather than by the failure of any of the system's components. **Always** check all wiring thoroughly before condemning an electrical component and work methodically to eliminate all other possibilities before deciding that a particular component is faulty.

2 The old practice of checking for a spark by holding the live end of an HT lead a short distance away from the engine is not recommended; not only is there a high risk of an electric shock, but the HT coil could be damaged. Similarly, **never** try to 'diagnose' misfires by pulling off one HT lead at a time, as well as the risk of electric shock, the test may be invalid in the ignition system fitted to later models has the capability to detect and temporarily disable HT lines that are open circuit.

Engine will not start

3 If the engine either will not turn over at all, or only turns very slowly, check the battery and starter motor. Connect a voltmeter across the battery terminals (meter positive probe to battery positive terminal), disconnect the ignition coil HT lead from the distributor cap and earth it, then note the voltage reading obtained while turning over the engine on the starter for (no more than) ten seconds. If the

reading obtained is less than approximately 9.5 volts, first check the battery, starter motor and charging systems (see Chapter 5, Part A).

4 If the engine turns over at normal speed but will not start, check the HT circuit by connecting a timing light (following the manufacturer's instructions) and turning the engine over on the starter motor; if the light flashes, voltage is reaching the spark plugs, so these should be checked first. If the light does not flash, check the HT leads themselves followed by the distributor cap, carbon brush and rotor arm using the information given in Chapter 1A.

5 If there is a spark, check the fuel system for faults referring to the relevant part of Chapter 4 for further information.

6 If there is still no spark, then the problem may lie within the ignition control/engine management system. In these cases, the vehicle should be referred to a Citroën dealer or automotive electrical specialist for assessment.

Engine misfires

7 An irregular misfire suggests either a loose connection or intermittent fault in the control system or primary circuit, or an HT fault on the coil side of the rotor arm (where applicable).

8 With the ignition switched off, check carefully through the system ensuring that all connections are clean and securely fastened. If the equipment is available, check the LT circuit as described above.

9 Check that the HT coil, the distributor cap and the HT leads are clean and dry. Check the leads themselves and the spark plugs (by substitution, if necessary), then check the distributor cap, carbon brush and rotor arm as described in Chapter 1A.

10 Regular misfiring is almost certainly due to a fault in the distributor cap (where applicable), coil, HT leads or spark plugs. Use a timing light (paragraph 4 above) to check whether HT voltage is present at all leads.

11 If HT voltage is not present on one particular lead, the fault will be in that lead or in the distributor cap. If HT is present on all leads, the fault will be in the spark plugs; check and renew them if there is any doubt about their condition.

12 If no HT voltage is present, check the HT coil(s); the secondary windings may be breaking down under load.

Models with Bosch Motronic engine management systems

13 If a fault appears in the engine management (fuel injection/ignition) system first ensure that the fault is not due to a poor electrical connection or poor maintenance; ie, check that the air cleaner filter element is clean, the spark plugs are in good condition and correctly gapped, that the engine breather hoses are clear and undamaged. Also check that the accelerator cable is correctly adjusted as described in the relevant part of Chapter 4. If the engine is running very

roughly, check the compression pressures and the valve clearances as described in Chapter 2A.

14 If these checks fail to reveal the cause of the problem the vehicle should be taken to a suitably equipped Citroën dealer for testing. A wiring block connector is incorporated in the engine management circuit into which a special electronic diagnostic tester can be plugged. The tester will locate the fault quickly and simply alleviating the need to test all the system components individually which is a time consuming operation that carries a high risk of damaging the ECU.

15 The only ignition system checks which can be carried out by the home mechanic are those described in Chapter 1A, relating to the spark plugs, and the ignition coil test described in this Chapter. If necessary, the system wiring and wiring connectors can be checked as described in Chapter 13 ensuring that the ECU wiring connector(s) have first been disconnected.

3 Ignition HT coil - removal, testing and refitting

Models with breakerless distributor ignition systems

Removal

1 The ignition HT coil is mounted on a bracket at the front of the engine, underneath the intake air ducting.

2 Disconnect the negative cable from the battery terminal.

3 Slacken the hose clips at either end and remove the section of air ducting that runs between the air flow meter and the throttle body.

4 Depress the retaining clip and disconnect the LT wiring connector from the HT coil.

5 Pull the HT king lead from the coil terminal. Pull on the cable end fitting, not the cable itself. Remove the securing screws and lift the coil away from is mounting bracket.

Testing

6 Testing of the coil consists of using a multimeter set to its resistance function, to check the resistance of the primary (LT '+' to '-' terminals) and secondary (LT '+' to HT lead terminal) windings. Compare the results obtained to those given in the Specifications at the start of this Chapter. Note the resistance of the coil windings will vary slightly according to the coil temperature, the results in the Specifications are approximate values for when the coil is at 20°C. A short circuit (zero ohms) or open circuit (no continuity) in either of the windings means that the coil is faulty.

7 Check that there is no continuity between the HT lead terminal and the coil body/ mounting bracket. If the coil is thought to be

5B

faulty, have your findings confirmed by a Citroën dealer before renewing the coil.

Refitting

8 Refitting is a reversal of the removal procedure, ensuring that the wiring connectors are securely reconnected.

Models with Bosch engine management systems

Removal

9 Disconnect the battery negative terminal. The ignition HT coil is mounted on the left-hand end of the cylinder head.
10 Depress the retaining clip and disconnect the wiring connector from the HT coil.
11 Make a note of the correct fitted positions of the HT leads then disconnect them from the coil terminals.
12 Undo the four retaining screws securing the coil to its mounting bracket and remove it from the engine compartment.

Testing

13 Testing of the coil is limited to checking the continuity of the primary (LT '+' to '-' terminals) and secondary (LT '+' to HT lead terminal) windings, using a multimeter. An open circuit (no continuity) in either of the windings means that the coil is faulty.
14 Check that there is no continuity between the HT lead terminal and the coil body/mounting bracket.
15 If the coil is thought to be faulty, have your findings confirmed by a Citroën dealer before renewing the coil.

Refitting

16 Refitting is a reversal of the relevant removal procedure ensuring that the wiring connectors are securely reconnected and, where necessary, the HT leads are correctly connected.

4 Ignition timing - checking and adjustment

Models with Bosch engine management systems

1 On all fuel-injected models, there are no timing marks on the flywheel or crankshaft pulley. The timing is constantly being monitored and adjusted by the engine management ECU, and nominal values cannot be given. Therefore, it is not possible for the home mechanic to check the ignition timing.
2 The only way in which the ignition timing can be checked is using dedicated electronic test equipment, connected to the engine management system diagnostic connector. Refer to your Citroën dealer, or a Bosch engine management system specialist for advice.

Models with breakerless distributor ignition systems

Checking

Note: *The engine should be at normal operating temperature to ensure correct results. Ideally the test should be completed before the coolant temperature exceeds 95ºC, or before the auxiliary cooling fan cuts in.*
3 Connect the stroboscopic timing light to the engine in accordance with the manufacturer's instructions, so that it is triggering from the No 1 cylinder HT lead.
4 Disconnect the vacuum advance hose from the diaphragm unit on the side of the distributor. Prevent air entering the manifold via the hose by plugging the open end of the hose.
5 Remove the cover from the timing mark access hole, located at the top of the transmission bellhousing, below the distributor **(see illustration)**. It may be necessary to push the adjacent coolant hoses to one side, to gain access to the cover.
6 Start the engine and run it at idling speed. Direct the beam from timing light at the timing mark access hole. The stroboscopic effect should 'freeze' the motion of the rotating flywheel and the timing mark stamped upon it. If the mark appears to be moving back and forth, this may be due erratic idling - check that all of the cars electrical accessories are switched off and that the auxiliary cooling fan is not running. The engine should be at normal operating temperature, but the idle speed may be become unstable if it is particularly hot day and the engine has been idling for some time.
7 Read off the ignition timing by observing the position of the mark on the flywheel in relation to the markings on the graduated plate. Compare your readings with the Specifications.

Ignition timing - adjusting

8 To adjust the basic ignition timing setting, first switch off the engine, to avoid the risk of a getting a shock from the HT voltage. Refer to Section 5 and slacken the distributor clamp bolt(s). **Note:** *It is good idea to mark the relationship between the distributor body and*

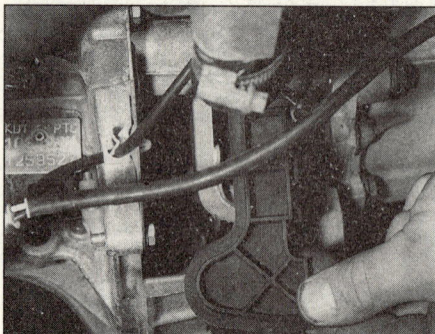

4.5 Remove the cover from the timing mark access hole, located at the top of the transmission bellhousing

the cylinder head with a dab of paint, before adjustment is attempted. This gives a reference point that can be reverted to if the timing setting is lost.
9 Turn the distributor *by a small amount* clockwise to advance (or anti-clockwise to retard) the ignition timing. Tighten the bolts lightly and re-check the timing using the stroboscopic light, as described in the previous sub-section. Repeat this process, until the timing setting is correct, then tighten the distributor clamp bolt securely and disconnect the stroboscopic light.

5 Distributor - removal and refitting

Removal

1 The distributor is mounted horizontally on the left hand end of the cylinder head and is driven directly by the camshaft.
2 Disconnect the battery negative cable and position it away from the terminal.
3 Set the engine to TDC on cylinder No 1, as follows. Remove the screws, lift off the distributor cap and move it to one side, with the HT leads still attached.
4 On the distributor body, find the alignment marking, which is a small recess cut into the upper edge of the distributor cap mating surface. Highlight it if required, using typists correction fluid or similar.
5 To bring any piston up to TDC, it will be necessary to rotate the crankshaft manually. This can be done by using a ratchet wrench and socket on the crankshaft bolt at the front of the engine.
6 Turn the crankshaft in its normal direction of rotation, until the distributor rotor arm electrode approaches the mark that was made on the distributor body.
7 With the cover removed from the timing mark access hatch (see Section 4), observe the timing mark on the flywheel. Continue rotating the crankshaft until the flywheel timing mark is aligned exactly with the '0' TDC notch on the graduated timing plate. Note: *Observe from directly above the pointer to ensure correct alignment.*
8 Check that the distributor rotor arm is now pointing to the marking on the distributor body. If it is offset by 180º, then No 1 cylinder is on its exhaust stroke; rotate the crankshaft through one complete revolution and repeat the steps in paragraphs 6 to 7 inclusive.
9 Cylinder No 1 will then be set at TDC. The distributor can now be removed and refitted without losing its alignment, provided that the engine is not rotated after the distributor body has been withdrawn from the crankcase.
10 Unplug the hose from the vacuum unit on the side of the distributor body.
11 Unplug the sensor wiring from the distributor, at the connector on the side of the distributor body.

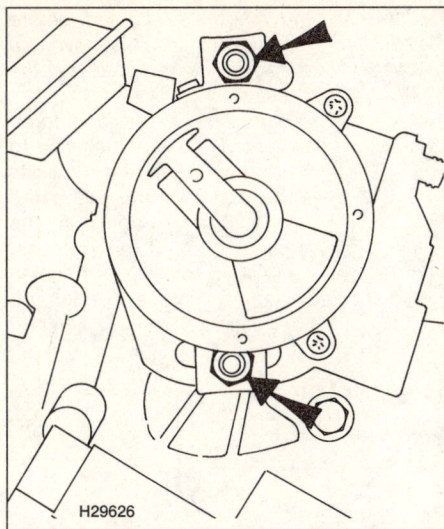
5.13 Distributor mounting nuts (arrowed)

12 Mark the relationship between the base of the distributor and the mounting flange, using a dab of paint or a scriber, to act as an alignment aid, during refitting.
13 Slacken and remove the securing nuts, then lift the distributor body from the cylinder head (see illustration).
14 Check the condition of the O-ring oil seal (where fitted); if it shows signs of deterioration or if there are signs of leakage, fit a new seal.

Refitting

15 If the engine has not been rotated from

TDC on cylinder No 1 since the removal of the distributor, then proceed from paragraph 17. If the alignment of the engine has been lost, then proceed from paragraph 16.
16 Remove No 1 spark plug with reference to Chapter 1. Place a thumb over the spark plug hole, then turn the crankshaft (as described during *Removal*) until the following conditions are true:
a) *Pressure can be felt building up at the spark plug hole.*
b) *The TDC marking on the flywheel is exactly aligned with the pointer on the bellhousing (see Removal for details).*
17 Insert the distributor into the cylinder head and engage the offset spigots at the end of the distributor shaft with the drive gear, such that:
a) *When the distributor body is flush with its mounting flange, the rotor arm is pointing at the alignment marking on the distributor body.*
b) *The securing bolt holes on the cylinder head are aligned with the centre of the elongated slots on the mounting flange at the base of the distributor.*
Turn the distributor body to achieve this, if necessary.
Note: *The distributor shaft drive spigots are offset, and can only be refitted the one way around.*
18 Insert the distributor securing bolts and tighten them securely.
19 Fit the distributor cap in position, then insert and tighten the securing screws. Use the alignment marks made on the base of the distributor body during removal, to give an approximate ignition timing setting.

20 Reconnect the vacuum unit hose and the sensor wiring.
21 On completion, check and if necessary adjust the ignition timing with reference to Section 4.

6 Rotor arm - removal and refitting

Removal

1 Remove the distributor cap as described in Section 5.
2 Pull the rotor arm from the end distributor shaft.
3 Check the condition of the rotor arms metal contact surfaces. If there are signs of corrosion, or severe pitting or wear, it should be renewed. Similarly, if the plastic body is cracked or shows signs or tracking, then the rotor arm must be renewed.

Refitting

4 Push the rotor arm firmly onto the end of the distributor shaft, ensuring that the lug on the inside of the rotor arm body engages with the recess in the end of the shaft.
5 Refit the distributor cap as described in Section 5.

7 Ignition control system components - removal and refitting

Refer to the information given in Chapter 4A.

5B

Notes

Chapter 5 Part C:
Preheating system (diesel models)

Contents

Degrees of difficulty

Easy, suitable for novice with little experience	Fairly easy, suitable for beginner with some experience	Fairly difficult, suitable for competent DIY mechanic	Difficult, suitable for experienced DIY mechanic	Very difficult, suitable for expert DIY or professional

Specifications

Torque wrench setting	Nm	lbf ft
Glow plugs	22	16

5C

1 Preheating system - description and testing

Description

1 Each swirl chamber has a heater plug (commonly called a glow plug) screwed into it. The plugs are electrically-operated before and during start-up when the engine is cold. Electrical feed to the glow plugs is controlled by a relay/timer unit, in conjunction with the diesel injection system electronic control unit (ECU).

2 On certain models, the glow plugs provide a 'post-heating' function, whereby the glow plugs remain switched on for a period after the engine has started. Once the starter has been switched off, the glow plugs begin a timed 3-minute 'post-heating' cycle. The operation of the plugs cannot be cancelled for the first 15 seconds, but after the first 15 seconds, the supply to the plugs will be interrupted by:

a) Operation of the accelerator pedal beyond a travel of 11 mm for a duration of more than 2.5 seconds.

b) A coolant temperature of more than 60°C.

3 A warning light in the instrument panel tells the driver that preheating is taking place. When the light goes out, the engine is ready to be started. The voltage supply to the glow plugs continues for several seconds after the light goes out. If no attempt is made to start, the timer then cuts off the supply, in order to avoid draining the battery and overheating the glow plugs.

Testing

4 If the system malfunctions, testing is ultimately by substitution of known good units, but some preliminary checks may be made as follows.

5 Connect a voltmeter or 12-volt test lamp between the glow plug supply cable and earth (engine or vehicle metal). Make sure that the live connection is kept clear of the engine and bodywork.

6 Have an assistant switch on the ignition, and check that voltage is applied to the glow plugs. Note the time for which the warning light is lit, and the total time for which voltage is applied before the system cuts out. Switch off the ignition.

7 At an under-bonnet temperature of 20°C, typical times noted should be 5 or 6 seconds for warning light operation, followed by a further 10 seconds supply after the light goes out. Warning light time will increase with lower temperatures and decrease with higher temperatures.

8 If there is no supply at all, the relay or associated wiring is at fault.

9 To locate a defective glow plug, remove the air intake ducting (Chapter 4B) and, where necessary, disconnect the breather hose from the engine oil filler tube. Disconnect the main supply cable and the interconnecting wire or strap from the top of the glow plugs. Be careful not to drop the nuts and washers.

10 Use a continuity tester, or a 12-volt test lamp connected to the battery positive terminal, to check for continuity between each glow plug terminal and earth. The resistance of a glow plug in good condition is very low (less than 1 ohm), so if the test lamp does not light or the continuity tester shows a high resistance, the glow plug is certainly defective.

11 As a final check, the glow plugs can be removed and inspected as described in the following Section.

2 Glow plugs - removal, inspection and refitting

Removal

Caution: If the preheating system has just been energised, or if the engine has been running, the glow plugs will be very hot.

1 Disconnect the battery negative lead. To improve access on 2.1 litre models, remove the air intake ducting (Chapter 4B) and, where necessary, disconnect the breather hose from the engine oil filler tube. To improve access on 2.5 litre models, remove the intercooler (Chapter 4B) and cooling system expansion tank (Chapter 3).

2 Unscrew the nut from the relevant glow plug terminal(s), and recover the washer(s).

2.2a Unscrew the nut (arrowed) from the relevant glow plug terminal - 2.1 litre model shown

2.2b Removing the glow plug supply wiring - 2.5 litre model shown

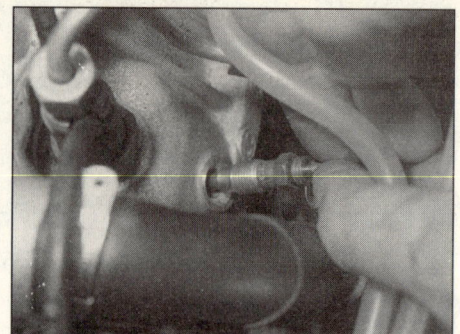

2.4a Unscrew the glow plug and remove it from the cylinder head - 2.1 litre model shown

The operation will be simplified if a universal or flexible joint can be attached between the wrench extension bar and socket. Note the main supply cable connection (usually to Number 2 cylinder glow plug) and the interconnecting wire fitted between the four plugs **(see illustrations)**.

3 Where applicable, carefully move any obstructing pipes or wires to one side to enable access to the relevant glow plug(s).

4 Unscrew the glow plug(s) and remove from the cylinder head **(see illustrations)**.

Inspection

5 Inspect each glow plug for physical damage. Burnt or eroded glow plug tips can be caused by a bad injector spray pattern. Have the injectors checked if this sort of damage is found.

6 If the glow plugs are in good physical condition, check them electrically using a 12 volt test lamp or continuity tester as described in the previous Section.

7 The glow plugs can be energised by applying 12 volts to them to verify that they heat up evenly and in the required time. Observe the following precautions.

a) *Support the glow plug by clamping it carefully in a vice or self-locking pliers. Remember it will become red-hot.*

b) *Make sure that the power supply or test lead incorporates a fuse or overload trip to protect against damage from a short-circuit.*

c) *After testing, allow the glow plug to cool for several minutes before attempting to handle it.*

8 A glow plug in good condition will start to glow red at the tip after drawing current for 5 seconds or so. Any plug which takes much longer to start glowing, or which starts glowing in the middle instead of at the tip, is defective.

Refitting

9 Refit by reversing the removal operations. Apply a smear of copper-based anti-seize compound to the plug threads and tighten the glow plugs to the specified torque. Do not overtighten, as this can damage the glow plug element.

2.4b Removing a glow plug - 2.5 litre model shown

---

3 Preheating system control unit - removal and refitting

Removal

1 The glow plug control unit is attached at the front of the battery holder **(see illustration)**.

2 Disconnect the battery negative lead.

3 Remove the securing screw and lift the control unit away from the battery holder.

4 Unplug the wiring at the connector.

Refitting

5 Refitting is a reversal of removal, ensuring that the wiring connectors are correctly engaged.

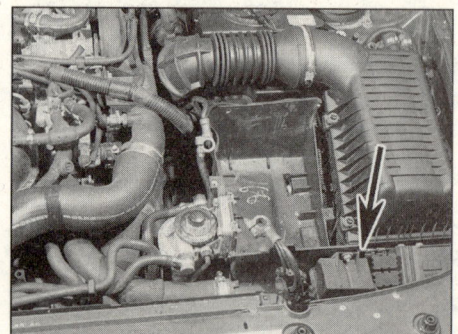

3.1 The glow plug control unit (arrowed) is attached at the front of the battery holder - 2.5 litre model shown

Chapter 6
Clutch

Contents

Degrees of difficulty

Easy, suitable for novice with little experience	**Fairly easy,** suitable for beginner with some experience	**Fairly difficult,** suitable for competent DIY mechanic	**Difficult,** suitable for experienced DIY mechanic	**Very difficult,** suitable for expert DIY or professional

Specifications

General
Type . . . .   Single dry plate with diaphragm spring, cable or hydraulic operation

Torque wrench setting

	Nm	lbf ft
Pressure plate retaining bolts	20	15

1 General information

The clutch consists of a friction plate, a pressure plate assembly, a release bearing and the release mechanism; all of these components are contained in the large cast-aluminium alloy bellhousing, sandwiched between the engine and the transmission. The release mechanism is mechanical, and is operated by a self-adjusting cable on all petrol models. On diesel models, the release mechanism is operated hydraulically by means of a master and slave cylinder and interconnecting hydraulic pipework.

The friction plate is fitted between the engine flywheel and the clutch pressure plate, and is allowed to slide on the transmission input shaft splines.

The pressure plate assembly is bolted to the engine flywheel. When the engine is running, drive is transmitted from the crankshaft, via the flywheel, to the friction plate (these components being clamped securely together by the pressure plate assembly) and from the friction plate to the transmission input shaft.

To interrupt the drive, the pressure plate spring fingers must be relaxed. On the models

covered in this manual, two different types of clutch release mechanism are used. The first is a conventional 'push-type' mechanism, where an independent clutch release bearing, fitted concentrically around the transmission input shaft, is pushed onto the pressure plate assembly; this type is fitted to all models except the 2.5 turbo diesel. The second is a 'pull-type' mechanism, where the clutch release bearing is an integral part of the pressure plate assembly, and is lifted away from the friction plate; this type is fitted to 2.5 litre turbo diesel models.

On models with the conventional 'push-type' mechanism, at the transmission end of the clutch cable, the outer cable is retained by a fixed mounting bracket, and the inner cable is attached to the release fork lever. Depressing the clutch pedal pulls the control cable inner wire, and this rotates the release fork by acting on the lever at the fork's upper end. The release fork then presses the release bearing against the pressure plate spring fingers. This causes the springs to deform and releases the clamping force on the pressure plate.

On 2.5 litre turbo diesel models with the 'pull-type' mechanism, the clutch pedal is connected to the clutch master cylinder by a pushrod. The master cylinder is mounted on the engine compartment bulkhead with the

slave cylinder mounted on the side of the transmission. A hydraulic fluid reservoir is also remotely mounted on the bulkhead and connected to the master cylinder by means of a fluid hose. Depressing the clutch pedal moves the piston in the master cylinder forwards, so forcing hydraulic fluid through the clutch hydraulic pipe to the slave cylinder. The piston in the slave cylinder moves forward under hydraulic pressure and actuates the release fork by means of a short pushrod. The release fork then lifts the release bearing, which is attached to the pressure plate springs, away from the friction plate, and releases the clamping force exerted at the pressure plate periphery.

The clutch master cylinder, fluid reservoir, slave cylinder, and interconnecting fluid pipework are all part of a maintenance-free, sealed assembly. The fluid reservoir cannot be opened and the fluid level does not require topping-up. In the event of fluid leakage or any malfunction of the hydraulic system, all the components must be renewed as a complete assembly.

On Diesel models, adjustment of the clutch to compensate for wear of the friction plate linings is automatically taken up by the hydraulic clutch components. For petrol models consult the following section for clutch cable adjustment.

6

2 Clutch cable -
removal, refitting and adjustment

Removal

1 Disconnect the battery negative terminal.
2 Slacken and withdraw the retaining screws, then remove the driver's side lower panel from the facia.
3 Remove the air cleaner housing and intake duct components as described in the relevant Part of Chapter 4.
4 Working in the engine compartment, release the inner cable and outer cable fittings from the clutch release lever and mounting bracket and free the cable from the transmission housing.
5 Working inside the vehicle, release the inner cable from the pedal. On some models it may be necessary to depress a plastic guide clip, at the top of the pedal, to release the cable.
6 Return to the engine compartment, then release the cable guide from the bulkhead and withdraw the cable forwards, releasing it from any relevant retaining clips and guides. Note its correct routing, and remove it from the vehicle. Note that on right hand drive vehicles, the cable has an integral right-angled bracket assembly, which directs the run of the cable across the bulkhead to the left hand side of the engine compartment.
7 Examine the cable, looking for worn end fittings or a damaged outer casing, and for signs of fraying of the inner wire. Renew the cable if it shows signs of excessive wear or any damage.

Refitting

8 Apply a thin smear of multi-purpose grease to the cable end fittings, then pass the cable through the engine compartment bulkhead.
9 Hold the clutch pedal in its raised position by wedging a suitable tool beneath it.
10 Guide the end of the cable into the pedal end (or plastic clip) making sure that it is fully engaged.
11 In the engine compartment refit the cable to the transmission housing and release lever. Re-secure the cable with any relevant retaining clips and guides.
12 Depress and release the clutch pedal at least thirty times, to allow the cable to stretch and settle. On completion, check the clutch adjustment.
13 Refit the lower panel to the facia then refit the air cleaner components and reconnect the battery.

Adjustment

14 The clutch cable has no integral self-adjusting mechanism; the pedal stroke can be adjusted as follows.
15 Using a tape measure, measure the distance between the clutch pedal rubber pad and the bottom of the steering wheel rim, when the pedal is at rest.
16 Now take the measurement again, but this time with the pedal fully depressed. Subtract the second measurement from the first.
17 Compare your results with the pedal travel figure given in the specifications of Chapters 1A or 1B. If the figure is incorrect, adjust the pedal travel by slackening the locknut at the transmission end of the clutch cable, and turning the adjuster nut. When the travel is correct, tighten the locknut securely.

3 Clutch pedal -
removal and refitting

Removal

1 Disconnect the battery negative terminal.
2 Slacken and withdraw the retaining screws and remove the driver's side lower panel from the facia.
3 On models with a cable-operated clutch, release the inner cable from the pedal. On some models it may be necessary to depress a plastic clip located just beneath the top of the pedal to release the cable. Where applicable, carefully release the pedal-assist spring assembly from the pedal (see **illustrations**).
4 On models with a hydraulically-operated clutch, release the master cylinder pushrod balljoint from the pedal and allow the pedal to rise until it reaches its stop.
5 Slacken and remove the pivot bolt and nut, and remove the clutch pedal from the vehicle. Slide the spacer out from the pedal pivot. Examine all components for signs of wear or damage, renewing them as necessary.

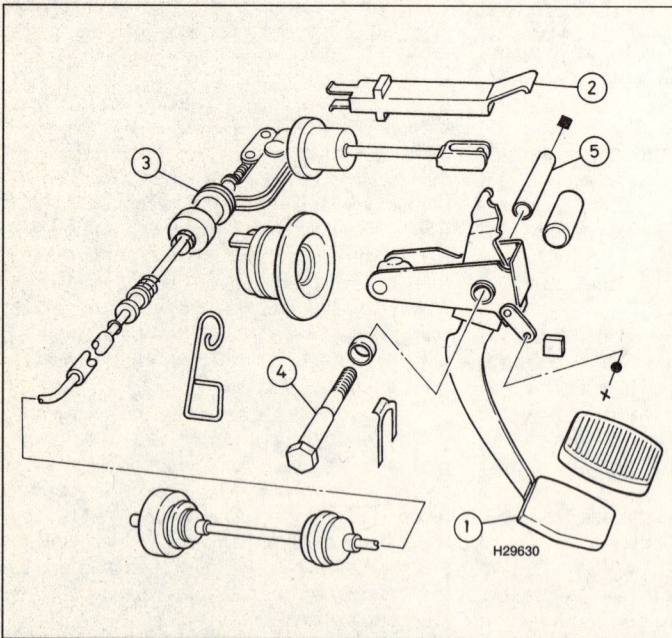

3.3a Clutch pedal and cable mounting details - right hand drive models

1	*Clutch pedal*	*3*	*Clutch cable assembly*	*4*	*Pivot bolt*
2	*Cable guide*			*5*	*Spacer*

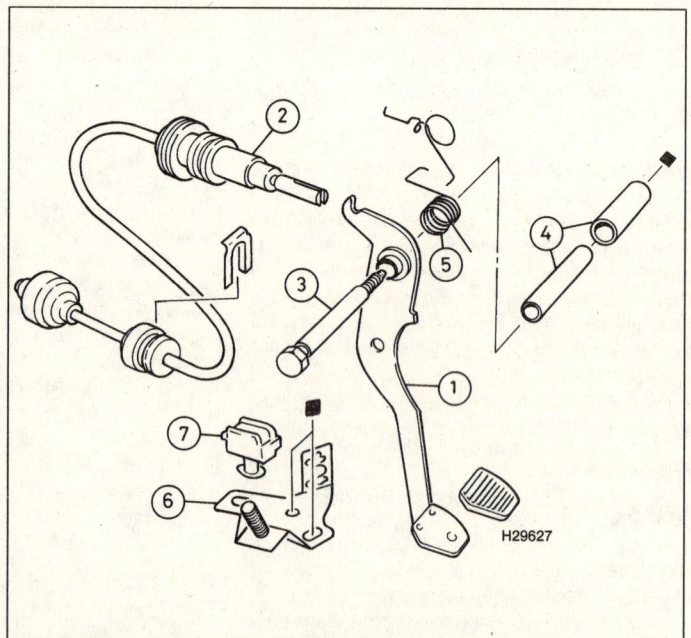

3.3b Clutch pedal and cable mounting details - left hand drive models

1	*Clutch pedal*	*4*	*Spacer*	*6*	*Mounting bracket*
2	*Cable guide*	*5*	*Return spring*	*7*	*Clip*
3	*Pivot bolt*				

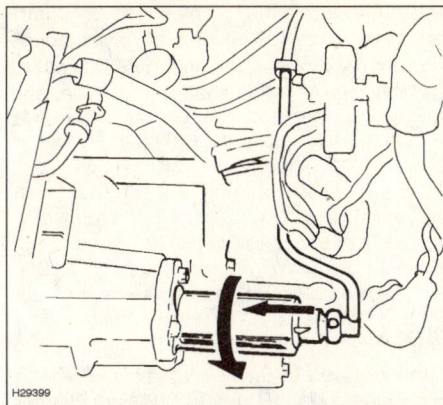

4.3a Release the clutch slave cylinder from the transmission by pushing it in by hand and at the same time turning it 90° anti-clockwise

4.3b Withdraw the slave cylinder, together with its pushrod from the transmission

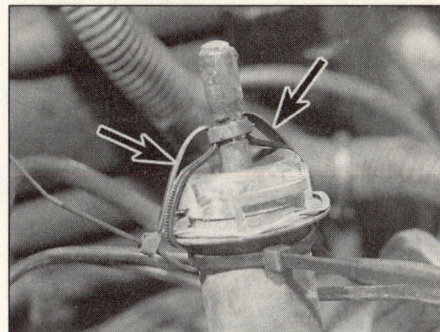

4.3c Retain the pushrod in place using cable ties (arrowed) or similar

Refitting

6 Apply a smear of multi-purpose grease to the spacer, and insert it into the pedal pivot bore.

7 Manoeuvre the pedal into position, and insert the pivot bolt. Refit the nut to the pivot bolt and tighten it securely.

8 Reconnect the clutch cable end/master cylinder pushrod (as applicable) to the pedal. Where applicable on cable-operated versions, refit the pedal assist spring assembly.

9 Depress and release the clutch pedal several times and check for correct operation.

10 Refit the lower panel to the facia and reconnect the battery.

4 Clutch hydraulic system components - removal and refitting

Note: *The hydraulic system components (master cylinder, slave cylinder, reservoir and pipework) are supplied as two sealed assemblies that cannot be dismantled.*

Note: *The master cylinder bulkhead union incorporates a quick-release fitting and Citroën recommend the use of a special slotted socket (tool 9040-TH) for this purpose. A suitable alternative can be fabricated from a folded strip of metal with a slot cut in it, to accommodate the hydraulic hose and union.*

Removal

Left hand drive models

1 Carry out the following to gain access to the slave cylinder:
a) *Disconnect the negative and positive battery cables, then remove the battery and battery tray.*
b) *Raise the front of the vehicle, support it securely on axle stands then remove the front left hand road wheel.*
c) *Extract the fixings and remove the lower section of the moulded wheel arch liner.*

d) *Remove the air cleaner housing and intake duct components as described in the relevant Part of Chapter 4.*

2 Unclip the clutch hydraulic pipe from the bodywork and the transmission casing.

3 Release the clutch slave cylinder from the transmission by pushing it in by hand and at the same time turning it 90° anti-clockwise. Withdraw the slave cylinder, together with its pushrod from the transmission. With the slave cylinder removed, retain the pushrod in place using cable ties or a similar arrangement **(see illustrations)**. **Do not** depress the clutch pedal with the slave cylinder removed or the push rod will be ejected. It is advisable to place a block of wood under the clutch pedal to prevent it being accidentally depressed.

4 Engage the Citroën special tool, or the home-made alternative over the master cylinder union and uncouple the master cylinder hydraulic union at the bulkhead **(see illustration)**.

5 Release the hydraulic pipework from the retaining clips and attachments in the engine compartment and remove it, together with the slave cylinder from the vehicle.

6 Working inside the car, remove the trim

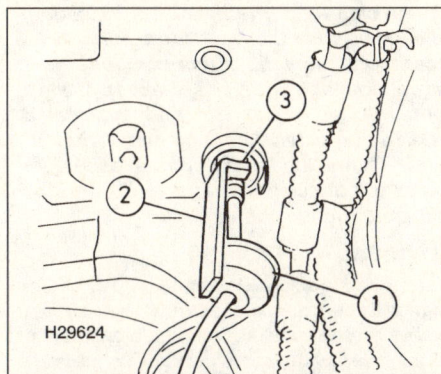

4.4 Engage the Citroën special tool, or the home-made alternative over the master cylinder union and uncouple the master cylinder hydraulic union at the bulkhead (left hand drive models)

1 Rubber protector
2 Special tool
3 Quick release union

panel from the underneath the drivers side of the facia.

7 Refer to Chapter 11 and carry out the following:
a) *Remove the drivers airbag (where applicable), paying close attention to the related warnings.*
b) *Remove the steering wheel.*
c) *Remove the lower section of the steering column shroud panel.*
d) *Remove the four securing nuts and lower the steering column support bracket away from its mountings.*

8 Slacken and remove the three nuts and detach the clutch pedal bracket from the bulkhead, complete with the clutch pedal and master cylinder.

9 Unclip the fluid reservoir from the pedal bracket, then unhook the master cylinder pushrod from the top of the clutch pedal.

10 Grasp the master cylinder and twist it anticlockwise through about quarter of a turn to release it from the clutch pedal bracket **(see illustration)**.

Right hand drive models

11 The removal/refitting procedure is essentially the same as that described for left hand drive models, with the exception that the quick release union is located on the right hand side of the passengers footwell, rather than at the engine compartment bulkhead. The clutch hydraulic pipe runs from the pedal

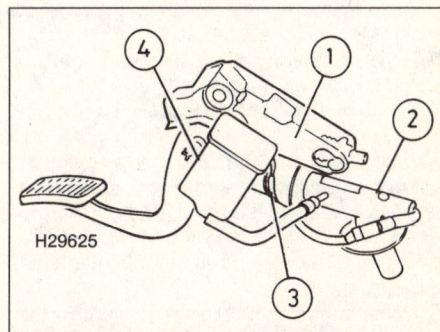

4.10 Clutch pedal, master cylinder and fluid reservoir assembly

1 Pedal bracket   3 Pushrod
2 Master cylinder 4 Fluid reservoir

6

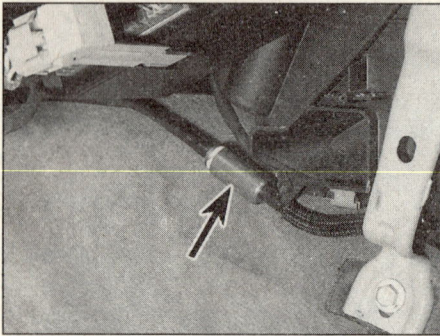

4.11 Location of the quick release union on right hand drive models

assembly along the underside of the facia and underneath the front of the centre console; access can be gained by removing the trim panels from the underside of the facia **(see illustration)**.

Refitting

Note: *On a new assembly, the slave cylinder pushrod is retained in the cylinder by a plastic collar which will automatically break off when the clutch pedal is depressed for the first time. Do not attempt to release this collar manually prior to fitting or the pushrod may be ejected.*

⚠️ **Warning: The hydraulic system is made up of two matched assemblies - the master cylinder assembly, and the pipework and slave cylinder assembly. Do not attempt to repair the system by renewing either of these assemblies individually.**

Left hand drive models

12 Locate the master cylinder in the clutch pedal mounting bracket, press and turn it clockwise through quarter of a turn to lock it in position.

13 Connect the end of the master cylinder pushrod to the clutch pedal by means of the clip.

14 Clip the fluid reservoir to the side of the pedal bracket.

15 Place the pedal bracket assembly in position on the bulkhead, ensuring that the union grommet locates correctly in the bulkhead aperture. Fit and tighten the bracket assembly securing nuts.

16 Refit the steering column support bracket, with reference to Chapter 10.

17 Remove the securing cable tie (where applicable) from the slave cylinder pushrod. Lubricate the end of the slave cylinder pushrod with molybdenum disulphide grease then locate the slave cylinder in the transmission. Push it in by hand and at the same time turn it clockwise through quarter of a turn to secure.

18 Reconnect the hydraulic pipework to the retaining clips and attachments in the engine compartment.

19 Connect the hydraulic pipework to the master cylinder, by pushing the two halves of the quick-release union together at the

bulkhead.

20 With the assembly installed, slowly depress the clutch pedal to the floor, then slowly lift it again by hand. Wait for ten seconds, then repeat this procedure twice. Depress the pedal again, release it and check that it rises correctly after being released.

21 The remainder of the refitting procedure is a reversal of the removal procedure, as follows:

a) *Refit the steering column shroud panel, steering wheel and airbag, with reference to Chapter 11.*

b) *Refit the facia lower trim panel.*

c) *Refit the air cleaner housing and intake duct components, as described in the relevant Part of Chapter 4.*

d) *Refit the battery tray and battery, then reconnect the positive and negative battery cables.*

e) *Refit the moulded wheel arch liner.*

f) *Refit the roadwheel, lower the car to the ground and tighten the wheel bolts to the specified torque.*

22 Note the hydraulic system is sealed and as such needs no bleeding or further maintenance.

Right hand drive models

23 Refer to the information given in paragraph 11.

---

5 Clutch assembly - removal, inspection and refitting

⚠️ **Warning: Dust created by clutch wear and deposited on the clutch components may contain asbestos, which is a health hazard. DO NOT blow it out with compressed air, or inhale any of it. DO NOT use petrol or petroleum-based solvents to clean off the dust. Brake system cleaner or methylated spirit should be used to flush the dust into a suitable receptacle. After the clutch components are wiped clean with rags, dispose of the contaminated rags and cleaner in a sealed, marked container. Although some friction materials may no longer contain asbestos, it is safest to assume that they do, and to take precautions accordingly.**

Removal

1 On all models, the installation of the engine/transmission assembly is such that removal of the transmission unit alone cannot be achieved without disturbing the subframe and the associated front suspension components.

2 Note that on petrol engined models and 2.1 litre diesel engined models only, it is possible to separate the engine and transmission in situ (without actually removing the transmission), to allow access to the clutch components, as described in Chapter 7A.

3 On 2.5 litre diesel engined models, it is

recommended that the engine and transmission are removed as a complete unit and then separated on the bench to gain access to the clutch components.

4 Before disturbing the clutch, mark the relationship of the pressure plate assembly to the flywheel, using a marker pen or similar.

5 Working in a diagonal sequence, slacken the pressure plate bolts by half a turn at a time, until spring pressure is released and the bolts can be unscrewed by hand.

6 Prise the pressure plate assembly off its locating dowels, and collect the friction plate, noting which way round the friction plate is fitted. Note that on 2.5 turbo diesel models with a 'pull-type' clutch, the release bearing remains attached to the centre of the pressure plate assembly.

Inspection

Note: *Due to the amount of work necessary to remove and refit clutch components, it is usually considered good practice to renew the clutch friction plate, pressure plate assembly and release bearing as a matched set, even if only one of these is actually worn enough to require renewal. It is also worth considering the renewal of the clutch components on a preventative basis if the engine and/or transmission have been removed for some other reason.*

7 Separate the pressure plate and friction plate and place them on the bench.

8 When cleaning clutch components, read first the warning at the beginning of this Section; remove dust using a clean, dry cloth, and working in a well-ventilated atmosphere.

9 Check the friction plate facings for signs of wear, damage or oil contamination. If the friction material is cracked, burnt, scored or damaged, or if it is contaminated with oil or grease (shown by shiny black patches), the friction plate must be renewed.

10 If the friction material is still serviceable, check that the centre boss splines are unworn, that the torsion springs are in good condition and securely fastened, and that all the rivets are tight. If any wear or damage is found, the friction plate must be renewed.

11 If the friction material is fouled with oil, this must be due to an oil leak from the crankshaft left-hand oil seal, from the sump-to-cylinder block joint, or from the transmission input shaft. Renew the seal or repair the joint, as appropriate, as described in Chapters 2A, 2B or 7A, before installing the new friction plate.

12 Check the pressure plate assembly for obvious signs of wear or damage; shake it to check for loose rivets or worn or damaged fulcrum rings, and check that the drive straps securing the pressure plate to the cover do not show signs of overheating (such as a deep yellow or blue discoloration). If the diaphragm spring is worn or damaged, or if its pressure is in any way suspect, the pressure plate assembly should be renewed.

13 Examine the machined bearing surfaces

5.15 On 2.5 litre turbo diesel models, if you intend to re-use the clutch assembly complete, it will be necessary to separate the release bearing from the pressure plate, by prising out the securing circlip with a small flat bladed screwdriver

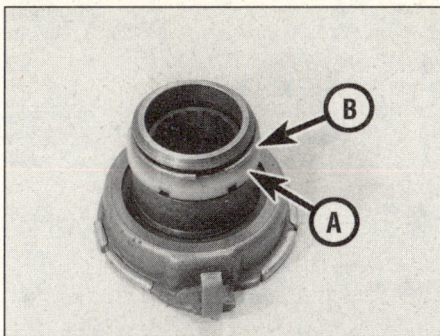

5.17a On 2.5 litre turbo diesel models, position the integral plastic collar (A) over the circlip (B) at the end of release bearing

5.17b Plastic collar in place over the circlip

of the pressure plate and of the flywheel; they should be clean, completely flat, and free from scratches or scoring. If either is discoloured from excessive heat, or shows signs of cracks, it should be renewed - although minor damage of this nature can sometimes be polished away using emery paper.

14 Check that the release bearing contact surface rotates smoothly and easily, with no sign of noise or roughness. Also check that the surface itself is smooth and unworn, with no signs of cracks, pitting or scoring. If there is any doubt about its condition, the bearing must be renewed.

15 On 2.5 litre turbo diesel models, if you intend to re-use the clutch assembly complete, it will be necessary to separate the release bearing from the pressure plate, by prising out the securing circlip with a small flat bladed screwdriver **(see illustration)**.

Refitting

16 On reassembly, ensure that the bearing surfaces of the flywheel and pressure plate are completely clean, smooth, and free from oil or grease. Use solvent to remove any protective grease from new components.

17 On 2.5 litre turbo diesel models, position the integral plastic collar over the circlip at the end of release bearing. Apply a smear of high temperature grease to the release fork arms, then slide the release bearing over the transmission input shaft, engaging the clips at the side of the bearing with the release fork arms **(see illustrations)**.

All models

18 Fit the friction plate so its spring hub faces away from the flywheel; there may also be a marking to show which way round the plate is to be refitted.

19 Refit the pressure plate assembly, aligning the marks made on dismantling (if the original pressure plate is re-used), and locating the pressure plate on its locating dowels. Fit the pressure plate bolts, but tighten them only finger-tight, so that the friction plate can still be moved.

20 The friction plate must now be

5.17c Apply a smear of high temperature grease to the release fork arms . . .

centralised, so that when the transmission is refitted, its input shaft will pass through the splines at the centre of the friction plate.

21 Centralisation can be achieved by passing a screwdriver or other long bar through the friction plate and into the hole in the crankshaft; the friction plate can then be moved around until it is centred on the crankshaft hole. Alternatively, a clutch-aligning tool can be used to eliminate the guesswork; these can be obtained from most accessory shops **(see illustration)**. A home-made aligning tool can be fabricated from a length of metal rod or wooden dowel which fits closely inside the crankshaft hole, and has insulating tape wound around it to match the diameter of the friction plate splined hole.

22 When the friction plate is centralised,

5.21 Using a clutch alignment tool to fit the clutch pressure and friction plate

5.17d then slide the release bearing over the transmission input shaft, engaging the clips at the side of the bearing with the release fork arms (arrowed)

tighten the pressure plate bolts evenly and in a diagonal sequence to the specified torque setting **(see illustration)**.

23 On 2.5 litre models, check that the release fork and bearing are still correctly engaged, and that bearing slide easily when the release fork is operated by hand.

24 Refit the transmission as described in Chapter 7A . On 2.5 litre diesel models refit the engine/transmission assembly as described in Chapter 2C.

25 On 2.5 litre turbo diesel models, the release bearing must be engaged with the pressure plate after the transmission has been refitted. Do this by hooking a piece of stout wire over the release arm and pulling on it

5.22 Tighten the pressure plate bolts evenly and in a diagonal sequence to the specified torque setting

6

sharply. It should be possible to hear the release bearing snap into position in the pressure plate **(see illustrations)**. Check the engagement by pushing on the release arm; if the release bearing is correctly engaged firm resistance to movement will be felt.

6 Clutch release mechanism - removal, inspection and refitting

Note: *Refer to the warning concerning the dangers of asbestos dust at the beginning of Section 5.*

Removal

1 Unless the complete engine/transmission is to be removed from the car and separated for major overhaul (see Chapter 2C), the clutch release mechanism can be reached by removing the transmission only, as described in Chapter 7A.

2.5 litre turbo diesel models

2 Separate the release bearing from the pressure plate, as described in Section 5.
3 Thread a suitable bolt into the upper end of the release fork shaft, at the top of the transmission casing. Extract the shaft by pulling on the bolt with a pair of grips. Catch the release fork as it slides off the lower end of the shaft.
4 Depress the retaining tabs and extract the upper and lower release shaft bushes from the transmission casing.

All other models

5 Unhook the release bearing from the fork, and slide it off the input shaft. Drive out the roll pin, and remove the release lever from the top of the release fork shaft. Discard the roll pin - a new one must be used on refitting.
6 On models with a cable operated clutch, depress the retaining tabs, then slide the upper bush off the end of the release fork

5.25a On 2.5 litre turbo diesel models, the release bearing can be engaged with the pressure plate using a hooked piece of wire

shaft. Disengage the shaft from its lower bush, and manoeuvre it out from the transmission. Depress the retaining tabs, and remove the lower pivot bush from the transmission housing **(see illustrations)**.

Inspection

7 Check the release mechanism, renewing any component which is worn or damaged. Carefully check all bearing surfaces and points of contact.
8 When checking the release bearing itself, note that it is often considered worthwhile to renew it as a matter of course. Check that the contact surface rotates smoothly and easily, with no sign of noise or roughness, and that the surface itself is smooth and unworn, with no signs of cracks, pitting or scoring. If there is any doubt about its condition, the bearing must be renewed.

Refitting

2.5 litre turbo diesel models

9 Fit the release shaft upper and lower bushes into place in the transmission casing. Apply a smear of molybdenum disulphide grease to each bush.
10 Slide the release shaft into position,

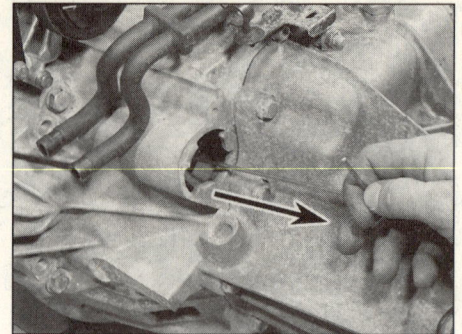

5.25b Hook the wire over the release arm and pull it sharply

engaging it with the release fork.
11 Slide the release bearing onto the input shaft, and engage it with the release fork. Apply a smear of molybdenum disulphide grease to the contact surfaces of the release fork.
12 Refit the transmission, then engage the release bearing with the clutch pressure plate, using the method described in Chapter 7A .

All other models

13 Apply a smear of molybdenum disulphide grease to the shaft pivot bushes and the contact surfaces of the release fork.
14 Locate the lower pivot bush in the transmission, ensuring that it is securely retained by its locating tangs, and refit the release fork. Slide the upper bush down the shaft, and clip it into position in the transmission housing.
15 Refit the release lever to the shaft. Align the lever with the shaft hole, and secure it in position by tapping a new roll pin fully into position. Slide the release bearing onto the input shaft, and engage it with the release fork.
16 Refit the transmission as described in Chapter 7A.

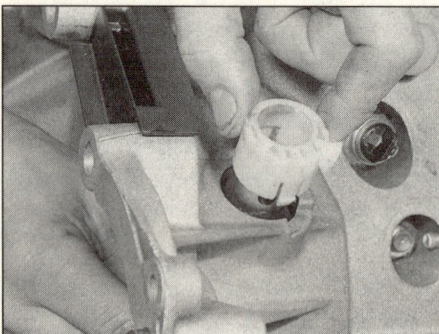

6.6a Removing the upper bush . . .

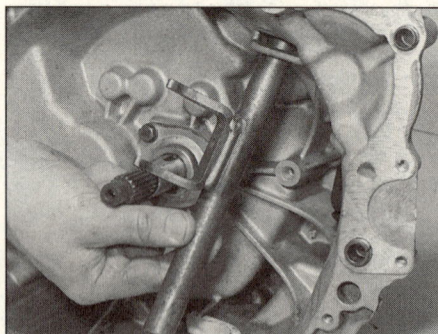

6.6b . . . release fork shaft . . .

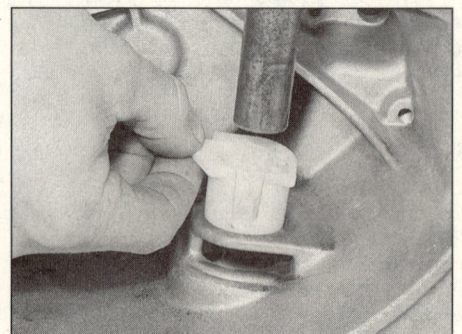

6.6c . . . and lower release bearing

Chapter 7 Part A:
Manual transmission

Contents

Degrees of difficulty

Easy, suitable for novice with little experience	Fairly easy, suitable for beginner with some experience	Fairly difficult, suitable for competent DIY mechanic	Difficult, suitable for experienced DIY mechanic	Very difficult, suitable for expert DIY or professional

Specifications

General

Type . . . . Manual, five forward speeds and reverse. Synchromesh on all forward speeds

Application:
Non-turbo petrol engine models BE3
Turbo petrol engine models ME5T
2.1 litre non-turbo diesel engine models BE3
2.1 litre turbo diesel ME5T
2.5 litre turbo diesel models MG5TB

Lubrication
Capacity:
BE3 . . . . 2.0 litres
ME5T . . . . 1.85 litres
MG5TB . . . . 2.2 litres

Torque wrench settings

	Nm	lbf ft
Roadwheel bolts	See Chapter 1A or 1B	
BE3 transmission		
Gearchange selector rod to lever pivot bolt	15	11
Gearchange linkage bellcrank pivot bolt	28	21
Oil filler/level plug	20	15
Oil drain plug	30	22
Clutch release bearing guide sleeve bolts	12	9
Reversing light switch	25	18
Right-hand driveshaft intermediate bearing retaining bolt nuts	10	7
Engine movement limiter-to- driveshaft intermediate bearing housing	50	37
Engine movement limiter-to- subframe	85	62
Left-hand engine/transmission mounting:		
Rubber mounting-to- bracket bolts	30	22
Mounting stud to transmission	60	44
Mounting stud bracket-to-transmission	60	44
Centre nut	65	48
Engine-to-transmission fixing bolts	45	33

7A

Torque wrench settings (continued)

	Nm	lbf ft
ME5T transmission		
Oil filler/level plug	20	15
Oil drain plug	30	22
Gearchange lever housing bolts	7	5
Clutch release bearing guide sleeve bolts	12	9
Reversing light switch	25	18
Right-hand driveshaft intermediate bearing retaining bolt nuts	10	7
Engine movement limiter-to-driveshaft intermediate bearing housing	50	37
Engine movement limiter-to-subframe	85	62
Left-hand engine/transmission mounting:		
Mounting to bracket	30	22
Mounting bracket-to-transmission	30	22
Centre nut	65	48
Engine-to-transmission fixing bolts	60	44
MG5TB transmission		
Engine-to-transmission fixing bolts	55	41
Oil filler/level plug	27	20
Oil drain plug	27	20
Gearchange lever housing bolts	7	5
Reversing light switch	25	18
Right-hand driveshaft intermediate bearing retaining bolt nuts	10	7

---

1 General information

The transmission is contained in a cast-aluminium alloy casing bolted to the engine's left-hand end, and consists of the gearbox and final drive differential. Three transmission types are fitted; refer to the application listing in the Specifications for details. All three transmission types are similar and operate as follows.

Drive is transmitted from the flywheel via the clutch to the input shaft, which has a splined extension to accept the clutch friction plate, and rotates in sealed ball-bearings. From the input shaft, drive is transmitted to the output shaft, which rotates in a roller bearing at its right-hand end, and a sealed ball-bearing at its left-hand end. From the output shaft, the drive is transmitted to the differential crownwheel, which rotates with the differential case and planetary gears, thus driving the sun gears and driveshafts. The rotation of the planetary gears on their shaft allows the inner roadwheel to rotate at a slower speed than the outer roadwheel when the car is cornering.

The input and output shafts are arranged side by side, parallel to the crankshaft and driveshafts, so that their gear pinion teeth are in constant mesh. In the neutral position, the relevant input shaft and output shaft gear pinions rotate freely, so that drive cannot be transmitted to the output shaft and crownwheel.

Gear selection is via a floor-mounted lever actuating a selector rod mechanism on BE3 and ME5T transmissions, or a selector cable mechanism on the MG5TB unit. The selector rod/cables cause the appropriate selector fork to move its respective synchro-sleeve along the shaft, to lock the gear pinion to the synchro-hub. Since the synchro-hubs are splined to the input and output shafts, this locks the pinion to the shaft, so that drive can be transmitted. To ensure that gear-changing can be carried out quickly and quietly, a synchro-mesh system is fitted to all forward gears, consisting of baulk rings and spring-loaded fingers, as well as the gear pinions and synchro-hubs. The synchro-mesh cones are formed on the mating faces of the baulk rings and gear pinions.

2 Manual transmission - draining and refilling

Note: *A suitable square section wrench may be required to undo the transmission filler/level and drain plugs on some models. These wrenches can be obtained from most motor factors or your Citroën dealer.*

1 This operation is much quicker and more efficient if the car is first taken on a journey of sufficient length to warm the engine/transmission up to operating temperature.

2 Park the car on level ground, switch off the ignition and apply the handbrake firmly. To ensure that the car remains level when refilling, jack up the front and rear of the car and support it securely on axle stands.

3 On models equipped with the BE3 transmission, remove the left-hand front roadwheel then release the screws and clips and remove the wheel arch liner from under the wing for access to the filler/level plug. On all models, remove the splash guard from under the engine.

4 Wipe clean the area around the filler/level plug. On the BE3 transmission, the filler/level plug is the largest bolt among those securing the end cover to the transmission. On the ME5T unit, the filler/level plug is located on the end face of the transmission, adjacent to the end cover. On the MG5TB unit, the filler/level plug is located on the engine side of the differential casing, to the rear of the right hand driveshaft. Unscrew the filler/level plug from the transmission and recover the sealing washer **(see illustration)**

5 Position a suitable container under the drain plug(s). On the BE3 and MG5TB units, there is one drain plug, situated on the underside of the final drive casing at the rear of the transmission. On the ME5T unit, there are two drain plugs - one on the underside of the differential casing and one on the under side of the transmission casing, adjacent to

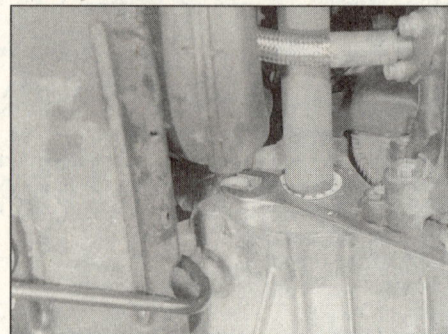

2.4 Unscrewing the filler/level plug - MG5TB transmission shown

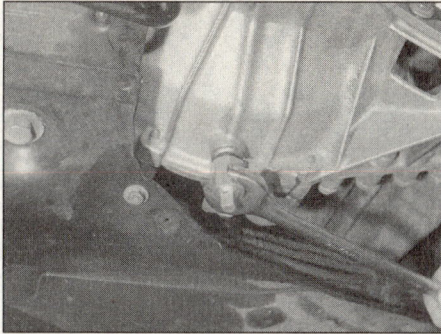

2.5 Unscrewing the drain plug - MG5TB transmission shown

the bellhousing mating surface. Unscrew the drain plug(s) **(see illustration)**.
6 Allow the oil to drain completely into the container. If the oil is hot, take precautions against scalding. Clean the filler/level and the drain plug(s), being especially careful to wipe any metallic particles off the magnetic inserts. Discard the original sealing washers; they should be renewed whenever they are disturbed.
7 When the oil has finished draining, clean the drain plug threads and those of the transmission casing, fit a new sealing washer and refit the drain plug, tightening it to the specified torque wrench setting.
8 Refilling the transmission is an extremely awkward operation. Above all, allow plenty of time for the oil level to settle properly before checking it. Note that the car must be level when checking the oil level.
9 Refill the transmission with the exact amount of the specified type of oil then check the oil level as described in the relevant Part of Chapter 1; if the correct amount was poured into the transmission and a large amount flows out on checking the level, refit the filler/level plug and take the car on a short journey so that the new oil is distributed fully around the transmission components, then check the level again on your return.
10 When the level is correct, fit a new sealing washer to the filler/level plug. Tighten the plug to the specified torque wrench setting. Wash off any spilt oil. Refit the wheel arch liner and splash guard, and secure with the retaining screws and clips. Refit the roadwheel (if removed) then lower the car to the ground.

3 Gearchange linkage (BE3, ME5T transmission) - removal and refitting

Removal

1 Remove the centre console (Chapter 12).
2 Chock the rear wheels, then jack up the front of the vehicle and support it on axle stands.
3 Refer to the relevant Part of Chapter 4 and remove the exhaust system and heat shields, as necessary for access to the gearchange linkage.

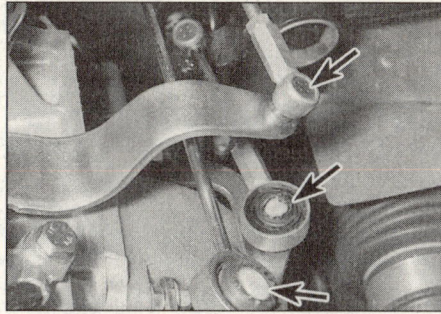

3.5 Disconnect the three gearchange linkage link rods (arrowed) from their transmission balljoints

4 Slacken and remove the nut, and withdraw the pivot bolt securing the selector rod to the base of the gearchange lever.
5 Using a flat-bladed screwdriver, carefully lever the three link rods off their balljoints on the transmission **(see illustration)**. Disengage the selector rod from the bellcrank pivot, and remove it from underneath the vehicle.
6 Carefully prise the plastic cap off the bolt securing the gearchange linkage bellcrank to the subframe.
7 Slacken and remove the bellcrank pivot bolt and washer, then manoeuvre the bellcrank and link rod out from under the vehicle, and recover the spacer and pivot bushes from the centre of the bellcrank.
8 Inspect all the linkage components for signs of wear or damage, paying particular attention to the pivot bushes and link rod balljoints, and renew worn components as necessary. If necessary, the gearchange lever can be removed and inspected as follows.
9 Slacken and remove the selector lever retaining nuts and lift off the retaining plate then lower the lever out from underneath the vehicle.
10 Peel back the lower gaiter from the base of the gearchange lever, then disengage the lever mounting plate, and slide the upper gaiter up the lever to gain access to the gearchange lever pivot ball. Examine the lever components for signs of wear or damage, paying particular attention to the rubber gaiters, and renew components as necessary. The lever can be separated from its baseplate after the retaining ring has been unclipped.

Refitting

11 Refitting is a reversal of the removal procedure, noting the following points:
a) Apply a smear of molybdenum disulphide grease to the gearchange lever pivot ball, the link rod balljoints and the bellcrank ball and pivot bushes.
b) Ensure that the gearchange lever rubber gaiters are correctly seated before refitting the lever assembly to the vehicle.
c) Ensure that the link rods are securely pressed onto their balljoints.
d) Refit the heat shields and exhaust components (Chapter 4C) and the centre console (Chapter 12).

4 Gearchange cables (MG5TB transmission) - removal and refitting

Removal

1 Remove the air cleaner assembly (Chapter 4B) and battery tray (Chapter 5A). If required, access can be further improved by removing the hydraulic fluid reservoir (Chapter 9).
2 Remove the centre console as described in Chapter 12.
3 Working in the engine compartment, carefully prise the two gearchange cable balljoints from the selector levers on the transmission. Access is extremely limited and to avoid damage to the ball joint assemblies, the fabrication of a special tool is recommended (see Tool Tip in Chapter 2B). Insert the forked tool under the balljoint head, then drive the balljoint off the selector levers by tapping the end of the tool with a mallet.
4 Chock the rear wheels, then jack up the front of the vehicle and support it on axle stands.
5 Extract the two horseshoe shaped clips securing the cables to the mounting bracket on the transmission. The clips can be accessed from above or below the engine.
6 Refer to Chapter 4C and detach the exhaust system and heat shields, as necessary for access to the cables and gearchange lever housing.
7 From inside the car, remove the sound-proofing shim then unscrew the bolts securing the lever housing to the floor. Release any clips or ties securing the gearchange cables, then remove the lever housing and gearchange cables as an assembly from under the car.

Refitting

8 Refitting is a reversal of the removal procedure, noting the following points:
a) Ensure that the sound-proofing shim is correctly positioned when refitting the lever housing.
b) Ensure that the cables are fitted to the correct selector levers on the transmission - the 13.0 mm diameter balljoint connects to the upper lever and the 10.0 mm diameter balljoint connects to the side lever.
c) Refit the heat shields, exhaust components and air cleaner assembly (Chapter 4C) and the centre console (Chapter 12).

5 Oil seals - renewal

Driveshaft oil seals

Note: A new suspension lower balljoint nut will be required on refitting.
1 Chock the rear wheels, then jack up the front of the car and support it on axle stands. Remove the appropriate front roadwheel, and

7A

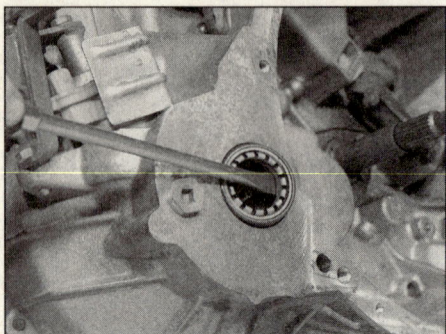

5.8 Use a large flat-bladed screwdriver to prise the driveshaft oil seals out of position

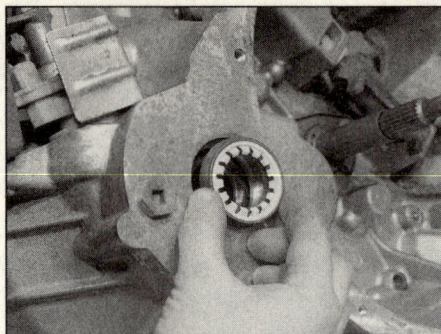

5.9a Fit the new seal to the transmission . . .

5.9b . . . and tap it into position using a tubular drift

remove the splash guard from under the engine.

2 Drain the transmission oil as described in Section 2.

3 On models equipped with ABS, remove the wheel sensor as described in Chapter 10.

4 Slacken and remove the nut securing the front suspension lower balljoint to the swivel hub, and free the balljoint from the lower arm (see Chapter 11). Discard the nut and remove the protector plate (if loose).

Right-hand seal

5 Loosen the two intermediate bearing retaining bolt nuts, then rotate the bolts through 90° so that their offset heads are clear of the bearing outer race.

6 Carefully pull the swivel hub assembly outwards, and pull on the inner end of the driveshaft to free the intermediate bearing from its mounting bracket.

7 Once the driveshaft end is free from the transmission, slide the dust seal off the inner end of the shaft, noting which way around it is fitted, and support the inner end of the driveshaft to avoid damaging the constant velocity joints or gaiters.

8 Carefully prise the oil seal out of the transmission, using a large flat-bladed screwdriver **(see illustration)**.

9 Remove all traces of dirt from the area around the oil seal aperture, then fill the space between the lips of the new oil seal with grease. Fit the new seal into its aperture, and drive it squarely into position using a suitable tubular drift (such as a socket) which bears only on the hard outer edge of the seal, until it contacts its locating shoulder. If the seal was supplied with a plastic protector sleeve, leave this in position until the driveshaft has been refitted **(see illustrations)**.

10 Thoroughly clean the driveshaft splines, then apply a thin film of grease to the oil seal lips and to the driveshaft inner end splines.

11 Slide the dust seal into position on the end of the shaft, ensuring that its flat surface is facing the transmission.

12 Carefully locate the inner driveshaft splines with those of the differential sun gear, taking care not to damage the oil seal, then align the intermediate bearing with its

mounting bracket, and push the driveshaft fully into position. If necessary, use a soft-faced mallet to tap the outer race of the bearing into position in the mounting bracket.

13 Ensure that the intermediate bearing is correctly seated, then rotate its retaining bolts back through 90° so that their offset heads are resting against the bearing outer race, and tighten the retaining nuts to the specified torque. Remove the plastic seal protector (where supplied), and slide the dust seal tight up against the oil seal.

14 Refit the protector plate (where removed) to the lower balljoint, then align the balljoint with the lower arm. Fit the new balljoint nut and tighten it to the specified torque setting (see Chapter 11).

15 Where necessary, refit the ABS wheel sensor as described in Chapter 10.

16 Refit the roadwheel and the engine splash guard, then lower the vehicle to the ground and tighten the roadwheel bolts to the specified torque.

17 Refill the transmission with the specified type and amount of fluid/oil, and check the level using the information given in Chapter 1A or 1B.

Left-hand seal

18 Pull the swivel hub assembly outwards and withdraw the driveshaft inner constant velocity joint from the transmission, taking care not to damage the driveshaft oil seal.

Support the driveshaft, to avoid damaging the constant velocity joints or gaiters.

19 On the BE3 and MG5TB transmissions, renew the oil seal as described in paragraphs 8 to 10. On ME5T transmissions, unbolt the differential bearing stop plate, and prise or drift the oil seal out of the stop plate. Also remove the sealing O-ring. Thoroughly clean the stop plate, then fill the space between the lips of the new oil seal with grease. Fit the new seal into its aperture, and drive it squarely into position using a suitable tubular drift (such as a socket) which bears only on the hard outer edge of the seal, until it is fully seated. Locate a new O-ring in position then refit the stop plate to the transmission.

20 Carefully locate the inner constant velocity joint splines with those of the differential sun gear, taking care not to damage the oil seal, and push the driveshaft fully into position. Where fitted, remove the plastic protector from the oil seal.

21 Carry out the operations described above in paragraphs 14 to 17.

Input shaft oil seal

22 Remove the transmission as described in Section 8 or 9 as applicable.

23 Undo the bolts securing the clutch release bearing guide sleeve in position, and slide the guide off the input shaft, along with its O-ring or gasket **(see illustrations)**. Recover any shims or thrustwashers which have stuck to

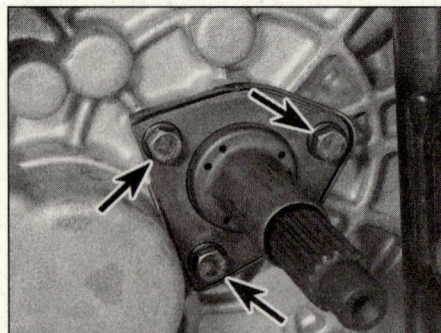

5.23a Clutch release bearing guide sleeve retaining bolts (arrowed) on the BE3 transmission . . .

5.23b . . . and on the ME5T transmission (arrowed)

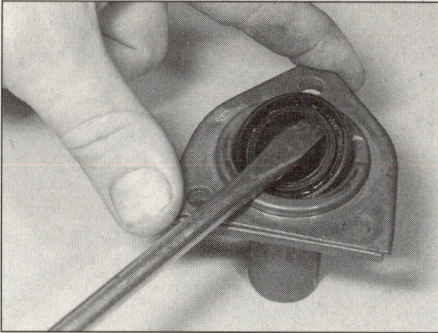

5.24 Removing the input shaft seal from the guide sleeve

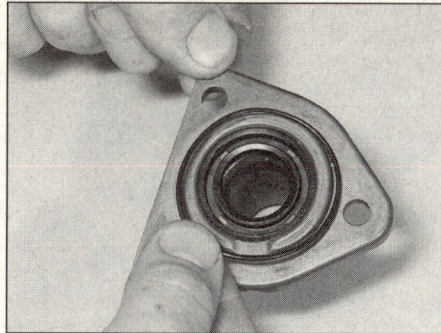

5.27 Fit a new O-ring/gasket (as applicable) to the guide sleeve

5.31a On the BE3 transmission, use a screwdriver to prise the selector shaft seal out of position . . .

the rear of the guide sleeve, and refit them to the input shaft.

24 Carefully lever the oil seal out of the guide using a suitable flat-bladed screwdriver **(see illustration)**.

25 Before fitting a new seal, check the input shaft's seal rubbing surface for signs of burrs, scratches or other damage, which may have caused the seal to fail in the first place. It may be possible to polish away minor faults of this sort using fine abrasive paper; however, more serious defects will require the renewal of the input shaft. Ensure the input shaft is clean and greased, to protect the seal lips on refitting.

26 Dip the new seal in clean oil, and fit it to the guide sleeve.

27 Fit a new O-ring or gasket (as applicable) to the rear of the guide sleeve, then carefully slide the sleeve into position over the input shaft **(see illustration)**. Refit the retaining bolts and tighten them to the specified torque.

28 Take the opportunity to inspect the clutch components if not already done (Chapter 6). Finally, refit the transmission (Section 8 or 9).

Selector shaft oil seal (BE3 transmission)

29 Park the car on level ground, chock the rear wheels, then jack up the front of the vehicle and support it on axle stands . Remove the left-hand front roadwheel, then release the screws and clips and remove the wheel arch liner from under the wing.

30 Using a large flat-bladed screwdriver,

lever the link rod balljoint off the transmission selector shaft, and disconnect the link rod.

31 Carefully prise the selector shaft seal out of the housing, and slide it off the end of the shaft **(see illustrations)**.

32 Before fitting a new seal, check the selector shaft's seal rubbing surface for signs of burrs, scratches or other damage, which may have caused the seal to fail in the first place. It may be possible to polish away minor faults of this sort using fine abrasive paper; however, more serious defects will require the renewal of the selector shaft.

33 Apply a smear of grease to the new seal's outer edge and sealing lip, then carefully slide the seal along the selector rod. Press the seal fully into position in the transmission housing.

34 Reconnect the link rod to the selector shaft, ensuring that its balljoint is pressed firmly onto the shaft.

35 Refit the wheel arch liner and secure it in position with its retaining screws and clips. Refit the roadwheel then lower the car to the ground.

---

6 Reversing light switch - testing, removal and refitting

Testing

1 The reversing light circuit is controlled by a

plunger-type switch that is screwed into the top of the transmission casing. If a fault develops in the circuit, first ensure that the circuit fuse has not blown.

2 To test the switch, remove the air cleaner components as required for access (Chapter 4A or 4B) then disconnect the wiring connector, and use a multimeter (set to the resistance function) or a battery-and-bulb test circuit to check that there is continuity across the switch terminals only when reverse gear is selected. If this is not the case, and there are no obvious breaks or other damage to the wires, the switch is faulty, and must be renewed.

Removal

3 Remove the air cleaner components as described in the relevant Part of Chapter 4.

4 Disconnect the wiring connector, then unscrew it from the transmission casing along with its sealing washer **(see illustrations)**.

Refitting

5 Fit a new sealing washer to the switch, then screw it back into position in the top of the transmission housing and tighten it to the specified torque setting. Reconnect the wiring connector, and test the operation of the circuit. Refit any components removed for access.

7A

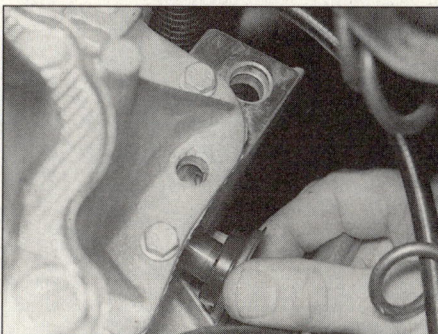

5.31b . . . then slide the seal off the shaft

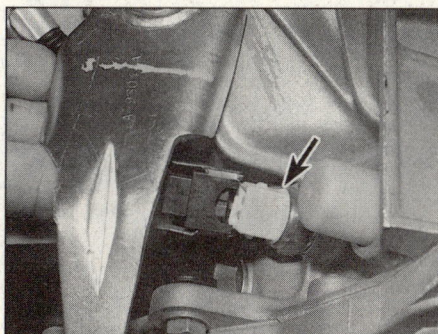

6.4a Disconnecting the wiring connector from the reversing light switch (arrowed)

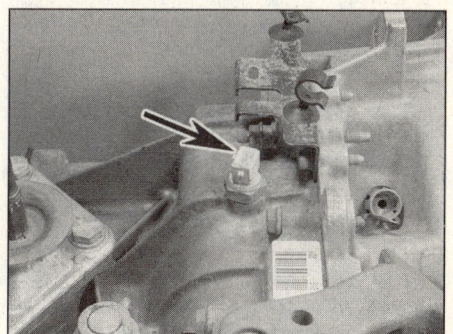

6.4b Location of the reversing light switch on the MG5TB transmission

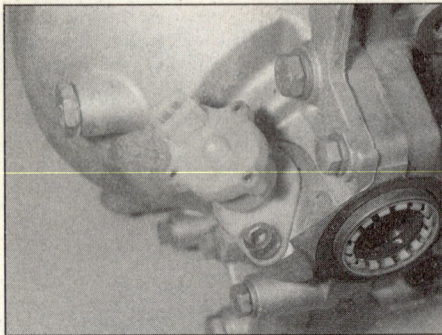
7.1a Speedometer transducer unit as fitted to the ME5T transmission

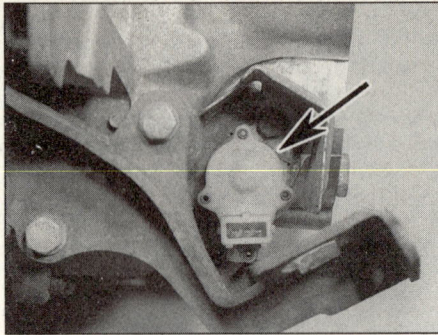
7.1b On MG5TB transmissions, the speedometer transducer (arrowed) is located at the top of the differential casing

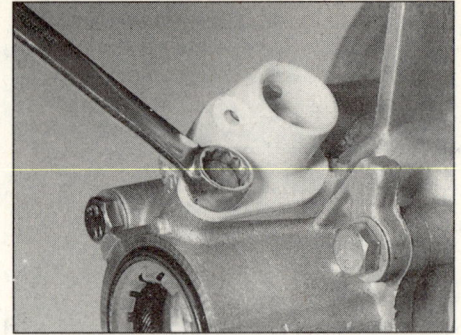
7.3a Slacken and remove the retaining bolt . . .

7 Speedometer drive - removal and refitting

Removal

1 Chock the rear wheels, then jack up the front of the car and support it on axle stands. Remove the splash guard from under the engine. The speedometer drive is situated on the rear of the transmission housing, next to the inner end of the right-hand driveshaft. According to model either a standard cable drive unit or transducer unit is fitted **(see illustrations)**.

7.3b . . . then withdraw the speedometer drive from the transmission (transmission removed for clarity)

2 On the cable type, pull out the speedometer cable retaining pin and disconnect the cable from the speedometer drive. Also disconnect the wiring connector (where applicable). On the transducer type, disconnect the wiring.

3 Slacken and remove the retaining bolt, along with the heat shield (where fitted), and withdraw the speedometer drive and driven pinion assembly from the transmission housing, along with its O-ring **(see illustrations)**.

4 If necessary, the pinion can be slid out of the housing, and the oil seal can be removed from the top of the housing. Examine the pinion for signs of damage, and renew if necessary. Renew the housing O-ring as a matter of course.

5 If the driven pinion is worn or damaged, also examine the drive pinion in the transmission housing for similar signs. The drive pinion on the BE3 transmission can be renewed as described below; on the ME5T transmission, major dismantling is necessary which must be entrusted to a dealer.

6 To remove the drive pinion on the BE3 transmission, first disengage the right-hand driveshaft from the transmission, as described in Section 5. Undo the three retaining bolts, and remove the speedometer drive housing from the transmission, along with its O-ring. Remove the drive pinion from the differential gear, and recover any adjustment shims from the gear **(see illustrations)**.

Refitting

7 Refit the adjustment shims to the differential gear, then locate the speedometer drive on the gear, ensuring it is correctly engaged in the gear slots **(see illustration)**. Fit a new O-ring to the rear of the speedometer drive housing, then refit the housing to the transmission and securely tighten its retaining bolts. Inspect the driveshaft oil seal for signs of wear, and renew if necessary. Refit the driveshaft to the transmission, using the information given in Section 5.

8 Apply a smear of grease to the lips of the seal and to the driven pinion shaft, and slide the pinion into position in the speedometer drive.

9 Fit a new O-ring to the speedometer drive and refit it to the transmission, ensuring that the drive and driven pinions are correctly engaged.

10 Refit the retaining bolt and the heat shield (where fitted), and tighten the bolt.

11 On the transducer type, reconnect the wiring. On the cable type, reconnect the wiring connector to the speedometer drive where applicable, then apply a smear of oil to the speedometer cable O-rings, reconnect the cable to the drive, and secure it in position with the rubber retaining pin.

12 Refit the splash guard under the engine then lower the vehicle to the ground.

7.6a Undo the three retaining bolts . . .

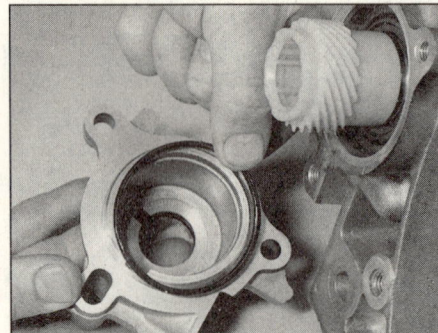
7.6b . . . and remove the housing, O-ring and drive pinion from the transmission (transmission removed for clarity)

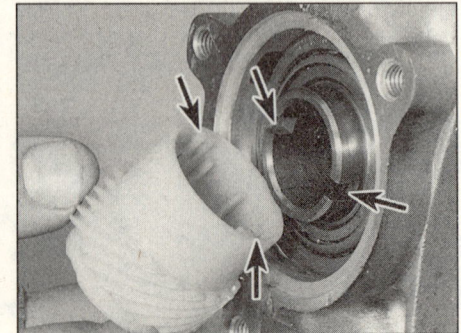
7.7 Ensure the drive pinion dogs are correctly engaged with the gear slots (arrowed)

8 Manual transmission - removal and refitting

General information

1 On all models, the installation of the engine/transmission assembly is such that removal of the transmission unit alone cannot be achieved without disturbing the subframe and the associated front suspension components. For this reason, it is recommended that the engine and transmission are removed as a complete assembly and separated on the bench; refer to Chapter 2C for details.

2 Note that on petrol engined models and 2.1 litre diesel engined models only, it is possible to separate the engine and transmission in situ, to allow access to the clutch components, as described in next sub-section.

Engine and transmission separation (in situ)

3 Refer to Chapter 9 and depressurise the hydraulic system. Disconnect the battery negative cable.

4 Raise the front of the car, support it securely on axle stands and remove the front wheels. Refer to Section 2 and drain the transmission.

5 Refer to Chapter 5A and remove the battery. Unbolt the battery holder from the bodywork and remove it from the engine compartment.

6 Refer to Chapter 4A or B (as applicable) and remove the air cleaner assembly and all associated intake air ducting from the engine compartment.

7 Remove the hydraulic fluid reservoir, with reference to Chapter 9.

8 With reference to the relevant Sections in this Chapter, carry out the following:
a) *Disconnect the gear change/selector rods from the transmission.*
b) *Disconnect the speedometer drive cable/sensor wiring from the transmission.*
c) *Disconnect the reversing light switch wiring from the transmission.*

9 Slacken the screws and withdraw the TDC sensor from the top of the transmission bellhousing.

10 Refer to Chapter 6 and disconnect the clutch cable from the release arm.

11 Unbolt the starter motor from the transmission bellhousing, with reference to Chapter 5A.

12 Refer to Chapter 9 and unbolt the pressure regulator from the front of the transmission casing.

13 Remove the securing screws and detach the plastic liners from the inner wheel arches. Unbolt and remove the wheel arch-to-subframe bracing bar.

14 Remove both driveshafts from the transmission, as described in Chapter 8.

15 Working underneath the car, unbolt and remove the flywheel protection plate from the underside of the transmission bellhousing.

16 Support the transmission from below using a trolley jack and an interposed block of wood. Attach an engine hoist, or lifting beam to the left hand end of the engine and raise the jib to just take the weight of the engine.

17 Refer to Chapter 2A or B (as applicable) and remove the left hand engine/transmission mounting. Note that it will be necessary to unscrew the transmission mounting stud from top of the transmission casing.

18 Check that nothing remains connected between the transmission and engine, then slacken and withdraw the transmission-to-engine mounting bolts and pull the transmission away from the engine on the trolley jack. With the end of the transmission casing drawn towards the left hand wheel arch, there should now be adequate access to the clutch components.

Refitting

19 Ensure the clutch plate and transmission input shaft splines are clean and dry. Do not apply grease to the splines as they have a special low-friction nickel coating.

20 Ensure the locating dowels are correctly positioned prior to installation and make sure the clutch release mechanism components are correctly fitted.

21 Carefully offer the transmission to the engine, until the locating dowels are engaged. Ensure that the weight of the transmission is not allowed to hang on the input shaft as it is engaged with the clutch friction disc.

22 Refit the transmission housing-to-engine bolts, ensuring that all the necessary brackets are correctly positioned, and tighten them to the specified torque setting.

23 The remainder of the refitting procedure is a reversal of removal, ensuring that the left hand engine/transmission mounting bolts are tightened to the specified torque (see Chapter 2A or 2B as applicable).

9 Manual transmission overhaul - general information

Overhauling a manual transmission is a difficult and involved job for the DIY home mechanic. In addition to dismantling and reassembling many small parts, clearances must be precisely measured and, if necessary, changed by selecting shims and spacers. Internal transmission components are also often difficult to obtain, and in many instances, extremely expensive. Because of this, if the transmission develops a fault or becomes noisy, the best course of action is to have the unit overhauled by a specialist repairer, or to obtain an exchange reconditioned unit.

Nevertheless, it is not impossible for the more experienced mechanic to overhaul the transmission, provided the special tools are available, and the job is done in a deliberate step-by-step manner, so nothing is overlooked.

The tools necessary for an overhaul include internal and external circlip pliers, bearing pullers, a slide hammer, a set of pin punches, a dial test indicator, and possibly a hydraulic press. In addition, a large, sturdy workbench and a vice will be required.

During dismantling of the transmission, make careful notes of how each component is fitted, to make reassembly easier and more accurate.

Before dismantling the transmission, it will help if you have some idea what area is malfunctioning. Certain problems can be closely related to specific areas in the transmission, which can make component examination and replacement easier. Refer to the Fault finding Section at the rear of this manual for more information.

7A

Notes

Chapter 7 Part B:
Automatic transmission

Contents

Degrees of difficulty

Easy, suitable for novice with little experience	**Fairly easy,** suitable for beginner with some experience	**Fairly difficult,** suitable for competent DIY mechanic	**Difficult,** suitable for experienced DIY mechanic	**Very difficult,** suitable for expert DIY or professional

Specifications

General

Type	Automatic, four forward speeds and reverse
Designation	4 HP 18

Lubrication

Recommended fluid	See *Lubricants and fluids*
Capacity (approximate):	
Drain and refill	2.5 to 3.0 litres approx.
Total capacity (including torque converter)	7.5 litres approx.

Torque wrench settings

	Nm	lbf ft
Transmission fluid drain plug	45	33
Filter base plate screws	10	7
Fluid cooler mounting bolts	50	37
Engine-to-transmission securing bolts	60	44
Torque converter screws	65	48

7B

1 General information

Automatic models are fitted with a four-speed, fully-automatic transmission, consisting of a torque converter, an epicyclic geartrain, and hydraulically-operated clutches and brakes.

Drive is taken from the engine to the transmission by a torque converter. The torque converter provides a fluid coupling between the engine and transmission, and acts as an automatic clutch, also providing a degree of torque multiplication when accelerating.

The epicyclic geartrain provides either of the four forward or one reverse gear ratios, according to which of its component parts are held stationary or allowed to turn. The components of the geartrain are held or released by brakes and clutches which are activated by a hydraulic control unit. A fluid pump within the transmission provides the necessary hydraulic pressure to operate the brakes and clutches.

Driver control of the transmission is by a multi-position selector lever. The transmission has a 'drive' position, and a 'hold' facility on the first three gear ratios. The 'drive' position 'D' provides automatic changing throughout the range of all four gear ratios, and is the one to select for normal driving. An automatic kick-down facility shifts the transmission down a gear if the accelerator pedal is fully depressed. This is useful when extra

acceleration is required. The 'hold' facility is very similar, but limits the number of gear ratios available - ie when the selector lever is in the '3' position, only the first three ratios can be selected; in the '2' position, only the first two can be selected, and so on. The lower ratio 'hold' is useful for providing engine braking when travelling down steep gradients, or for preventing unwanted selection of top gear on twisty roads.

Due to the complexity of the automatic transmission, any repair or overhaul work must be left to a Citroën dealer with the necessary special equipment for fault diagnosis and repair. The contents of the following Sections are therefore confined to supplying general information, and any service information and instructions that can be used by the owner.

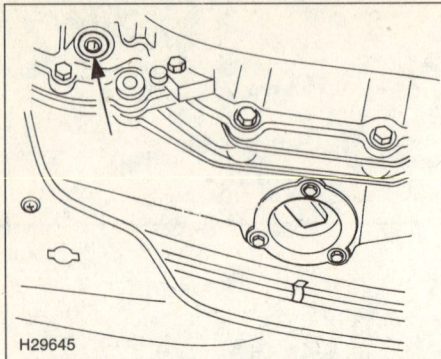

2.3 Location of transmission fluid drain plug (arrowed)

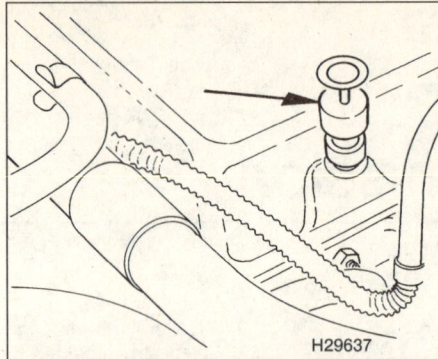

2.6 Fill the transmission via the dipstick tube (arrowed)

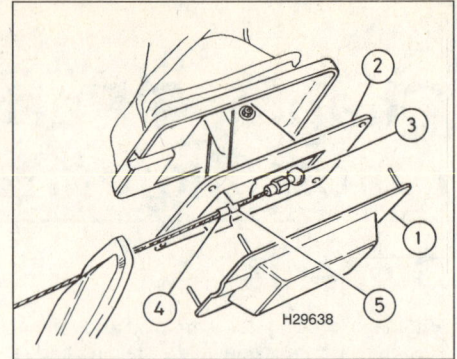

3.9 Selector cable and lever components

1 Lower casing	4 Spring clip
2 Support bracket	5 Locknut
3 Ball joint	

2 Automatic transmission fluid - draining and refilling

1 Take the vehicle on a short run, to warm the transmission up to normal operating temperature.

2 Park the vehicle on level ground, switch off the ignition and apply the parking brake firmly. For improved access, jack up the front of the vehicle and support it securely on axle stands. Note that the vehicle must be lowered to the ground and be level, to ensure accuracy when refilling and checking the fluid level.

3 Remove the dipstick, then position a suitable container under the transmission drain plug. The drain plug is located under the transmission on the differential casing (see illustration).

4 Unscrew the drain plug and allow the fluid to drain completely into the container. Note that only approximately 2.5 to 3.0 litres will drain out as it is not possible to completely drain the torque converter. If the fluid is hot, take precautions against scalding. Clean the drain plug, being especially careful to wipe any metallic particles off the magnetic insert. Discard the original sealing washer which should be renewed whenever it is disturbed.

5 When the fluid has finished draining, clean the drain plug threads and those of the transmission casing, fit a new sealing washer to the drain plug and refit it to the transmission, tightening securely. If the vehicle was raised for the draining operation, lower it to the ground.

6 Refilling the transmission is an extremely awkward operation, adding the specified type of fluid to the transmission a little at a time via the dipstick tube (see illustration). Alternatively, use the filler cap (breather) having cleaned around the area first. Use a funnel with a fine mesh gauze, to avoid spillage and to ensure that no foreign matter enters the transmission. Allow plenty of time for the fluid level to settle properly before checking. Note that the vehicle must be parked on flat level ground when checking the fluid level.

7 Add approximately 2.5 to 3.0 litres and check the level on the dipstick continuously as the last litre is added. Once the level is up to the MAX mark on the dipstick, refit the dipstick then start the engine and allow it to idle for a few minutes. Recheck the level with the engine still turning and top-up if necessary. Take the vehicle on a short run to fully distribute the new fluid around the transmission, then recheck the fluid level.

3 Selector cable - removal, refitting and adjustment

Removal

1 Remove the air cleaner assembly as described in Chapter 4A.

2 Move the gear selector lever to the 'P' position.

3 Working in the engine compartment, carefully prise the selector cable balljoint from the selector lever swivel plate on the transmission unit.

4 Remove the two screws and lift off the flange securing the cable to the mounting bracket on the transmission.

5 With reference to Chapter 13, remove the selector lever control grid illumination bulbholder, the control grid and the sound proofing gaiter. Unscrew the nuts securing the selector lever housing to the floor.

6 Chock the rear wheels, then jack up the front of the car and support it on axle stands.

7 Refer to Chapter 4C and remove the exhaust system and heat shields, as necessary for access to the cable and gear selector lever housing.

8 Work back along the selector cable, releasing it from any relevant retaining clips, and noting its correct routing.

9 Release any remaining clips or ties, then remove the selector lever housing and cable as an assembly from under the car (see illustration),

10 Detach the lower casing from the selector lever assembly and withdraw the lever support bracket.

11 Remove the spring clip and unscrew the locknut from the cable end.

12 Prise the ball joint from the base of the selector lever.

Refitting

13 Refitting is the reverse of removal, ensuring that the selector lever is in the 'P' position and the cable is correctly routed and retained with any relevant clips and ties.

14 At the transmission unit end of the cable, rotate the ball joint end fitting to extend the cable, so that the joint can be pressed onto the swivel plate ball - ensure that the selector lever is in the 'P' position as you do this.

Adjustment

15 Provided that no other components have been disturbed, the selector cable should not need further adjustment. If alignment between the selector mechanism components has been lost, carry out the following adjustments before refitting the selector cable.

16 Inside the car, place the selector lever in position 'P'.

17 Working underneath the transmission, rotate the selector lever swivel plate fully clockwise, to set the transmission in 'Park'.

18 Locate the linkrod that connects the selector lever swivel plate to the selector level on the transmission unit. Check that its length is 73 mm. If not, prise it off with a small screwdriver and rotate one of the balljoint end fittings to achieve the correct length (see illustration).

19 At the end of the selector cable, rotate the ball joint end fitting to extend the cable, so that the joint can be pressed onto the selector lever swivel plate.

20 Finally, check the operation of the transmission as follows:

a) Check that the car is immobilised when 'P' is selected.

b) Check that the car can be reversed when 'R' is selected.

c) Check that the starter motor can only be turned when either 'P' or 'N' is selected.

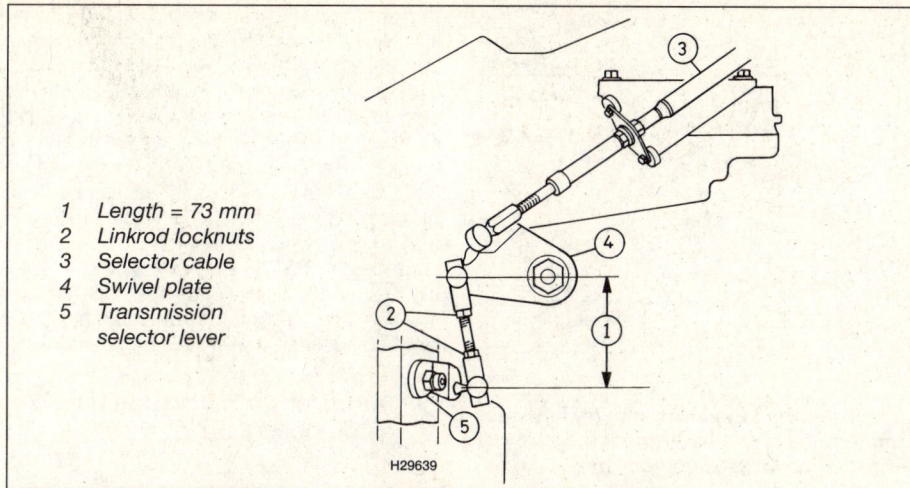

1 Length = 73 mm
2 Linkrod locknuts
3 Selector cable
4 Swivel plate
5 Transmission selector lever

H29639

3.18 Checking the length of the selector cable linkrod

4 Oil seals - renewal

Only the driveshaft oil seals can be renewed without extensive dismantling. The procedure is essentially the same as that described for manual transmission models, in Chapter 7A .

5 Speedometer drive - removal and refitting

Removal

1 Chock the rear wheels, then jack up the front of the car and support it on axle stands. Remove the splash guard from under the engine. The speedometer drive is situated on the rear of the transmission housing, next to the inner end of the right-hand driveshaft.
2 Disconnect the wiring connector/drive cable (as applicable) from the speedometer drive housing.
3 Slacken and remove the retaining bolt, along with the heat shield, and withdraw the speedometer drive and driven pinion assembly from the transmission housing, along with its O-ring. As the drive is withdrawn, hold the pinion assembly in place as there is a possibility that it can be dislodged and fall into the transmission casing.
4 If necessary, the pinion can be slid out of the housing, and the oil seal can be removed from the top of the housing. Renew the housing O-ring as a matter of course.

Refitting

5 Fit a new O-ring to the speedometer drive and refit it to the transmission, ensuring the drive and driven pinions are correctly engaged.
6 Refit the retaining bolt and the heat shield and tighten the bolt. Reconnect the wiring.

7 Refit the splash guard under the engine then lower the vehicle to the ground.

6 Fluid cooler - removal and refitting

Removal

1 The fluid cooler is mounted on the front of the transmission housing. To gain access to the fluid cooler, carry out the following:
a) Remove the air cleaner assembly and intake ducts as described in Chapter 4A or 4B.
b) Remove the battery and battery holder, with reference to Chapter 5A.
2 Unclip the wiring connector from the support bracket located just above the fluid cooler, then remove the support bracket.
3 Using hose clamps or similar, clamp both the fluid cooler coolant hoses to minimise coolant loss during the following operations.
4 Disconnect both coolant hoses from the

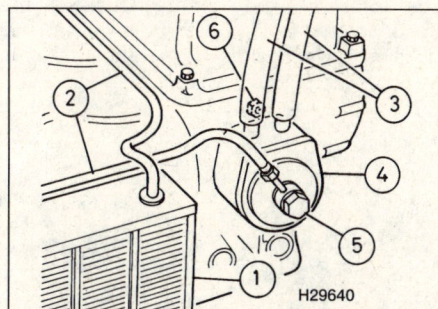

6.5a Transmission fluid cooler details

1 Air-to-transmission fluid heat exchanger
2 Transmission fluid lines
3 Coolant pipes
4 Coolant-to-transmission fluid heat exchanger
5 Banjo union bolt
6 Brake band adjustment screw

fluid cooler being prepared for some coolant spillage. Wash off any spilt coolant immediately with cold water, and dry the surrounding area before proceeding further.
5 Slacken and remove the union bolts, and remove the cooler from the transmission. Remove the seals from the mounting bolts, and the two seals fitted to the base of the cooler, and discard them; new ones must be used on refitting **(see illustrations)**.

Refitting

6 Lubricate the new seals with clean automatic transmission fluid, then fit the two new seals to the base of the fluid cooler, and a new seal to each mounting bolt.
7 Locate the fluid cooler on the top of transmission housing, taking care not to obscure the brake band adjustment screw behind it. Refit the union bolts, and tighten securely.
8 Reconnect the coolant hoses to the fluid cooler and remove the hose clamps.
9 Refit the support bracket and wiring connector, then refit the battery and its holder (Chapter 5A) and the disturbed intake duct/air cleaner components (see Chapter 4A).
10 On completion, top-up and bleed the cooling system and check the automatic transmission fluid level as described in Chapters 1A or 1B.

7 Kick-down cable - general information and adjustment

General information

1 The removal of the kick-down cable requires the removal of the transmission hydraulic valve block. This is an extremely complex task that should be entrusted to an automatic transmission expert or a Citroën dealer.

Adjustment

2 Check the accelerator cable adjustment as described in Chapter 4A or 4B.

7B

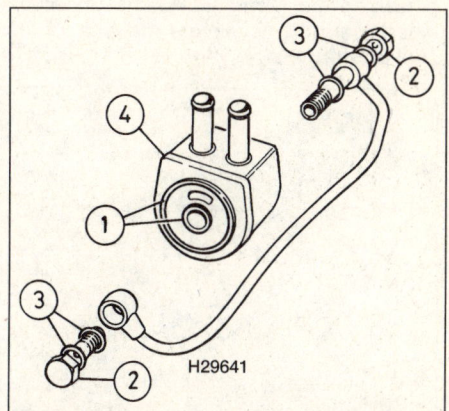

H29641

6.5b Coolant-to-transmission fluid heat exchanger assembly

1 Seals
2 Banjo union bolts
3 Sealing washers
4 Heat exchanger

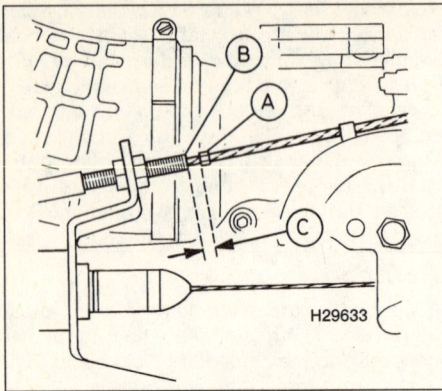

7.3 Check that there is a clearance (C) of 0.5 to 1.0 mm between the threaded section of the kick-down cable outer (B) and the ferrule (A) crimped to the cable inner, when the accelerator pedal is at rest

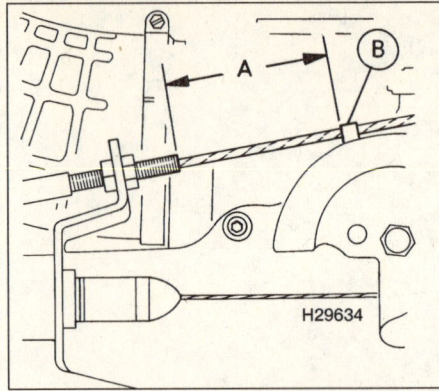

7.4 Measuring the clearance (A) between the ferrule (B) and the threaded section of the kick-down cable outer - petrol model shown

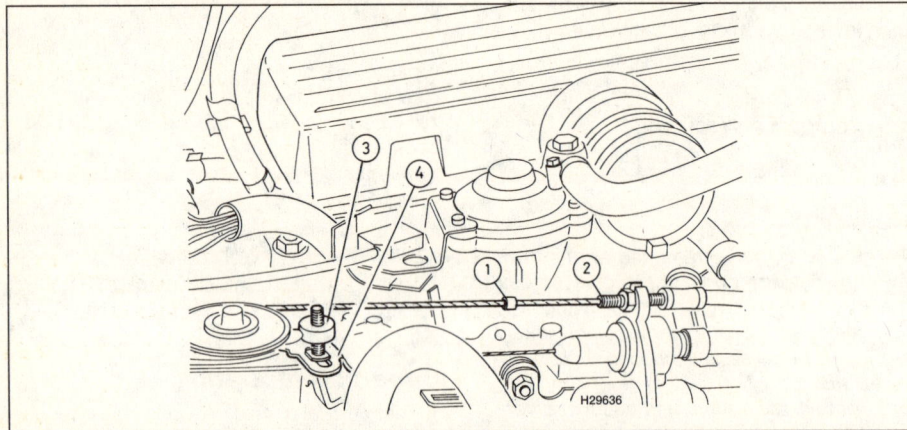

7.6 Measuring the clearance between the ferrule and the threaded section of the kick-down cable outer - 2.1 litre diesel model shown

1 Ferrule   2 Kick-down cable outer   3 Accelerator cable pivot   4 Accelerator lever

3 At the throttle housing (petrol models) or fuel injection pump (diesel models), check that there is a clearance of 0.5 to 1.0 mm between the threaded section of the kick-down cable outer and the ferrule crimped to the cable inner, when the accelerator pedal is at rest **(see illustration)**.

4 Operate the accelerator lever slowly by hand. At a point roughly three quarters of the way through the lever's total travel, slight resistance will be felt as the kick-down cable start to operate the cam mechanism on the transmission. Hold the throttle still at this position and measure the distance between

the end of the threaded section of the kick-down cable outer and the crimped ferrule. This should be 39 mm **(see illustration)**.

5 On petrol models, if adjustment is required, slacken and reposition the locknuts on either side of the kick-down cable mounting bracket, to achieve the correct cable length.

6 On diesel models, turn the accelerator lever fully (until it reaches its end stop) and repeat the above measurement - the distance should now be 45 mm. Adjust the cable if necessary by slackening the locknut and moving the accelerator cable pivot along the slotted accelerator lever **(see illustration)**.

8 Automatic transmission fluid filter - renewal

1 Jack up the front of the car and rest it securely on axle stands.

2 Refer to Section 2 of this Chapter and drain the automatic transmission fluid.

3 Working under the transmission, unscrew the three screws and lower the filter housing away from the transmission **(see illustration)**. Be prepared for some transmission fluid loss and take precautions against scalding.

4 Detach the filter cartridge from the base plate and clean it thoroughly. Clean the magnet at the centre of the base plate, taking care to remove all metallic particles.

5 Fit a new seal to the base plate and the filter cartridge. Press the filter into place on the base plate **(see illustration)**.

6 Refit the base plate to the transmission, then insert the securing screws and tighten them to the specified torque.

7 On completion, refill the transmission as described in Section 2.

9 Automatic transmission - removal and refitting

1 The transmission cannot be removed with the engine *in situ* - both units must be lifted from the engine compartment as an assembly and separated on the bench. This procedure is described in Chapter 2C.

10 Automatic transmission overhaul - general information

In the event of a fault occurring with the transmission, it is first necessary to determine whether it is of an electrical, mechanical or hydraulic nature, and to do this, special test equipment is required. It is therefore essential to have the work carried out by a Citroën dealer if a transmission fault is suspected.

Do not remove the transmission from the car for possible repair before professional fault diagnosis has been carried out, since most tests require the transmission to be in the vehicle.

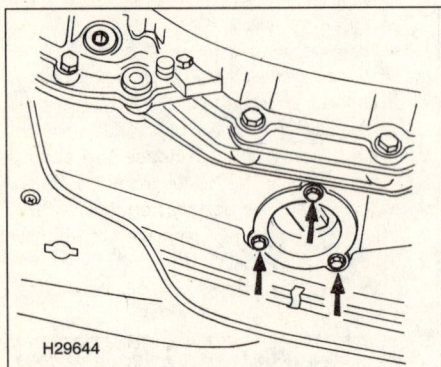

8.3 Unscrew the three screws (arrowed) and lower the filter housing away from the transmission

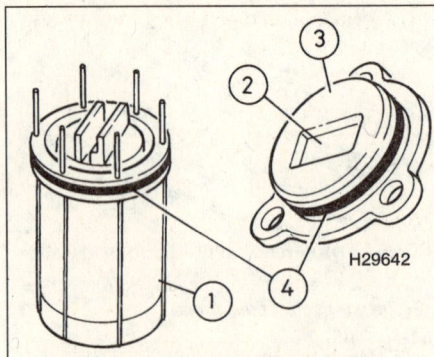

8.5 Transmission fluid filter assembly

1 Filter cartridge   3 Cover plate
2 Magnet   4 Seals

Chapter 8
Driveshafts

Contents

Degrees of difficulty

Easy, suitable for novice with little experience	**Fairly easy,** suitable for beginner with some experience	**Fairly difficult,** suitable for competent DIY mechanic	**Difficult,** suitable for experienced DIY mechanic	**Very difficult,** suitable for expert DIY or professional

Specifications

Torque wrench settings

	Nm	lbf ft
Driveshaft retaining nut	320	236
Right-hand driveshaft intermediate bearing retaining bolt nuts	10	7
Front suspension lower balljoint nut	45	33
Roadwheel bolts	90	66

1 General information

Drive is transmitted from the differential to the front wheels by means of two solid-steel driveshafts of unequal length.

Both driveshafts are splined at their outer ends, to accept the wheel hubs, and are threaded so that each hub can be secured to the end of the driveshaft by a large nut. The inner end of each driveshaft is splined to accept the differential sun gear.

Constant velocity (CV) joints are fitted to each end of the driveshafts, to ensure the smooth and efficient transmission of power through all suspension and steering angles. The inner constant velocity joints are of the tripod type, and the outer joints are of the ball-and-cage type.

On the right-hand side, due to the length of the driveshaft, the inner constant velocity joint is situated approximately halfway along the shaft's length, and an intermediate support bearing is mounted in the engine/transmission rear mounting bracket.

2 Driveshaft –
removal and refitting

Note: *A balljoint separator tool will be required for this operation. A new suspension lower balljoint nut will be required on refitting.*

Removal

1 Chock the rear wheels of the car, then jack up the front of the car and support it on axle stands (see *Jacking and Vehicle Support*). Remove the appropriate front roadwheel. On models where access to the driveshaft nut

can be obtained by removing the wheel trims, before jacking up the vehicle, loosen the driveshaft nut as follows.

a) *Chock the front wheels, and remove the wheel trim.*
b) *Apply the parking brake firmly.*
c) *Proceed as described in paragraph 6.*
d) *Loosen the driveshaft nut using a socket and extension.*

2 Using a screwdriver, release the securing clip, and remove the cover from the battery, then disconnect the battery negative lead.

3 Where applicable, remove the engine undershield.

4 Drain the transmission oil or fluid as described in Chapter 7A or 7B.

5 On models equipped with ABS, remove the ABS wheel sensor as described in Chapter 10.

6 If the driveshaft nut has been loosened, proceed to paragraph 8, otherwise withdraw

2.6a Withdraw the R-clip (arrowed) . . .

2.6b . . . and remove the locking cap

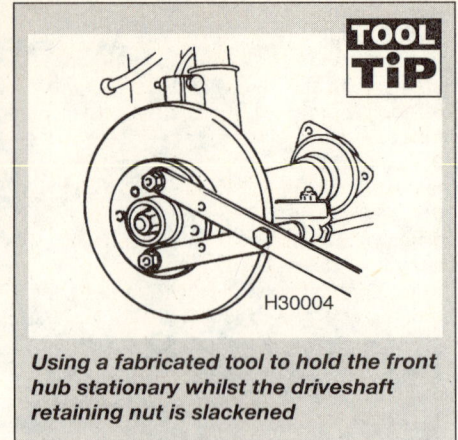

TOOL TiP

H30004

Using a fabricated tool to hold the front hub stationary whilst the driveshaft retaining nut is slackened

the R-clip and remove the locking cap from the driveshaft retaining nut **(see illustrations)**.
7 Refit at least two roadwheel bolts to the front hub, and tighten them securely. Have an assistant firmly depress the brake pedal to prevent the front hub from rotating, then using a socket and a long extension bar, slacken and remove the driveshaft retaining nut. Alternatively, a tool can be fabricated from two lengths of steel strip (one long, one short) and a nut and bolt; the nut and bolt forming the pivot of a forked tool. Bolt the tool to the hub using two wheel bolts, and hold the tool to prevent the hub from rotating as the driveshaft retaining nut is undone **(see Tool Tip)**. This nut is very tight; make sure that there is no risk of pulling the car off the axle stands. (If the roadwheel trim allows access to the driveshaft nut, the initial slackening can be done with the wheels chocked and on the ground.)
8 Slacken and partially unscrew the suspension lower balljoint nut (unscrew the nut as far as the end of the threads on the balljoint to prevent damage to the threads as the joint is released), then release the balljoint using a balljoint separator tool **(see illustration)**. Remove the nut.

Left-hand driveshaft

9 Carefully pull the hub carrier assembly outwards, and withdraw the driveshaft outer constant velocity joint from the hub assembly. If necessary, the shaft can be tapped out of the hub using a soft-faced mallet.
10 Support the driveshaft, then withdraw the inner constant velocity joint from the

transmission, taking care not to damage the driveshaft oil seal. Remove the driveshaft from the vehicle.

Right-hand driveshaft

11 Loosen the two intermediate bearing retaining bolt nuts, then rotate the bolts through 90°, so that their offset heads are clear of the bearing outer race **(see illustration)**.
12 Carefully pull the hub carrier assembly outwards, and withdraw the driveshaft outer constant velocity joint from the hub assembly. If necessary, the shaft can be tapped out of the hub using a soft-faced mallet.
13 Support the outer end of the driveshaft, then pull on the inner end of the shaft to free the intermediate bearing from its mounting bracket.
14 Once the driveshaft end is free from the transmission, remove the driveshaft from the vehicle.

Refitting

15 Before installing the driveshaft, examine the driveshaft oil seal in the transmission for signs of damage or deterioration and, if necessary, renew it, with reference to Chapter 7A or 7B. (Having got this far it is worth renewing the seal as a matter of course.)
16 Thoroughly clean the driveshaft splines, and the apertures in the transmission and hub assembly. Apply a thin film of grease to the oil seal lips, and to the driveshaft splines and shoulders. Check that all gaiter clips are securely fastened.

2.8 Releasing the suspension lower balljoint using a balljoint separator tool

2.11 Right-hand driveshaft intermediate bearing retaining bolt nut (arrowed)

Left-hand driveshaft

17 Offer up the driveshaft, and locate the joint splines with those of the differential sun gear, taking great care not to damage the oil seal. Push the joint fully into position.
18 Locate the outer constant velocity joint splines with those of the hub, and slide the joint back into position in the hub.
19 Align the suspension lower balljoint with the lower arm, then refit and tighten a new retaining nut to the specified torque setting.
20 Lubricate the inner face and threads of the driveshaft retaining nut with clean engine oil, and refit it to the end of the driveshaft. Use the method employed on removal to prevent the hub from rotating, and tighten the driveshaft retaining nut to the specified torque. Check that the hub rotates freely.

HAYNES HiNT *On models where access to the driveshaft nut can be obtained by removing the wheel trims, the driveshaft nut can be tightened with the parking brake applied, and the vehicle resting on its wheels.*

21 Engage the locking cap with the driveshaft nut so that one of its cut-outs is aligned with the driveshaft hole. Secure the cap in position with the R-clip.
22 Where applicable, refit the ABS wheel sensor, with reference to Chapter 10.
23 Where applicable, refit the engine undershield.
24 Refit the roadwheel, then lower the vehicle to the ground and tighten the roadwheel bolts to the specified torque.
25 Reconnect the battery negative lead, and refit the battery cover.
26 Refill the transmission with the specified type and amount of fluid/oil, and check the level using the information given in Chapters 1A or 1B and 7A or 7B.

Right-hand driveshaft

27 Check that the intermediate bearing rotates smoothly, without any sign of roughness or undue free play between its inner and outer races. If necessary, renew the bearing as described in Section 5. Examine the dust seal for signs of damage or deterioration, and renew if necessary.

28 Apply a smear of grease to the outer race of the intermediate bearing.

29 Pass the inner end of the shaft through the bearing mounting bracket.

30 Carefully locate the inner driveshaft splines with those of the differential sun gear, taking care not to damage the oil seal. Align the intermediate bearing with its mounting bracket, and push the driveshaft fully into position. If necessary, use a soft-faced mallet to tap the outer race of the bearing into position in the mounting bracket.

31 Locate the outer constant velocity joint splines with those of the hub, and slide the joint back into position in the hub.

32 Ensure the intermediate bearing is correctly seated, then rotate its retaining bolts back through 90°, so that their offset heads are resting against the bearing outer race. Tighten the retaining nuts to the specified torque.

33 Carry out the operations described above in paragraphs 19 to 26.

3 Driveshaft inner joint gaiter – renewal

1 Remove the driveshaft from the vehicle as described in Section 2.

2 Secure the driveshaft in a vice equipped with soft jaws, and release the two rubber gaiter retaining clips. If necessary, the gaiter retaining clips can be cut to release them.

3 Slide the rubber gaiter down the shaft, to expose the outer constant velocity joint. Scoop out the excess grease.

4 Using a hammer and suitable soft metal drift, sharply strike the inner member of the outer joint to drive it off the end of the shaft. The joint is retained on the driveshaft by a circlip, and striking the joint in this manner forces the circlip into its groove, so allowing the joint to slide off.

5 Once the joint assembly has been removed, remove the circlip from the groove in the driveshaft splines, and discard it. A new circlip must be fitted on reassembly.

6 Tape over the splines on the driveshaft, and carefully remove the outer constant velocity joint rubber gaiter, and the gaiter inner end plastic bush. It is recommended that the outer joint gaiter is also renewed, regardless of its apparent condition.

7 Release the retaining clips, then slide the inner gaiter off the shaft, and remove its plastic bush. As the gaiter is released, the joint outer member will also be freed from the end of the shaft **(see illustrations)**.

8 Thoroughly clean the joint using paraffin, or a suitable solvent, and dry it thoroughly. Check the tripod joint bearings and joint outer member for signs of wear, pitting or scuffing on their bearing surfaces. Check that the bearing rollers rotate smoothly and easily around the tripod joint, with no traces of roughness.

9 If on inspection, the tripod joint or outer member reveal signs of wear or damage, it will be necessary to renew the complete driveshaft assembly, since the joint is not available separately. If the joint is in satisfactory condition, obtain a repair kit consisting of a new gaiter, retaining clips, and the correct type and quantity of grease. Although not strictly necessary, it is also recommended that the outer constant velocity joint gaiter is renewed, regardless of its apparent condition.

10 On reassembly, pack the inner joint with the grease supplied in the gaiter kit. Work the grease well into the bearing tracks and rollers, while twisting the joint.

11 Clean the shaft, using emery cloth to remove any rust or sharp edges which may damage the gaiter, then slide the plastic bush and inner joint gaiter along the driveshaft. Locate the plastic bush in its recess on the shaft, and seat the inner end of the gaiter on top of the bush.

12 Fit the outer member over the end of the shaft, and locate the gaiter in the groove on the joint outer member. Push the outer member onto the joint, so that its spring-loaded plunger is compressed, then lift the outer edge of the gaiter to equalise air pressure in the gaiter. Fit both the inner and outer retaining clips, securing them in position using the information given in Section 4, paragraph 11. Ensure the gaiter retaining clips are securely tightened, then check that the joint moves freely in all directions.

13 Refit the outer constant velocity joint components using the information given in Section 4, paragraph 11.

4 Driveshaft outer joint gaiter – renewal

1 Remove the driveshaft from the car as described in Section 2.

2 Secure the driveshaft in a vice equipped with soft jaws, and release the two rubber gaiter retaining clips. If necessary, the gaiter retaining clips can be cut to release them.

3 Slide the rubber gaiter down the shaft, to expose the outer constant velocity joint. Scoop out the excess grease.

4 Using a hammer and suitable soft metal drift, sharply strike the inner member of the outer joint to drive it off the end of the shaft. The joint is retained on the driveshaft by a circlip, and striking the joint in this manner forces the circlip into its groove, so allowing the joint to slide off.

5 Once the joint assembly has been removed, remove the circlip from the groove in the driveshaft splines, and discard it. A new circlip must be fitted on reassembly.

6 Withdraw the rubber gaiter from the driveshaft, and slide off the gaiter inner end plastic bush.

7 With the constant velocity joint removed from the driveshaft, thoroughly clean the joint using paraffin, or a suitable solvent, and dry it thoroughly. Carry out a visual inspection of the joint.

8 Move the inner splined driving member from side to side, to expose each ball in turn at the top of its track. Examine the balls for cracks, flat spots, or signs of surface pitting.

9 Inspect the ball tracks on the inner and outer members. If the tracks have widened, the balls will no longer be a tight fit. At the same time, check the ball cage windows for wear or cracking between the windows.

3.7a Release the inner gaiter retaining clips, and remove the joint outer member

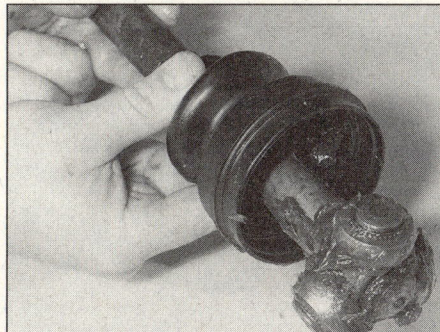

3.7b Slide the gaiter off the end of the driveshaft . . .

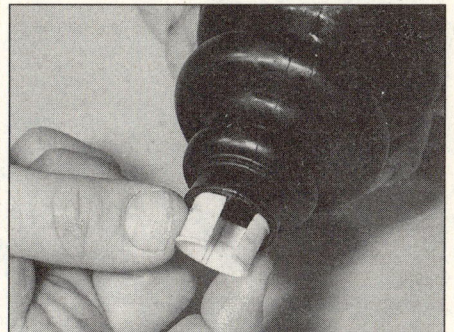

3.7c . . . and remove the plastic bush

8

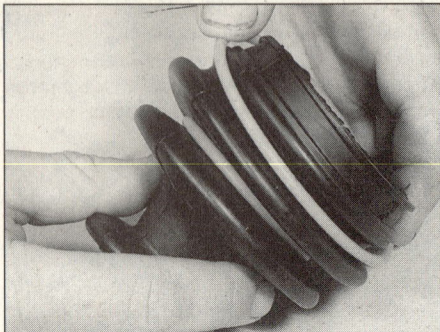

4.11a Fit the hard plastic rings to the outer CV joint gaiter . . .

4.11b . . . then slide on the new plastic bush (arrowed), and seat it in its recess in the shaft. Slide the gaiter onto the shaft . . .

4.11c . . . and seat the gaiter inner end on top of the plastic bush

4.11d Fit the new circlip to its groove in the driveshaft splines . . .

4.11e . . . then locate the joint outer member on the splines, and slide it into position over the circlip. Ensure that the joint is retained by the circlip before proceeding

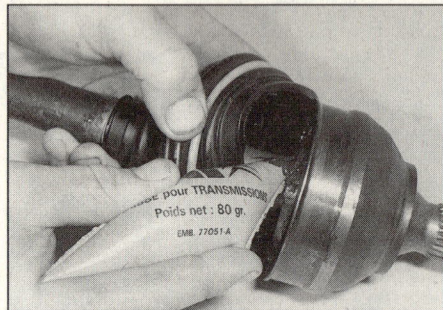

4.11f Pack the joint with the grease, working it into the ball tracks while twisting the joint, then locate the gaiter outer lip in its groove on the outer member

4.11g Fit the outer gaiter retaining clip and, using a hook fabricated out of a welding rod and a pair of pliers, pull the clip tightly to remove all slack

4.11h Bend the clip end back over the buckle, then cut the excess clip

10 If on inspection, any of the constant velocity joint components are found to be worn or damaged, it will be necessary to renew the complete joint assembly (where available), or even the complete driveshaft (where no joint components are available separately). Refer to your Citroën dealer for further information on parts availability. If the joint is in satisfactory condition, obtain a repair kit consisting of a new gaiter, circlip, retaining clips, and the correct type and quantity of grease.

11 To install the new gaiter, refer to the accompanying illustrations, and perform the operations shown **(see illustrations 4.11a to 4.11k)**. Be sure to stay in order, and follow the

4.11i Fold the clip end underneath the buckle . . .

4.11j . . . then fold the buckle firmly down onto the clip to secure the clip in position

4.11k Carefully lift the gaiter inner end to equalise air pressure in the gaiter, then secure the inner gaiter retaining clip in position using the same method

captions carefully. Note that the hard plastic rings are not fitted to all gaiters, and the gaiter retaining clips supplied with the repair kit may be different to those shown in the sequence. To secure this other type of clip in position, lock the ends of the clip together, then remove any slack in the clip by carefully compressing the raised section of the clip using a pair of side cutters.

12 Check that the constant velocity joint moves freely in all directions, then refit the driveshaft to the car as described in Section 2.

5 Right-hand driveshaft intermediate bearing – inspection and renewal

Note: *A suitable bearing puller will be required, to draw the bearing and collar off the driveshaft end.*

1 Remove the right-hand driveshaft as described in Section 2.

2 Check that the bearing outer race rotates smoothly and easily, without any signs of roughness or undue free play between the inner and outer races. If necessary, renew the bearing as follows.

3 Using a long-reach universal bearing puller, carefully draw the collar and intermediate bearing off the driveshaft inner end **(see illustration)**. Apply a smear of grease to the inner race of the new bearing, then fit the

bearing over the end of the driveshaft. Using a hammer and suitable piece of tubing which bears only on the bearing inner race, tap the new bearing into position on the driveshaft, until it abuts the constant velocity joint outer member. Once the bearing is correctly positioned, tap the bearing collar onto the shaft until it contacts the bearing inner race.

4 Check that the bearing rotates freely, then refit the driveshaft as described in Section 2.

6 Driveshaft overhaul – general information

1 If any of the checks described in Chapter 1 reveal wear in any driveshaft joint, first remove the roadwheel trim or centre cap (as appropriate).

2 If the R-clip is fitted, the driveshaft nut should be correctly tightened; if in doubt, remove the R-clip and locking cap, and use a torque wrench to check that the nut is securely fastened. Once tightened, refit the locking cap and R-clip, then refit the centre cap or trim. Repeat this check on the remaining driveshaft nut.

3 Road test the vehicle, and listen for a metallic clicking from the front as the vehicle is driven slowly in a circle on full-lock. If a clicking noise is heard, this indicates wear in the outer constant velocity joint. This means

that the joint must be renewed; reconditioning is not possible.

4 If vibration, consistent with road speed, is felt through the car when accelerating, there is a possibility of wear in the inner constant velocity joints.

5 To check the joints for wear, remove the driveshafts, then dismantle them as described in Sections 3 and 4; if any wear or free play is found, the affected joint must be renewed. In the case of the inner joints (and on some models, the outer joints), this means that the complete driveshaft assembly must be renewed, as the joints are not available separately. Refer to your Citroën dealer for information on the availability of driveshaft components.

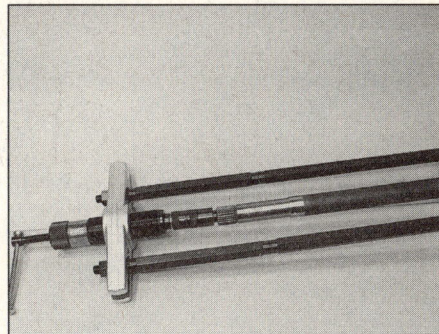

5.3 Using a long-reach bearing puller to remove the intermediate bearing from the right-hand driveshaft

8

Notes

Chapter 9
Hydraulic system

Contents

Degrees of difficulty

Easy, suitable for novice with little experience	Fairly easy, suitable for beginner with some experience	Fairly difficult, suitable for competent DIY mechanic	Difficult, suitable for experienced DIY mechanic	Very difficult, suitable for expert DIY or professional

Specifications

Pressure regulator
Cut-out pressure . . . . 170 ± 5 bars
Cut-in pressure . . . . 145 ± 5 bars

Torque wrench settings	Nm	lbf ft
Hydraulic pipe unions:		
3.5 and 4.5 mm diameter pipes	8	6
6.35 mm diameter pipes:		
With sleeve seal	10	7
Without sleeve seal	13	10
10 mm diameter pipes	30	22

1 General information and precautions

General information

The hydraulic suspension, the braking system, and where applicable the power steering system are pressurised by a common hydraulic system.

Hydraulic fluid is drawn from the hydraulic reservoir, mounted in the engine compartment, and is delivered under pressure to the hydraulic pressure regulator. The hydraulic system is pressurised by a belt-driven pump, which is driven by the engine crankshaft pulley.

From the pressure regulator, fluid passes to the security valve, which is connected to the compensator control valve and the front and rear suspension height corrector units.

Fluid from the suspension height corrector units flows to the suspension hydraulic unit cylinders. From the suspension hydraulic unit cylinders, the low pressure return fluid is returned to the hydraulic reservoir.

The height correctors maintain the suspension at the manually-selected height by admitting fluid to, and releasing fluid from, the suspension cylinders according to the movement of the front and rear anti-roll bars to which they are connected.

The hydraulic fluid flow in the suspension system is controlled by an electronic control unit, via solenoid valves. The control unit receives signals from various sensors, and regulates the hydraulic fluid pressure and flow within the suspension system to suit the prevailing driving conditions.

Hydraulic pressure for the braking system is supplied from the compensator control valve with separate front and rear circuits. The front circuit is supplied direct from the compensator control valve, whilst the rear brake circuits operate in conjunction with hydraulic circuits to the rear suspension. This arrangement results in the braking effort being biased in favour of the front brakes, and at the same time regulates the braking effort on the rear wheels according to the load on the rear suspension - the greater the load, the greater the pressure on the rear suspension, thus the more braking effort. On models with ABS, the hydraulic pressure to all four brakes is controlled by a hydraulic modulator – further details can be found in Chapter 10.

A flow distributor is fitted between the high pressure pump and the pressure regulator unit. The purpose of the flow distributor is to control the hydraulic pressure between the steering circuit and the suspension and brake circuits.

Procedures for the basic system components are given in this Chapter. The overall hydraulic system comprises a number of valves, regulators and actuators). Any work involving components not covered in this Chapter or Chapter 11 should be referred to a Citroën dealer or qualified specialist. Similarly, system checking and fault diagnosis work should be entrusted to a Citroën dealer or specialist.

9

Precautions

⚠️ *Warning: The fluid used in the XM hydraulic system is LHM mineral fluid, which is green in colour. The use of any other type of fluid will damage the system rubber seals and hoses. Keep the fluid carefully sealed in its original container.*

In an *emergency*, SAE 10 or SAE 20 engine oil (no other type of fluid) may be used in the system, but in this case the complete hydraulic system **must** be drained, and fresh LHM fluid substituted at the earliest opportunity.

If there is any possibility of fluid other than genuine LHM fluid being in the system, drain the complete hydraulic system, and fill it with the special rinsing solution obtainable from Citroën dealers. Bleed the system and leave the solution in the circuit for approximately 600 miles (1000 km), then drain it out and fill with LHM fluid. If the rubber seals are damaged by the incorrect fluid, it will also be necessary to renew these items at the same time (it is wise to entrust this task to a Citroën dealer or specialist).

Use only genuine spare parts. Components are identified by their white or green colour, and are of a special quality for use with LHM fluid.

Cleanliness is of the utmost importance when working on the hydraulic system and its components. Clean all adjacent areas before disconnecting components. After removal, blank off all orifices, and ensure that components, pipes and hoses do not get contaminated.

Use only petrol to clean hydraulic components.

Before carrying out any work on hydraulic system components, depressurise the system (see Section 2), and then disconnect the battery negative lead.

2 Hydraulic system – depressurising, pressurising and priming

Depressurising

Note: *Refer to the precautions given at the beginning of Section 1 before proceeding. The following operation must be carried out with the engine running. The system can be depressurised with the engine stopped, but special equipment is required, and the operation should therefore be entrusted to a Citroën dealer or a qualified specialist.*

1 The hydraulic system is depressurised using the pressure regulator pressure release screw. The release screw is located on the main accumulator assembly mounted at the front of the transmission **(see illustration)**.
2 Ensure that the pressure regulator pressure release screw is fully closed (see paragraph 1).

2.1 Hydraulic pressure regulator pressure release screw (arrowed) – viewed from under front of vehicle

3 Move the suspension mode switch to the 'Auto' (models up to 1992) or 'Normal' (models from 1993) position, as applicable **(see illustration)**.
4 With the engine running, set the suspension height control lever to the 'Low' position.
5 Allow the engine to run for approximately one minute, whilst the vehicle suspension sinks down. **Do not** move the steering wheel.
6 When the suspension has stopped sinking, stop the engine, then unscrew the pressure regulator pressure release screw by one turn (see paragraph 1). It should be possible to hear a whistling sound, which indicates that hydraulic fluid under pressure is flowing and returning to the reservoir.

⚠️ *Warning: Do not remove the pressure release screw, as the sealing ball beneath the screw is easily lost.*

Pressurising

7 On completion of work, to pressurise the system, tighten the pressure regulator pressure release screw, then move the suspension height control lever to the 'Maximum' position.
8 Start the engine, and allow the vehicle suspension to rise to its maximum height. Operate the height control lever through its full range of movement several times to check the operation of the hydraulic system.

Priming

9 Normally, the system will prime automatically when the engine is started,

2.11 High pressure fluid hose (arrowed)

2.3 Moving the suspension mode switch to the 'Normal' position

however sometimes it may be necessary to assist priming of the high pressure pump as follows.
10 Ensure that the pressure regulator pressure release screw has been slackened (see paragraph 1).
11 Disconnect the high pressure fluid hose from the top of the reservoir **(see illustration)**.
12 Pour LHM hydraulic fluid directly into the hose.
13 Start the engine.
14 Reconnect the hose as soon as the fluid level in the hose falls.
15 Once the pump has been primed, slacken and then tighten the pressure regulator pressure release screw several times to bleed the air from the system.
16 Move the suspension height control lever to the 'Maximum' position, then top up the level in the fluid reservoir.

3 Hydraulic pipes – renewal

Note: *Refer to the precautions given at the beginning of Section 1 before proceeding. New pipe seals must be used on refitting.*

1 Depressurise the hydraulic system as described in Section 2.
2 Before disconnecting a pipe, thoroughly clean the area around the union.
3 If a complete pipe section is to be removed, release the pipe from any retaining clips and mountings. Avoid distorting or damaging the pipe as it is removed.
4 Plug all openings to prevent the entry of dirt into the system.
5 To ensure that a perfect seal exists when hydraulic pipe joints are assembled, the following procedure must be carried out.
6 Clean the relevant fluid port, the hydraulic pipe, the union nut, and the rubber seal, and lightly lubricate them with LHM fluid.
7 Slide the rubber seal onto the end of the pipe until the pipe protrudes from it.
8 Reconnect the pipe so that the end of the pipe enters the relevant fluid port. Ensure that the sealing rubber fully enters its location in

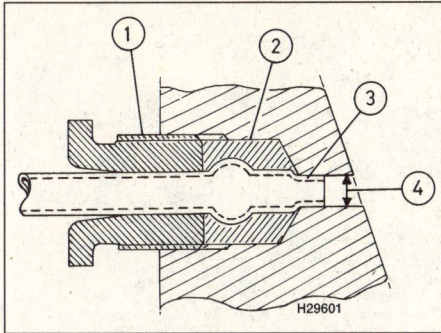

3.8 Hydraulic pipe end fitting details

1 *Union nut*   3 *Pipe end*
2 *Rubber seal*   4 *Fluid bore*

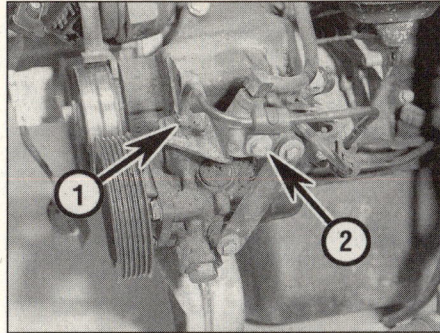

4.5 Disconnect the hydraulic pump fluid feed pipe (1) and unbolt the pipe bracket (2) – 2.5 litre diesel engine model

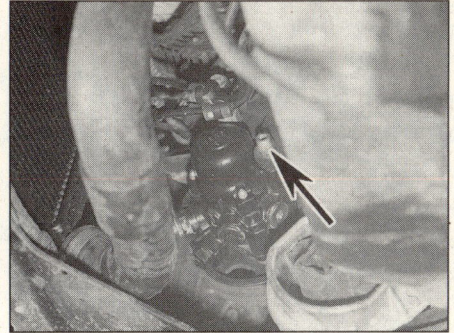

4.7 Hydraulic pump lower through-bolt (arrowed) – 2.5 litre diesel engine model

the fluid port **(see illustration)**. Also check that the end of the pipe is seated centrally in the port.

9 Screw the union nut into position, whilst keeping the pipe stationary. Do not overtighten the union nut.

10 On completion, check that the hydraulic system pipes do not touch each other or surrounding components which may stress or chafe the pipes.

11 Pressurise and if necessary prime the hydraulic system as described in Section 2.

12 Check and if necessary top up the hydraulic fluid level as described in *"Weekly checks"*.

4 High pressure (hydraulic fluid) pump – removal and refitting

Note: *Refer to the precautions given at the beginning of Section 1 before proceeding. New seals will be required when reconnecting the fluid pipes, and a new hose clip will be required.*

Removal

1 The pump is located at the right-hand side of the engine, and shares a drivebelt with the alternator. On most models, the pump is easily accessible from the top of the engine compartment, but on the 2.5 litre diesel engine, the pump is mounted below the alternator, and access must be obtained from underneath the vehicle.

2 Depressurise the hydraulic system as described in Section 2.

3 On the 2.5 litre diesel engine , chock the rear wheels, then jack up the front of the vehicle and support securely on axle stands (see *Jacking, and Vehicle Support*). Similarly, if necessary, to improve access, remove the right-hand front roadwheel, and the wheel arch liner.

4 Remove the auxiliary drivebelt as described in Chapter 1A or 1B.

5 Place a container beneath the pump to catch escaping hydraulic fluid, then unscrew the union nuts, and disconnect the fluid pipes

from the pump. Be prepared for fluid spillage, and plug the open ends of the pipes and pump. If necessary, unbolt the pipe support brackets to allow the pipes to be moved clear **(see illustration)**.

6 Similarly, disconnect the fluid hose from the pump. Discard the hose clip, and use a new one on refitting.

7 Working at the rear of the pump, unscrew the lower through-bolt and nut securing the pump to the mounting bracket **(see illustration)**.

8 Where applicable, unscrew the bolts securing the pipe bracket to the rear of the pump.

9 Unscrew the remaining mounting bolt, and withdraw the pump.

Refitting

10 Refitting is a reversal of removal, bearing in mind the following points.

a) *Use a new clip when reconnecting the fluid hose.*

b) *Where applicable, use new seals when reconnecting the hydraulic fluid pipes (see Section 3).*

c) *Refit and tension the auxiliary drivebelt as described in Chapter 1A or 1B.*

d) *On completion, pressurise, and if necessary prime the pump as described in Section 2.*

e) *Check and if necessary top up the hydraulic fluid level as described in "Weekly checks".*

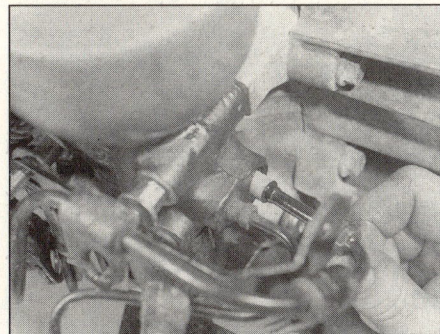

5.3 Disconnect the fluid pipes from the top of the pressure regulator

5 Hydraulic pressure regulator unit – removal and refitting

Note: *Refer to the precautions given at the beginning of Section 1 before proceeding. New pipe seals and a new hose clip will be required on refitting.*

Removal

1 Depressurise the hydraulic system as described in Section 2.

2 Place a container under the regulator unit (to catch escaping hydraulic fluid), then working under the unit, release the securing clip, and disconnect the fluid hose. Be prepared for fluid spillage, and plug the open ends of the hose and regulator.

3 Working at the top of the unit, unscrew the unions and disconnect the two fluid pipes **(see illustration)**. Again, be prepared for fluid spillage, and plug the open ends of the pipes and regulator.

4 Where applicable, unscrew the nut and bolt securing the pipe clamps at the top of the unit, to allow the pipes to be moved clear.

5 Unscrew the bolt, and separate the lower hose bracket from the regulator bracket **(see illustration)**.

6 Unscrew the securing nut, and separate the pipe brackets from the regulator unit front securing stud.

9

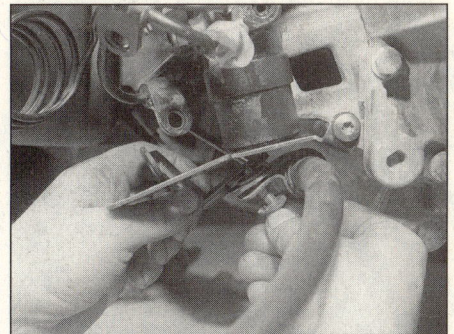

5.5 Unscrew the bolt and separate the lower hose bracket from the regulator bracket

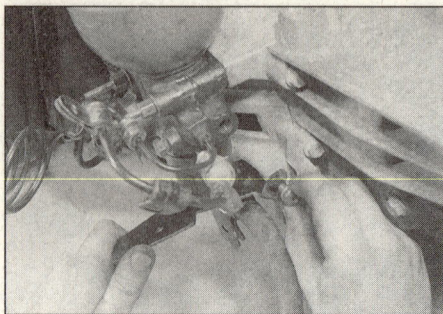

5.7 Unscrew the bolt securing the regulator unit mounting bracket to the transmission

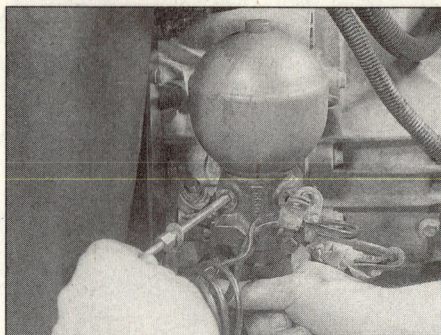

5.8 Removing the regulator securing stud

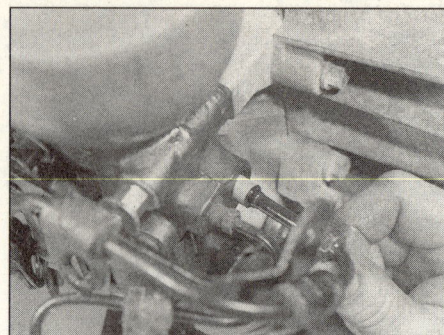

5.9 Use new seals when reconnecting the fluid pipes

7 Unscrew the bolt securing the regulator unit mounting bracket to the transmission **(see illustration)**.

8 Unscrew the bolt and stud securing the front of the regulator to the transmission, and withdraw the unit **(see illustration)**.

Refitting

9 Refitting is a reversal of removal, bearing in mind the following points **(see illustration)**.

a) *Ensure that the fluid pipe brackets are correctly refitted, and that none of the pipes are strained.*

b) *Where applicable, use new seals when reconnecting the fluid pipes.*

c) *On completion, pressurise the hydraulic system as described in Section 2.*

d) *Check and if necessary top up the hydraulic fluid level as described in "Weekly checks".*

6 Hydraulic fluid reservoir – removal and refitting

Removal

1 Depressurise the hydraulic system as described in Section 2.

2 Remove the air cleaner cover, and the air cleaner air intake trunking, as described in Chapter 4A or 4B.

3 Disconnect the wiring plug from the hydraulic fluid level indicator, and pull the two fluid return hoses from the reservoir centre section **(see illustrations)**.

4 Release the clip securing the centre section to the top of the reservoir **(see illustration)**.

5 Carefully lift the centre section out of the

reservoir, then cover it and place it to one side, clear of the reservoir – take care not to damage the filters as the centre section is withdrawn **(see illustration)**. Ideally, the centre section should be placed in a suitable container to prevent any possibility of dirt entry.

> **HAYNES HINT** *Cut the top off a large plastic bottle, and store the reservoir centre section in the bottle.*

6 Remove the reservoir securing screw, and release the securing clip, then release the reservoir from the locating guide, and withdraw the reservoir from the engine compartment **(see illustrations)**. Empty the contents of the reservoir into a suitable container.

6.3a Disconnect the wiring plug from the hydraulic fluid level indicator . . .

6.3b . . . and pull the fluid return hoses from the reservoir centre section

6.4 Release the reservoir centre section securing clip . . .

6.5 . . . then lift the centre section from the reservoir

6.6a Hydraulic fluid reservoir securing screw . . .

6.6b . . . and securing clip (arrowed)

Refitting

7 Refitting is a reversal of removal, but make sure that the base of the reservoir is correctly engaged with the locating guide.

8 Refill the hydraulic system with fluid, as described in Chapter 1A or 1B.

---

7 Hydraulic fluid reservoir bulbs ("spheres") - removal and refitting

Front suspension units

Removal

Note: *A strap wrench will be required for this operation. A new seal will be required on refitting.*

1 Depressurise the hydraulic system as described in Section 2.

2 Using a strap wrench, loosen the bulb, then unscrew the bulb from the relevant hydraulic unit **(see illustration)**. The bulbs are self-sealing, but be prepared for fluid spillage from the hydraulic unit. Plug the open end of the hydraulic unit to prevent dirt ingress.

Refitting

3 Grease the contact face of the bulb, and refit the bulb using a new seal. Tighten the bulb by hand only.

4 Re-pressurise the hydraulic system as described in Section 2. Check and if necessary top up the hydraulic fluid level as described in *"Weekly checks"*.

Rear suspension units

5 In theory, the procedure for removing the bulbs from the rear suspension units is identical to that for the front units. However, the rear units are exposed to the elements, and therefore suffer from corrosion over time, making unscrewing the bulbs extremely difficult.

6 If the bulbs have recently been serviced, or do not appear to have suffered corrosion, jack up and support the rear of the car, and proceed as described in paragraphs 1 and 2.

6.6c Removing the hydraulic fluid reservoir

If excessive effort is required to loosen the bulb, to the point where the suspension unit itself turns (risking damage to the fluid pipe), proceed as follows.

⚠ *Warning: The following procedure involves working without depressurising the hydraulic system, and there is a risk of fluid being released under pressure. Take adequate precautions before proceeding (wear goggles and gloves), since the operating pressure is sufficient to cause serious injury. Further, this procedure involves working under the car with the suspension fully raised - make sure the car cannot drop significantly, and be sure to have an assistant on hand. If you doubt your ability to carry out his work, leave it to a Citroen dealer or independent Citroen specialist.*

7 Apply the handbrake fully, ensure the transmission is in neutral (or Park), and chock the front and rear wheels.

8 Start the engine, and allow it to idle. Set the suspension height control lever to the "Maximum" position, and let the car rise to its maximum height.

9 Place axle stands directly below, but not touching, the rear suspension crosstube. The stands are a safety precaution, should the car drop significantly as a result of hydraulic pressure loss - make sure they are securely

7.2 Using a strap wrench to unscrew a front suspension unit hydraulic reservoir bulb

located. For this procedure to be effective, the car must be supported only by its suspension.

10 Switch off the engine. Before moving under the car, confirm that the ride height has stabilised.

11 Clean the bulb to be loosened, and the hydraulic unit, and if necessary position an inspection light so that the bulb can clearly be seen. The purpose of this procedure is to just start to loosen the bulb, **not** to remove it completely, and it must be possible to see clearly when the bulb starts to turn.

12 Using a strap wrench if necessary, start to loosen the bulb from the hydraulic unit. Ensure that the hydraulic unit itself does not turn, since this risks damaging the fluid supply pipe. Once you are sure the bulb has turned relative to the hydraulic unit, the car can be lowered as follows.

⚠ *Warning: Only loosen the bulb very slightly. Loosening the bulb more than just a fraction could result in fluid spraying out at high pressure, and the car dropping as hydraulic pressure is lost.*

13 Remove the stands below the rear axle. Start the engine, and set the suspension height control lever to the "Normal" position. Allow the car to settle at its normal ride height, then switch off the engine.

14 Jack up and support the rear of the car, then proceed as described in paragraphs 1 to 4.

9

Notes

Chapter 10
Braking system

Contents

Degrees of difficulty

Easy, suitable for novice with little experience	Fairly easy, suitable for beginner with some experience	Fairly difficult, suitable for competent DIY mechanic	Difficult, suitable for experienced DIY mechanic	Very difficult, suitable for expert DIY or professional

Specifications

General
System type . . . . Dual hydraulic circuit, front-rear split. Anti-lock braking system available on some models. Front and rear disc brakes fitted to all models. Hydraulic pressure provided by main hydraulic system (see Chapter 9). Cable-operated parking brake operating on front wheels.

Front brakes
Type . . . . Disc with single piston sliding caliper
Disc diameter:
  Models up to February 1991 . . . . 276.0 mm
  Models from March 1991 . . . . 283.0 mm
Disc thickness:
  New:
    Models up to February 1991 . . . . 22.0 mm
    Models from March 1991 . . . . 26.0 mm
  Minimum thickness:
    Models up to February 1991 . . . . 20.0 mm
    Models from March 1991 . . . . 24.0 mm
Maximum disc run-out . . . . 0.05 mm
Maximum surface distortion . . . . 0.01 mm
Minimum disc pad friction material thickness (not including
  backing plate) . . . . 3.0 mm

10

Rear brakes

Type . . . . Disc, with twin piston fixed calipers
Disc diameter:
    Hatchback models . . . . 224.0 mm
    Estate models . . . . 251.0 mm
Disc thickness:
  New:
    Hatchback models . . . . 9.0 mm
    Estate models . . . . 12.0 mm
  Minimum thickness:
    Hatchback models . . . . 7.0 mm
    Estate models . . . . 10.0 mm
Maximum disc run-out . . . . 0.05 mm
Maximum surface distortion . . . . 0.01 mm
Minimum disc pad friction material thickness (not including
  backing plate) . . . . 2.0 mm

Torque wrench settings

	Nm	lbf ft
Front caliper bracket bolts	105	77
Rear caliper securing bolts:		
Hatchback models	45	33
Estate models	70	52
Brake control valve securing bolts	20	15
Parking brake cable adjuster locknuts	20	15
ABS wheel sensor securing bolt	10	7
ABS wheel sensor clamp bolt	3	2

1 General information and precautions

General information

The dual circuit hydraulic system, with disc brakes fitted to all four wheels, is hydraulically operated from the main hydraulic system (see Chapter 9) via the brake control valve, which replaces the master cylinder found in a conventional braking system.

The hydraulic pressure to the front brakes is supplied directly from the main system (via the pressure regulator), but the pressure to the rear brakes is supplied from the rear suspension system. This arrangement favours the front brakes, and imposes a braking effort limitation on the rear axle in relation to the load on the suspension.

The brake pedal acts on the brake control valve.

The front brake calipers are of single piston floating type, operating on ventilated discs.

The rear brake calipers are of twin piston fixed type, operating on solid discs.

The mechanical parking brake mechanism is operated by a foot pedal and a lever on the facia, and acts on the front calipers via flexible cables.

An anti-lock braking system (ABS) is fitted as standard on certain models, and is available as an option on others. The system is described in more detail in Section 15.

Precautions

⚠ **Warning: The fluid used in the XM hydraulic system is LHM mineral fluid, which is green in colour. The use of any other type of fluid will damage the system rubber seals and hoses. Keep the fluid carefully sealed in its original container.**

In an *emergency*, SAE 10 or SAE 20 engine oil (no other type of fluid) may be used in the system, but in this case the complete hydraulic system **must** be drained, and fresh LHM fluid substituted at the earliest opportunity.

If there is any possibility of fluid other than genuine LHM fluid being in the system, drain the complete hydraulic system, as described in Chapter 9, and fill it with the special rinsing solution obtainable from Citroën dealers. Bleed the system and leave the solution in the circuit for approximately 600 miles (1000 km), then drain it out and fill with LHM fluid. If the rubber seals are damaged by the incorrect fluid, it will also be necessary to renew these items at the same time (it is wise to entrust this task to a Citroën dealer).

Use only genuine spare parts. Components are identified by their white or green colour, and are of a special quality for use with LHM fluid.

Cleanliness is of the utmost importance when working on the hydraulic system and its components. Clean all adjacent areas before disconnecting components. After removal, blank off all orifices, and ensure that components, pipes and hoses do not get contaminated.

2 Brake hydraulic system – bleeding

Note: *Refer to the precautions at the end of Section 1 before proceeding.*

1 The brake hydraulic system must be bled after renewing and refitting any components, brake pipe of hose. If this procedure is not carried out, air will be trapped in the hydraulic circuit, and the brakes will not function correctly.

2 Before starting work, check all the brake lines, unions, hoses and connections for possible leakage.

3 If there is any possibility of fluid other than genuine LHM fluid being in the system, drain the complete hydraulic system, as described in Chapter 9, and fill it with the special rinsing solution obtainable from Citroën dealers. Bleed the system and leave the solution in the circuit for approximately 600 miles (1000 km), then drain it out and fill with LHM fluid. If the rubber seals are damaged by the incorrect fluid, it will also be necessary to renew these items at the same time.

4 The brake bleeding procedure for the front and rear brakes differs. Proceed as follows.

5 With the engine running, operate the suspension several times, switching between the 'Low' and 'Maximum' height positions.

6 Set the height control to the 'Maximum' height position, then switch off the ignition.

7 Jack up the vehicle and support on axle stands with all four wheels clear of the ground (see *Jacking and Vehicle Support*). Remove the roadwheels.

2.8 ABS hydraulic valve block bleed screws (arrowed)

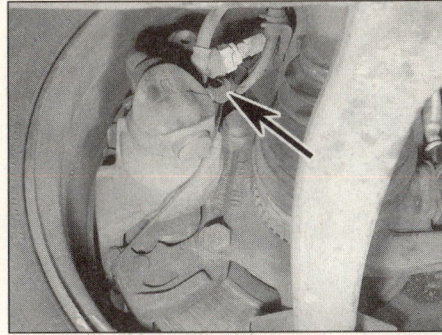

2.13a Front brake caliper bleed screw location (arrowed)

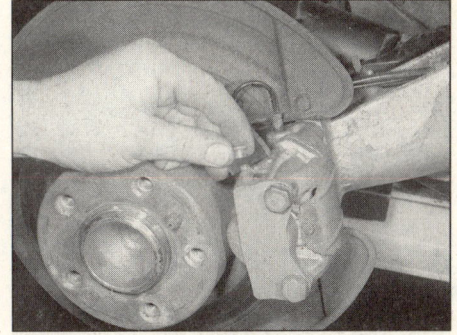

2.13b Removing the dust cap from a rear caliper bleed screw

8 On models with ABS, proceed as follows **(see illustration)**.
a) *Working under the front left-hand wheel arch, remove the wheel arch liner for access to the ABS hydraulic valve block.*
b) *Start the engine and allow it to run at idle speed.*
c) *Slacken the bleed screws on the valve block using a spanner.*
d) *Depress the brake pedal slightly, which will allow the hydraulic fluid to flow through the valve block return circuit. Keep the pedal depressed for a few seconds, then release it.*
e) *Tighten the bleed screws on the valve block.*

9 If not already done, start the engine, and allow it to run at idle speed.
10 If the system has been only partially disconnected, and suitable precautions were taken to minimise fluid loss, it should be necessary only to bleed that part of the system. If the complete system is being bled, bleed the brakes in the following order.
a) *Right-hand rear.*
b) *Left-hand rear.*
c) *Right-hand front.*
d) *Left-hand front.*

11 Collect a clean glass jar, a suitable length of plastic or rubber tubing, which is a tight fit over the bleed screw, and a ring spanner to fit the screw. The help of an assistant will also be required.
12 Check that the level in the hydraulic fluid reservoir is maintained at least above the 'MIN' mark (see Chapter 1A or 1B).
13 Remove the dust cap from the relevant caliper bleed screw, then fit the spanner and the tube to the screw **(see illustrations)**. Place the other end of the tube in the jar.
14 Have the assistant depress the brake pedal **slightly**.
15 Loosen the bleed screw to allow fluid to flow into the jar. Continue until the fluid emerging is free from air bubbles.
16 When no more air bubbles appear, tighten the bleed screw securely, remove the tube and spanner, and refit the dust cap. Do not overtighten the bleed screw.
17 Where applicable, repeat the procedure

on the remaining screws in the sequence, until all air is removed from the system.
18 Stop the engine.
19 Where applicable, refit the mud shields, then refit the roadwheels, and lower the vehicle to the ground.
20 Check the hydraulic fluid level, and top up if necessary as described in Chapter 1A or 1B.

3 Hydraulic pipes and hoses – renewal

Note: *Refer to the precautions at the end of Section 1 before proceeding.*
1 If any pipe or hose is to be renewed, minimise fluid loss from the hoses using a proprietary brake hose clamp; metal brake pipe unions can be plugged (if care is taken not to allow dirt into the system) or capped immediately they are disconnected. Place a wad of rag under any union that is to be disconnected, to catch any spilt fluid.
2 If a flexible hose is to be disconnected, unscrew the brake pipe union nut before removing the spring clip which secures the hose to its mounting bracket **(see illustration)**.
3 To unscrew the union nuts, it is preferable to obtain a brake pipe spanner of the correct size; these are available from most large motor accessory shops. Failing this, a close-

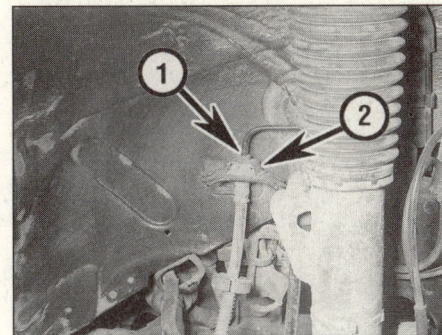

3.2 Brake pipe union nut (1) and spring clip (2)

fitting open-ended spanner will be required, though if the nuts are tight or corroded, their flats may be rounded-off if the spanner slips. In such a case, a self-locking wrench is often the only way to unscrew a stubborn union, but it follows that the pipe and the damaged nuts must be renewed on reassembly. Always clean a union and surrounding area before disconnecting it. If disconnecting a component with more than one union, make a careful note of the connections before disturbing any of them.
4 If a brake pipe is to be renewed, it can be obtained, cut to length and with the union nuts and end flares in place, from Citroën dealers. All that is then necessary is to bend it to shape, following the line of the original, before fitting it to the car. Alternatively, most motor accessory shops can make up brake pipes from kits, but this requires very careful measurement of the original, to ensure that the replacement is of the correct length. The safest answer is usually to take the original to the shop as a pattern.
5 On refitting, do not overtighten the union nuts. It is not necessary to exercise brute force to obtain a sound joint.
6 Ensure that the pipes and hoses are correctly routed, with no kinks, and that they are secured in the clips or brackets provided. After fitting, remember to remove the brake hose clamps and bleed the hydraulic system as described in Section 2. Wash off any spilt fluid, and check carefully for fluid leaks.

4 Front brake pads – renewal

10

⚠️ *Warning: Renew both sets of front brake pads at the same time - never renew the pads on only one wheel, as uneven braking may result. Note that the dust created by wear of the pads may contain asbestos, which is a health hazard. Never blow it out with compressed air, and don't inhale any of it. An approved filtering mask*

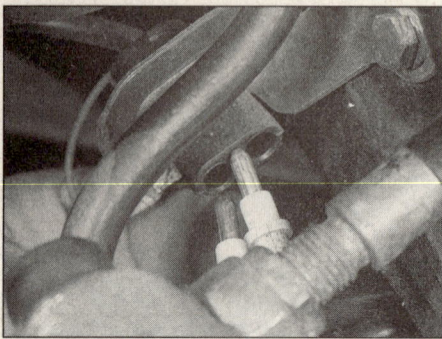

4.2 Disconnecting the brake pad wear sensor wiring connectors

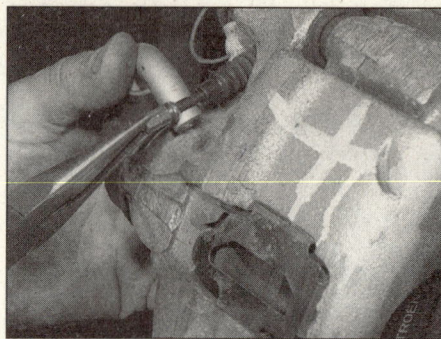

4.3 Disconnecting the parking brake cable from the caliper

4.4a Extract the spring clip . . .

should be worn when working on the brakes. DO NOT use petrol or petroleum-based solvents to clean brake parts; use brake cleaner or methylated spirit only.

1 Chock the rear wheels, then jack up the front of the vehicle and support securely on axle stands (see *Jacking and Vehicle Support*). Remove the front roadwheels.

2 Disconnect the pad wear sensor wiring connectors **(see illustration)**.

3 Disconnect the cable end from the parking brake operating lever. Pull the cable outer from the lug on the caliper, and move the cable to one side, clear of the caliper **(see illustration)**.

4 Using pliers, extract the small spring clip from the caliper retaining pin, then slide the retaining pin from the caliper. If necessary,

carefully tap the pin free using a suitable pin-punch **(see illustrations)**.

5 Release the pad wear sensor wiring from the clip at the bottom of the caliper **(see illustration)**. Take care not to lose the clip.

6 Swivel the caliper upwards, taking care not to strain the fluid hose **(see illustration)**.

7 Withdraw the pads from the caliper **(see illustration)**. Note how the lug on the inboard pad locates into the notch in the caliper piston.

8 First measure the thickness of each brake pad's friction material. If either pad is worn at any point to the specified minimum thickness or less, all four pads must be renewed **(see illustration)**. Also, the pads should be renewed if any are fouled with oil or grease; there is no satisfactory way of degreasing

friction material, once contaminated. If any of the brake pads are worn unevenly, or are fouled with oil or grease, trace and rectify the cause before reassembly. New brake pads and retaining pin kits are available from Citroën dealers.

9 If the brake pads are still serviceable, carefully clean them using a clean, fine wire brush or similar, paying particular attention to the sides and back of the metal backing. Pick out any large embedded particles of dirt or debris. Carefully clean the pad locations in the caliper body/mounting bracket.

10 Prior to fitting the pads, check that the guide pin is free to slide easily in the caliper body/mounting bracket, and check that the rubber guide pin gaiter is undamaged **(see illustration)**. Brush the dust and dirt from the

4.4b . . . then remove the caliper retaining pin

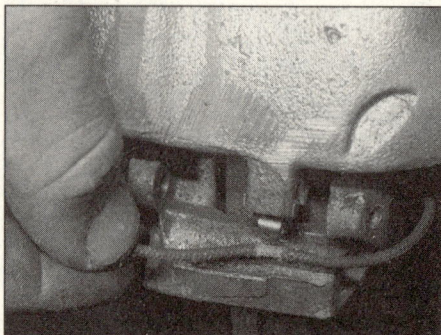

4.5 Release the pad wear sensor wiring from the clip . . .

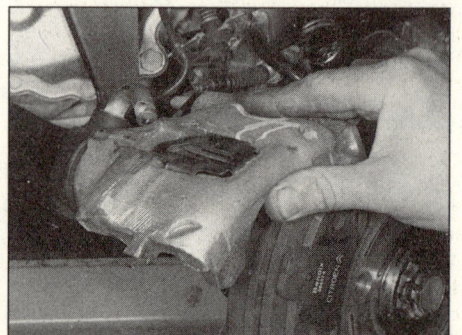

4.6 . . . and swivel the caliper upwards

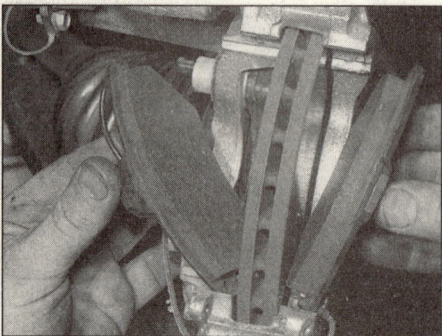

4.7 Withdrawing the brake pads from the caliper

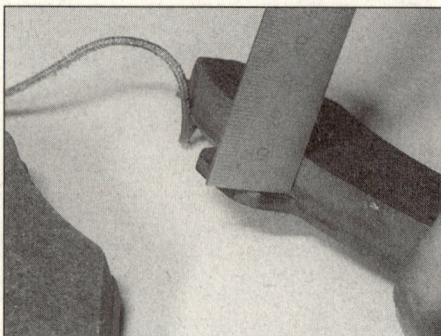

4.8 Measuring the thickness of a front brake pad

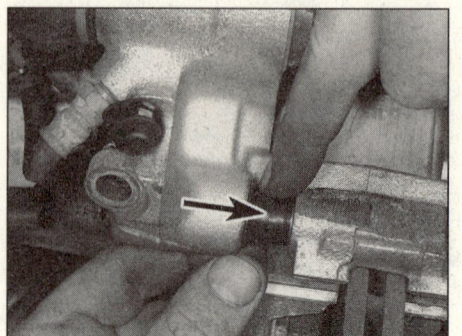

4.10 Check the condition of the guide pin gaiter (arrowed)

4.11 Using a length of flat bar to retract the caliper piston

4.12 Make sure that the locating lug (arrowed) engages with one of the piston notches

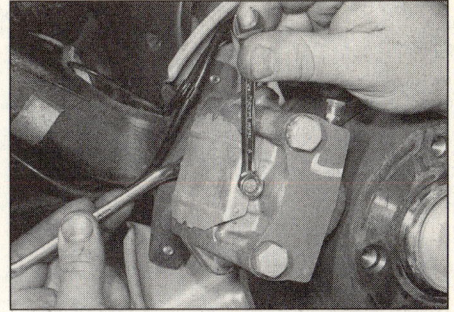

5.2a Counterhold the pad retaining pin, and unscrew the nut from the end of the pin . . .

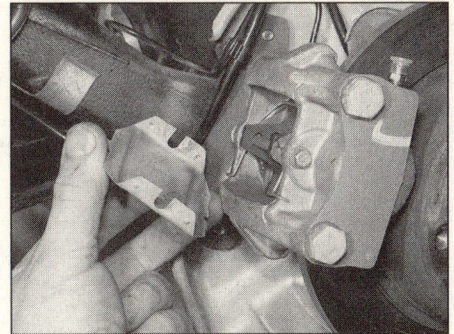

5.2b . . . then withdraw the pad shield

caliper and piston, but *do not* inhale it, as it is injurious to health. Inspect the dust seal around the piston for damage, and the piston for evidence of fluid leaks, corrosion or damage. If attention to any of these components is necessary, refer to Section 7.

11 If new pads are to be fitted, the caliper piston must be pushed back into the cylinder to make room for them. In order to retract the piston, the piston must be turned clockwise as it is pushed into the caliper. Citroën tool 9011-T is designed for this purpose, but a short length of flat bar or any similar tool can be used instead **(see illustration)**. Push the piston fully into the cylinder bore. **Note:** *Rotate the piston so that one of the notches in the piston is opposite the opening in the front of the caliper.*

12 If new pads are being fitted, where applicable, remove the protective paper from the pad backing plates. Fit the pads, ensuring that the friction material is against the brake disc. Make sure that the locating lug on the inboard pad engages with one of the notches in the piston **(see illustration)**.

13 Move the caliper into position over the brake disc.

14 Route the pad wear sensor wiring through the clip at the bottom of the caliper, then reconnect the sensor wiring connectors.

15 Slide the caliper retaining pin into position in the caliper bracket/body, with the hole for the spring clip at the inner end of the pin, then fit the spring clip.

16 Slide the parking brake cable into position in the caliper, then reconnect the end of

the cable to the operating lever on the caliper.

17 Repeat the procedure on the remaining front brake caliper.

18 Depress the brake pedal several times to operate the automatic wear adjustment mechanism in the calipers. Release the brake pedal, and check the parking brake adjustment as described in Section 11.

19 Make a final check on the operation of the brake calipers and the parking brake, then refit the roadwheels and lower the vehicle to the ground.

20 Check the hydraulic fluid level as described in *"Weekly checks"*.

21 New pads will not give full braking efficiency until they have bedded in. Be prepared for this, and avoid hard braking as far as possible for the first few hundred miles or so after pad renewal.

5 Rear brake pads – renewal

⚠ *Warning: Renew both sets of rear brake pads at the same time - never renew the pads on only one wheel, as uneven braking may result. Note that the dust created by wear of the pads may contain asbestos, which is a health hazard. Never blow it out with compressed air, and don't inhale any of it. An approved filtering mask should be worn when working on the*

brakes. DO NOT use petrol or petroleum-based solvents to clean brake parts; use brake cleaner or methylated spirit only.

1 Apply the parking brake and chock the front wheels, then jack up the rear of the vehicle and support securely on axle stands (see *Jacking and Vehicle Support*). Remove the rear roadwheels.

2 Working at the inboard end of the caliper, unscrew the securing nut from the end of the pad retaining pin, then withdraw the pad shield **(see illustrations)**.

3 Slide the pad retaining pin from the outboard end of the caliper, then lift out the anti-rattle spring, noting how the spring locates over the pads and caliper **(see illustrations)**.

4 Slide the pads from the caliper, noting the locations and orientation of the shims, where applicable (shims are not fitted to all models) **(see illustration)**.

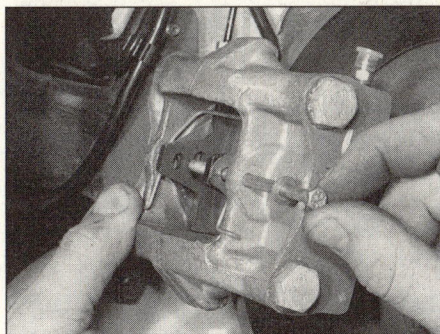

5.3a Slide out the pad retaining pin . . .

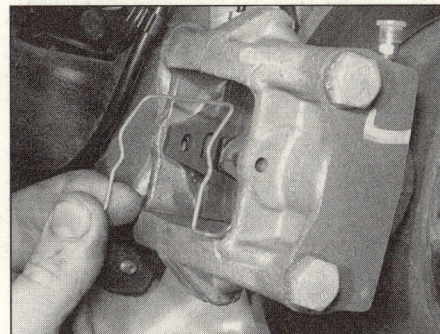

5.3b . . . and lift out the anti-rattle spring

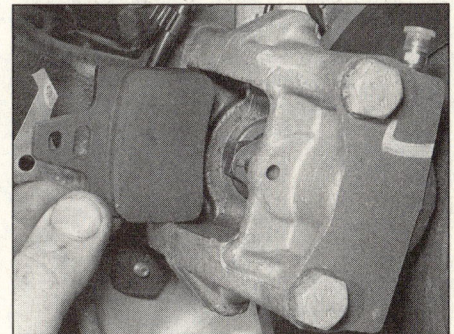

5.4 Sliding the inboard pad from the caliper

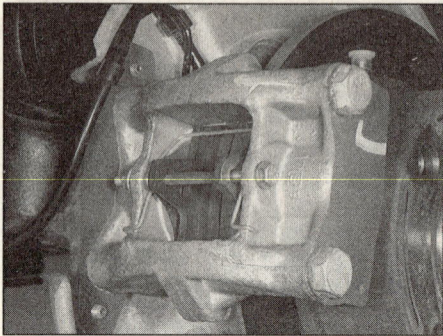

5.11 View of caliper showing anti-rattle spring and pad retaining pin correctly fitted

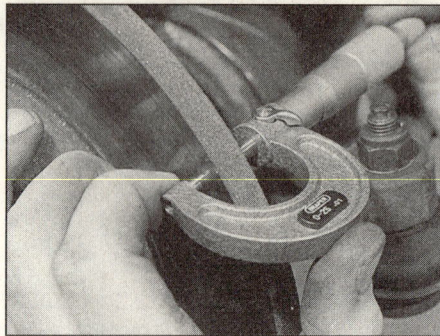

6.3 Using a micrometer to measure disc thickness

6.4 Checking disc run-out using a dial gauge

5 First measure the thickness of each brake pad's friction material. If either pad is worn at any point to the specified minimum thickness or less, all four pads must be renewed. Also, the pads should be renewed if any are fouled with oil or grease; there is no satisfactory way of degreasing friction material, once contaminated. If any of the brake pads are worn unevenly, or are fouled with oil or grease, trace and rectify the cause before reassembly. New brake pads are available from Citroën dealers.

6 If the brake pads are still serviceable, carefully clean them using a clean, fine wire brush or similar, paying particular attention to the sides and back of the metal backing. Pick out any large embedded particles of dirt or debris. Carefully clean the pad locations in the caliper body/mounting bracket.

7 Clean the end of each piston with petrol, then dry the pistons, and pour a few drops of fresh LHM fluid onto the end of each piston.

8 If new brake pads are to be fitted, proceed as follows.
a) *Refit the old brake pads, and slide the pad retaining pin into position.*
b) *Push each brake pad away from the disc, in order to push the pistons fully into their cylinder bores.*
c) *Remove the retaining pin and the pads, and again clean the pad locations in the caliper.*

9 Fit the (new, where applicable) pads into position in the caliper. Make sure that the friction material is against the brake disc. Where applicable, make sure that the shims are fitted as noted before removal.

10 Fit the anti-rattle spring, ensuring that it is located correctly over the pads and caliper, as noted before removal.

11 Fit the pad retaining pin, ensuring that it passes in front of the anti-rattle spring, and the securing nut, but do not tighten the nut at this stage **(see illustration)**.

12 Refit the pad shield, then tighten the retaining pin securing nut.

13 Repeat the procedure on the remaining rear brake caliper.

14 Refit the roadwheels and lower the vehicle to the ground.

15 With the engine running, depress the brake pedal several times to bring the pads into contact with the discs.

16 Check the hydraulic fluid level as described in *"Weekly checks"*.

17 New pads will not give full braking efficiency until they have bedded in. Be prepared for this, and avoid hard braking as far as possible for the first few hundred miles or so after pad renewal.

6 Brake discs - inspection, removal and refitting

Front brake disc
Inspection

⚠ *Warning: If either front disc requires renewal, BOTH should be renewed at the same time, to ensure even and consistent braking. New brake pads should also be fitted. Note that the dust created by wear of the pads may contain asbestos, which is a health hazard. Never blow it out with compressed air, and don't inhale any of it. An approved filtering mask should be worn when working on the brakes. DO NOT use petrol or petroleum-based solvents to clean brake parts; use brake cleaner or methylated spirit only.*

1 Chock the rear wheels, then jack up the front of the car and support it on axle stands (see *Jacking and Vehicle Support*). Remove the appropriate front roadwheel.

2 Slowly rotate the brake disc so that the full area of both sides can be checked; remove the brake pads if better access is required to the inboard surface (see Section 4). Light scoring is normal in the area swept by the brake pads, but if heavy scoring or cracks are found, the disc must be renewed.

3 It is normal to find a lip of rust and brake dust around the disc's perimeter; this can be scraped off if required. If, however, a lip has formed due to excessive wear of the brake pad swept area, then the disc's thickness must be measured using a micrometer **(see illustration)**. Take measurements at several places around the disc, at the inside and

outside of the pad swept area; if the disc has worn at any point to the specified minimum thickness or less, the disc must be renewed.

4 If the disc is thought to be warped, it can be checked for run-out. Either use a dial gauge mounted on any convenient fixed point, while the disc is slowly rotated, or use feeler gauges to measure (at several points all around the disc) the clearance between the disc and a fixed point, such as the caliper mounting bracket **(see illustration)**. If the measurements obtained are at the specified maximum or beyond, the disc is excessively warped, and must be renewed; however, it is worth checking first that the hub bearing is in good condition (Chapters 1A, section 7 or 1B, section 6 and/or Chapter 11). Also try the effect of removing the disc and turning it through 180°, to reposition it on the hub; if the run-out is still excessive, the disc must be renewed.

5 Check the disc for cracks, especially around the wheel bolt holes, and any other wear or damage, and renew if necessary.

Removal

6 Remove the front brake pads as described in Section 4.

7 Unscrew the two bolts securing the caliper bracket to the hub carrier, and move the caliper to one side, taking care not to strain the fluid hose. Suspend the caliper to one side using wire or string. **Do not** remove the caliper from the guide pin.

8 Unscrew the two brake disc securing screws, then withdraw the disc from the hub.

Refitting

9 Ensure that the mating faces of the disc and the hub are absolutely clean, then fit the disc to the hub. Refit and tighten the disc securing screws.

10 Refit the caliper, and tighten the bracket securing bolts to the specified torque.

11 Refit the brake pads as described in Section 4.

Rear brake disc
Inspection

⚠ *Warning: If either rear disc requires renewal, BOTH should be renewed at the same time, to ensure even and consistent braking. New brake pads should also be fitted. Note that the dust created by wear of the pads may contain asbestos, which is a health hazard. Never blow it out with compressed air, and don't inhale any of it. An approved filtering mask should be worn when working on the brakes. DO NOT use petrol or petroleum-based solvents to clean brake parts; use brake cleaner or methylated spirit only.*

12 Apply the parking brake and chock the front wheels, then jack up the rear of the car and support securely on axle stands (see *Jacking and Vehicle Support*).

13 Proceed as described for the front brake disc in paragraphs 2 to 5, but refer to Section 5 if the brake pads are to be removed.

Removal

14 Jack up the vehicle and support securely on axle stands with the wheels clear of the ground (see *Jacking and Vehicle Support*).

15 Remove the relevant roadwheel.

16 Depressurise the hydraulic system as described in Chapter 9.

17 Remove the brake pads as described in Section 5.

18 Refit the pad retaining pin to the caliper, then refit and tighten the pin securing nut to hold the two halves of the caliper together.

19 Unscrew the two bolts securing the brake caliper to the trailing arm.

20 Move the caliper clear of the disc, taking care not to strain the fluid pipe. Suspend the caliper using wire or string - if desired, the caliper can be temporarily secured using the upper securing bolt once the disc has been removed.

21 Remove the disc securing screw, then withdraw the disc from the hub.

Refitting

22 Where applicable, remove the upper caliper securing bolt, and again move the caliper to one side to facilitate refitting of the disc.

23 Ensure that the mating faces of the disc and the hub are absolutely clean, then fit the disc to the hub. Refit and tighten the disc securing screw.

24 Lubricate the threads and heads of the caliper securing bolts, then refit the caliper,

and tighten the bolts to the specified torque.

25 Unscrew the securing nut, and remove the pad retaining pin from the caliper.

26 Refit the brake pads as described in Section 5.

27 Refit the roadwheel.

28 Pressurise the hydraulic system as described in Chapter 9.

29 Lower the vehicle to the ground.

7 Front brake caliper - removal, overhaul and refitting

Note: *Refer to the precautions at the end of Section 1 before proceeding.*

Removal

1 Remove the brake pads as described in Section 4.

2 Place a suitable container under the fluid hose union on the caliper, to catch any escaping hydraulic fluid. Unscrew the union and disconnect the hose from the caliper. Plug the open ends of the hose and caliper to prevent dirt ingress and further fluid loss.

3 Remove the two securing bolts, and withdraw the caliper/bracket assembly from the hub carrier **(see illustrations)**.

Overhaul

4 The caliper can be overhauled after obtaining the relevant repair kit from a Citroën dealer, however is strongly recommended that overhaul is entrusted to a Citroën dealer, as various special tools are required to fully dismantle the caliper. If the caliper is to be overhauled, note the following points.

a) *Ensure that the correct repair kit is obtained for the caliper being worked on.*
b) *Note the locations of all components to ensure correct refitting.*
c) *Lubricate the new seals using LHM.*
d) *Follow the assembly instructions supplied with the repair kit.*

Refitting

5 Offer the caliper/bracket assembly into position, then refit the securing bolts and tighten to the specified torque.

6 Reconnect the fluid hose, and tighten the union.

7 Refit the brake pads as described in Section 4.

8 Bleed the brake hydraulic system as described in Section 2.

9 Refit the roadwheels and lower the vehicle to the ground.

8 Rear brake caliper – removal, overhaul and refitting

Note: *Refer to the precautions at the end of Section 1 before proceeding. A new fluid pipe seal will be required on refitting.*

Removal

1 Jack up the vehicle and support securely on axle stands with the wheels clear of the ground (see *Jacking and Vehicle Support*).

2 Remove the relevant roadwheel.

3 Depressurise the hydraulic system as described in Chapter 9.

4 Remove the brake pads as described in Section 5.

5 Refit the pad retaining pin to the caliper, then refit and tighten the pin securing nut to hold the two halves of the caliper together.

6 Place a container under the caliper fluid pipe union, to catch any escaping fluid. Smear a little LHM fluid on the joint between the pipe and union, and progressively unscrew the union. If the pipe turns, retighten the union slightly before loosening again, to free the pipe. Once the pipe is disconnected from the caliper, plug the pipe and caliper fitting to prevent fluid loss.

7 Unscrew the two securing bolts, and withdraw the caliper **(see illustration)**.

Overhaul

8 The caliper can be overhauled after obtaining the relevant repair kit from a Citroën dealer, bearing in mind the following points.

a) *Do not separate the two halves of the caliper.*
b) *Ensure that the correct repair kit is obtained for the caliper being worked on.*
c) *Note the locations of all components to ensure correct refitting.*

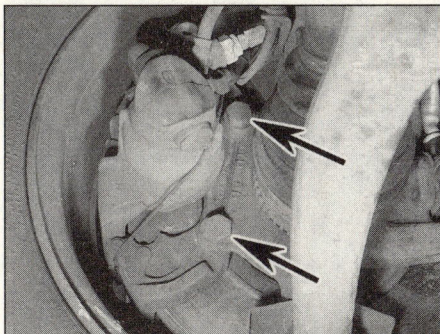

7.3a Remove the two securing bolts (arrowed) . . .

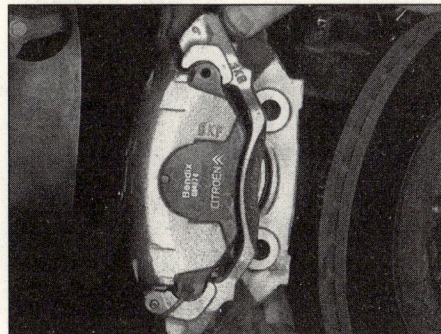

7.3b . . . and withdraw the caliper

8.7 Rear caliper securing bolts (1). Note that the pad retaining pin (2) must be refitted before removing the caliper

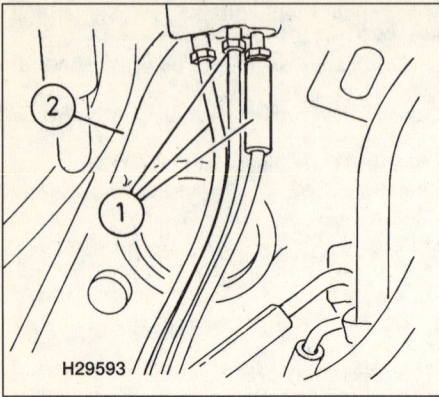

9.6 Disconnect the three fluid pipes (1) and the fluid return hose (2) from the brake control valve/compensator – left-hand-drive models

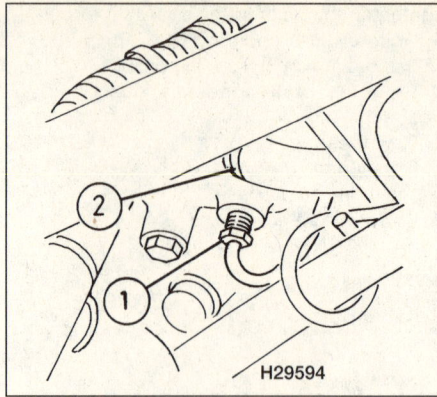

9.7 Slacken the remaining pipe union (1) and the remaining fluid return hose clip (2) – left-hand-drive models

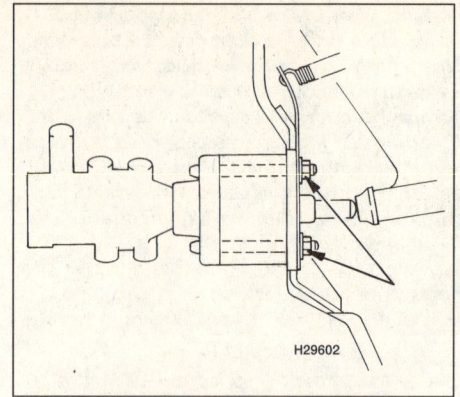

9.8 Brake control valve/compensator securing nuts (arrowed) – left-hand-drive models

d) *Lubricate the new seals using LHM.*
e) *Follow the assembly instructions supplied with the repair kit.*

Refitting

9 Lubricate the threads of the caliper securing bolts, then refit the caliper, and tighten the bolts to the specified torque.
10 Reconnect the fluid pipe to the caliper, using a new seal, and tighten the union.
11 Tighten the securing nut, and refit the pad retaining pin from the caliper.
12 Refit the brake pads as described in Section 5.
13 Refit the roadwheel.
14 Pressurise the hydraulic system as described in Chapter 9.
15 Lower the vehicle to the ground.

9 Brake control valve/compensator - removal, overhaul and refitting

Note: *Refer to the precautions given at the end of Section 1 before proceeding. New hose clips, and new fluid pipe seals will be required on refitting.*

Left-hand-drive models

Removal

1 Chock the rear wheels, then jack up the front of the car and support it on axle stands (see *Jacking and Vehicle Support*). Remove the left-hand front roadwheel, then remove the mud shield from under the wheel arch.
2 Depressurise the hydraulic system as described in Chapter 9.
3 Using a screwdriver, release the securing clip, and remove the cover from the battery, then disconnect the battery negative lead.
4 Remove the air cleaner assembly as described in Chapter 4A or 4B.
5 Remove the hydraulic fluid reservoir as described in Chapter 9.
6 Unscrew the unions, and disconnect the

three fluid pipes shown from the valve. Also disconnect the fluid return hose **(see illustration)**. Note the locations of the pipes and hose to aid refitting. Be prepared for fluid spillage, and plug the open ends of the pipes, hose and valve to prevent dirt entry and further fluid loss.
7 Slacken the union securing the remaining fluid pipe, and the clip securing the remaining fluid return hose **(see illustration)**.
8 Unscrew the two brake control valve/compensator securing nuts, and unclip the clamp from the top of the valve/compensator **(see illustration)**.
9 Carefully lift the valve/compensator from the bulkhead, and recover the spacer and seal.
10 Disconnect the remaining fluid pipe and return hose, and withdraw the assembly from the engine compartment. Again, note the locations of the pipe and hose to aid refitting. Be prepared for fluid spillage, and plug the open ends of the pipe, hose and valve to prevent dirt entry and further fluid loss.

Overhaul

11 The valve/compensator can be overhauled after obtaining the relevant repair kit from a Citroën dealer. Note the locations of all components to ensure correct refitting, and lubricate the new seals using LHM. Follow the assembly instructions supplied with the repair kit.

Refitting

12 Refitting is a reversal of removal, bearing in mind the following points.
a) *Make sure that the spacer and seal are in place when refitting the valve/compensator assembly. Examine the seal, and renew if necessary.*
b) *Use new seals when reconnecting the fluid pipes.*
c) *Ensure that the fluid pipes and hoses are correctly reconnected as noted before removal.*
d) *Tighten the valve/compensator securing nuts to the specified torque.*

e) *Refit the hydraulic fluid reservoir as described in Chapter 9.*
f) *On completion, pressurise the hydraulic system as described in Chapter 9, and bleed the brake hydraulic circuit as described in Section 2.*

Right-hand-drive models

Removal

13 Depressurise the hydraulic system as described in Chapter 9.
14 Remove the front scuttle trim panel as described in Chapter 12.
15 Unclip both bonnet support struts, and move the bonnet to its fully-raised position. Ensure that the bonnet is held securely in position (using a wooden prop or similar).
16 Disconnect the windscreen wiper motor wiring plug. If desired, to improve access, remove the windscreen wiper motor as described in Chapter 13.
17 Slacken the four unions securing the fluid pipes to the valve/compensator.
18 Disconnect the fluid return hose from the top of the valve. Note the location of the hose to aid refitting **(see illustration)**. Be prepared for fluid spillage, and plug the open ends of the hose and valve to prevent dirt entry and further fluid loss.

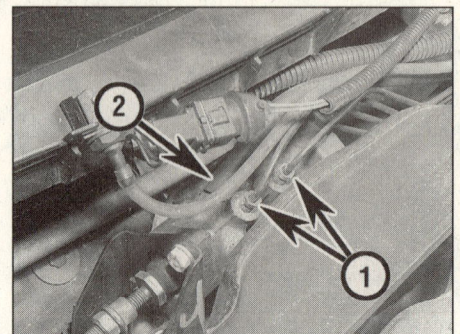

9.18 Brake control valve compensator top fluid pipe connections (1) and fluid return hose (2) – right-hand-drive models

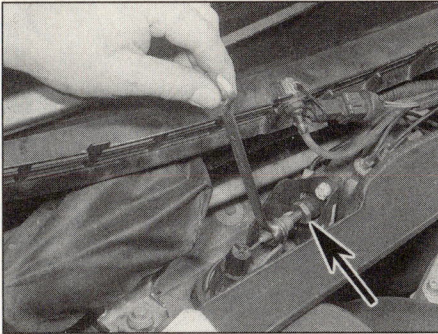
9.19 Slacken the locknut, and screw the pedal adjustment stop (arrowed) in as far as possible – right-hand-drive models

19 Slacken the locknut, and screw brake pedal adjustment stop in as far as possible **(see illustration)**.
20 Unscrew the two valve/compensator securing bolts **(see illustration)**.
21 Disconnect the four fluid pipes from the valve/compensator. Note the locations of the pipes to aid refitting. Be prepared for fluid spillage, and plug the open ends of the pipes and valve to prevent dirt entry and further fluid loss.
22 Swivel the valve/compensator round through 180° for access to the fluid return hose at the bottom of the assembly.
23 Disconnect the remaining fluid return hose (again, be prepared for fluid spillage, and plug the valve/compensator and hose), and withdraw the valve/compensator from the scuttle.

Overhaul
24 Refer to paragraph 11.

Refitting
25 Refitting is a reversal of removal, bearing in mind the following points.
a) Use new seals when reconnecting the fluid pipes.
b) Ensure that the fluid pipes and hoses are correctly reconnected as noted before removal.
c) Tighten the valve/compensator securing nuts to the specified torque.

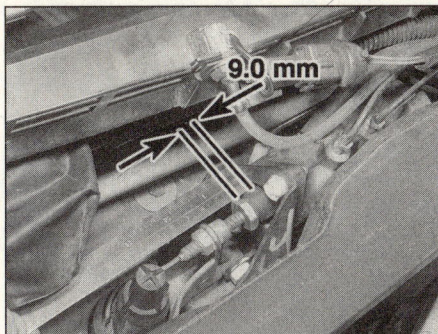
10.3 Measure the distance between the end face of the brake control valve/compensator mounting bracket and the brake pedal stop

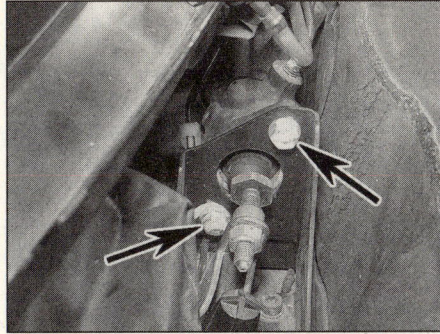
9.20 Unscrew the two valve/compensator securing bolts (arrowed) – right-hand-drive models

d) Adjust the brake pedal adjustment stop to give the specified brake pedal height (see Section 10), then tighten the locknut.
e) Refit the front scuttle trim panel with reference to Chapter 12.
f) On completion, pressurise the hydraulic system as described in Chapter 9, and bleed the brake hydraulic circuit as described in Section 2.

10 Brake pedal – adjustment

Left-hand-drive models
1 No adjustment of the brake pedal is possible.

Right-hand-drive models
2 Remove the front scuttle trim panel as described in Chapter 12.
3 With the brake pedal in the rest position, measure the distance between the end face of the brake control valve/compensator mounting bracket and the brake pedal stop. The distance should be 9.0 mm **(see illustration)**.
4 If adjustment is required, slacken the brake pedal adjustment stop locknut, and turn the stop as necessary to give the specified distance **(see illustration)**. Tighten the locknut when adjustment is complete.
5 On completion, refit the front scuttle trim panel with reference to Chapter 12.

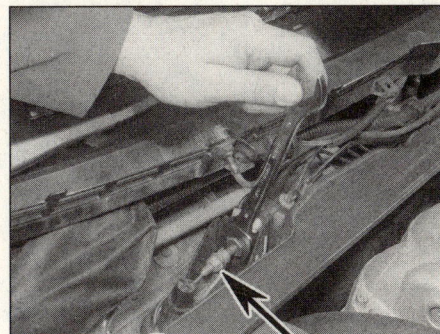
10.4 Slacken the locknut (arrowed) and turn the stop to give the specified distance

11 Parking brake – adjustment

Automatic wear adjustment
1 The front brake calipers incorporate an automatic adjustment mechanism, which compensates for the clearance in the parking brake operating mechanism created by brake pad wear. The adjustment mechanism is operated by hydraulic pressure as the brakes are applied.
2 The following operation should be carried out if the front brake pads have been removed, or if work has been carried out on the front calipers, in order to initially set the adjustment mechanism.
3 Start the engine, and allow it to run at idle.
4 Ensure that the parking brake lever is in the 'released' position.
5 Depress the brake pedal several times to operate the adjustment mechanism.
6 Release the brake pedal, and check that the parking brake can be fully applied by depressing the parking brake pedal between 4 to 12 clicks. Stop the engine. If necessary, adjust the parking brake cables as described in the following paragraphs.

Cable adjustment
7 Chock the rear wheels, then jack up the front of the car and support it on axle stands (see Jacking and Vehicle Support). Remove the front roadwheels.
8 Fully release the parking brake.
9 Depress the brake pedal to bring the brake pads into contact with the disc, then release the pedal.
10 Push in the parking brake lever on the facia to the 'locked' position.
11 Depress the parking brake pedal to the 4th notch of its travel.
12 Working on one of the calipers, loosen the cable adjuster locknut at the caliper, then turn the adjuster nut to obtain a balance between the cable lengths at the equaliser to within 1.5 mm **(see illustration)**.
13 If necessary, repeat the procedure on the remaining caliper.

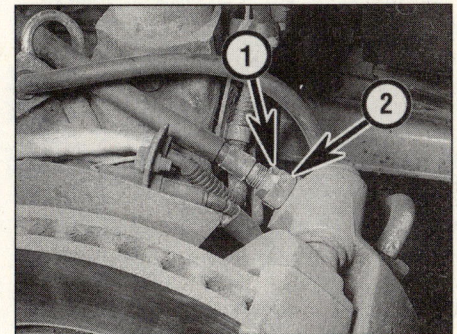
11.12 Parking brake cable adjuster locknut (1) and adjuster nut (2)

10

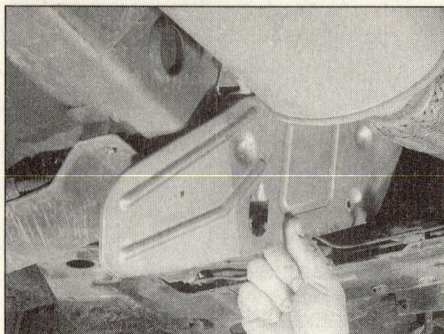

12.3 Removing the parking brake cable equaliser cover – right-hand-drive model

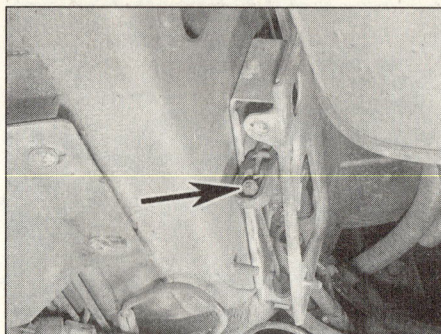

12.4 Unhook the end of the cable (arrowed) from the equaliser – right-hand-drive model

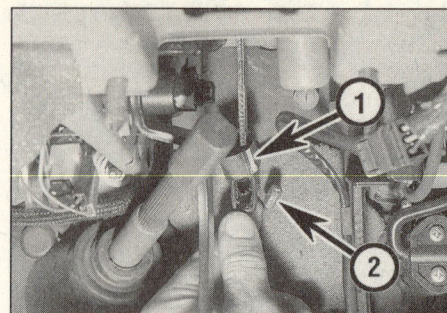

12.6 Slide off the plastic cover (1), then unhook the end of the parking brake cable (2) from the lever – viewed with steering column removed for clarity

14 Pull the parking brake lever on the facia to the 'released' position. The parking brake pedal should return to its rest position.

15 With the caliper parking brake operating levers at rest (parking brake released), the levers should not be pulled by the cables whatever the steering lock angle and the ride height of the vehicle. Check this, and re-adjust if necessary.

16 Tighten the cable adjuster locknuts.

17 Depress the parking brake pedal several times, and make sure that it returns to its rest position.

18 Operate the parking brake mechanism several times, and check that the parking brake can be fully applied by depressing the pedal between 6 to 12 clicks.

12 Parking brake cables - removal and refitting

Pedal-to-equaliser cable

Removal

1 Jack up the vehicle and support securely on axle stands (see *Jacking and Vehicle Support*).

2 Fully release the parking brake.

3 Working under the vehicle, unbolt the parking brake cable equaliser cover plate **(see illustration)**.

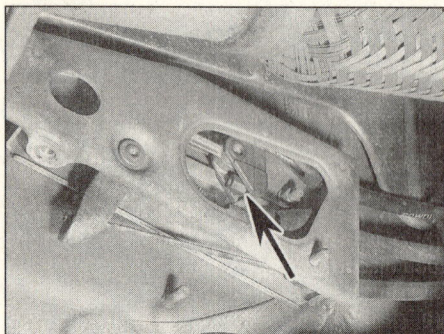

12.13 Unhook the end of the equaliser cable (arrowed) from the equaliser – right-hand-drive model

4 Unhook the end of the cable from the equaliser, then release the cable from the locating brackets under the vehicle, noting its routing **(see illustration)**.

5 Working at the engine compartment bulkhead, unscrew the two nuts securing the parking brake cable plate to the bulkhead – access to these nuts is extremely difficult.

6 Working in the driver's footwell, slide the plastic cover from the end of the parking brake lever, then unhook the end of the cable from the lever **(see illustration)**.

7 Working in the engine compartment, pull the cable through the bulkhead, then withdraw it from the engine compartment.

Refitting

8 Refitting is a reversal of removal, but ensure that the cable is routed as noted before removal and, on completion, check the parking brake adjustment as described in Section 11.

Equaliser-to-caliper cable

Removal

9 Jack up the vehicle and support securely on axle stands (see *Jacking and Vehicle Support*). Remove the relevant front roadwheel.

10 Fully release the parking brake.

11 Disconnect the cable end from the parking brake operating lever. Pull the cable outer from the lug on the caliper, and move the cable to one side, clear of the caliper.

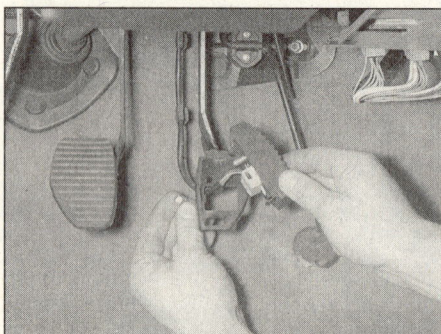

13.6 Unscrew the securing nut and remove the pedal rubber/switch assembly

12 Working under the vehicle, unbolt the parking brake cable equaliser cover plate.

13 Unhook the end of the relevant cable from the equaliser, then release the cable from the locating brackets, and withdraw it from the vehicle **(see illustration)**.

Refitting

14 Refitting is a reversal of removal but, on completion, check the parking brake adjustment as described in Section 11.

13 Stop-light switch - removal and refitting

Removal

1 The switch is integral with the brake pedal rubber.

2 Using a screwdriver, release the securing clip, and remove the cover from the battery, then disconnect the battery negative lead.

3 Remove the securing clips and screws, and remove the driver's footwell carpet trim panel from the underside of the facia.

4 Trace the wiring up from the pedal, and separate the two halves of the stop light switch wiring connector.

5 Unclip the wiring from the pedal.

6 Working behind the pedal, unscrew the securing nut, then remove the pedal rubber/switch assembly **(see illustration)**.

Refitting

7 Refitting is a reversal of removal.

14 Parking brake 'on' warning light switch – removal and refitting

Removal

1 The switch is mounted in the parking brake pedal mounting bracket in the driver's footwell.

2 Using a screwdriver, release the securing clip, and remove the cover from the battery, then disconnect the battery negative lead.

3 Pull back the footwell carpet trim panel for access to the switch.
4 Disconnect the wiring plugs from the switch.
5 Slacken the locknut, then unscrew the switch from the pedal bracket. Note the number of turns necessary to unscrew the switch, to aid refitting

Refitting

6 Refitting is a reversal of removal, but screw the switch into the bracket by the number of turns noted before removal and, if necessary, adjust the position of the switch to achieve satisfactory operation. Tighten the locknut when adjustment is complete.

15 Anti-lock braking system (ABS) - general information

An anti-lock braking system is available as standard equipment on some models, and as an option on certain others.

To prevent wheel locking, the system provides pressure modulation in the braking circuits. To achieve this, sensors mounted each wheel monitor the rotational speeds of the wheels, and are able to detect when there is a risk of wheel locking (low rotational speed). Solenoid valves are positioned in the brake circuits to each wheel, and the solenoid valves are incorporated in a modulator assembly, which is controlled by an electronic control unit (ECU). The ECU controls modulation of the braking effort applied to each wheel, according to the information supplied by the wheel sensors.

Should a fault develop in the system, a self-diagnostic facility is incorporated in the ECU, which can be used in conjunction with special diagnostic equipment available to a Citroën dealer, to determine the nature of the fault.

The braking system components used on models with ABS are similar to those used on models with a conventional braking system.

16 Anti-lock braking system components - removal and refitting

Hydraulic valve block

Removal

1 Depressurise the hydraulic system as described in Chapter 9.
2 Chock the rear wheels, then jack up the front of the vehicle and support securely on axle stands (see *Jacking and Vehicle Support*). Remove the left-hand front roadwheel.
3 Working under the front left-hand wheel arch, remove the wheel arch liner for access to the ABS hydraulic valve block.
4 Remove the air cleaner assembly as described in Chapter 4A or 4B.
5 Remove the battery, with reference to Chapter 5A if necessary.
6 Working around the battery tray, unbolt all electrical units, fuseboxes, etc, and move them to one side to allow the battery tray to be removed.
7 Release the wiring harnesses, hoses and cables, as applicable, from any clips and brackets attached to the battery tray. It may be necessary to disconnect certain wiring connectors, in which case note their locations.
8 Check that all relevant components have been moved clear, then unscrew the securing bolts and nuts, and remove the battery tray assembly. Note that on some models, the air intake trunking passes through the battery tray, and it will be necessary to disconnect the trunking to allow removal of the battery tray.
9 Move any remaining wiring harnesses, hoses and cables to one side (noting their routing to aid refitting) to provide adequate access to the ABS hydraulic valve block.
10 Disconnect the two wiring plugs from the side of the valve block, then release the wiring harnesses from the clips attached to the assembly.
11 Release the clip, and disconnect the hydraulic fluid return hose from the top of the valve block – be prepared for fluid spillage.

12 Unscrew the unions, and disconnect the hydraulic fluid pipes from the left-hand side of the valve block – note the locations of the pipes to ensure correct refitting. Again be prepared for fluid spillage, and position a container beneath the assembly to catch escaping fluid.
13 Unscrew the front and rear securing nuts, and the lower securing nut, and withdraw the valve block assembly from its mounting bracket **(see illustrations)**.

Refitting

14 Refitting is a reversal of removal, bearing in mind the following points.
a) *Ensure that the valve block fluid pipes are correctly reconnected as noted before removal.*
b) *Ensure that all wiring harnesses, hoses and cables are correctly routed and reconnected a noted before removal.*
c) *On completion, pressurise the hydraulic system as described in Chapter 9, then bleed the brake hydraulic system, as described in Section 2.*

Front wheel sensor

Removal

15 Using a screwdriver, release the securing clip, and remove the cover from the battery, then disconnect the battery negative lead.
16 Chock the rear wheels, then jack up the front of the vehicle, and support securely on axle stands (see *Jacking and Vehicle Support*). Remove the relevant roadwheel.
17 Trace the wiring back from the sensor, and separate the two halves of the connector. Note the routing of the wiring to aid correct refitting.
18 Where applicable, unclip the wiring from the brackets and/or clips.
19 Unscrew the sensor clamp screw, and the sensor securing bolt, and slide the sensor from the hub carrier **(see illustration)**.

Refitting and adjustment

20 Ensure that the mating faces of the sensor and housing are clean, and lightly grease the sensor body.
21 If a new sensor is being fitted, it will normally be supplied with a setting shim fitted to the end of the sensor to aid fitting.

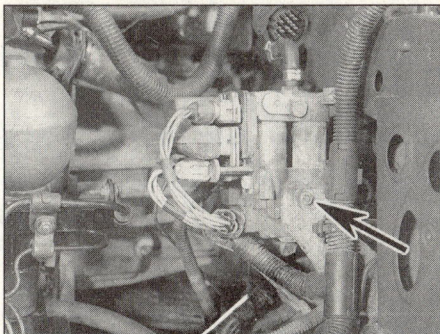

16.13a **ABS hydraulic valve block front securing nut (arrowed) – viewed with body front panel removed**

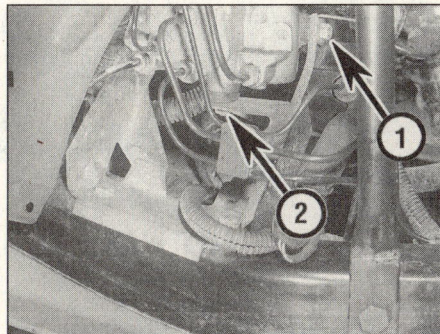

16.13b **ABS hydraulic valve block rear (1) and lower (2) securing nuts – viewed through left-hand wheel arch**

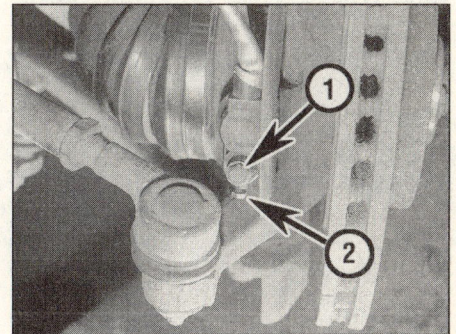

16.19 **ABS front wheel sensor securing bolt (1) and clamp screw (2)**

10

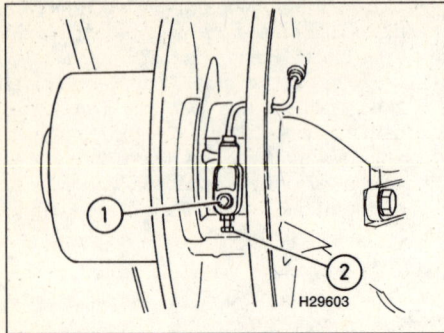

16.34 ABS rear wheel sensor securing bolt (1) and clamp screw (2)

16.44 Unclip the cover from the control unit housing . . .

16.45 . . . then withdraw the ABS electronic control unit

22 Ensure that the sensor clamp screw in the hub carrier is loosened.

23 Fit the sensor, together with its shim, where applicable, into the housing in the hub carrier.

24 Screw in the sensor securing bolt, without tightening fully.

25 If the sensor has a setting shim fitted, move the sensor until the setting shim contacts the toothed sensor wheel. Do not turn the hub during this procedure.

26 If the sensor is not fitted with a setting shim, insert a 0.5 mm feeler gauge between the end of the sensor and the toothed sensor wheel, and adjust the position of the sensor until the feeler blade is a stiff sliding fit.

27 Hold the sensor in position, then tighten the sensor securing bolt, followed by the clamp screw.

28 Reconnect the sensor wiring connector, and clip the wiring into position, ensuring that it is routed as noted before removal.

29 Refit the roadwheel, then lower the vehicle to the ground and reconnect the battery negative lead.

Rear wheel sensor

Removal

30 Using a screwdriver, release the securing clip, and remove the cover from the battery, then disconnect the battery negative lead.

31 Apply the parking brake and chock the front wheels, then jack up the front of the vehicle, and support securely on axle stands (*see Jacking and Vehicle Support*). Remove the relevant roadwheel.

32 Trace the wiring back from the sensor, and separate the two halves of the connector. Note the routing of the wiring to aid correct refitting.

33 Where applicable, unclip the wiring from the brackets and/or clips.

34 Unscrew the sensor clamp screw, and the sensor securing bolt, and slide the sensor from the trailing arm **(see illustration)**.

Refitting and adjustment

35 Ensure that the sensor clamp screw and the sensor securing bolt are loose.

36 Ensure that the mating faces of the sensor and the bore in the trailing arm are clean, and lightly grease the sensor body.

37 Using a depth gauge inserted into the sensor bore in the trailing arm, measure the distance between the top of one of the sensor wheel teeth, and the upper surface of the sensor bore.

38 Fit the sensor, and slide it into the bore to give a clearance between the upper surface of the sensor bore and the end of the sensor equal to the measurement taken in paragraph 37, less 0.5 mm. This can be achieved by measuring the distance from the end of the

sensor to the lower face of the flange at the top of the sensor, and calculating the appropriate clearance between the lower face of the flange on the sensor, and the upper surface of the sensor bore.

39 Hold the sensor in position, then tighten the sensor securing bolt, followed by the clamp screw.

40 Reconnect the sensor wiring connector, and clip the wiring into position, ensuring that it is routed as noted before removal.

41 Refit the roadwheel, then lower the vehicle to the ground and reconnect the battery negative lead.

Electronic control unit

Removal

42 The Electronic control unit (ECU) is located in the control unit housing at the right-hand side of the engine compartment.

43 Using a screwdriver, release the securing clip, and remove the cover from the battery, then disconnect the battery negative lead.

44 Unclip the cover from the control unit housing **(see illustration)**.

45 Withdraw the ECU, then release the metal securing clip, and disconnect the wiring connector **(see illustration)**.

Refitting

46 Refitting is a reversal of removal.

Chapter 11
Suspension and steering

Contents

Degrees of difficulty

Easy, suitable for novice with little experience	Fairly easy, suitable for beginner with some experience	Fairly difficult, suitable for competent DIY mechanic	Difficult, suitable for experienced DIY mechanic	Very difficult, suitable for expert DIY or professional

Specifications

Front wheel alignment
Front wheel toe-setting . . . . 0 to 3.0 mm toe-out

Vehicle ride height (see text)
Models up to 1994:
    Front . . . . 144.0 +10.0 mm/ -7.0 mm
    Rear . . . . 431.0 +10.0 mm/ -7.0 mm
Models from 1995:
    Front . . . . 141.5 mm
    Rear . . . . 136.5 mm

Torque wrench settings

	Nm	lbf ft
Front suspension		
Driveshaft retaining nut	320	236
Hub carrier-to-suspension strut clamp nut and bolt	55	41
Lower balljoint-to-lower arm nut	40	30
Lower arm balljoint-to-hub carrier	250	185
Anti-roll bar drop link securing nuts	30	22
Anti-roll bar drop link adjuster locknut	30	22
Lower arm rear securing bolts	35	26
Lower arm front pivot bolt and nut	85	63
Hydraulic unit-to-body bolts	25	18
Suspension strut-to-hydraulic unit nut*	65	48
Rear suspension		
Rear hub nut	380	280
Rear hub assembly securing bolts	32	24
Anti-roll bar-to-trailing arm bolts:		
Hatchback models	80	59
Estate models	100	74
Trailing arm pivot shaft nut	130	96
Rear suspension front securing bolts	60	44
Rear suspension rear securing bolts	40	30
Rear suspension rear mounting rubber-to-body nuts	30	22
Steering		
Steering wheel nut	30	22
Track-rod balljoint nut	45	33
Steering column universal joint pinch-bolt	20	15
Steering column nuts	20	15
Intermediate shaft-to-steering gear pinion pinch-bolt	20	15
Steering gear-to-subframe bolts	70	52
Track-rod balljoint locknut	45	33
Hydraulic pipe unions		
3.5 and 4.5 mm diameter pipes	8	6
6.35 mm diameter pipes:		
With sleeve seal	10	7
Without sleeve seal	13	10
10.0 mm diameter pipes	30	22
Roadwheels		
Roadwheel bolts	90	66

*Use thread/locking compound

1 General information

The suspension is of an independent hydraulic type, exclusive to Citroën.

The front suspension comprises a vertically-mounted hydraulic suspension strut unit, a lower arm, and an anti-roll bar. The lower arms and the anti-roll bar are mounted on the front subframe. The front suspension hydraulic units are supplied with hydraulic fluid from the main hydraulic system via the front height corrector which is actuated by the front anti-roll bar. The anti-roll bar is attached to the suspension struts by drop links.

A trailing arm rear suspension is used, and the rear hydraulic units are supplied with hydraulic fluid from the main system via the rear height corrector. The height corrector is actuated by the rear anti-roll bar.

The ground clearance of the vehicle may be adjusted using a lever mounted inside the vehicle. The lever is connected via operating rods to the front and rear height correctors.

Automatic damping is incorporated in the hydraulic units, which take the place of the coil springs and dampers found in a conventional suspension system.

Certain models have 'Hydractive' suspension system, which allows the driver to switch between 'Normal' (or 'Auto' on early models) and 'Sport' suspension settings. This system is electronically controlled via various sensors. The 'Normal' (or 'Auto') position provides maximum comfort, and adapts automatically to driving and road conditions. The 'Sport' setting provides a stiffer suspension more suited to a sporting driving style, particularly on winding roads.

The steering is of rack-and-pinion type, mounted on a crossmember attached to the front subframe. The steering column incorporates a universal joint and coupling. Power steering is fitted to all models, and hydraulic fluid is supplied from the main hydraulic system (see Chapter 9), and the pressure is controlled by a flow distributor unit and a control valve. The power assistance is provided via a hydraulic ram mounted on the steering gear.

2 Front hub carrier assembly – removal and refitting

Note: *A balljoint separator tool will be required for this operation. A new suspension lower balljoint nut, a new track-rod balljoint nut, and a new hub carrier-to-strut nut will be required on refitting.*

Removal

1 Depressurise the suspension hydraulic system, as described in Chapter 9.
2 Chock the rear wheels, then jack up the front of the car and support it on axle stands (see *Jacking and Vehicle Support*). Remove the appropriate front roadwheel. On models

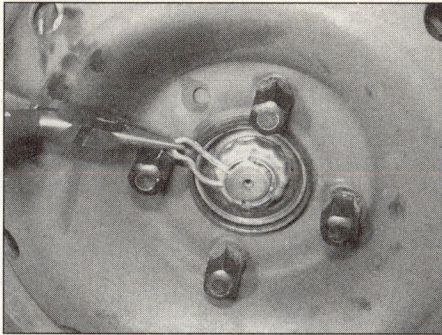
2.2a Withdraw the R-clip . . .

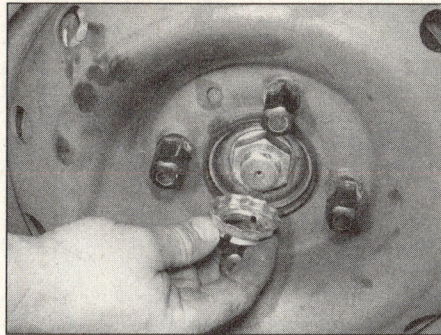
2.2b . . . and remove the locking cap, then loosen the driveshaft nut

2.3 Releasing the track-rod balljoint using a balljoint separator tool

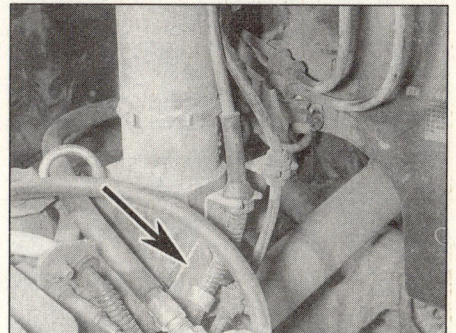
2.4 Unbolt the wiring/hose bracket (arrowed) from the hub carrier

where access to the driveshaft nut can be obtained by removing the wheel trims, before jacking up the vehicle, loosen the driveshaft nut as follows (see illustrations).
a) Chock the front wheels, and remove the wheel trim.
b) Apply the parking brake firmly.
c) Withdraw the R-clip and remove the locking cap from the driveshaft nut..
d) Loosen the driveshaft nut using a socket and extension.

3 Slacken and partially unscrew the track-rod balljoint nut (unscrew the nut as far as the end of the threads on the balljoint to prevent damage to the threads as the joint is released), then release the balljoint using a balljoint separator tool (see illustration). Remove the nut.

4 Unbolt the wiring/hose bracket(s) from the hub carrier, and move to one side, taking care not to strain the wiring or hose (see illustration).

5 On models with ABS, remove the wheel sensor as described in Chapter 10, section 16, and unbolt the sensor shield.

6 Remove the brake disc, as described in Chapter 10.

7 Disconnect the outboard end of the driveshaft from the hub carrier, as described in Chapter 8, noting the following points.
a) There is no need to drain the transmission oil.
b) Do not disconnect the inboard end of the driveshaft from the transmission, and

when working on the right-hand driveshaft, there is no need to release the intermediate bearing.
c) Support the free, outboard end of the driveshaft by suspending it using wire or string - do not allow the end of the driveshaft to hang down under its own weight.

8 Unscrew and remove the clamp nut and bolt securing the hub carrier to the suspension strut (see illustrations).

9 Engage an 8.0 mm Allen key or hexagon bit in the slot in the hub carrier, and turn the key/bit through a quarter turn to spread the slot (see illustration). Simultaneously pull the hub carrier down to release it from the strut.

Refitting

10 Commence refitting by using the Allen key to spread the slot in the hub carrier, as during removal, if not already done.

11 Engage the hub carrier with the strut, noting that the raised positioning boss on the strut must engage with the hub carrier slot.

12 Push the hub carrier onto the strut until the top surface of the hub carrier rests against the shoulder on the strut.

13 Refit the clamp bolt, noting that the bolt fits from the rear of the strut, then fit a new nut, and tighten to the specified torque.

14 Refit the brake disc and the caliper, then reconnect the parking brake cable and check the adjustment, as described in Chapter 10.

15 Reconnect the driveshaft to the hub

carrier, and tighten the driveshaft retaining nut, as described in Chapter 8. Note that on models where access to the driveshaft nut can be obtained by removing the wheel trims, the driveshaft nut can be tightened with the parking brake applied, and the vehicle resting on its wheels.

16 Once the driveshaft nut has been tightened, where applicable refit the ABS wheel sensor shield, then refit and adjust the ABS wheel sensor as described in Chapter 10.

17 Refit the wiring/hose bracket(s) to the hub carrier, and secure with the bolt(s).

18 Reconnect the track-rod end to the hub carrier, and secure using a new balljoint nut, tightened to the specified torque.

2.8a Unscrew the clamp nut . . .

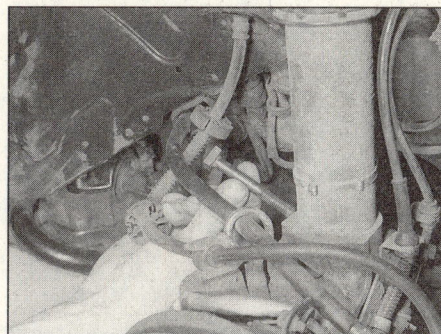
2.8b . . . and remove the bolt securing the hub carrier to the suspension strut

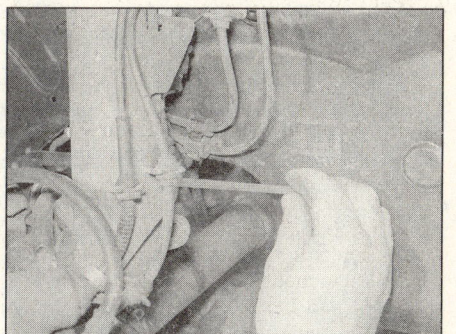
2.9 Engage an 8.0 mm Allen key in the slot in the hub carrier

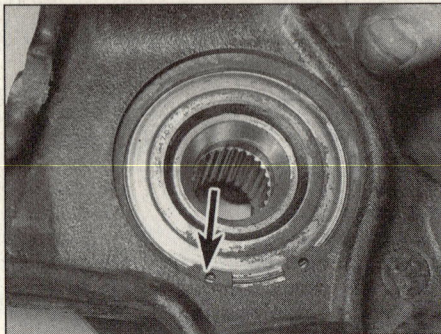

3.3 Front wheel bearing retaining circlip (arrowed)

19 Refit the roadwheel, and lower the vehicle to the ground.
20 Pressurise the hydraulic system as described in Chapter 9.
21 On completion, have the front wheel alignment checked, with reference to Section 17.

3 Front hub bearings – renewal

Note: *The bearing is a sealed, pre-adjusted and pre-lubricated, double-row roller type, and is intended to last the car's entire service life without maintenance or attention. Never overtighten the driveshaft nut beyond the specified torque wrench setting in an attempt to adjust the bearing.*

Note: *A press will be required to dismantle and rebuild the assembly; if such a tool is not available, a large bench vice and spacers (such as large sockets) will serve as an adequate substitute. The bearing's inner races are an interference fit on the hub; if the inner race remains on the hub when it is pressed out of the hub carrier, a knife-edged bearing puller will be required to remove it. A new bearing retaining circlip should be used on refitting.*

1 Remove the hub carrier assembly as described in Section 2.
2 Support the hub carrier securely on blocks or in a vice. Using a tubular spacer which bears only on the inner end of the hub flange, press the hub flange out of the bearing. If the bearing's outboard inner race remains on the hub, remove it using a bearing puller (see note above).
3 Extract the bearing retaining circlip from the inner end of the hub carrier assembly **(see illustration)**.
4 Where necessary, refit the inner race back in position over the ball cage, and securely support the inner face of the hub carrier. Using a tubular spacer which bears only on the inner race, press the complete bearing assembly out of the hub carrier.
5 Thoroughly clean the hub and hub carrier, removing all traces of dirt and grease, and polish away any burrs or raised edges which might hinder reassembly. Check both for

cracks or any other signs of wear or damage, and renew them if necessary. Renew the circlip, regardless of its apparent condition.
6 On reassembly, apply a light film of oil to the bearing outer race and hub flange shaft, to aid installation of the bearing.
7 Securely support the hub carrier, and locate the bearing in the hub. Press the bearing fully into position, ensuring that it enters the hub squarely, using a tubular spacer which bears only on the bearing outer race.
8 Once the bearing is correctly seated, secure the bearing in position with the new circlip, ensuring that it is correctly located in the groove in the hub carrier.
9 Securely support the outer face of the hub flange, and locate the hub carrier bearing inner race over the end of the hub flange. Press the bearing onto the hub, using a tubular spacer which bears only on the inner race of the hub bearing, until it seats against the hub shoulder. Check that the hub flange rotates freely, and wipe off any excess oil or grease.

4 Front suspension lower arm – removal, overhaul and refitting

Note: *A balljoint separator tool will be required for this operation. New lower arm securing nuts, a new lower balljoint nut, and a new anti-roll bar drop link nut must be used on refitting.*

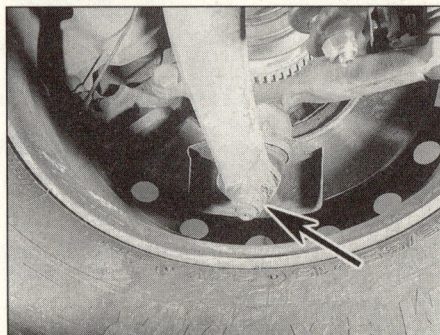

4.3 Unscrew the lower balljoint nut (arrowed)

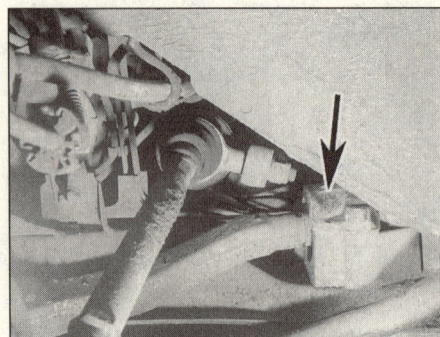

4.5b . . . and lift off the securing clamp (arrowed)

Removal

1 Move the suspension height control lever to the 'Low' position.
2 Chock the rear wheels, then jack up the front of the vehicle and support securely on axle stands (*see Jacking and Vehicle Support*). Remove the relevant roadwheel.
3 Slacken and partially unscrew the suspension lower balljoint nut (unscrew the nut as far as the end of the threads on the balljoint to prevent damage to the threads as the joint is released), then release the balljoint using a balljoint separator tool **(see illustration)**. Remove the nut.
4 Similarly, partially unscrew the nut securing the anti-roll bar drop link to the end of the anti-roll bar. Counterhold the end of the drop link pin using a 5.0 mm Allen key whilst unscrewing the nut. Remove the nut.
5 Unscrew and remove the lower arm rear securing bolts (recover the washers), and lift off the securing clamp **(see illustrations)**.
6 Unscrew the nut and front pivot bolt securing the front of the lower arm to the subframe, then withdraw the lower arm **(see illustration)**.

Overhaul

7 Thoroughly clean the lower arm and the area around the arm mountings, removing all traces of dirt and underseal if necessary, then check carefully for cracks, distortion or any other signs of wear or damage, paying particular attention to the pivot bushes, and

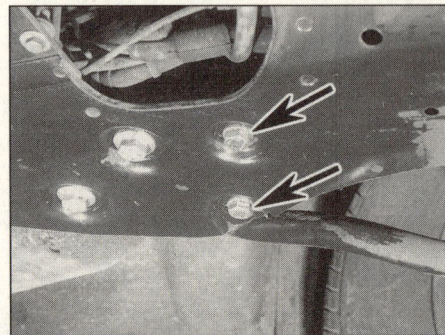

4.5a Unscrew the lower arm rear securing bolts (arrowed) . . .

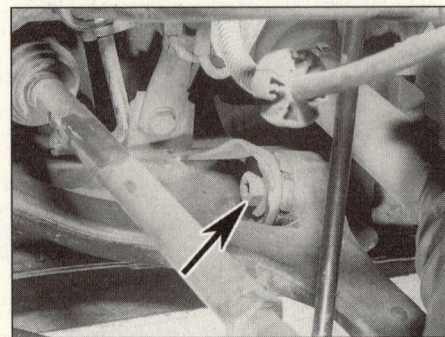

4.6 Lower arm front pivot bolt (arrowed)

renew components as necessary. Check with a Citroën dealer regarding the availability of spares.

8 Examine the shank of the pivot bolt for signs of wear or scoring, and renew if necessary.

9 It is possible to renew the lower arm bushes, but the bushes are an extremely tight fit, and a hydraulic press and special adapter tools are required to remove the old bushes and fit the new bushes successfully. Bush renewal should be referred to a Citroën dealer, or a suitably-qualified specialist.

Refitting

10 Offer the lower arm into position, then loosely refit the front pivot bolt, and a new nut. Do not fully tighten the nut and bolt at this stage.

11 Refit the clamp, then refit the two lower arm rear securing bolts, ensuring that the washers are in place. Again, do not fully tighten the bolts at this stage.

12 Before tightening the lower arm fixings, the lower edge of the outboard end of the lower arm should be positioned approximately level with the lower surface of the suspension subframe.

13 With the arm positioned as described in the previous paragraph, tighten the fixings to the specified torque.

14 Reconnect the suspension lower balljoint, then fit a new securing nut, and tighten to the specified torque.

15 Reconnect the anti-roll bar drop link to the end of the anti-roll bar, then fit a new securing nut and tighten to the specified torque. Hold the balljoint stem in position using a 5.0 mm Allen key.

16 Refit the roadwheel and lower the vehicle to the ground.

17 On completion, set the suspension height control to the 'Normal' position.

---

5 Front suspension lower balljoint – renewal

Note: *Citroën special tool 7103-T or an equivalent, and an impact wrench (such as a Facom Dynapact wrench) will be required to unscrew and tighten the balljoint. If these tools are not available, the task should be entrusted to a Citroën dealer. **Do not** attempt the work using improvised tools. A new balljoint nut must be used on refitting. Take care not to pull the inner end of the driveshaft from the transmission during this procedure, as this will result in loss of transmission oil/fluid.*

Removal

1 Move the suspension height control lever to the 'Low' position.

2 Chock the rear wheels, then jack up the front of the vehicle and support securely on

5.4 Tapping off the lower balljoint dust shield

axle stands (*see Jacking and Vehicle Support*). Remove the relevant roadwheel.

3 Slacken and partially unscrew the suspension lower balljoint nut (unscrew the nut as far as the end of the threads on the balljoint to prevent damage to the threads as the joint is released), then release the balljoint using a balljoint separator tool. Remove the nut.

4 Tap the dust shield from the balljoint, using a drift **(see illustration)**.

5 Fit the special tool 7103-T to the balljoint, engaging the tool with the cut-outs in the balljoint, and secure it by screwing the tool locknut onto the threaded section of the balljoint **(see illustration)**. Engage the impact wrench with the tool, and unscrew the balljoint.

Refitting

6 Refitting is a reversal of removal, bearing in mind the following points.

a) Tighten the balljoint as far as possible by hand before finally tightening to the specified torque using the special tools.

b) Take care not to damage the balljoint rubber gaiter during fitting.

c) Lock the dust shield in position by staking it in the cut-outs in the bottom of the hub carrier.

d) After lowering the vehicle to the ground, set the suspension height control to the Normal position.

6.4 Disconnect the anti-roll bar drop link (arrowed) from the strut

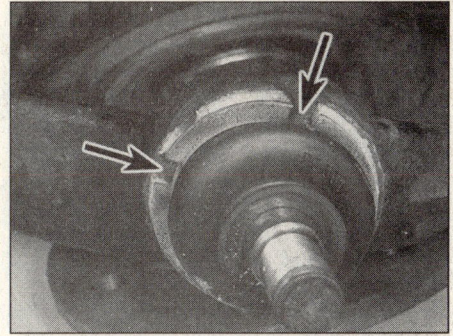
5.5 Engage the tool with the cut-outs (arrowed) in the balljoint

6 Front suspension hydraulic unit/strut assembly – removal and refitting

Removal

Note: *All Nyloc self-locking nuts must be renewed on refitting. Where applicable use new seals when reconnecting the hydraulic fluid pipes.*

1 Depressurise the hydraulic system as described in Chapter 9.

2 Jack up the vehicle, and support securely on axle stands, with the roadwheels clear of the ground (*see Jacking and Vehicle Support*). Remove the relevant front roadwheel.

3 Drain as much fluid as possible from the suspension hydraulic unit by compressing the suspension, pushing up on the lower arm.

4 Unscrew the securing nut, and disconnect the anti-roll bar drop link from the suspension strut. Counterhold the end of the drop link pin using a 5.0 mm Allen key whilst unscrewing the nut. If necessary, use a balljoint separator tool to free the drop link from the strut, but in this case, screw the nut onto the end of the drop link, to prevent damage to the thread **(see illustration)**.

5 Unscrew and remove the clamp nut and bolt securing the hub carrier to the suspension strut **(see illustration)**.

6.5 Removing the hub carrier-to-suspension strut bolt

11

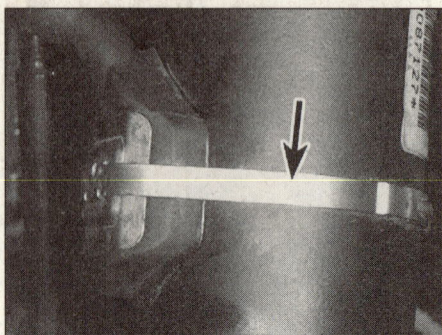

6.7 Suspension strut fluid return pipe securing clip (arrowed)

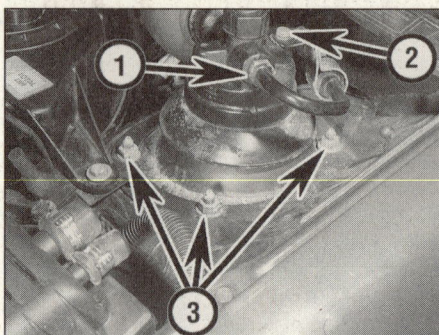

6.10 Disconnect the fluid feed pipe (1), unbolt the pipe bracket (2), then unscrew the nuts (3) securing the hydraulic unit

6.12 Prise the suspension strut gaiter from the hydraulic unit

6 Engage an 8.0 mm Allen key in the slot in the hub carrier, and turn the Allen key through a quarter turn to spread the slot. Simultaneously pull the hub carrier down to release it from the strut.

7 Disconnect the hydraulic fluid return pipe from the suspension strut. To disconnect the pipe, cut the securing clip, and pull the end of the pipe from the strut **(see illustration)**. Be prepared for fluid spillage, and plug the open ends of the pipe and strut to prevent dirt entry.

8 Proceed as follows according to whether the complete hydraulic unit/strut assembly, or just the strut is to be removed.

Complete hydraulic unit/strut assembly

9 Working in the engine compartment, unscrew the union and disconnect the hydraulic fluid feed pipe from the hydraulic unit. Plug the open ends of the pipe and hydraulic unit to prevent dirt entry and reduce fluid loss. Unbolt the pipe bracket(s) from the hydraulic unit.

10 Unscrew the four nuts securing the hydraulic unit to the body **(see illustration)**.

11 Lower the assembly, and manipulate it down from under the wheel arch.

Suspension strut

12 Working under the wheel arch, prise the suspension strut gaiter from the base of the hydraulic unit **(see illustration)**.

13 Where applicable, prise off the plastic cover, then slacken the nut securing the top of the strut to the hydraulic unit. If necessary, counterhold the strut piston using a splined key **(see illustrations)**.

14 Spray penetrating oil, or a lubricating fluid onto the top of the suspension strut cone (below the nut), to help to free the strut from the hydraulic unit.

15 Remove the nut, then withdraw the strut downwards and remove it from under the wheel arch.

Refitting

Complete hydraulic unit/strut assembly

16 Offer the assembly into position, taking care not to trap the hydraulic fluid return pipe, then refit the hydraulic unit securing nuts, and tighten to the specified torque.

17 Reconnect the hydraulic fluid feed pipe to the hydraulic unit, using a new seal where applicable, and tighten the union.

18 Proceed to paragraph 22.

Suspension strut

19 Lubricate the top of the strut cone and the sealing surfaces with clean LHM fluid.

20 Coat the threads at the top of the strut with thread-locking compound, then offer the assembly into position from under the wheel arch. Refit the securing nut, and tighten to the specified torque. If necessary, counterhold the strut piston using a splined key. Where applicable, refit the plastic cover.

21 Secure the strut gaiter to the base of the hydraulic unit.

Strut with or without hydraulic unit

22 Commence refitting by using the Allen key to spread the slot in the hub carrier, as during removal, if not already done.

23 Engage the hub carrier with the strut, noting that the raised positioning boss on the strut must engage with the hub carrier slot.

24 Push the hub carrier onto the strut until the top surface of the hub carrier rests against the shoulder on the strut.

25 Refit the clamp bolt, noting that the bolt fits from the rear of the strut, then fit a new nut, and tighten to the specified torque.

26 Reconnect the anti-roll bar drop link to the suspension strut, and tighten a new nut to the specified torque. If necessary, counterhold

6.13a Prise off the plastic cover . . .

6.13b . . . then slacken the nut securing the strut to the hydraulic unit

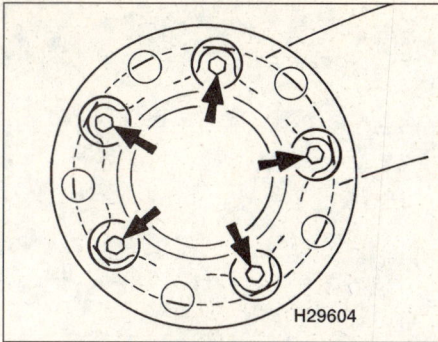

7.2 Rear hub assembly securing bolts (arrowed)

the end of the drop link pin using a 5.0 mm Allen key.

27 Reconnect the hydraulic fluid return pipe to the suspension strut, using a new seal where applicable, and secure with a new clip.

28 Refit the roadwheel, and lower the vehicle to the ground.

29 Pressurise the hydraulic system as described in Chapter 9, and check for fluid leaks around the disturbed components.

30 Check and if necessary top up the hydraulic fluid level as described in *Weekly Checks*.

7 Rear hub assembly – removal and refitting

Removal

1 Remove the rear brake disc as described in Chapter 10.

2 Working through the holes in the hub flange, unscrew the five bolts securing the hub assembly to the trailing arm, and remove the hub assembly **(see illustration)**.

Refitting

3 Refit the hub assembly to the trailing arm, and tighten the securing bolts to the specified torque.

4 Refit the brake disc as described in Chapter 10.

8 Rear hub bearings – renewal

The rear hub bearing is integral with the rear hub mounting plate. To renew the bearing/mounting plate assembly, it is necessary to separate the hub flange from the bearing/mounting plate assembly. This operation requires the use of Citroën special tools, and should be entrusted to a Citroën dealer or a suitably-equipped specialist.

9 Rear suspension trailing arm – removal and refitting

Removal

Note: *A new trailing arm pivot shaft nut must be used on refitting.*

1 Remove the relevant rear suspension hydraulic unit as described in Section 10.

2 Remove the securing screws, and withdraw the brake backplate from the trailing arm.

3 On models with ABS, proceed as follows:

a) *Unscrew the securing bolt and remove the wheel sensor. Move the sensor to one side, taking care not to strain the wiring.*

b) *Remove the wheel sensor heat shield.*

4 Disconnect the fluid pipe from the brake caliper. Smear a little LHM fluid on the joint between the pipe and union, and progressively unscrew the union. If the pipe turns, retighten the union slightly before loosening again, to free the pipe. Once the pipe is disconnected from the caliper, plug the pipe and caliper fitting to prevent fluid loss.

5 Release the hydraulic fluid pipe and the ABS sensor wiring from the clips on the trailing arm and the rear axle assembly. Unbolt the hydraulic pipe bracket from the trailing arm, and move it aside without straining the pipes.

6 Set the suspension height control to the 'Normal' position.

7 Working on the side from which the trailing arm is to be removed, unscrew the first of two

bolts securing the anti-roll bar to the trailing arm **(see illustration)**. Recover the shims, noting their locations.

8 To remove the second anti-roll bar bolt, raise the suspension unit (using a jack and block of wood) until the bolt is accessible through the hole in the suspension unit housing, using a socket and long extension bar. To improve access, lower the spare wheel carrier. Unscrew and remove the bolt, and recover the shims.

9 Unscrew the nut from the inner end of the trailing arm pivot shaft **(see illustration)**.

10 Withdraw the trailing arm pivot shaft, then manipulate the arm out from under the vehicle **(see illustration)**.

Refitting

11 Refitting is a reversal of removal, bearing in mind the following points:

a) *Grease the entire length of the trailing arm pivot shaft.*

b) *Use a new pivot shaft nut, and tighten all fixings to the specified torque.*

c) *Ensure that the shims are in place on the anti-roll bar securing bolts as noted on removal.*

d) *Reconnect the fluid pipe to the brake caliper using a new seal.*

e) *Refit the hydraulic unit as described in Section 10.*

f) *Refit and adjust the ABS wheel sensor as described in Chapter 10.*

g) *Bleed the rear brake hydraulic system as described in Chapter 10, Section 2.*

10 Rear suspension hydraulic unit – removal and refitting

Note: *A new fluid reservoir bulb seal and a new fluid pipe seal will be required on refitting.*

Removal

1 Remove the relevant hydraulic fluid reservoir bulb as described in Chapter 9, Section 7. Also remove the relevant rear wheel.

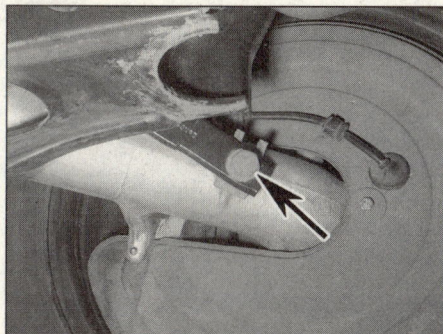

9.7 Anti-roll bar-to-trailing arm securing bolt (arrowed)

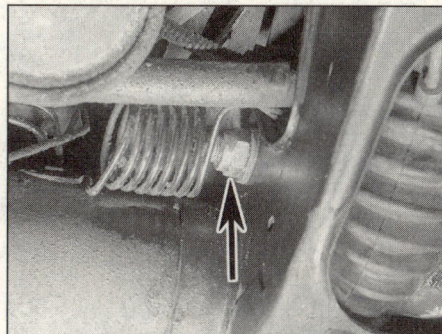

9.9 Unscrew the nut (arrowed) from the trailing arm pivot shaft . . .

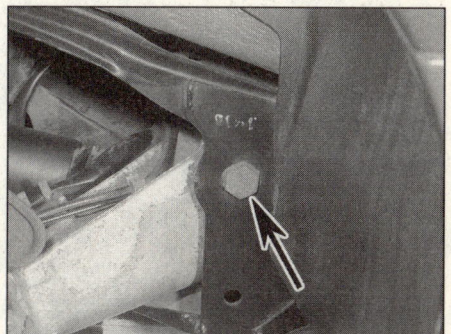

9.10 . . . then withdraw the pivot shaft (arrowed)

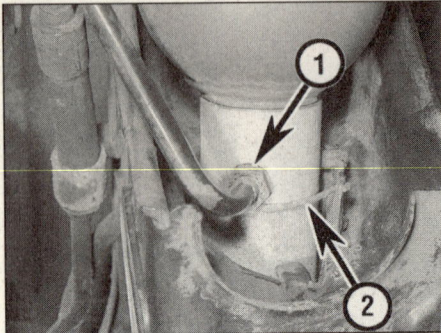

10.3 Rear suspension hydraulic unit fluid supply pipe (1) and hydraulic unit retaining clip (2)

10.4 Suspension link rod spring clip (arrowed)

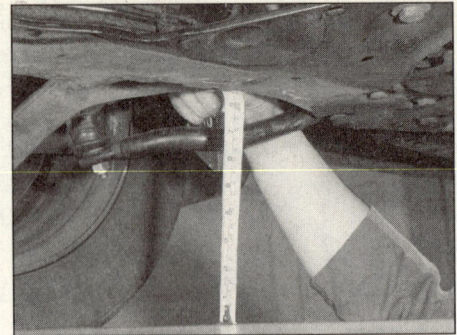

11.7 Measuring the front ride height

2 Drain as much fluid as possible from the suspension hydraulic unit by compressing the suspension, pushing up on the trailing arm.

3 Disconnect the hydraulic fluid supply pipe from the hydraulic unit. Plug the open ends of the hydraulic unit and the pipe to prevent dirt entry **(see illustration)**.

4 Remove the spring clip from the suspension link rod **(see illustration)**.

5 Remove the hydraulic unit retaining spring clip.

6 Manipulate the hydraulic unit out from the rear suspension assembly, and disconnect the hydraulic fluid return pipe and the vent pipe.

Refitting

7 Refitting is a reversal of removal, bearing in mind the following points.

   a) *Ensure that the fluid return pipe and the vent pipe are securely reconnected, and make sure that they are not trapped.*

   b) *Use a new seal when reconnecting the fluid supply pipe.*

   c) *Grease the contact face of the hydraulic fluid reservoir bulb, and refit the reservoir bulb using a new seal. Tighten the reservoir bulb by hand only.*

   d) *Pressurise the hydraulic system as described in Chapter 9, and check for leaks from the disturbed components.*

   e) *On completion, check and if necessary top up the hydraulic fluid level as described in "Weekly checks".*

11 Vehicle ride height –
checking and adjustment

Checking

Note: *After each movement of the bodyshell, and each measurement during the following procedure, move the vehicle backwards and forwards slightly by turning one of the roadwheels. This will relieve any stress in the suspension components.*

1 Check the tyre pressures, and adjust if necessary.

2 Raise the vehicle on a four-post lift.

3 Set the suspension height control to the 'Normal' position.

4 Release the parking brake.

5 Start the engine.

6 Working under the vehicle, push the vehicle up as far as possible by hand, then let go. The vehicle will drop, then rise and settle.

7 Measure the front ride height at each side, by measuring the distance between the bed of the four-post lift (the flat area on which the roadwheels are resting), and the lower surface of the subframe longitudinal member on each side of the vehicle **(see illustration)**.

8 Push the vehicle down as far as possible by hand, then let go. The vehicle will rise, then drop and settle.

9 Again measure the front ride height as

described in paragraph 7. Take an average of the two measurements to give the front ride height for each side of the vehicle.

10 Repeat the procedure in paragraphs 6 to 9 to measure the rear ride height but, on models up to 1994, take the measurements between the bed of the four-post lift (the flat area on which the roadwheels are resting), and the point on the body directly behind the rear suspension rear mounting point. On models from 1995, take the measurement between the flat area on which the roadwheels are resting, and the lower surface of the rear suspension crosstube **(see illustrations)**. Again, take an average of the two measurements to give the rear ride height for each side of the vehicle.

11 Check the ride height measurements with the figures given in the Specifications.

Adjustment

12 The ride height is adjusted by rotating the height corrector control rod clamps on the anti-roll bars. This is a trial-and-error process.

13 To adjust the position of the relevant control rod clamp (one on the front and one on the rear anti-roll bars), slacken the clamp bolt, and pivot the clamp as necessary, then tighten the clamp bolt, and re-check the ride height as described previously in this Section **(see illustrations)**.

14 Note that when carrying out adjustment, there should be approximately 1.0 mm freeplay between the height corrector rod

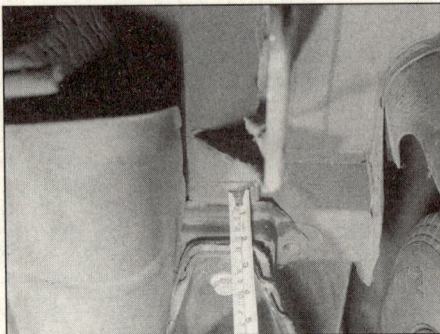

11.10a Measuring the rear ride height – models up to 1994

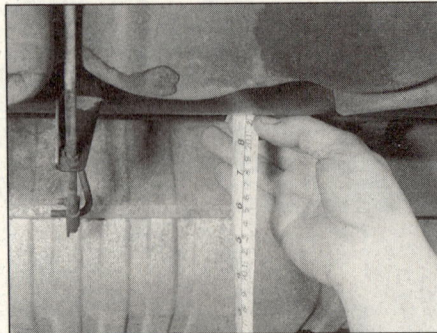

11.10b Measuring the rear ride height – models from 1995

11.13a Front anti-roll bar clamp bolt (arrowed)

11.13b Rear anti-roll bar clamp bolt (arrowed)

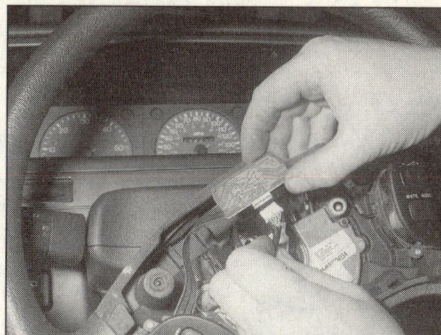

12.7 Disconnect the radio/cassette player remote control switch wiring plug

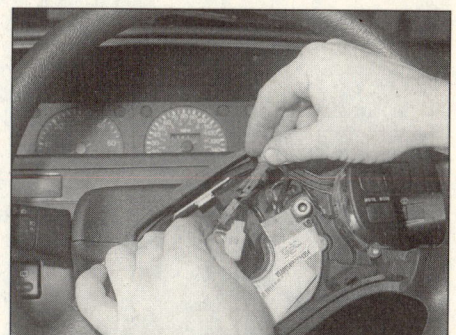

12.8 Separate the two halves of the air bag electronic control unit wiring connector

balljoint and the height corrector control rod lever.

15 The front ride heights on each side of the vehicle must be equal to within ± 2.0 mm. On early models, if necessary, once the overall front ride height has been set, any difference in height from side-to-side can be eliminated by adjusting the length of the anti-roll bar drop links. To adjust a drop link, unscrew the locknut towards the top of the drop link, and rotate the drop link upper section whilst holding the lower section still. Tighten the locknut on completion.

12 Steering wheel – removal and refitting

Models without air bag

Removal

1 Using a screwdriver, release the securing clip, and remove the cover from the battery, then disconnect the battery negative lead.
2 Carefully prise the centre pad from the steering wheel.
3 Where applicable, disconnect the steering wheel-mounted radio/cassette player remote control switch wiring plug(s).
4 Unscrew the steering wheel securing nut, and lift off the washer, then withdraw the steering wheel. Where applicable, feed the wiring through the steering wheel as it is withdrawn.

Refitting

5 Refitting is a reversal of removal but, where applicable, make sure that the wiring is correctly routed through the wheel, and tighten the steering wheel securing nut to the specified torque.

Models with air bag

Note: *The air bag electronic control unit is integral with the steering wheel. Take care not to damage the unit during removal, and store the wheel carefully once removed.*

Removal

6 Remove the air bag unit as described in Chapter 13.

7 Where applicable, disconnect the radio/cassette player remote control switch wiring plug **(see illustration)**.
8 Unclip the air bag electronic control unit wiring connector from the steering wheel, and separate the two halves of the connector **(see illustration)**.
9 Unscrew the steering wheel securing nut, and recover the washer, then withdraw the steering wheel **(see illustration)**. Feed the wiring through the steering wheel as it is withdrawn.
10 Store the wheel carefully, taking care not to damage the air bag electronic control unit.

Refitting

11 Refitting is a reversal of removal, bearing in mind the following points.
a) Make sure that the wiring is correctly routed through the wheel.
b) Tighten the steering wheel securing nut to the specified torque.
c) Refit the air bag unit as described in Chapter 13.

13 Steering column – removal, inspection and refitting

Removal

1 Using a screwdriver, release the securing clip, and remove the cover from the battery, then disconnect the battery negative lead.
2 Set the roadwheels in the straight-ahead position.

12.9 Removing the steering wheel securing nut

3 Remove the steering wheel as described in Section 12.
4 Remove the securing clips and screws, and remove the driver's footwell carpet trim panel from the underside of the facia.
5 Working on the driver's side of the centre console, remove the side trim panel as follows.
a) Unscrew the two screws securing the front air vent, and pull out the air vent.
b) Unscrew the screw securing the rear air vent, and withdraw the vent.
c) Working through the slit in the carpet trim, unscrew the side trim panel rear securing nut.
d) Withdraw the trim panel.
6 Remove the securing screws, and withdraw the driver's side lower facia panel.
7 On models up to 1994, disconnect the end of the parking brake lever cable from the parking brake linkage balljoint.
8 Remove the steering column shrouds, as described in Chapter 12, section 23.
9 On left-hand-drive models, where applicable, remove the securing screw from the lower end of the fusebox, then unclip the upper end of the fusebox, and lower the assembly for access to the steering column switch wiring connectors.
10 Trace the wiring back from the switches mounted on the steering column, and disconnect the wiring connectors. Note that on models from 1995, it may be necessary to unclip the plastic wiring ducts from the steering column in order to allow the wiring to be disconnected **(see illustration)**.

13.10 Unclipping a wiring duct from the steering column – models from 1995

11

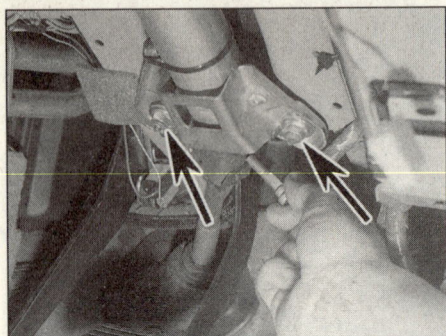

13.11 Removing the steering column universal joint clamp bolt – lower steering column securing nuts arrowed

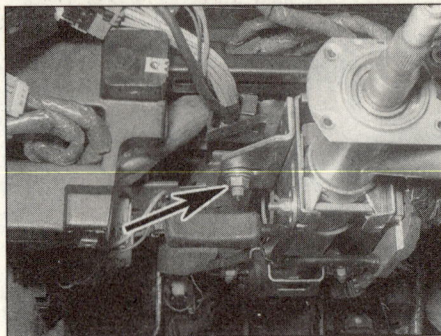

13.12 Steering column upper securing nut (arrowed)

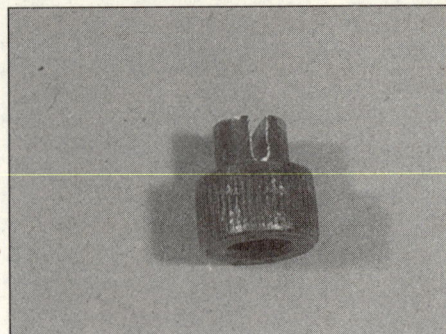

13.16 Improvised tool (made from M7 bolt) for centring steering column bearing

11 Unscrew the clamp nut and bolt, and, then carefully prise out the legs of the safety clip, and disconnect the steering column universal joint from the intermediate shaft **(see illustration)**.

12 Unscrew the four steering column securing nuts, then withdraw the column assembly. Note the locations of any washers under the nuts **(see illustration)**.

Inspection

13 The steering column incorporates a telescopic safety feature. In the event of a front-end crash, the shaft collapses and prevents the steering wheel injuring the driver. Before refitting the steering column, examine the column and mountings for signs of damage and deformation, and renew as necessary.

14 Check the steering shaft for signs of free play in the column bushes, and check the universal joints for signs of damage or roughness in the joint bearings. If any damage or wear is found on the steering column universal joints or shaft bushes, the column must be renewed as an assembly.

Refitting

15 Commence refitting by offering the steering column into position, and loosely refitting the securing nuts. Make sure that any washers noted during removal are in place.

16 If the column tube has a red plastic stop sleeve attached to it, unclip the sleeve, and

engage the lug on the sleeve with the bearing slot in the column tube. If no stop sleeve is fitted to the column, a suitable alternative can be improvised as follows **(see illustration)**.

a) *Obtain an M7 bolt, and cut off the threaded section of the bolt to leave the shank and head.*

b) *Cut a 2.0 mm wide slot in the centre of the end of the shank.*

17 Fit the stop sleeve or the improvised tool to the slot in the steering column tube, making sure that the slot in the sleeve or tool engages with the steering column bearing plate **(see illustration)**. If desired, the improvised tool can be temporarily held in place using a cable-tie.

18 Reconnect the steering column universal joint to the intermediate shaft, and ensure that the safety clip is correctly fitted, then refit the clamp nut and bolt, and tighten to the specified torque.

19 Tighten the steering column securing nuts to the specified torque, then remove the stop sleeve or improvised tool (as applicable) from the slot in the column tube.

20 Further refitting is a reversal of removal, bearing in mind the following points.

a) *Refit the steering column shrouds with reference to Chapter 12, section 23.*

b) *Where applicable, reconnect the parking brake cable, and check the parking brake*

operation and adjustment as described in Chapter 10.

c) *Refit the steering wheel as described in Section 12.*

21 On completion, adjust the steering column as follows **(see illustration)**.

a) *Unlock the steering column adjuster lever, then pull the column out and push it down.*

b) *Slacken the column adjuster locknut.*

c) *Turn the adjuster nut so that the steering column slides horizontally and vertically without resistance.*

d) *Tighten the locknut.*

e) *Lock the adjuster lever, and check that the steering column is firmly locked in position – if not, repeat the adjustment procedure.*

14 Ignition switch/steering column lock – removal and refitting

Right-hand-drive models

Removal

1 Using a screwdriver, release the securing clip, and remove the cover from the battery, then disconnect the battery negative lead.

2 Remove the lower steering column shroud, as described in Chapter 12.

3 Trace the wiring back from the ignition switch, and disconnect the wiring connectors under the steering column **(see illustration)**.

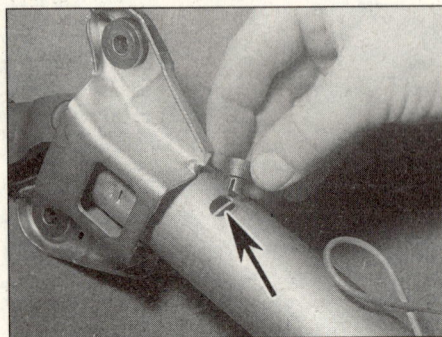

13.17 The slot in the tool must engage with the bearing plate (arrowed)

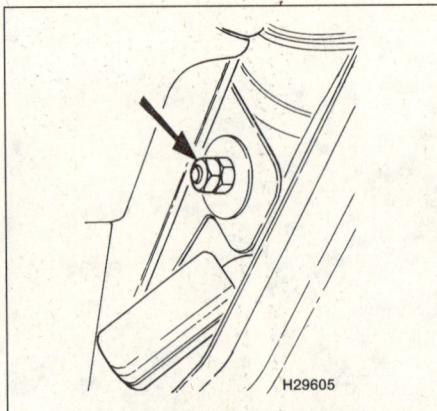

13.21 Steering column adjuster nut (arrowed)

14.3 Disconnecting an ignition switch wiring connector

14.5 Removing the lock cylinder securing screw

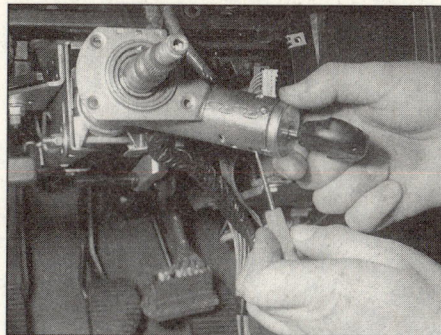

14.7a Depress the lock securing lug . . .

14.7b . . . then pull the lock cylinder from the housing

On later models, it may be necessary to unclip the plastic wiring housing from the steering column, in order to release the wiring – the housing can be opened to allow the wiring to be released or, alternatively, it may be possible to feed the wiring through the housing (in this case, refitting may be difficult).

4 Release the ignition switch wiring from any clips and brackets, taking note of its routing.

5 Unscrew the lock cylinder securing screw from the housing on the steering column **(see illustration)**.

6 Insert the ignition key into the lock, and turn the key to the position between 'A' and 'S'.

7 Using a screwdriver, depress the lock securing lug, then pull the lock cylinder from the housing using the ignition key **(see illustrations)**. Feed the wiring connectors through the hole in the steering column.

8 If desired, the ignition switch can be separated from the steering column lock after unscrewing the securing grub screw – take care, as a spring will be ejected as the switch and lock cylinder are separated.

Refitting

9 Refitting is a reversal of removal, bearing in mind the following points.

a) *Where applicable, make sure that the spring engages with the pin when refitting the ignition switch to the steering column lock.*

b) *Make sure that the lock cylinder securing lug engages securely.*

c) *Ensure that the wiring is routed as noted before removal.*

Left-hand-drive models

Removal

10 Using a screwdriver, release the securing clip, and remove the cover from the battery, then disconnect the battery negative lead.

11 Remove the securing clips and screws, and remove the driver's footwell carpet trim panel from the underside of the facia.

12 Working on one side of the centre console, remove the side trim panel as follows.

a) *Unscrew the two screws securing the front air vent, and pull out the air vent.*

b) *Unscrew the screw securing the rear air vent, and withdraw the vent.*

c) *Working through the slit in the carpet trim, unscrew the side trim panel rear securing nut.*

d) *Withdraw the trim panel.*

13 Remove the securing screws, and withdraw the driver's side lower facia panel.

14 On models up to 1994, disconnect the end of the parking brake lever cable from the parking brake linkage balljoint.

15 Remove the lower steering column shroud, with reference to Chapter 12 if necessary.

16 Remove the securing screw from the lower end of the fusebox, then unclip the upper end of the fusebox, and lower the assembly for access to the steering column switch wiring connectors.

17 Disconnect the steering column switch wiring connectors from the rear of the fusebox, then move the fusebox assembly to one side, taking care not to strain the wiring. Take careful note of the routing of the ignition switch wiring.

18 Proceed as described in paragraphs 5 to 8.

Refitting

19 Refer to paragraph 9.

15 Steering gear assembly – removal, overhaul and refitting

Removal

Note: *A balljoint separator tool will be required for this operation. New track-rod balljoint nuts, and a new intermediate shaft-to-steering gear pinion pinch-bolt will be required on refitting. New hydraulic fluid pipe seals will be required.*

1 Jack up the vehicle, and support on axle stands with the roadwheels clear of the ground (see *Jacking and Vehicle Support*). Remove the front roadwheels.

2 Slacken the pressure release screw on the hydraulic pressure regulator – see Chapter 9.

3 Turn the steering from lock-to-lock to drain the steering gear hydraulic ram.

4 Remove the securing clips and screws, and remove the driver's footwell carpet trim panel from the underside of the facia.

5 Unscrew the clamp nut and bolt, and, then carefully prise out the legs of the safety clip, and disconnect the steering column universal joint from the intermediate shaft.

6 Unscrew the clamp nut and bolt, and disconnect the steering intermediate shaft flexible coupling from the steering gear pinion. Pull up on the intermediate shaft to disconnect the coupling from the pinion.

7 On manual transmission models, disconnect the two gearchange link rods from their balljoints, then move the gearchange control rod to one side.

8 Disconnect the exhaust front section from the intermediate section with reference to Chapter 4C, section 7. Move the exhaust sections clear of the working area, and if necessary support them using wire or string.

9 Disconnect the steering gear hydraulic fluid return pipe, then unscrew the union nuts and disconnect the two fluid feed pipes – disconnect the pipes from the unions, **do not** unscrew the unions from the steering gear. Be prepared for fluid spillage, and plug the open ends of the pipes and steering gear to prevent dirt entry and further fluid loss.

10 Working on each side of the vehicle in turn, slacken and partially unscrew the track-rod balljoint nut (unscrew the nut as far as the end of the threads on the balljoint to prevent damage to the threads as the joint is released), then release the balljoint using a balljoint separator tool **(see illustration)**. Remove the nut.

15.10 Slackening the track-rod balljoint nut

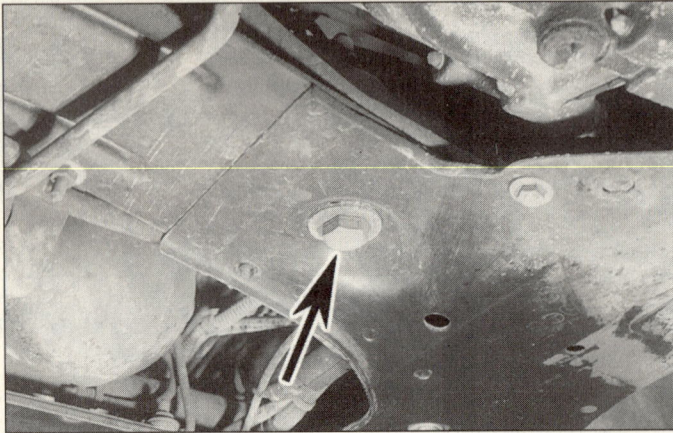

15.12 Steering gear securing bolt (arrowed)

16.4 Hold the track-rod and unscrew the track-rod end locknut

11 Unbolt and remove the steering gear heat shield.

12 Unscrew the two steering gear securing bolts from the subframe, then manipulate the steering gear out from under the right-hand wheel arch **(see illustration)**.

13 Recover the two thrust washers, which fit under the steering gear.

Overhaul

14 Examine the steering gear assembly for signs of wear or damage, and check that the rack moves freely throughout the full length of its travel, with no signs of roughness or excessive free play between the steering gear pinion and rack. It is possible to overhaul the steering gear assembly, but this task should be entrusted to a Citroën dealer. The only components which can be renewed easily by the home mechanic are the track-rod ends (see Section 16).

15 Inspect all the steering gear fluid unions for signs of leakage, and check that all union nuts are securely tightened. Also examine the steering gear hydraulic ram for signs of fluid leakage or damage, and if necessary renew it.

Refitting

16 Refitting is a reversal of removal, bearing in mind the following points.

a) *Centralise the rack so that the steering is effectively in the straight-ahead position before refitting the steering gear.*

b) *Ensure that the thrust washers are in position between the subframe and the steering gear, and that the washers are in position under the steering gear securing bolt heads.*

c) *Use new nuts when reconnecting the track-rod balljoints, and use new nuts when refitting the steering universal joint and flexible coupling clamp-bolts.*

d) *Use new seals when reconnecting the fluid pipes to the steering gear.*

e) *Ensure that the steering wheel is in the straight-ahead position, and that the steering gear is centralised before*

reconnecting the intermediate shaft to the steering gear and column.

f) *Before lowering the vehicle to the ground, retighten the hydraulic pressure regulator bleed screw.*

g) *On completion, check and if necessary adjust the front wheel alignment as described in Section 17.*

16 Track-rod end – removal and refitting

Removal

Note: *A new balljoint-to-hub carrier nut will be required on refitting.*

1 Chock the rear wheels, then jack up the front of the vehicle, and support securely on axle stands (see *Jacking and Vehicle Support*). Remove the relevant roadwheel.

2 Slacken and partially unscrew the track-rod balljoint nut (unscrew the nut as far as the end of the threads on the balljoint to prevent damage to the threads as the joint is released), then release the balljoint using a balljoint separator tool. Remove the nut.

3 If the track-rod end is to be re-used, use a straight-edge and a scriber, or similar, to mark its relationship to the track-rod.

4 Hold the track-rod, using a spanner on the flats provided, and unscrew the track-rod end locknut by a quarter of a turn **(see illustration)**. Do not move the locknut from this position, as it will serve as a handy reference mark on refitting.

5 Counting the **exact** number of turns necessary to do so, unscrew the track-rod end from the track-rod.

6 Count the number of exposed threads between the end of the track-rod end and the locknut, and record this figure. If a new track-rod end is to be fitted, unscrew the locknut from the old track-rod end.

7 Carefully clean the track-rod end and the

threads. Renew the track-rod end if the balljoint movement is sloppy or too stiff, if excessively worn, or if damaged in any way; carefully check the balljoint taper and threads. If the balljoint gaiter is damaged, the complete track-rod end assembly must be renewed; it is not possible to obtain the gaiter separately.

Refitting

8 If a new track-rod end is to be fitted, screw the locknut onto its threads, and position it so that the same number of exposed threads are visible, as was noted on the old track-rod end prior to removal.

9 Screw the track-rod end into the track-rod by the number of turns noted on removal. This should bring the locknut to within a quarter of a turn from the shoulder on the track-rod, with the alignment marks that were made on removal (if applicable) lined up.

10 Reconnect the track-rod balljoint to the hub carrier, then fit a new nut and tighten to the specified torque.

11 Refit the roadwheel, then lower the vehicle to the ground and tighten the roadwheel bolts to the specified torque.

12 Check and, if necessary, adjust the front wheel alignment as described in Section 17, then securely tighten the balljoint locknut.

17 Wheel alignment and steering angles – general information

General

1 A car's steering and suspension geometry is defined in four basic settings - all angles are expressed in degrees (toe settings are also expressed as a measurement); the relevant settings are camber, castor, steering axis inclination, and toe-setting. With the exception of front wheel toe-setting, none of these settings are adjustable.

Front wheel toe setting - checking and adjustment

2 Due to the special measuring equipment necessary to accurately check the wheel alignment, and the skill required to use it properly, checking and adjustment is best left to a Citroën dealer or similar expert. Note that most tyre-fitting shops now possess sophisticated checking equipment. The following is provided as a guide, should the owner decide to carry out a DIY check.

3 The front wheel toe setting is checked by measuring the distance between the front and rear inside edges of the roadwheel rims. Proprietary toe measurement gauges are available from motor accessory shops. Adjustment is made by screwing the track-rod ends in or out of their track-rods, to alter the effective length of the track-rod assemblies.

4 For **accurate** checking, the vehicle **must** be at the kerb weight, ie unladen and with a full tank of fuel, and the ride height must be correct (see Section 11).

5 Before starting work, check first that the tyre sizes and types are as specified, then check the tyre pressures and tread wear, the roadwheel run-out, the condition of the hub bearings, the steering wheel free play, and the condition of the front suspension components (Chapter 1A or 1B). Correct any faults found.

6 Park the vehicle on level ground, check that the front roadwheels are in the straight-ahead position, then rock the rear and front ends to settle the suspension. Release the parking brake, and roll the vehicle backwards 1 metre, then forwards again, to relieve any stresses in the steering and suspension components.

7 Measure the distance between the front edges of the wheel rims and the rear edges of the rims. Subtract the rear measurement from the front measurement, and check that the result is within the specified range.

8 If adjustment is necessary, chock the rear wheels, then jack up the front of the vehicle and support it securely on axle stands (*see Jacking and Vehicle Support*). Turn the steering wheel onto full-left lock, and record the number of exposed threads on the right-hand track-rod end. Now turn the steering onto full-right lock, and record the number of threads on the left-hand side. If there are the same number of threads visible on both sides, then subsequent adjustment should be made equally on both sides. If there are more threads visible on one side than the other, it will be necessary to compensate for this during adjustment. **Note:** *It is most important that after adjustment, the same number of threads are visible on each track-rod end.*

9 First clean the track-rod threads; if they are corroded, apply penetrating fluid before starting adjustment. Release the steering gear rubber gaiter outboard clips (where necessary), and peel back the gaiters; apply a smear of grease to the inside of the gaiters, so that both are free, and will not be twisted or strained as their respective track-rods are rotated.

10 Use a straight-edge and a scriber or similar to mark the relationship of each track-rod to its track-rod end then, holding each track-rod in turn (use a spanner on the flats provided), unscrew its track-rod end locknut fully.

11 Alter the length of the track-rods, bearing in mind the note made in paragraph 8. Screw them into or out of the track-rod ends, rotating the track-rod using an open-ended spanner fitted to the flats provided on the track-rod. Shortening the track-rods (screwing them into their track-rod ends) will reduce toe-in/increase toe-out.

12 When the setting is correct, hold the track-rods and securely tighten the track-rod end locknuts. Check that the balljoints are seated correctly in their sockets, and count the exposed threads to check the length of both track-rods. If they are not the same, then the adjustment has not been made equally, and problems will be encountered with tyre scrubbing in turns; also, the steering wheel spokes will no longer be horizontal when the wheels are in the straight-ahead position.

13 If the track-rod lengths are the same, lower the vehicle to the ground and re-check the toe setting; re-adjust if necessary. When the setting is correct, securely tighten the track-rod end locknuts. Ensure that the steering gear rubber gaiters are seated correctly, and are not twisted or strained, and secure them in position with new retaining clips (where necessary).

11

Notes

Chapter 12
Bodywork and fittings

Contents

Degrees of difficulty

Easy, suitable for novice with little experience	Fairly easy, suitable for beginner with some experience	Fairly difficult, suitable for competent DIY mechanic	Difficult, suitable for experienced DIY mechanic	Very difficult, suitable for expert DIY or professional

1 General information

The bodyshell is of five-door Hatchback or Estate configuration, and is made of pressed steel sections. Most components are welded together, but some use is made of structural adhesives. The front wings are bolted on.

The bonnet, doors and some other vulnerable panels are made of zinc-coated metal, and are further protected by being coated with an anti-chip primer prior to being sprayed.

Extensive use is made of plastic materials, mainly in the interior, but also in exterior components. The front and rear bumpers and the front grille are injection-moulded from a synthetic material which is very strong, and yet light. Plastic components such as wheel arch liners are fitted to the underside of the vehicle, to improve the body's resistance to corrosion.

2 Maintenance – bodywork and underframe

The general condition of a vehicle's bodywork is the one thing that significantly affects its value. Maintenance is easy, but needs to be regular. Neglect, particularly after minor damage, can lead quickly to further deterioration and costly repair bills. It is important also to keep watch on those parts of the vehicle not immediately visible, for instance the underside, inside all the wheel arches, and the lower part of the engine compartment.

The basic maintenance routine for the bodywork is washing - preferably with a lot of water, from a hose. This will remove all the loose solids which may have stuck to the vehicle. It is important to flush these off in such a way as to prevent grit from scratching the finish. The wheel arches and underframe need washing in the same way, to remove any accumulated mud which will retain moisture and tend to encourage rust. Paradoxically enough, the best time to clean the underframe and wheel arches is in wet weather, when the mud is thoroughly wet and soft. In very wet weather, the underframe is usually cleaned of large accumulations automatically, and this is a good time for inspection.

Periodically, except on vehicles with a wax-based underbody protective coating, it is a good idea to have the whole of the underframe of the vehicle steam-cleaned, engine compartment included, so that a thorough inspection can be carried out to see what minor repairs and renovations are necessary. Steam-cleaning is available at many garages, and is necessary for the removal of the accumulation of oily grime, which sometimes is allowed to become thick in certain areas. If steam-cleaning facilities are not available, there are one or two excellent grease solvents available, which can be brush-applied; the dirt can then be simply hosed off. Note that these methods should not be used on vehicles with wax-based underbody protective coating, or the coating will be removed. Such vehicles should be inspected annually, preferably just prior to Winter, when the underbody should be washed down, and any damage to the wax coating repaired. Ideally, a completely fresh coat should be applied. It would also be worth considering the use of such wax-based protection for injection into door panels, sills, box sections, etc, as an additional safeguard against rust damage, where such protection is not provided by the vehicle manufacturer.

After washing paintwork, wipe off with a chamois leather to give an unspotted clear finish. A coat of clear protective wax polish will give added protection against chemical pollutants in the air. If the paintwork sheen has dulled or oxidised, use a cleaner/polisher combination to restore the brilliance of the shine. This requires a little effort, but such dulling is usually caused because regular washing has been neglected. Care needs to be taken with metallic paintwork, as special non-abrasive cleaner/polisher is required to

12

avoid damage to the finish. Always check that the door and ventilator opening drain holes and pipes are completely clear, so that water can be drained out. Brightwork should be treated in the same way as paintwork. Windscreens and windows can be kept clear of the smeary film which often appears, by the use of proprietary glass cleaner. Never use any form of wax or other body or chromium polish on glass.

3 Maintenance – upholstery and carpets

Mats and carpets should be brushed or vacuum-cleaned regularly, to keep them free of grit. If they are badly stained, remove them from the vehicle for scrubbing or sponging, and make quite sure they are dry before refitting. Seats and interior trim panels can be kept clean by wiping with a damp cloth. If they do become stained (which can be more apparent on light-coloured upholstery), use a little liquid detergent and a soft nail brush to scour the grime out of the grain of the material. Do not forget to keep the headlining clean in the same way as the upholstery. When using liquid cleaners inside the vehicle, do not over-wet the surfaces being cleaned. Excessive damp could get into the seams and padded interior, causing stains, offensive odours or even rot. If the inside of the vehicle gets wet accidentally, it is worthwhile taking some trouble to dry it out properly, particularly where carpets are involved. *Do not leave oil or electric heaters inside the vehicle for this purpose.*

4 Minor body damage – repair

Repairs of minor scratches in bodywork

If the scratch is very superficial, and does not penetrate to the metal of the bodywork, repair is very simple. Lightly rub the area of the scratch with a paintwork renovator, or a very fine cutting paste, to remove loose paint from the scratch, and to clear the surrounding bodywork of wax polish. Rinse the area with clean water.

Apply touch-up paint to the scratch using a fine paint brush; continue to apply fine layers of paint until the surface of the paint in the scratch is level with the surrounding paintwork. Allow the new paint at least two weeks to harden, then blend it into the surrounding paintwork by rubbing the scratch area with a paintwork renovator or a very fine cutting paste. Finally, apply wax polish.

Where the scratch has penetrated right through to the metal of the bodywork, causing the metal to rust, a different repair technique is required. Remove any loose rust from the bottom of the scratch with a penknife, then apply rust-inhibiting paint, to prevent the formation of rust in the future. Using a rubber or nylon applicator, fill the scratch with bodystopper paste. If required, this paste can be mixed with cellulose thinners, to provide a very thin paste which is ideal for filling narrow scratches. Before the stopper-paste in the scratch hardens, wrap a piece of smooth cotton rag around the top of a finger. Dip the finger in cellulose thinners, and quickly sweep it across the surface of the stopper-paste in the scratch; this will ensure that the surface of the stopper-paste is slightly hollowed. The scratch can now be painted over as described earlier in this Section.

Repairs of dents in bodywork

When deep denting of the vehicle's bodywork has taken place, the first task is to pull the dent out, until the affected bodywork almost attains its original shape. There is little point in trying to restore the original shape completely, as the metal in the damaged area will have stretched on impact, and cannot be reshaped fully to its original contour. It is better to bring the level of the dent up to a point which is about 3 mm below the level of the surrounding bodywork. In cases where the dent is very shallow anyway, it is not worth trying to pull it out at all. If the underside of the dent is accessible, it can be hammered out gently from behind, using a mallet with a wooden or plastic head. Whilst doing this, hold a suitable block of wood firmly against the outside of the panel, to absorb the impact from the hammer blows and thus prevent a large area of the bodywork from being 'belled-out'.

Should the dent be in a section of the bodywork which has a double skin, or some other factor making it inaccessible from behind, a different technique is called for. Drill several small holes through the metal inside the area - particularly in the deeper section. Then screw long self-tapping screws into the holes, just sufficiently for them to gain a good purchase in the metal. Now the dent can be pulled out by pulling on the protruding heads of the screws with a pair of pliers.

The next stage of the repair is the removal of the paint from the damaged area, and from an inch or so of the surrounding 'sound' bodywork. This is accomplished most easily by using a wire brush or abrasive pad on a power drill, although it can be done just as effectively by hand, using sheets of abrasive paper. To complete the preparation for filling, score the surface of the bare metal with a screwdriver or the tang of a file, or alternatively, drill small holes in the affected area. This will provide a really good 'key' for the filler paste.

To complete the repair, see the Section on filling and respraying.

Repairs of rust holes or gashes in bodywork

Remove all paint from the affected area, and from an inch or so of the surrounding 'sound' bodywork, using an abrasive pad or a wire brush on a power drill. If these are not available, a few sheets of abrasive paper will do the job most effectively. With the paint removed, you will be able to judge the severity of the corrosion, and therefore decide whether to renew the whole panel (if this is possible) or to repair the affected area. New body panels are not as expensive as most people think, and it is often quicker and more satisfactory to fit a new panel than to attempt to repair large areas of corrosion.

Remove all fittings from the affected area, except those which will act as a guide to the original shape of the damaged bodywork (eg headlamp shells etc). Then, using tin snips or a hacksaw blade, remove all loose metal and any other metal badly affected by corrosion. Hammer the edges of the hole inwards, in order to create a slight depression for the filler paste.

Wire-brush the affected area to remove the powdery rust from the surface of the remaining metal. Paint the affected area with rust-inhibiting paint; if the back of the rusted area is accessible, treat this also.

Before filling can take place, it will be necessary to block the hole in some way. This can be achieved by the use of aluminium or plastic mesh, or aluminium tape.

Aluminium or plastic mesh, or glass-fibre matting is probably the best material to use for a large hole. Cut a piece to the approximate size and shape of the hole to be filled, then position it in the hole so that its edges are below the level of the surrounding bodywork. It can be retained in position by several blobs of filler paste around its periphery.

Aluminium tape should be used for small or very narrow holes. Pull a piece off the roll, trim it to the approximate size and shape required, then pull off the backing paper (if used) and stick the tape over the hole; it can be overlapped if the thickness of one piece is insufficient. Burnish down the edges of the tape with the handle of a screwdriver or similar, to ensure that the tape is securely attached to the metal underneath.

Bodywork repairs - filling and respraying

Before using this Section, see the Sections on dent, deep scratch, rust holes and gash repairs.

Many types of bodyfiller are available, but generally speaking, those proprietary kits which contain a tin of filler paste and a tube of resin hardener are best for this type of repair. A wide, flexible plastic or nylon applicator will be found invaluable for imparting a smooth and well-contoured finish to the surface of the filler.

Mix up a little filler on a clean piece of card or board - measure the hardener carefully (follow the maker's instructions on the pack), otherwise the filler will set too rapidly or too slowly. Using the applicator, apply the filler paste to the prepared area; draw the applicator across the surface of the filler to achieve the correct contour and to level the surface. As soon as a contour that approximates to the correct one is achieved, stop working the paste - if you carry on too long, the paste will become sticky and begin to 'pick-up' on the applicator. Continue to add thin layers of filler paste at 20-minute intervals, until the level of the filler is just proud of the surrounding bodywork.

Once the filler has hardened, the excess can be removed using a metal plane or file. From then on, progressively-finer grades of abrasive paper should be used, starting with a 40-grade production paper, and finishing with a 400-grade wet-and-dry paper. Always wrap the abrasive paper around a flat rubber, cork, or wooden block - otherwise the surface of the filler will not be completely flat. During the smoothing of the filler surface, the wet-and-dry paper should be periodically rinsed in water. This will ensure that a very smooth finish is imparted to the filler at the final stage.

At this stage, the 'dent' should be surrounded by a ring of bare metal, which in turn should be encircled by the finely 'feathered' edge of the good paintwork. Rinse the repair area with clean water, until all of the dust produced by the rubbing-down operation has gone.

Spray the whole area with a light coat of - this will show up any imperfections in the surface of the filler. Repair these imperfections with fresh filler paste or bodystopper, and once more smooth the surface with abrasive paper. If bodystopper is used, it can be mixed with cellulose thinners, to form a really thin paste which is ideal for filling small holes. Repeat this spray-and-repair procedure until you are satisfied that the surface of the filler, and the feathered edge of the paintwork, are perfect. Clean the repair area with clean water, and allow to dry fully.

The repair area is now ready for final spraying. Paint spraying must be carried out in a warm, dry, windless and dust-free atmosphere. This condition can be created artificially if you have access to a large indoor working area, but if you are forced to work in the open, you will have to pick your day very carefully. If you are working indoors, dousing the floor in the work area with water will help to settle the dust which would otherwise be in the atmosphere. If the repair area is confined to one body panel, mask off the surrounding panels; this will help to minimise the effects of a slight mis-match in paint colours. Bodywork fittings (eg chrome strips, door handles etc) will also need to be masked off. Use genuine masking tape, and several thicknesses of newspaper, for the masking operations.

Before commencing to spray, agitate the aerosol can thoroughly, then spray a test area (an old tin, or similar) until the technique is mastered. Cover the repair area with a thick coat of primer; the thickness should be built up using several thin layers of paint, rather than one thick one. Using 400 grade wet-and-dry paper, rub down the surface of the primer until it is really smooth. While doing this, the work area should be thoroughly doused with water, and the wet-and-dry paper periodically rinsed in water. Allow to dry before spraying on more paint.

Spray on the top coat, again building up the thickness by using several thin layers of paint. Start spraying in the centre of the repair area, and then, using a circular motion, work outwards until the whole repair area and about 2 inches of the surrounding original paintwork is covered. Remove all masking material 10 to 15 minutes after spraying on the final coat of paint.

Allow the new paint at least two weeks to harden, then, using a paintwork renovator or a very fine cutting paste, blend the edges of the paint into the existing paintwork. Finally, apply wax polish.

Plastic components

With the use of more and more plastic body components by the vehicle manufacturers (eg bumpers. spoilers, and in some cases major body panels), rectification of more serious damage to such items has become a matter of either entrusting repair work to a specialist in this field, or renewing complete components. Repair of such damage by the DIY owner is not really feasible, owing to the cost of the equipment and materials required for effecting such repairs. The basic technique involves making a groove along the line of the crack in the plastic, using a rotary burr in a power drill. The damaged part is then welded back together, using a hot air gun to heat up and fuse a plastic filler rod into the groove. Any excess plastic is then removed, and the area rubbed down to a smooth finish. It is important that a filler rod of the correct plastic is used, as body components can be made of a variety of different types (eg polycarbonate, ABS, polypropylene).

Damage of a less serious nature (abrasions, minor cracks etc) can be repaired by the DIY owner using a two-part epoxy filler repair. Once mixed in equal, this is used in similar fashion to the bodywork filler used on metal panels. The filler is usually cured in twenty to thirty minutes, ready for sanding and painting.

If the owner is renewing a complete component himself, or if he has repaired it with epoxy filler, he will be left with the problem of finding a suitable paint for finishing which is compatible with the type of plastic used. At one time, the use of a universal paint was not possible, owing to the complex range of plastics encountered in body component applications. Standard paints, generally speaking, will not bond to plastic or rubber satisfactorily, but suitable paints to match any plastic or rubber finish, can be obtained from dealers. However, it is now possible to obtain a plastic body parts finishing kit which consists of a pre-primer treatment, a primer and coloured top coat. Full instructions are normally supplied with a kit, but basically, the method of use is to first apply the pre-primer to the component concerned, and allow it to dry for up to 30 minutes. Then the primer is applied, and left to dry for about an hour before finally applying the special-coloured top coat. The result is a correctly-coloured component, where the paint will flex with the plastic or rubber, a property that standard paint does not normally possess.

5 Major body damage – repair

Where serious damage has occurred, or large areas need renewal due to neglect, it means that complete new panels will need welding-in, and this is best left to professionals. If the damage is due to impact, it will also be necessary to check completely the alignment of the bodyshell, and this can only be carried out accurately by a Citroën dealer using special jigs. If the body is left misaligned, it is primarily dangerous, as the car will not handle properly, and secondly, uneven stresses will be imposed on the steering, suspension and possibly transmission, causing abnormal wear, or complete failure, particularly to such items as the tyres.

6 Bumpers – removal and refitting

Front bumper

Removal

1 To improve access, chock the rear wheels, then jack up the front of the car and support it securely on axle stands (see *Jacking and Vehicle Support*). Remove the front roadwheels.

2 Where applicable, unclip the access panels from the front of the wheel arch liners to allow access to the bumper side securing bolts. If no access panels are provided, detach the wheel arch liners as described in Section 20. On models fitted with headlight washers, the right-hand wheel arch liner must be pulled back to allow the headlight washer fluid hose to be disconnected.

12

6.3 Disconnecting the headlight washer fluid hose from the T-piece

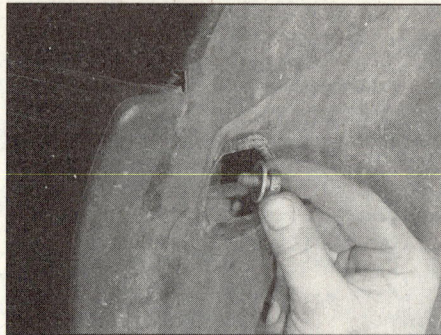

6.4 Removing a bumper side securing bolt, working through the access hole in the wheel arch liner

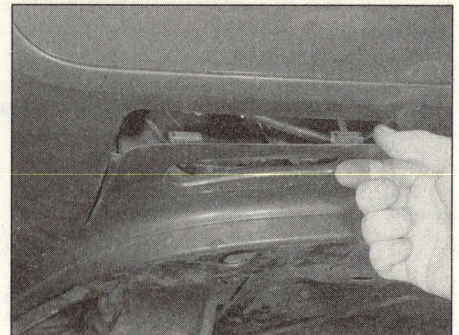

6.5a Pull off the access cover . . .

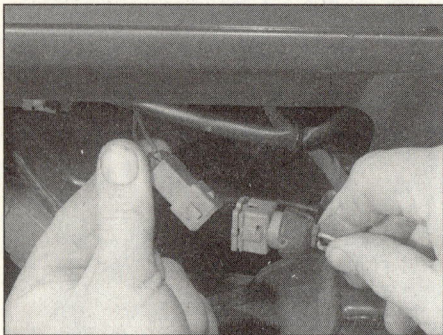

6.5b . . . and disconnect the foglight wiring connector

6.7 Bumper upper securing screws (arrowed)

3 Where applicable, working behind the right-hand side of the bumper, disconnect the headlight washer fluid hose from the T-piece **(see illustration)**.

4 Unscrew the bumper side securing bolts, one on each side of the vehicle **(see illustration)**.

5 Working underneath the front bumper, pull off the access covers, then separate the two halves of each foglight wiring connector, and disconnect the wiring plugs from the direction indicator light bulbholders **(see illustrations)**.

6 Remove the front grille panel as described in Section 20.

7 Working at the top of the bumper, unscrew the three upper securing screws (exposed by removal of the front grille panel) **(see illustration)**.

8 Remove the number plate, and unscrew the two front bumper securing screws **(see illustration)**.

9 Working under the front of the bumper, unscrew the two lower bumper securing screws, then withdraw the bumper from the vehicle **(see illustrations)**.

10 If desired, the bumper impact absorber components can now be pulled from their locations at the front of the vehicle **(see illustrations)**.

Refitting

11 Refitting is a reversal of removal.

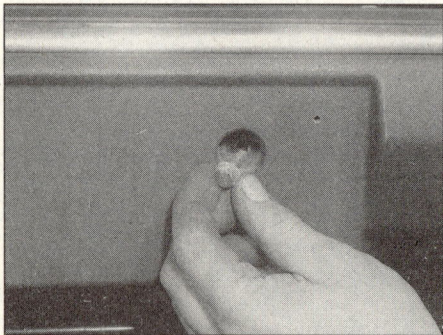

6.8 Removing a front bumper securing screw

6.9a Unscrew the lower bumper securing screws . . .

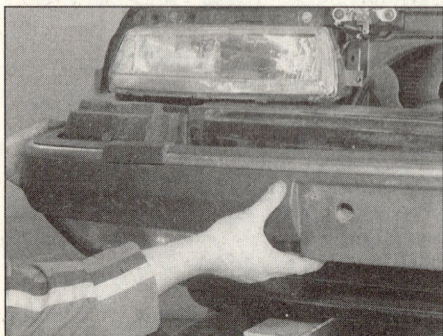

6.9b . . . and withdraw the bumper

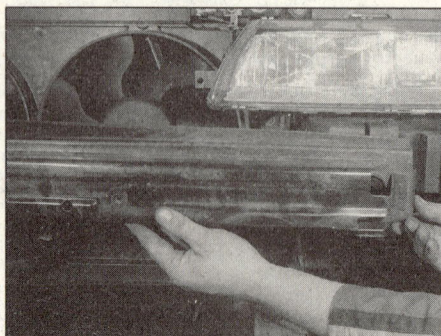

6.10a Pull off the metal impact absorber bar . . .

6.10b . . . the foam blocks . . .

6.10c . . . and the mounting tubes

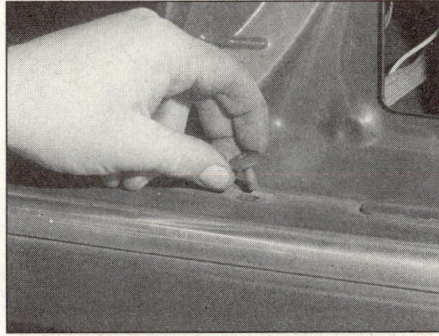
6.14 Unscrew the two upper bumper securing screws (one each side) – Hatchback model

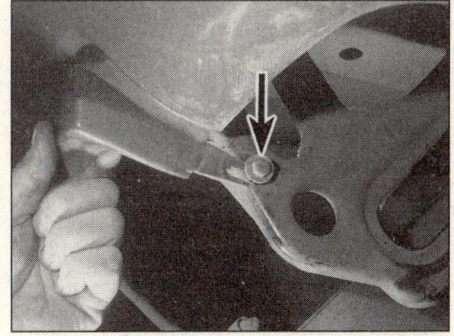
6.15 Slacken the nut and bolt (arrowed) securing the lower end of the bumper – Hatchback model

Rear bumper – Hatchback models

Removal

12 To improve access, apply the parking brake, then jack up the rear of the vehicle and support securely on axle stands (see *Jacking and Vehicle Support*).
13 Remove the rear light assemblies as described in Chapter 13, section 7.
14 Unscrew the two upper bumper securing screws, exposed by removal of the light units (see illustration).

15 Working on each side of the bumper in turn, slacken the nut and bolt securing the lower end of the bumper to the rear towing brackets (see illustration).
16 Again working on each side of the bumper in turn, reach up under the rear corners of the bumper and remove the air vent covers, then unscrew the bumper side securing bolts (see illustrations).
17 Withdraw the bumper from the rear of the vehicle (see illustration).
18 If desired, the bumper impact absorbers can be removed by unscrewing the two nuts

securing each impact absorber to the body panel (see illustration).

Refitting

19 Refitting is a reversal of removal.

Rear bumper – Estate models

Removal

20 Proceed as described in paragraphs 12 to 14.
21 Unscrew the remaining two upper securing screws from the top surface of the bumper (see illustration).

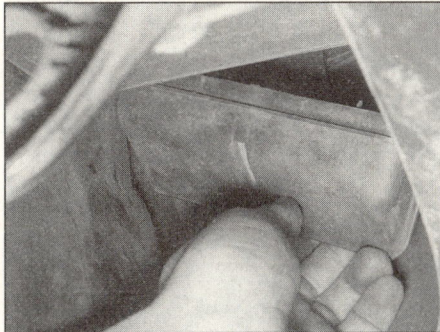
6.16a Remove the air vent covers . . .

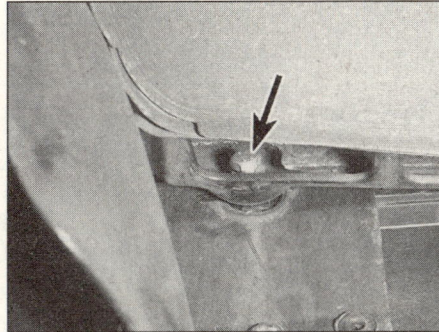
6.16b . . . and unscrew the bumper side securing bolts (arrowed) – Hatchback model

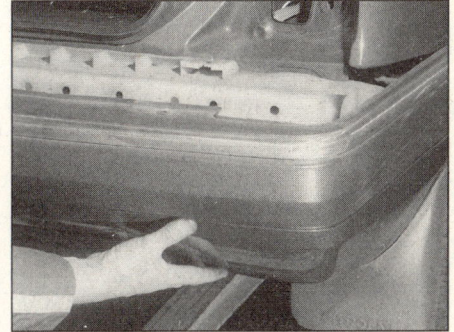
6.17 Withdrawing the rear bumper – Hatchback model

6.18 Rear bumper impact absorber securing nut (arrowed) – Hatchback model

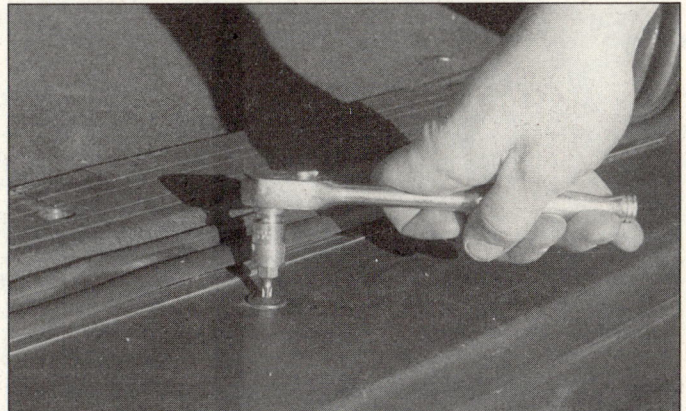
6.21 Unscrewing a rear bumper upper securing screw – Estate model

12

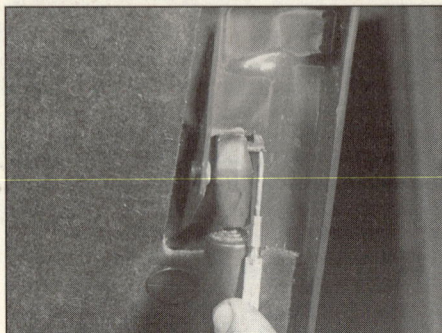

7.2 Prising out a bonnet support strut locking clip

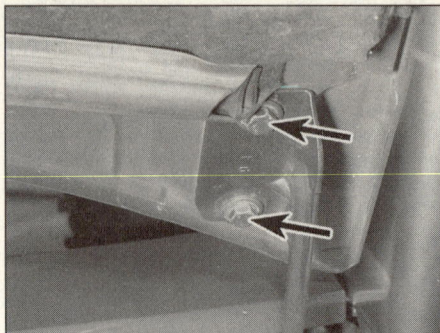

7.3 Bonnet retaining bolts (arrowed)

8.4 Unhook the bonnet release cable (arrowed) from the bonnet lock

22 Working underneath the bumper, unscrew the four lower bumper securing screws.

23 Working under the rear of the vehicle, reach up behind the wing panels, and unscrew the bumper side securing bolts (one on each side of the bumper), accessible through the cut-outs in the body panels.

24 Pull the bumper rearwards from the vehicle.

Refitting

25 Refitting is a reversal of removal.

7 Bonnet and support struts – removal, refitting and adjustment

Bonnet

Removal

1 Open the bonnet and have an assistant support it, then, using a pencil or felt tip pen, mark the outline position of each bonnet hinge relative to the bonnet, to use as a guide on refitting.

2 Using a screwdriver, carefully prise out the locking clips, and pull the bonnet support struts from the studs on the bonnet (see illustration).

3 Unscrew the bonnet retaining bolts and, with the help of the assistant, carefully lift the bonnet from the vehicle (see illustration). Store the bonnet out of the way in a safe place.

Refitting and adjustment

4 With the aid of an assistant, offer up the bonnet and loosely fit the retaining bolts. Align the hinges with the marks made on removal, then tighten the retaining bolts securely, and reconnect the bonnet support struts.

5 Close the bonnet, and check for alignment with the adjacent panels. If necessary, slacken the hinge bolts and re-align the bonnet to suit. Once the bonnet is correctly aligned, tighten the hinge bolts.

6 Once the bonnet is correctly aligned, check that the bonnet fastens and releases in a satisfactory manner. If adjustment is necessary, slacken the bonnet lock retaining

bolts, and adjust the position of the lock(s) to suit. Once the lock operation is satisfactory, securely tighten the retaining bolts.

Support struts

Removal

7 Support the bonnet in the open position, with the help of an assistant or using a suitable wooden prop.

8 Using a flat-bladed screwdriver, prise out the locking clip, then pull the support strut from its balljoint on the bonnet.

9 Similarly, release the strut from the balljoint on the body, and withdraw the strut from the vehicle.

Refitting

10 Refitting is a reversal of removal, but ensure that the locking clips are securely engaged.

8 Bonnet release cable and lock components – removal and refitting

Bonnet release cable

General

1 The bonnet release cable consists of two sections, joined by a connector plate positioned in the engine compartment.

2 The front and rear sections of the cable can be renewed individually.

Front cable section

3 During this procedure, take careful note of the routing of the cable to aid refitting.

4 Working in the engine compartment, unhook the cable end from the bonnet lock (see illustration).

5 Again working in the engine compartment, unbolt and move to one side any items necessary to allow access to the release cable securing clips. Release the cable from the securing clips.

6 Locate the cable connector plate, which is covered by a plastic shield, and is located to the side of the left-hand suspension turret, in the engine compartment.

7 Where applicable, release the cable connector plate shield from the securing clip, then withdraw the shield, and disconnect the front section of the cable from the connector plate.

8 Withdraw the front section of the cable, noting its routing.

9 Fit the new cable section using a reversal of the removal procedure. Ensure that the cable is routed as noted before removal.

Rear cable section

10 Working inside the vehicle, remove the sill trim panel as described in Section 23.

11 Unhook the end of the release cable from the release lever, then free the cable from the clips inside the vehicle.

12 To aid routing on refitting, tie a length of string to the end of the cable inside the vehicle.

13 Proceed as described in paragraphs 5 to 7, but disconnect the rear section of the cable from the connector plate.

14 Pull the cable through the bulkhead, into the engine compartment. Untie the string from the end of the cable, and leave it in place to aid refitting. Withdraw the rear cable section from the vehicle.

15 Fit the new cable section by using the string to pull it into position. Make sure that the cable is routed as noted during removal, and ensure that the cable is fitted free from kinks, and clear of surrounding components.

16 Secure the cable in position using the clips, then reconnect the cable to the connector plate and release lever, and check the lock operation.

17 Finally, refit the sill trim panel.

Bonnet lock

Removal

18 If removing the lock which is operated directly by the bonnet release cable, unhook the cable and release the lock interconnecting rod from the lock.

19 If removing the lock operated by the interconnecting rod, release the clip and disconnect the interconnecting rod from the lock.

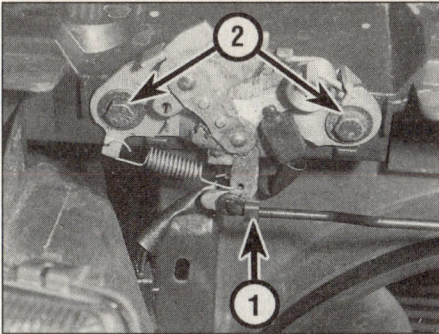

8.20 Disconnect the interconnecting rod (1), then unscrew the securing bolts (2) and withdraw the lock

9.2 Disconnect the door wiring connector

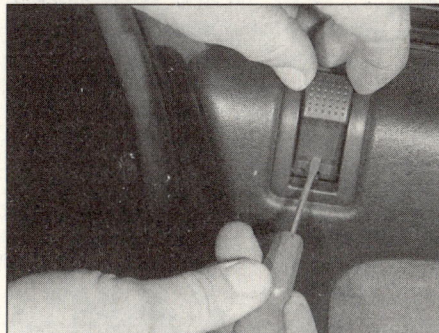

9.5 Front door hinge pin (arrowed)

20 Unscrew the two securing bolts, and withdraw the lock **(see illustration)**.

Refitting

21 Refitting is a reversal of removal, but if necessary adjust the position of the lock to achieve satisfactory operation.

9 Door – removal and refitting

Removal

1 Using a screwdriver, release the securing clip, and remove the cover from the battery, then disconnect the battery negative lead.
2 Open the door and, working at the front edge of the door, twist the locking ring anti-clockwise, and disconnect the door wiring connector **(see illustration)**.
3 Unbolt the door check strap from the body pillar.
4 Have an assistant support the door or, alternatively, support the lower edge of the door using wooden blocks, with pads to protect the paintwork.
5 Using a slide hammer and a suitable adapter, or a hammer and punch, tap out the door hinge pins **(see illustration)**.
6 Carefully lift the door from the vehicle.

Refitting

7 Refitting is a reversal of removal, but check

the condition of the hinge pins, and renew if necessary.

10 Door inner trim panel – removal and refitting

Front door

Removal

1 Using a screwdriver, release the securing clip, and remove the cover from the battery, then disconnect the battery negative lead.
2 Lift up the inner door lock operating button then, using a small screwdriver, depress the retaining tab, and slide off the button **(see illustration)**.
3 On models with manually-operated window

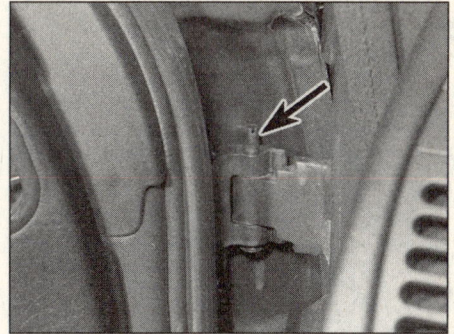

10.2 Depress the retaining tab and slide off the lock operating button

regulators, pull the winder handle off the spindle, then remove the spindle trim plate.
4 Unclip the cover panel from the door-mounted loudspeaker, then remove the securing screws, withdraw the loudspeaker, and disconnect the wiring plugs **(see illustration)**.
5 Where applicable, carefully prise the electric window control switch from the armrest, and disconnect the wiring plug **(see illustration)**. Alternatively, where applicable, prise out the blanking plate from the armrest.
6 Unclip the surround from the door interior handle **(see illustration)**.
7 Unscrew the following trim panel securing screws **(see illustrations)**.
a) Unscrew the single screw from the electric window control switch (or blanking plate) aperture.

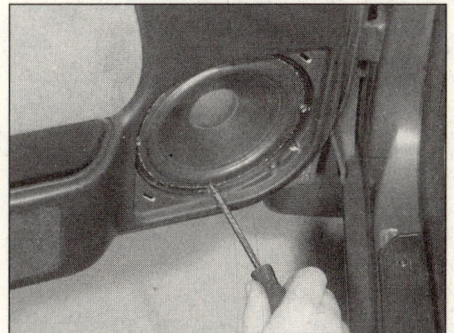

10.4 Unscrewing a loudspeaker securing screw – front door

10.5 Removing the electric window control switch – front door

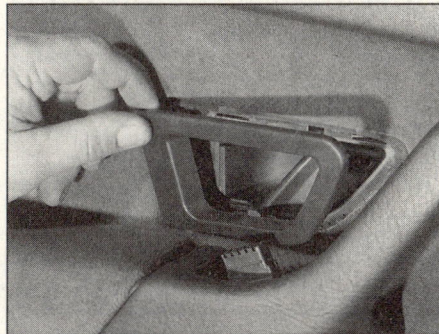

10.6 Unclip the surround from the door interior handle – front door

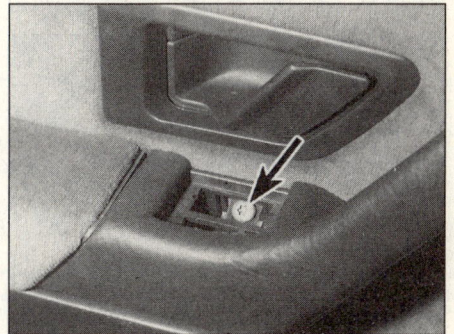

10.7a Unscrew the screws from the electric window control switch aperture ...

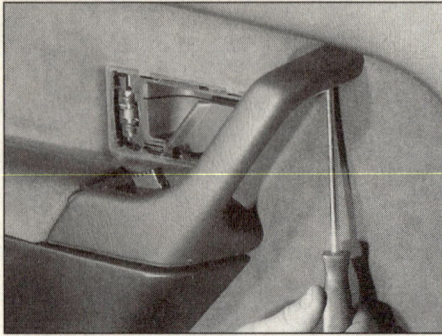

10.7b . . . the top edge of the armrest . . .

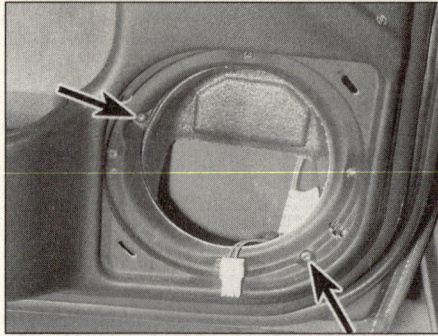

10.7c . . . and the loudspeaker aperture (arrowed) – front door

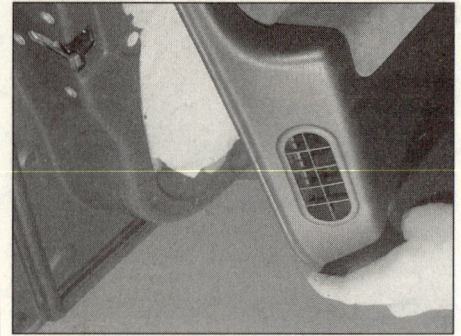

10.8a Withdraw the trim panel . . .

b) Unscrew the single screw from the top edge of the armrest.
c) Unscrew the two screws from the loudspeaker aperture.

8 Work around the edge of the panel using a suitable forked tool, and release the securing clips, then carefully withdraw the trim panel from the door. Disconnect the wiring from the kerb light switch mounted in the trim panel and, where applicable, disconnect the wiring plug from the electronic control unit mounted on the rear of the trim panel. Where applicable, feed the window switch wiring through the aperture in the door as the panel is removed **(see illustrations)**.

9 If work is to be carried out on the components inside the door, the sealing sheet must be removed as follows.

a) Where applicable, carefully prise the wiring connector clips from the door.
b) Unclip the door interior handle, and disconnect the lock operating rod from the handle.
c) Carefully peel the sealing sheet from the door – it should be possible to remove the sheet in one piece if the adhesive is carefully cut using a sharp knife.

Refitting

10 Refitting is a reversal of removal, bearing in mind the following points.

a) Before refitting, check whether any of the trim panel retaining clips were broken on removal, and renew them as necessary.
b) Ensure that all wiring is correctly routed.
c) To refit the inner door lock operating

button, first lock the door, to ensure that the link rod is in its lowest position. Position the button locating tab in the lower of its two holes, then firmly push the button onto the rod, until it clips into position and the retaining tab appears in the upper hole **(see illustrations)**.

Rear door

11 Proceed as described previously in this Section for the front door, noting the following differences **(see illustration)**.

a) Ignore the references to removing the loudspeaker and the two screws from the loudspeaker aperture.
b) Ignore the reference to disconnecting the wiring plug from the electronic control unit mounted on the trim panel.

---

11 Door handle and lock components – removal and refitting

Door interior handle

Removal

1 Remove the door inner trim panel as described in Section 10.

2 Slide the interior handle forwards, and unclip the door interior handle, and disconnect the lock operating rod from the handle **(see illustration)**.

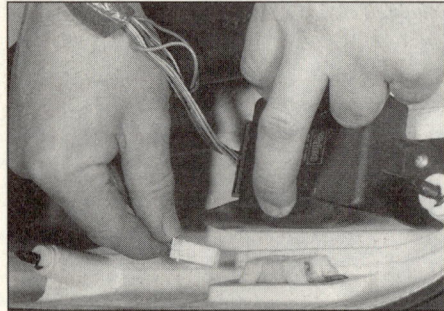

10.8b . . . and disconnect the wiring from the kerb light and the electronic control unit – front door

10.10a Position the lock button locating tab in the lower position

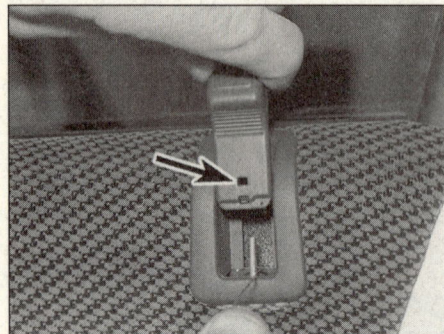

10.10b Push the button onto the rod until the retaining tab appears in the upper hole (arrowed)

10.11 Unscrewing the door trim panel securing screw from the electric window switch aperture – rear door

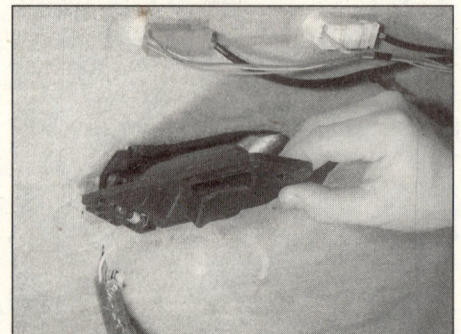

11.2 Removing a front door interior handle

11.6a Pull the rubber grommet from the outside of the door . . .

11.6b . . . slacken the three screws (arrowed) . . .

11.6c . . . and withdraw the inner part of the handle

Refitting

3 Refitting is a reversal of removal.

Front door exterior handle

Removal

4 Remove the door lock cylinder as described later in this Section.

5 Slide the exterior part of the handle assembly towards the rear of the car and remove it from the door.

6 Once the exterior part of the handle assembly has been removed, if desired, the inner part of handle assembly can be removed as follows (see illustrations).

a) *Remove the door inner trim panel and the plastic sealing sheet, as described in Section 10.*

b) *Pull the handle rubber grommet from the outside of the door.*

c) *Working outside the door, slacken the three screws securing the inner part of the handle, then withdraw the inner part of the handle from inside door.*

Refitting

7 Refitting is a reversal of removal, noting the following points (see illustrations).

a) *When refitting the exterior part of the handle, ensure that the lever on the exterior part of the handle engages with the spring and the lever on the inner part of the handle - pull the spring forwards*

using a length of hooked wire to allow it to engage with the lever.

b) *Where applicable, refit the plastic sealing sheet and the door inner trim panel as described in Section 10.*

c) *Refit the door lock cylinder as described later in this Section.*

Rear door exterior handle

8 Proceed as described previously in this Section for the front door, noting that a plastic block is fitted to the handle assembly in place of the lock cylinder. The screw securing the plastic block can be accessed through the hole in the rear edge of the door (remove the grommet) using a long-reach Allen key (see illustration).

11.7a Pull the spring (arrowed) forwards . . .

Front door lock cylinder

Removal

9 Working at the rear edge of the door, prise out the grommet for access to the lock cylinder securing screw.

10 Unscrew the securing screw using a suitable Allen key or hexagon bit, then pull the lock cylinder assembly from the exterior handle (see illustrations).

Refitting

11 Refitting is a reversal of removal.

Front door lock

Removal

12 Remove the door inner trim panel and the plastic sealing sheet, as described in Section 10.

11.7b . . . to allow it to engage with the lever on the exterior part of the handle

11.8 Unscrewing the rear door handle plastic block securing screw

11.10a Unscrew the securing screw . . .

11.10b . . . then pull the lock cylinder assembly from the exterior handle

12

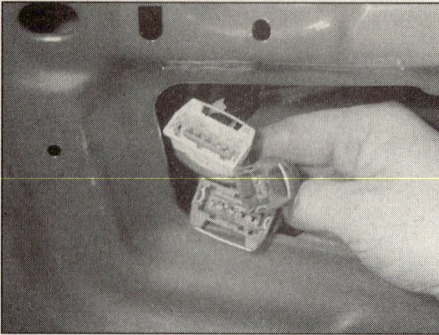
11.15 Disconnect the wiring plugs from the door lock assembly

11.16 Unscrew the three lock securing screws (arrowed) . . .

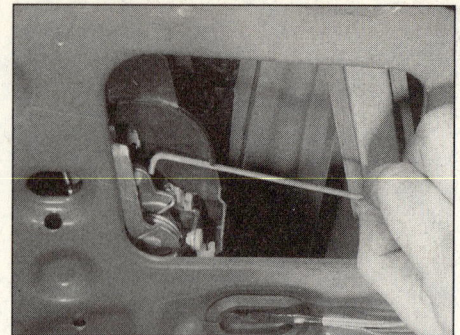
11.17a . . . then disconnect the operating rods . . .

13 Remove the door interior handle, as described previously in this Section.
14 Remove the exterior section of the door exterior handle, as described previously in this Section.
15 Reach in through door aperture, and disconnect wiring plugs from lock assembly. Unclip the lock assembly wiring harnesses from the door **(see illustration)**.
16 Working at the rear edge of the door, remove the three door lock securing screws **(see illustration)**.
17 Manipulate the lock assembly from the

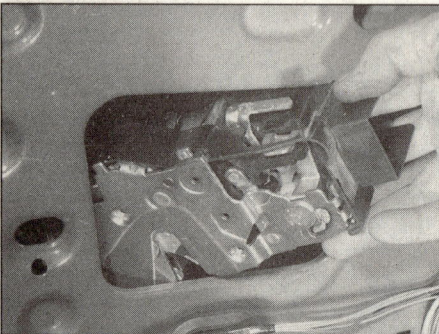
11.17b . . . and withdraw the lock assembly through the upper door aperture

rear of the door, and disconnect the lock button operating rod and the interior handle operating rod from lock, then withdraw the lock assembly through the upper door aperture **(see illustrations)**.

Refitting

18 Refitting is a reversal of removal, bearing in mind the following points.
a) Make sure that the lock button and interior handle operating rods are reconnected to the lock before the lock assembly is manoeuvred into its final position.
b) Refit the exterior section of the door exterior handle, and the door interior handle as described previously in this Section.
c) Refit the plastic sealing sheet and the door inner trim panel as described in Section 10.

Rear door lock

19 Proceed as described previously in this Section for the front door lock, but note that there is no need to remove the exterior section of the door exterior handle, and note the interior handle operating rod must be unclipped from the inside of the door before it can be disconnected from the lock **(see illustration)**.

12 Door window glass and regulator – removal and refitting

Fixed door window glass

1 These areas of glass are bonded in position with a special adhesive. Renewal of such fixed glass is a difficult, messy and time-consuming task, which is considered beyond the scope of the home mechanic. The task carries a high risk of breakage. In view of this, owners are strongly advised to have this sort of work carried out by one of the many specialist windscreen fitters.

Front door sliding window glass

Removal

2 Fully lower the window, then remove the door inner trim panel and the plastic sealing sheet as described in Section 10.
3 Working at the rear edge of the window aperture, unscrew the nut securing the inner window aperture plastic trim panel **(see illustration)**. Pull the panel from the door to release the securing clips – there is no need to release the panel from the front edge of the door.
4 Pull the lower edge of the inner weatherstrip from the window aperture **(see illustration)**.
5 Working inside the door, pull off the plastic

11.19 Removing a rear door lock

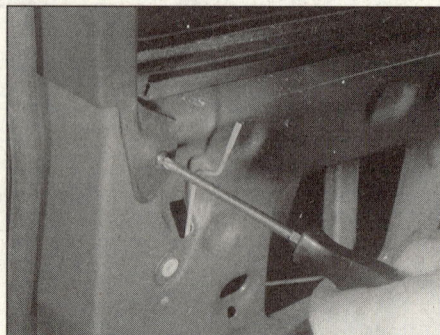
12.3 Unscrewing the inner window aperture trim panel securing nut – front door

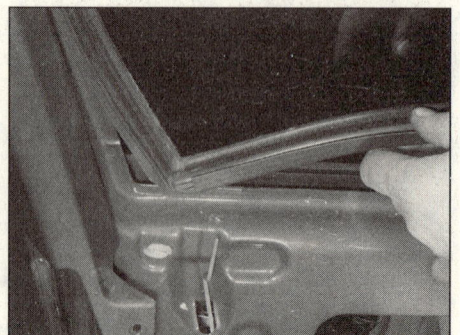
12.4 Pull the inner weatherstrip from the window aperture – front door

12.5a Pull off the plastic clip securing the window glass to the regulator – front door

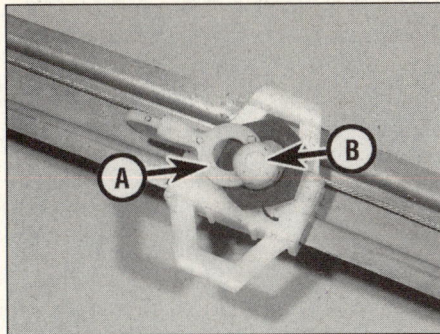

12.5b View of window glass plastic clip in position with regulator mechanism removed from door. Clip (A) engages with peg (B)

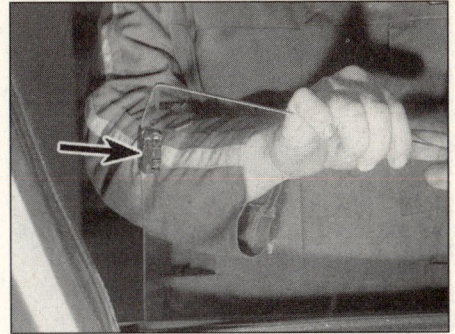

12.7 Withdraw the glass from the front door – note the peg (arrowed) which must slide out from the window guide rail as the glass is lowered

clip securing the window glass to the regulator mechanism **(see illustrations)**.

6 Lower the glass down to the bottom of the door, until the peg attached to the rear of the glass slides out from the bottom of the window guide rail.

7 Lift the glass up, and withdraw it through the outside of the window aperture **(see illustration)**.

Refitting

8 Refitting is a reversal of removal, but ensure that the peg attached to the rear of the glass engages with the window guide rail. Check the operation of the window mechanism before refitting the plastic sealing sheet and the door inner trim panel with reference to Section 10.

Rear door sliding window glass
Removal

9 Fully lower the window, then remove the door inner trim panel and the plastic sealing sheet as described in Section 10.

10 Remove the door lock as described in Section 11.

11 Working at the front edge of the window aperture, unscrew the nut securing the inner window aperture plastic trim panel **(see illustration)**. Pull the panel from the door to release the securing clips – there is no need to release the panel from the rear edge of the door.

12 Pull the lower and front edges of the inner

weatherstrip from the window aperture **(see illustration)**.

13 Working inside the door, pull off the plastic clip securing the window glass to the regulator mechanism (see paragraph 5).

14 Tilt the front edge of the glass down until the peg attached to the front of the glass slides out from the bottom of the window guide rail.

15 Lift the glass up, and withdraw it through the inside of the window aperture **(see illustration)**.

Refitting

16 Refitting is a reversal of removal, but ensure that the peg attached to the front of the glass engages with the window guide rail.

12.11 Unscrewing the inner window aperture trim panel securing nut – rear door

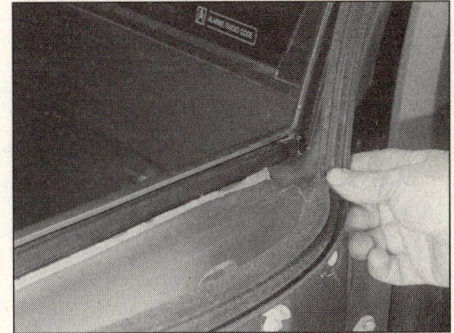

12.12 Pull the inner weatherstrip from the window aperture – rear door

Check the operation of the window mechanism before refitting the plastic sealing sheet and the door inner trim panel with reference to Section 10.

Window regulator assembly
Removal

17 Remove the door sliding window glass as described previously in this Section.

18 Where applicable, reach in through the door aperture, and disconnect the wiring plug from the regulator motor **(see illustration)**.

19 Unscrew the three nuts securing the regulator motor assembly to the door, and the two nuts securing the regulator rail **(see illustration)**.

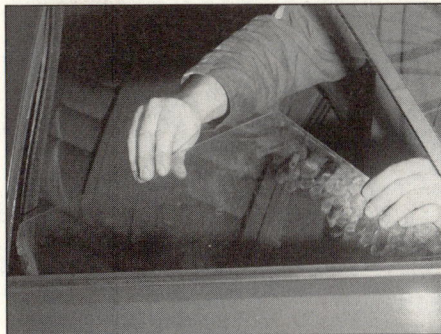

12.15 Withdrawing the rear door window glass

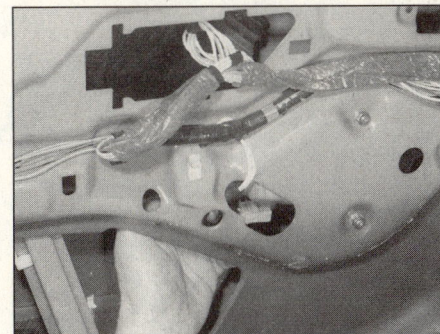

12.18 Disconnecting the front door window regulator motor wiring plug

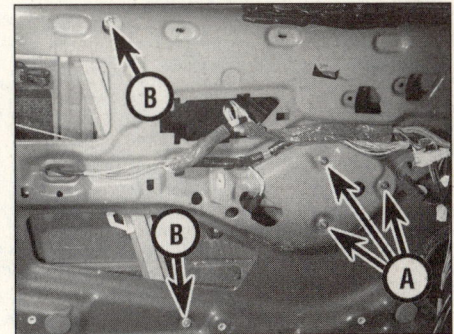

12.19 Unscrew the three nuts (A) securing the motor, and the two nuts (B) securing the regulator rail – front door

12

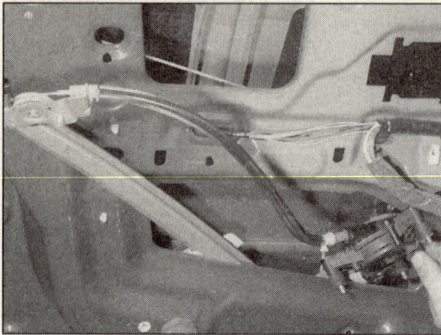

12.20 Manipulate the front door window regulator assembly out through the lower door aperture

20 Manipulate the assembly out through lower door aperture **(see illustration)**.

Refitting

21 Refitting is a reversal of removal, but refit the door sliding window glass as described previously in this Section.

13 Tailgate and support struts – removal, refitting and adjustment

Tailgate - Hatchback models
Removal

1 Using a screwdriver, release the securing clip, and remove the cover from the battery, then disconnect the battery negative lead.

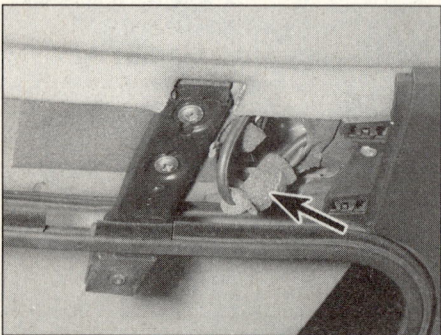

13.3b . . . and the washer fluid hose connector (arrowed) – Hatchback model

13.2 Prising off the rear roof trim panel – Hatchback model

2 Open the tailgate, then working inside the vehicle, carefully prise off the rear roof trim panel **(see illustration)**.

3 Working through the holes in each side of the upper rear body panel, disconnect the tailgate wiring harness connectors, and the washer fluid connector **(see illustrations)**. If necessary, prise the foam insulation from the connectors.

4 Unscrew the bolt securing the tailgate wiring harness earth lead to the upper rear body panel **(see illustration)**.

5 Pull the tailgate wiring harness grommets from the holes in the body panel, then withdraw the wiring harnesses through the body panels, noting the harness routing.

6 Support the tailgate in the open position using a wooden prop or similar tool.

7 Prise out the securing clips, then carefully prise the upper ends of the tailgate support struts from the studs on the tailgate.

8 Working at the top edges of the tailgate, unscrew the bolts securing the tailgate to the hinges **(see illustration)**.

9 With the aid of an assistant, carefully lift the tailgate from the vehicle.

Refitting

10 Refitting is a reversal of removal, bearing in mind the following points.
a) *Ensure that the wiring harnesses are routed as noted before removal.*
b) *Before fully tightening the tailgate hinge nuts, temporarily close the tailgate and*

check the alignment with surrounding body panels.

Tailgate - Estate models
Removal

11 Proceed as described in paragraphs 1 to 7.

12 Working at the top of the tailgate, carefully prise the locking clips from the tailgate hinge pins **(see illustration)**.

13 Ensure that the tailgate is adequately supported, then tap out the hinge pins and, with the aid of an assistant, lift the tailgate from the vehicle.

Refitting

14 Refitting is a reversal of removal, but ensure that the wiring harnesses are routed as noted before removal.

Support struts

15 Support the tailgate in the open position, with the help of an assistant or using a suitable wooden prop.

16 Using a flat-bladed screwdriver, prise out the locking clip, then pull the support strut from its balljoint on the tailgate.

17 Similarly, release the strut from the balljoint on the body, and withdraw the strut from the vehicle.

Refitting

18 Refitting is a reversal of removal, but ensure that the locking clips are securely engaged.

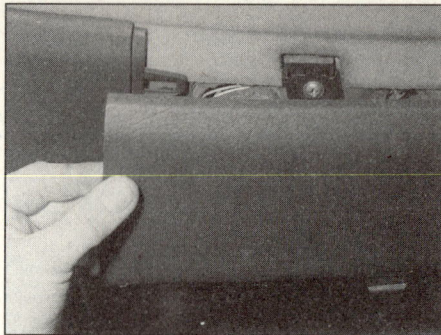

13.3a Disconnect the tailgate wiring harness connectors (arrowed) . . .

13.4 Unscrew the bolt (arrowed) securing the wiring harness earth lead – Hatchback model

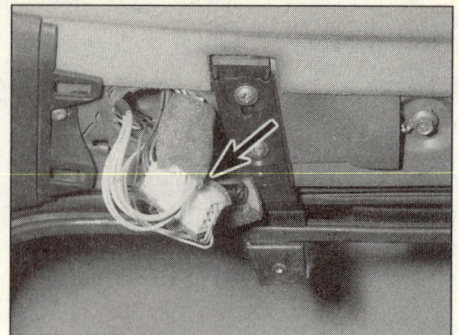

13.8 Tailgate-to-hinge securing bolts (arrowed) – Hatchback model

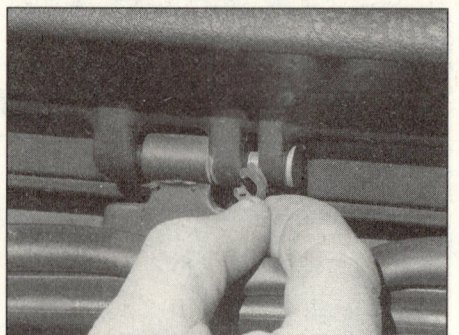

13.12 Prising the locking clip from a tailgate hinge pin – Estate model

14.3 Removing a tailgate trim panel securing screw – Hatchback model

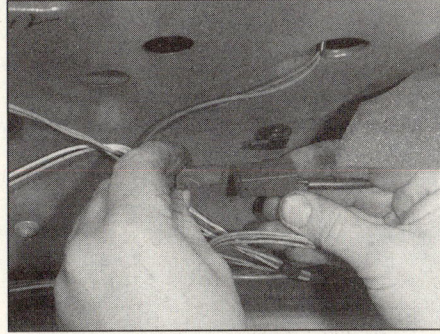

14.6 Separate the two halves of the lock wiring connector – Hatchback model

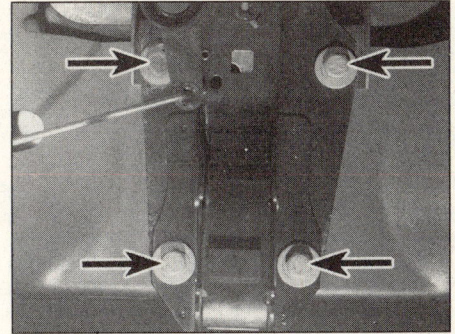

14.7a Unscrew the four securing bolts (arrowed) . . .

14 Tailgate lock components – removal and refitting

Lock assembly – Hatchback models

Removal

1 Using a screwdriver, release the securing clip, and remove the cover from the battery, then disconnect the battery negative lead.
2 Open the tailgate, and unclip the tailgate-mounted rear light covers.
3 Remove the securing screws, and withdraw the tailgate inner trim panel (see illustration).
4 Reach in through the tailgate aperture, and disconnect the lock operating rods from the lock assembly.
5 Where applicable, pull the foam pad from the top of the lock assembly.
6 Trace the wiring back from the lock assembly, and separate the two halves of the wiring connector (see illustration).
7 Unscrew the four lock securing bolts, and withdraw the lock assembly from the tailgate (see illustrations).

Refitting

8 Refitting is a reversal of removal,

Lock cylinder – Hatchback models

Removal

9 Proceed as described in paragraphs 1 to 3.

10 Where applicable, trace the wiring back from the switch on the lock, then separate the two halves of the switch wiring connector.
11 Unscrew the two lock cylinder securing screws, then disconnect the lock operating rod, and withdraw the lock cylinder from the tailgate (see illustrations).
12 If desired, the switch can be removed from the lock cylinder after unscrewing the two securing screws.

Refitting

13 Refitting is a reversal of removal.

Lock striker – Hatchback models

Removal

14 Working at the rear of the luggage compartment, prise out the cover plugs, then unscrew the four screws securing the luggage compartment rear trim panel. Lift off the panel, and disconnect the wiring from the luggage compartment light switch.
15 Unscrew the securing bolt, and remove the lock striker.

Refitting

16 Refitting is a reversal of removal but, if necessary, adjust the position of the striker to achieve satisfactory lock operation.

Lock assembly - Estate models

Removal

17 Using a screwdriver, release the securing clip, and remove the cover from the battery, then disconnect the battery negative lead.

18 Remove the securing screws, and withdraw the handle from the tailgate trim panel. Work around the trim panel and remove the securing screws, then withdraw the trim panel.
19 Trace the wiring back from the lock, then unclip the wiring connector from the tailgate, and separate the two halves of the connector (see illustration).
20 Unscrew the two lock securing bolts, then withdraw the lock from the tailgate and disconnect the lock operating rod.

Refitting

21 Refitting is a reversal of removal, but before refitting the tailgate trim panel, close the tailgate and check the operation of the lock. If necessary, move the lock within its elongated bolt holes to achieve satisfactory lock operation.

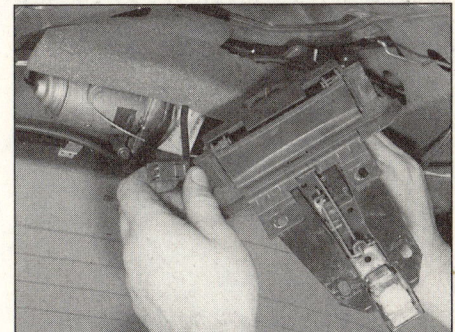

14.7b . . . and withdraw the lock assembly – Hatchback model

14.11a Unscrew the two lock cylinder securing screws (arrowed) . . .

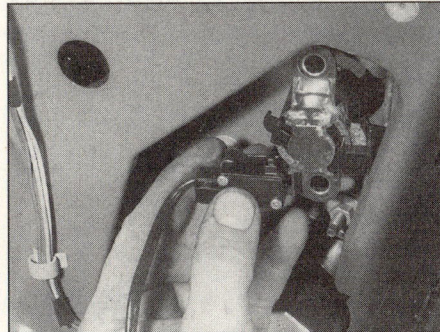

14.11b . . . and withdraw the lock cylinder – Hatchback model

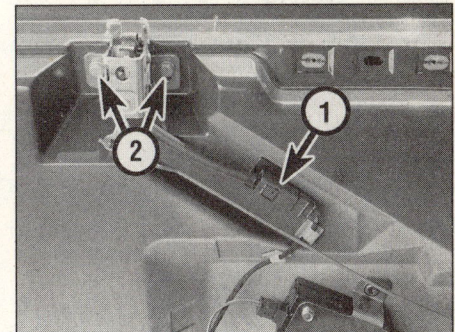

14.19 Tailgate lock wiring connector (1) and lock securing bolts (2) – Estate models

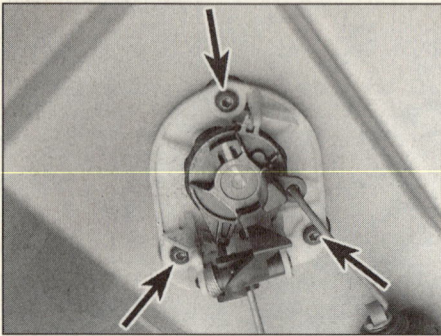

14.23 Tailgate lock cylinder securing screws (arrowed) – Estate models

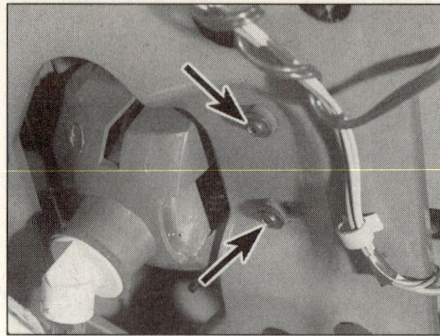

15.5a Unscrew the two securing screws (arrowed) . . .

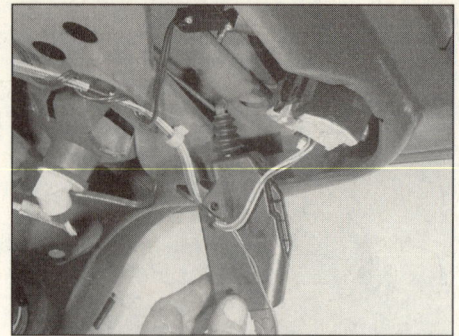

15.5b . . . and withdraw the tailgate lock motor – Hatchback model

Lock cylinder – Estate models

Removal

22 Proceed as described in paragraphs 17 and 18.
23 Working inside the tailgate, unscrew the three securing screws, then lift the lock cylinder assembly away from the tailgate, and disconnect the lock operating rods (see illustration).

Refitting

24 Refitting is a reversal of removal, but check the operation of the lock mechanism before refitting the tailgate trim panel.

Lock striker – Estate models

Removal

25 Unclip the striker cover from the luggage compartment rear trim panel, then unscrew the securing bolts and withdraw the striker.

Refitting

26 Refitting is a reversal of removal.

15 Central locking components – removal and refitting

Door lock motor

1 The lock motors are integral with the door locks, and cannot be renewed separately.

15.11 Tailgate lock motor wiring plug (1) and motor bracket securing bolts (2) – Estate model

Tailgate lock motor – Hatchback models

Removal

2 Using a screwdriver, release the securing clip, and remove the cover from the battery, then disconnect the battery negative lead.

3 Open the tailgate, and unclip the tailgate-mounted rear light covers.
4 Remove the securing screws, and withdraw the tailgate inner trim panel.
5 Unscrew the two securing screws, then lower the motor from the tailgate. Disconnect the lock operating rod, and the wiring plug, and withdraw the motor (see illustrations).

Refitting

6 Refitting is a reversal of removal.

Tailgate lock motor – Estate models

Removal

7 Using a screwdriver, release the securing clip, and remove the cover from the battery, then disconnect the battery negative lead.
8 Remove the securing screws, and withdraw the handle from the tailgate trim panel.
9 Work around the trim panel and remove the securing screws, then withdraw the trim panel.
10 Working inside the tailgate, disconnect the wiring plug from the lock motor.
11 Remove the two screws securing the lock motor mounting bracket to the tailgate, then

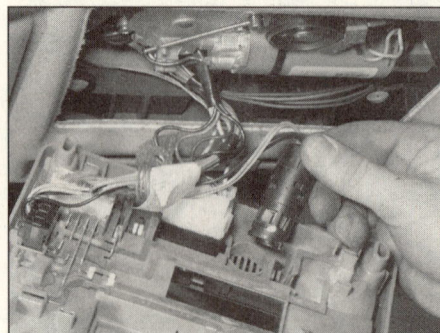

15.16 Removing the remote control receiver unit from the roof console

withdraw the lock motor assembly, and disconnect the lock operating rod (see illustration). If desired, the lock motor can be unbolted from the mounting bracket.

Refitting

12 Refitting is a reversal of removal, but check the operation of the lock mechanism before refitting the tailgate trim panel.

Remote control receiver unit

Removal

13 Using a screwdriver, release the securing clip, and remove the cover from the battery, then disconnect the battery negative lead.

14 Carefully prise the lens from the front courtesy light in the roof console.
15 Unscrew the now-exposed securing screws, and lower the roof console.
16 Unclip the receiver unit from the console, then disconnect the wiring plug, and remove the receiver unit (see illustration).

Refitting

17 Refitting is a reversal of removal.

Remote control transmitter batteries - renewal

18 Remove the securing screw, and withdraw the end cover from the key assembly.
19 Withdraw the old batteries, noting their orientation.
20 Fit the new batteries, ensuring that they are fitted the correct way round, as noted before removal, then refit the cover and tighten the securing screw.

16 Electric window components – removal and refitting

Electric window motors

1 The motors are integral with the regulator assemblies, and cannot be removed separately. Removal and refitting of the regulator assemblies is described in Section 12.

Electric window switches

2 Refer to Chapter 13.

17 Exterior mirrors and associated components – removal and refitting

Door mirror glass - renewal

1 Using a small flat-bladed screwdriver, working at the lower edge of the glass, locate the hook in the glass retaining clip, and lever the clip down to release the glass **(see illustrations)**.

2 To fit the glass, ensure that the clip is fully engaged with the lugs on the rear of the mirror glass, then push the glass firmly into position. The clip should click into position as it engages with the mirror.

Manually-operated door mirror

Removal

3 Pull the mirror adjuster trim plate from the door, then unclip the trim plate from the end of the adjuster.

4 Working at the front edge of the door, unscrew the two mirror securing screws, then carefully withdraw the mirror from the door, and feed the adjuster mechanism through the grommet in the door.

Refitting

5 Refitting is a reversal of removal.

Electrically-operated door mirrors

Removal

6 On models with electric mirrors, using a screwdriver, release the securing clip, and remove the cover from the battery, then disconnect the battery negative lead.

7 Remove the door inner trim panel as described in Section 10.

8 Working inside the door, locate the mirror wiring connectors, then separate the two halves of each connector. Note the routing of the wiring harnesses.

9 Working at the front edge of the door, unscrew the two mirror securing screws, then carefully withdraw the mirror from the door, and feed the wiring harnesses through the grommet inside the door **(see illustrations)**.

17.1a Release the door mirror glass retaining clip . . .

17.1b . . . and remove the glass – engage the screwdriver with the hook (arrowed) in the clip to release the glass

17.9a Unscrew the two securing screws . . .

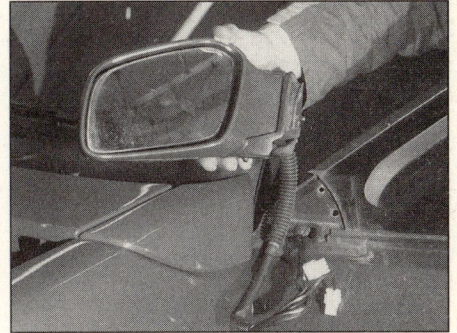

17.9b . . . and withdraw the mirror from the door

10 Pull the wiring grommet from the outside of the door, and withdraw the mirror/wiring harness assembly.

Refitting

11 Refitting is a reversal of removal, but ensure that the wiring harnesses are routed as noted before removal.

18 Windscreen, tailgate and fixed window glass – general information

These areas of glass are bonded in position with a special adhesive. Renewal of such fixed glass is a difficult, messy and time-consuming task, which is considered beyond the scope of the home mechanic. It is difficult, unless one has plenty of practice, to obtain a secure, waterproof fit. Furthermore, the task carries a high risk of breakage; this applies especially to the laminated glass windscreen. In view of this, owners are strongly advised to have this sort of work carried out by one of the many specialist windscreen fitters.

19 Sunroof – general information

1 The factory-fitted sunroof is of the electric tilt/slide type.

2 Due to the complexity of the sunroof mechanism, considerable expertise is required to repair, replace or adjust the sunroof components successfully. Removal of the sunroof first requires the headlining to be removed, which is a tedious operation, and not a task to be undertaken lightly. Therefore, any problems with this type of sunroof should be referred to a Citroën dealer.

3 Refer to Chapter 13 for details of sunroof switch removal.

4 If the sunroof motor fails, and the roof panel is stuck in the open or closed position, the panel can be moved manually as follows **(see illustrations)**.

a) Open the fusebox cover, and unclip the sunroof crank handle from its securing bracket.

19.4a Unclip the sunroof crank handle . . .

19.4b . . . move the release lever (arrowed), then use the crank handle to move the sunroof panel

12

b) *Carefully prise the lens from the courtesy light assembly in the roof console.*
c) *Using the cross-head end of the crank handle, unscrew the four roof console securing screws, then lower the console, taking care not to strain the wiring.*
d) *Engage the hexagon end of the crank handle with the motor spindle.*
e) *Move the crank mechanism release lever in the direction shown by the arrow on the motor, then use the crank handle to turn the mechanism and move the sunroof panel.*

20 Body exterior fittings – removal and refitting

Front grille panel

Removal

1 Open the bonnet.
2 On early models, where applicable, unscrew the securing screws, or prise out the clips, as applicable, and remove the grille centre panel.
3 Remove the front sidelights as described in Chapter 13.
4 Unscrew the three upper securing bolts, and the single bolt accessible through each front sidelight aperture, then withdraw the front grille panel **(see illustrations)**.

Refitting

5 Refitting is a reversal of removal.

Front scuttle trim panel

Removal

6 Working on each side of the vehicle in turn, carefully prise off the side trim panel from the scuttle. The panel should be pushed downwards to release the locating hooks, then pulled up from the scuttle – take care as the clips are easily broken **(see illustration)**.
7 Unscrew the two now-exposed scuttle trim panel securing screws on each side of the vehicle, then lift the scuttle panel to disengage the lugs from the cut-outs in the body panel, and withdraw the scuttle panel **(see illustration)**.

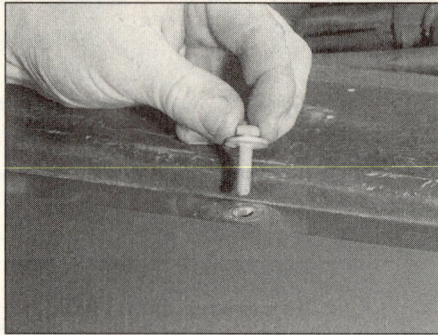
20.4a Removing a front grille panel upper securing bolt

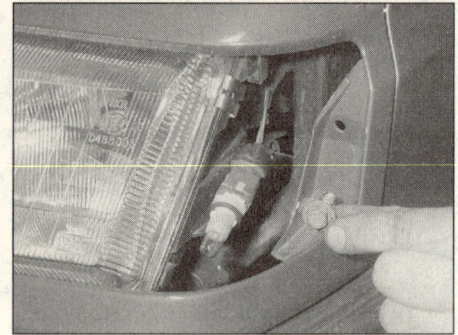
20.4b Remove the single bolt from each sidelight aperture . . .

20.4c . . . and withdraw the front grille panel

Refitting

8 Refitting is a reversal of removal, but take care not to damage the clips when refitting the scuttle side trim panels.

Wheel arch liners

9 The various plastic covers fitted to the underside of the vehicle are secured in position by a mixture of screws and retaining clips, and the method of removal will be obvious on inspection. Work methodically around the panel, removing its retaining screws and releasing the retaining clips until the panel is free **(see illustration)**. Some of the plastic clips may consist of two parts – where this is the case, the centre pin should be pushed out to free the main part of the clip. Note that on some models, panels may be

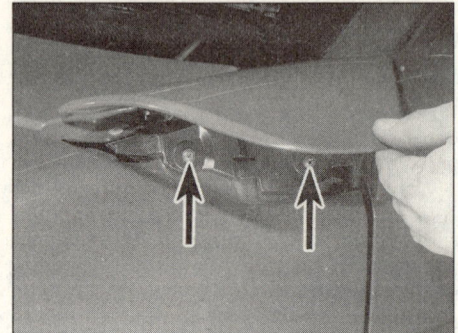
20.6 Prise the side trim panel from the scuttle to expose the two scuttle trim panel securing screws

secured by pop-rivets, which must be drilled out.

21 Seats – removal and refitting

Front seats

Removal

1 Slide the seat fully forwards then, working at the rear of the seat rails, remove the securing screws, and withdraw the trim plates from the rear of the seat rails **(see illustration)**.

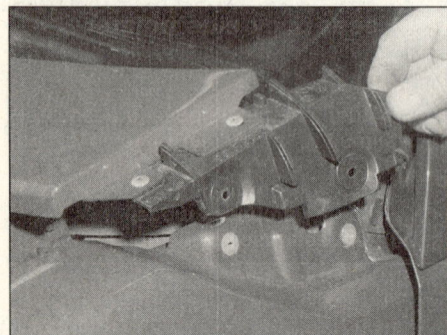
20.7 Withdrawing the scuttle trim panel

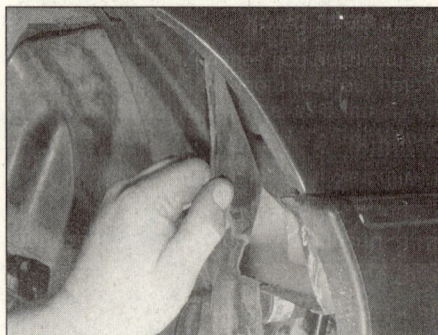
20.9 Removing a front wheel arch liner

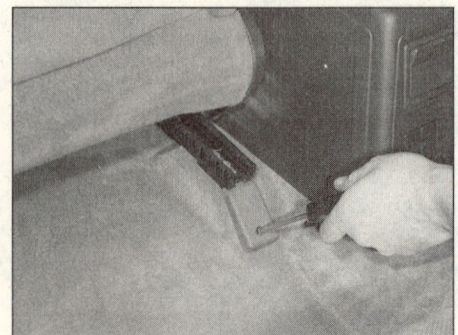
21.1 Unscrew the securing screws, and remove the trim plate from the rear of the front seat rails . . .

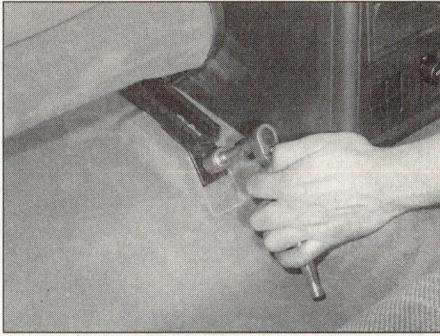

21.2 . . . then unscrew the bolts securing
the rear of the seat rails to the floor

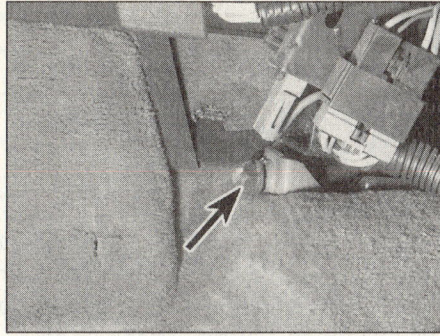

21.4 Front seat rail securing nut
(arrowed)

21.6 Disconnect the seat wiring plugs

2 Unscrew the bolts (one on each side) securing the rear of the seat rails to the floor **(see illustration)**.

3 Slide the seat fully rearwards.

4 Working at the front of the seat, unscrew the nuts (one on each side) securing the front of the seat rails to the floor **(see illustration)**.

5 On models fitted with front seat belt tensioners, electrically adjustable seats and/or heated front seats, release the securing clip, and remove the cover from the battery, then disconnect the battery negative lead.

⚠️ **Warning: On models with front seat belt tensioners and/or an airbag, wait for ten minutes before proceeding further – this will ensure that the seat belt tensioners and the airbag are de-activated.**

6 Where applicable, reach under the seat, and disconnect the seat wiring plug(s) **(see illustration)**. Similarly, where applicable, reach up under the seat side trim panel and disconnect the wiring from the outboard seat adjustment switch.

7 Lift the seat from the floor, and tilt it to one side, for access to the outboard side trim panel securing screws. There are three securing screws, one at the back, one at the side, and one underneath at the front.

8 Remove the securing screws, and withdraw the seat side trim panel, then unscrew the bolt securing the seat belt to the seat **(see illustration)**.

21.8 Unscrew the bolt (arrowed) securing
the seat belt to the seat

9 Withdraw the seat from the vehicle.

Refitting

10 Refitting is a reversal of removal, but tighten the seat mounting bolts, and the seat belt mounting bolt securely.

Rear seat back

Removal

11 Tilt the rear seat cushion forwards, then unscrew the now-exposed hinge bolts **(see illustration)**.

12 Fold the rear seat back down, then lift up the carpet trim panel for access to the remaining seat back hinge bolts **(see illustration)**.

13 Unscrew the remaining seat back hinge bolts, and remove the seat back.

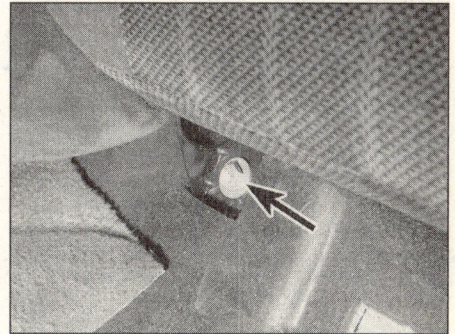

21.11 Rear seat back hinge bolt (arrowed)
– Hatchback model

Refitting

14 Refitting is a reversal of removal, but ensure that the hinge bolts are securely tightened.

Rear seat back side bolster

Removal

15 Remove the parcel shelf side support panel as described in Section 23.

16 Unscrew the side bolster upper securing nut **(see illustration)**.

17 Fold the seat back down, then unscrew the now-exposed side bolster securing clip **(see illustration)**.

18 Pull the side bolster up, and withdraw it from the locating lugs.

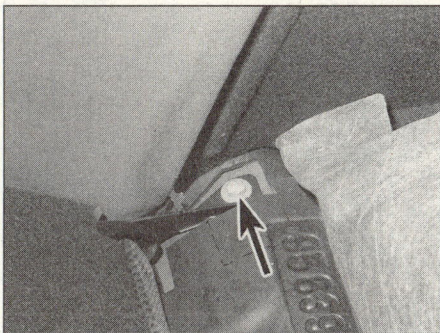

21.12 Fold the seat back down for access
to the remaining seat back hinge bolts
(arrowed) – Hatchback model

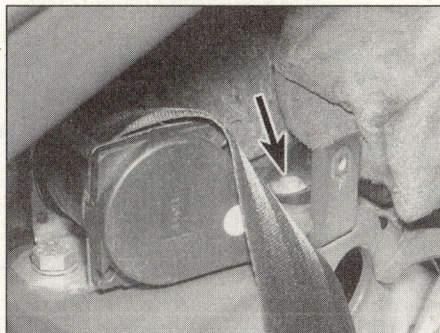

21.16 Unscrew the side bolster upper
securing nut (arrowed) . . .

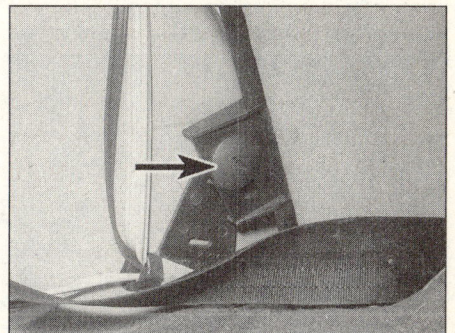

21.17 . . . and the securing clip (arrowed) –
Hatchback model

12

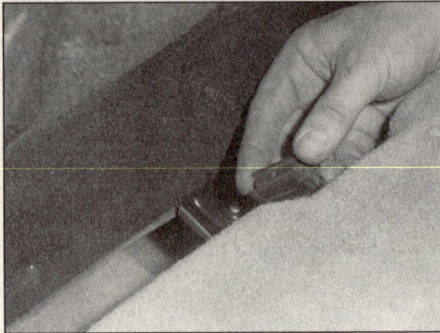

21.20 Removing a rubber retaining block from a rear seat cushion hinge – Hatchback model

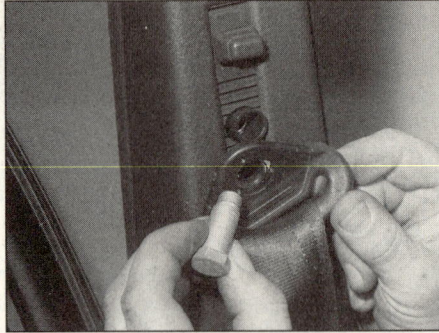

22.4 Removing a front seat belt upper anchor bolt

22.5a Prise out the trim plate . . .

Refitting

19 Refitting is a reversal of removal.

Rear seat cushion

Removal

20 Tilt the seat cushion forwards, then reach down behind each hinge, and pull the rubber retaining block from the rear of the hinge (see illustration).
21 Once the rubber blocks have been removed, the seat cushion can be lifted from the vehicle.

Refitting

22 Refitting is a reversal of removal, but make sure that the rubber blocks are correctly located in the hinge assemblies.

22 Seat belt components – removal and refitting

Front seat belt

Removal

1 Remove the relevant front seat, as described in Section 21.
2 Open the front and rear doors, and carefully prise the weatherstrips from the sill trim panel, and from the centre pillar trim panel.
3 Remove the relevant sill trim panel as described in Section 23.
4 Pull off the trim plate, and unbolt the seat belt upper anchor bolt (see illustration).

5 Prise out the trim plate from the centre of the centre pillar trim panel, and unscrew the now-exposed trim panel securing screw (see illustrations).
6 Carefully prise away the roof side trim panels from the top of the centre pillar trim panel to expose the two centre pillar trim panel upper securing screws (see illustration). Remove the upper securing screws.
7 Using a small flat-bladed screwdriver, prise out the centre pin, and remove the seat belt height adjuster knob, then withdraw the centre pillar trim panel (see illustrations).
8 Feed the seat belt webbing through the slot in the bracket attached to the centre pillar, the unbolt the inertia reel, and withdraw the seat belt assembly (see illustrations).

22.5b . . . and unscrew the centre pillar trim panel securing screw

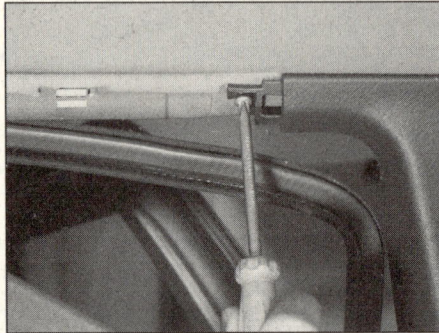

22.6 Unscrew the centre pillar trim panel upper securing screws

22.7a Remove the seat belt height adjuster knob . . .

22.7b . . . then withdraw the centre pillar trim panel

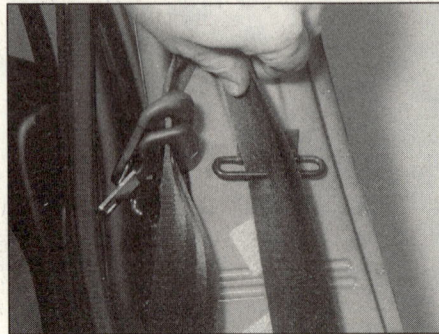

22.8a Feed the seat belt webbing through the bracket . . .

22.8b . . . then unbolt the inertia reel assembly

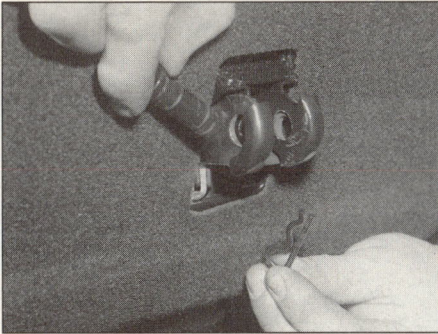

22.11 Pull off the circlip, and pull out the seat belt pivot pin . . .

22.12a . . . then pull out the inertia reel . . .

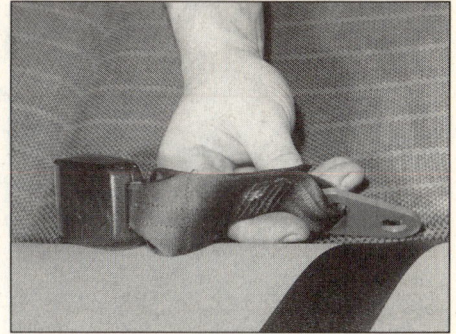

22.12b . . . or the buckle

Refitting

9 Refitting is a reversal of removal, but make sure that the seat belt anchor bolts are tightened securely, and refit the front seat with reference to Section 21.

Rear seat belt buckles and centre inertia reel

Removal

10 The rear seat belt buckles and centre inertia reel are attached to the rear seat cushion.

11 Fold the rear seat cushion forwards then, working under the seat cushion, pull off the large circlip, and pull out the seat belt pivot pin **(see illustration)**.

12 Pull the inertia reel, or buckle, as applicable, out through the top of the seat cushion, noting its orientation **(see illustrations)**.

Refitting

13 Refitting is a reversal of removal, but make sure that the components are correctly refitted to the seat cushion as noted before removal, and ensure that the circlip is securely refitted.

Rear side seat belt assembly – Hatchback models

Removal

14 Fold the rear seat cushion forwards, then

pull the carpet trim panel away from the seat belt lower anchor bolt, and unscrew the anchor bolt **(see illustration)**.

15 Remove the parcel shelf side support panel as described in Section 23.

16 Unscrew the inertia reel securing bolt, then withdraw the seat belt assembly **(see illustration)**.

Refitting

17 Refitting is a reversal of removal, but tighten the seat belt anchor bolt securely.

Rear side seat belt assembly – Estate models

Removal

18 Fold the rear seat cushion forwards, then pull the carpet trim panel away from the seat belt lower anchor bolt, and unscrew the anchor bolt.

19 Unclip the luggage compartment cover and the rear parcel shelf, and remove them from the vehicle.

20 Remove the screws securing the relevant parcel shelf support panel to the side of the luggage compartment.

21 Pull the parcel shelf support panel upwards to release the securing clips, then withdraw the panel and disconnect the loudspeaker wiring plug **(see illustration)**.

22 Unscrew the inertia reel securing bolt, then withdraw the seat belt assembly **(see illustration)**.

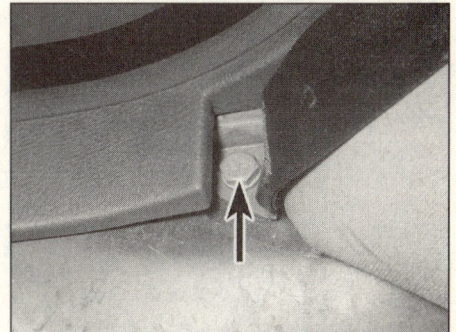

22.14 Unscrew the lower rear side seat belt lower anchor bolt (arrowed)

Refitting

23 Refitting is a reversal of removal, but tighten the seat belt anchor bolt securely.

23 Interior trim – removal and refitting

Steering column shrouds – models up to 1994

Removal

1 Using a screwdriver, release the securing clip, and remove the cover from the battery, then disconnect the battery negative lead.

22.16 Unscrewing a rear side seat belt inertia reel securing bolt – Hatchback model

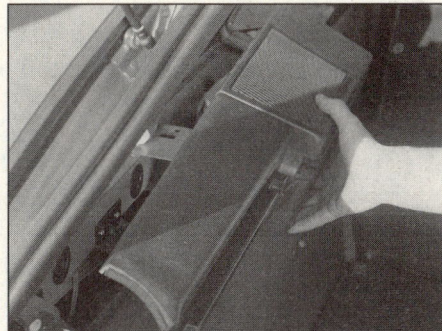

22.21 Removing the parcel shelf support panel – Estate model

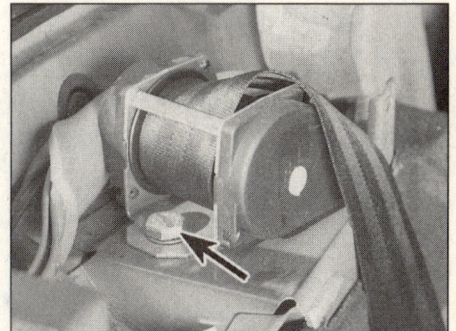

22.22 Rear side seat belt inertia reel securing bolt (arrowed) – Estate model

12

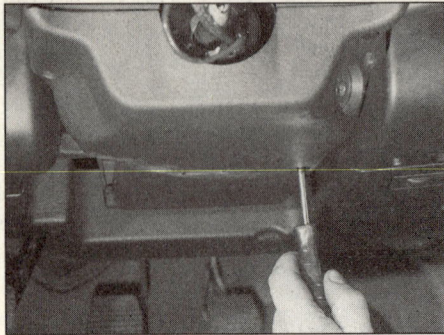

23.14a Unscrew the five securing screws . . .

23.14b . . . then unclip the illumination bulb and remove the lower steering column shroud (viewed with steering wheel removed for clarity)

23.17 Depress the securing tabs and remove the rotary switch assembly – models from 1995

2 Working under the lower steering column shroud, unscrew the five screws securing the shroud, then lower the lower shroud from the steering column. Disconnect the wiring plug from the lighting switch (mounted in the shroud) and, where applicable, unclip the steering lock illumination bulb from the lower shroud. Remove the shroud.

3 Working at each side of the instrument panel in turn, prise out the cover plate, and unscrew the instrument panel surround securing screw.

4 Withdraw the Instrument panel surround from the facia.

5 Carefully pull the knobs from the heater/ventilation controls.

6 Unscrew the two securing screws (located in the blower motor and air temperature control apertures), and remove the heater/ventilation control unit trim panel.

7 Unscrew the four screws securing the heater/ventilation control unit to the facia.

8 Move the adjuster to fully lower the steering column.

9 Working underneath the facia, on either side of the steering column, unscrew the two lower switch/heater control/display trim panel securing screws.

10 Working at the front of the instrument panel aperture, unscrew the two upper switch/heater control/display trim panel securing screws.

11 Pull the top of the switch/heater control/display trim panel forwards from the facia to release the securing clips then, working at the rear of the panel, disconnect the wiring plugs from the panel-mounted components, noting the locations of the plugs, and withdraw the panel.

12 Carefully manipulate the upper steering column shroud out and withdraw it from the steering column.

Refitting

13 Refitting is a reversal of removal.

Steering column shrouds – models from 1985-on

Removal

Note: *During this procedure, take note of the routing of all wiring to aid correct refitting.*

14 Working under the lower steering column shroud, unscrew the five screws securing the shroud, then lower the lower shroud from the steering column. Where applicable, unclip the steering lock illumination bulb from the lower shroud, and remove shroud (see illustrations).

15 Remove the steering wheel as described in Chapter 11.

16 Move the steering column height adjuster to fully lower the steering column.

17 Where applicable, depress the three securing tabs, then pull the rotary switch assembly from the centre of the steering

column – take care not to strain the wiring (see illustration).

18 Unscrew the three screws securing the stalk switch assembly to the steering column, then lift up the upper column shroud as far as possible (see illustration).

19 Remove the screw securing the wiring connector block to the bottom of the stalk switch assembly (see illustration).

20 Release the securing clips and disconnect the wiring plugs from the rear of the stalk switches.

21 Pull the stalk switch assembly forwards, complete with the steering wheel position sensor, where applicable, then disconnect wiring connector halves at the block under the column (which was secured by the previously removed screw – see paragraph 19).

22 Reach up under the upper steering column shroud, and cut the cable-ties securing the wiring harnesses to the upper column shroud.

23 Carefully manipulate the upper steering column shroud out and withdraw it from the steering column (see illustration).

Refitting

24 Refitting is a reversal of removal, bearing in mind the following points.

a) *Ensure that all wiring is correctly reconnected and routed, as noted before removal.*

23.18 Unscrew the three screws (arrowed) securing the stalk switch assembly – models from 1995

23.19 Remove the screw securing the wiring connector block – models from 1995

23.23 Removing the upper steering column shroud – models from 1995

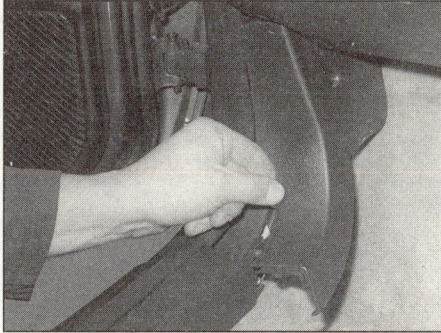
23.27 Removing the footwell side trim panel

23.28a Unscrew the securing screws . . .

23.28b . . . and remove the sill trim panel

b) *Refit the steering wheel as described in Chapter 11.*

Sill trim panel
Removal
25 Remove the relevant front seat, as described in Section 21.

26 Open the front and rear doors, and carefully prise the weatherstrips from the sill trim panel.

27 Remove the securing screw, and withdraw the relevant footwell side trim panel **(see illustration)**.

28 Unscrew the sill trim panel securing screws and, where applicable, unbolt the bonnet release lever from the sill, then withdraw the sill trim panel **(see illustrations)**.

Refitting
29 Refitting is a reversal of removal, but refit the front seat with reference to Section 21.

Rear parcel shelf side support panel – Hatchback models
Removal
30 Using a screwdriver, release the securing clip, and remove the cover from the battery, then disconnect the battery negative lead.
31 Fold the rear seat cushion forwards.
32 Working in the luggage compartment, pull the carpet trim panel away from the relevant side of the luggage compartment.
33 Pull off the trim covers, and unscrew the

two tie-down brackets from the side of the luggage compartment **(see illustration)**.
34 Pull out the cover plate, then unbolt the seat back catch striker from the side of the luggage compartment **(see illustrations)**.
35 Unscrew the securing screw, and remove the air vent from the side of the luggage compartment side trim panel **(see illustration)**.
36 Release the upper clip, and remove the luggage compartment side trim panel **(see illustration)**. Where applicable, disconnect the wiring from the luggage compartment light as the panel is withdrawn.
37 Carefully unclip the loudspeaker cover from the parcel shelf side support panel **(see illustration)**.

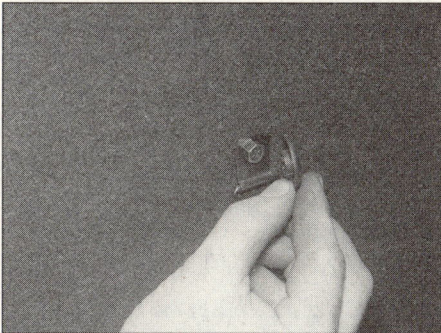
23.33 Pull off the covers to expose the tie-down bracket securing bolts

23.34a Pull out the cover plate . . .

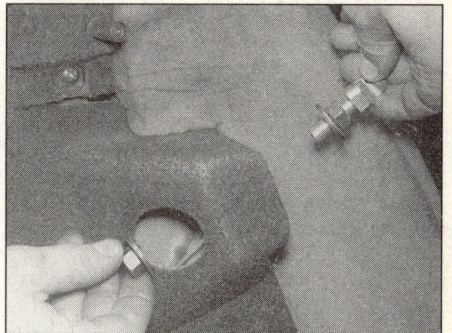
23.34b . . . then unbolt the seat back catch striker

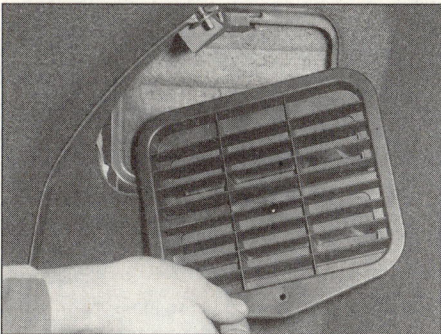
23.35 Unscrew the securing screw and remove the air vent

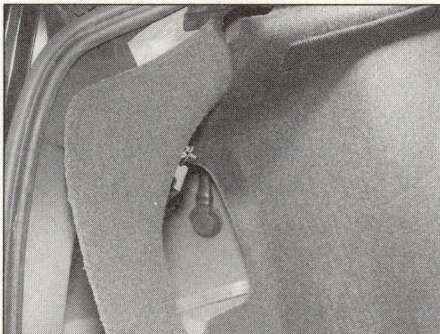
23.36 Remove the luggage compartment side trim panel

23.37 Unclip the loudspeaker cover

12

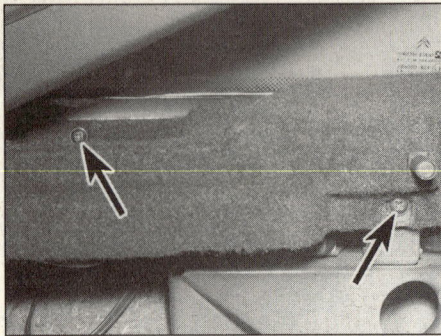

23.38a Remove the two screws (arrowed) . . .

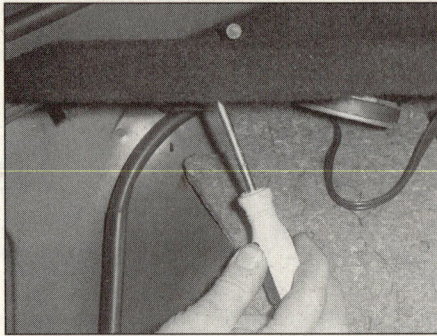

23.38b . . . and the securing clip . . .

23.38c . . . and lift out the parcel shelf side support panel

38 Remove the two screws and the securing clip, and lift out the parcel shelf side support panel **(see illustrations)**. Disconnect the wiring from the loudspeaker as the panel is withdrawn.

Refitting

39 Refitting is a reversal of removal.

Rear parcel shelf side support panel – Estate models

Removal

40 Unclip the luggage compartment cover and the rear parcel shelf, and remove them from the vehicle.
41 Remove the screws securing the relevant parcel shelf support panel to the side of the luggage compartment.

42 Pull the parcel shelf support panel upwards to release the securing clips, then withdraw the panel and disconnect the loudspeaker wiring plug.

Refitting

43 Refitting is a reversal of removal.

Rear passenger compartment glass shield

Removal

44 Open the tailgate, then hold the glass panel in the raised position.
45 Carefully tap out the panel hinge pins using a suitable punch, then unhook the panel support cord, and lift out the panel.

Refitting

46 Refitting is a reversal of removal, but ensure that the hinge pins are tapped securely into position.

24 Centre console – removal and refitting

Removal

1 Using a screwdriver, release the securing clip, and remove the cover from the battery, then disconnect the battery negative lead.
2 Working on one side of the centre console, remove the side trim panel as follows **(see illustrations)**.
a) *Unscrew the two screws securing the front air vent, and pull out the air vent.*
b) *Unscrew the screw securing the rear air vent, and withdraw the vent.*
c) *Working through the slit in the carpet trim, unscrew the side trim panel rear securing nut.*
d) *Withdraw the trim panel.*
3 Repeat the procedure to remove the side trim panel on the remaining side of the centre console.
4 Working at the front of the centre console, unscrew the screw on each side securing the heater ducts to the console **(see illustration)**.

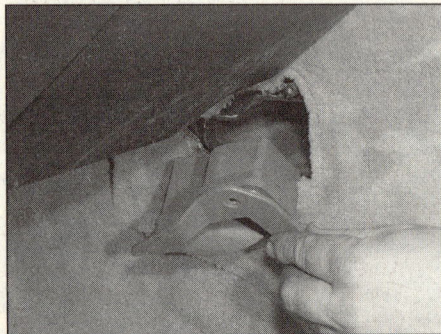

24.2a Remove the front air vent . . .

24.2b . . . unscrew the screw (arrowed) and remove the rear air vent . . .

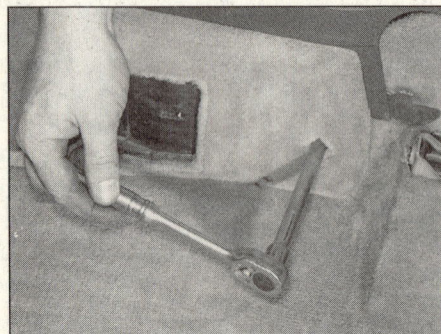

24.2c . . . then unscrew the side trim panel rear securing nut . . .

24.2d . . . and withdraw the trim panel

24.4 Unscrewing a heater duct securing screw

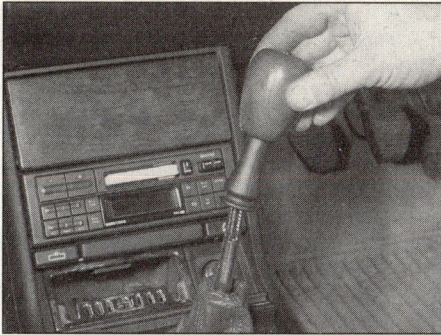

24.5 Pulling off the gear lever selector knob

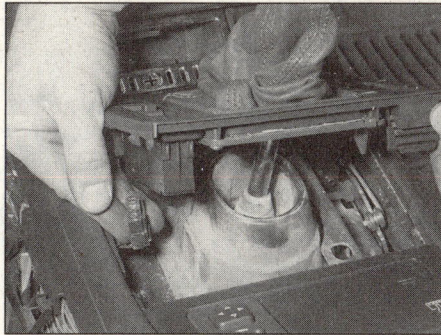

24.7 Unclip the selector lever surround from the console

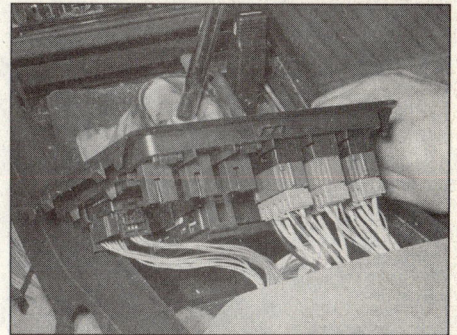

24.9 Prise the anti-theft alarm key pad from the console and disconnect the wiring plugs

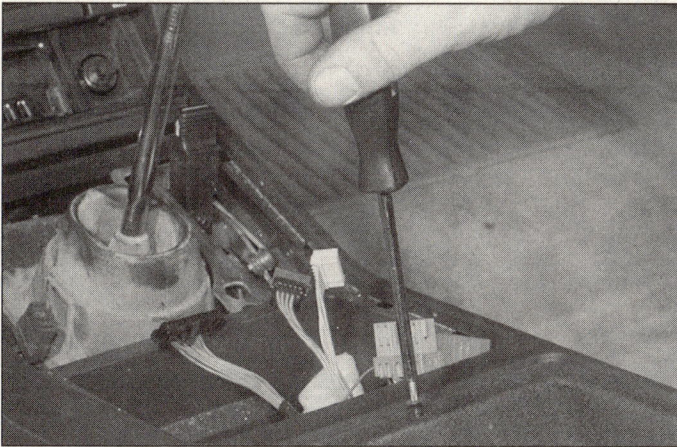

24.10a Unscrew the securing screws . . .

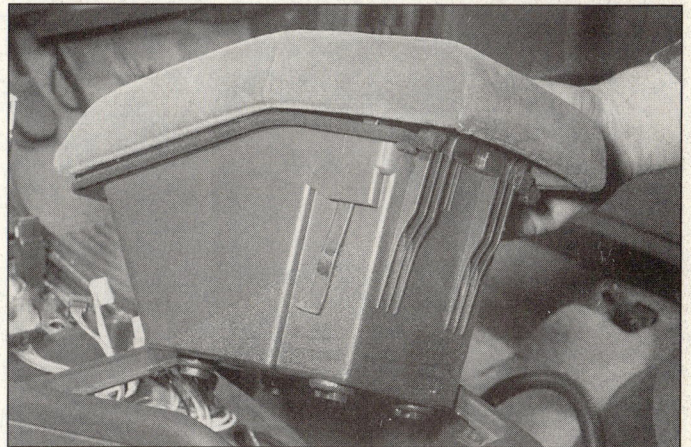

24.10b . . . and withdraw the armrest/storage tray assembly

5 On models with manual transmission, pull off the gear selector lever knob **(see illustration)**.

6 On models with automatic transmission, proceed as follows.

a) Unclip the top of the gear selector lever gaiter from the lever, and push the gaiter down to expose the gear selector lever knob securing screws.

b) Unscrew the two gear lever knob securing screws.

c) Lift the selector lever knob by approximately 10 mm, then rotate the knob through a quarter-turn anti-clockwise, and lift by approximately 7 mm.

d) Press the button on the side of the selector lever knob, then rotate the knob back through a quarter-turn clockwise, and lift off the knob.

7 Unclip the selector lever surround from the console, complete with the suspension control switch(es). Where applicable, disconnect the wiring plug from the Normal/Sport switch **(see illustration)**.

8 Slide the surround (complete with the rubber gaiter) up the gear selector lever, and withdraw the assembly.

9 Where applicable, carefully prise the trip computer or anti-theft alarm key pad/mirror switch/electric window switch panel from the centre console, and disconnect the wiring plugs from the components in the panel **(see illustration)**. Alternatively, prise the blanking plate from the centre console.

10 Open the lid of the armrest/storage tray assembly, then unscrew the three securing screws, and withdraw the assembly from the centre console **(see illustrations)**.

11 Working at the rear of the centre console, pull out the ashtray, then unclip the ashtray surround. Disconnect the wiring from the cigarette lighter, the ashtray illumination bulb and, where applicable, the switches mounted in the ashtray surround, then withdraw the surround **(see illustration)**.

12 Where applicable, disconnect the wiring plug from the relay attached to the rear of the centre console.

13 Disconnect the two earth connectors located at the bottom of the ashtray aperture.

14 Unscrew the four securing nuts, two accessible through the ashtray aperture, and two through the trip computer or anti-theft alarm key pad/mirror switch/electric window switch panel aperture (or blanking plate aperture) **(see illustrations)**.

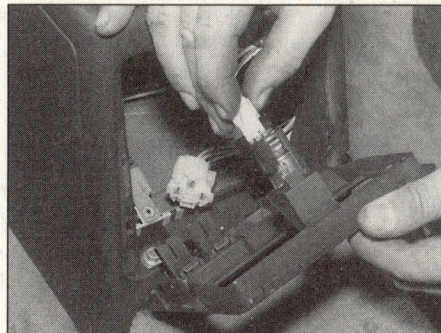

24.11 Unclip the rear ashtray surround and disconnect the wiring

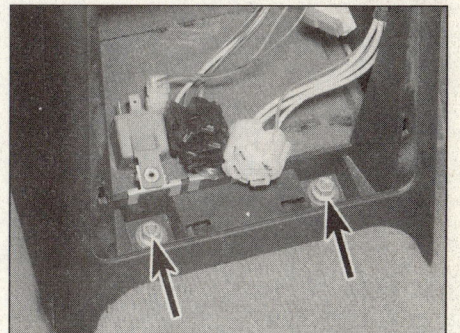

24.14a Unscrew the two nuts (arrowed) from the rear ashtray aperture . . .

12

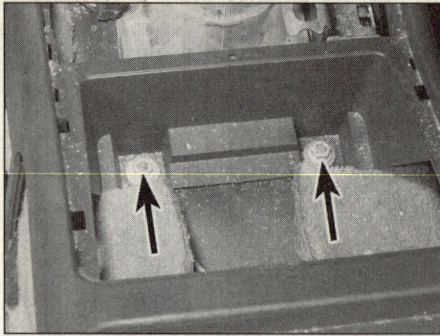
24.14b . . . and the two nuts (arrowed) from the anti-theft alarm key pad aperture

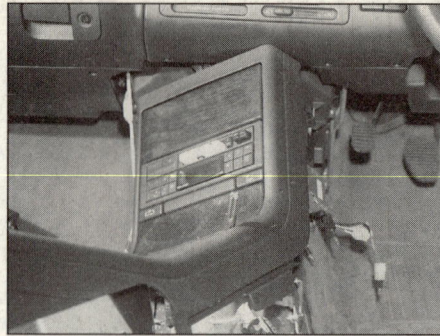
24.18 Withdrawing the centre console

15 Unclip the heater ducts from the front of the console.

16 Carefully pull the console assembly rearwards, and lift the assembly up over the suspension height control lever.

17 Working at the front of the console, disconnect the wiring plugs and the aerial lead from the rear of the radio/cassette player, and disconnect the wiring plugs from the switch illumination bulbs, ashtray and cigarette lighter, as applicable.

18 Withdraw the centre console from the vehicle **(see illustration)**.

Refitting

19 Refitting is a reversal of removal, but ensure that all wiring plugs are correctly reconnected.

Chapter 13
Body electrical system

Contents

Degrees of difficulty

Easy, suitable for novice with little experience	Fairly easy, suitable for beginner with some experience	Fairly difficult, suitable for competent DIY mechanic	Difficult, suitable for experienced DIY mechanic	Very difficult, suitable for expert DIY or professional

Specifications

General

System type . . . . 12-volt negative earth

Fuses

Refer to wiring diagrams

Bulbs

	Type	Wattage
Headlight:		
Main beam	H1	55
Dipped beam	H1	55
Front sidelight	Push-fit	5
Front direction indicator light	Bayonet-fit	21
Front direction indicator side repeater light	Push-fit	5
Front foglight	H3	55
Tail light	Bayonet-fit	5
Stop light	Bayonet-fit	21
Rear direction indicator light	Bayonet-fit	21
Reversing light	Bayonet-fit	21
Rear foglight	Bayonet-fit	21
Number plate light	Push-fit	5
Kerb lights	Festoon	10
Courtesy light	Push-fit	5
Map reading lights	Push-fit	5
Luggage compartment light	Push-fit	5
Glovebox light	Push-fit	5
Vanity mirror light	Festoon	3

13

1 General information and precautions

⚠️ **Warning: Before carrying out any work on the electrical system, read through the precautions given in Safety first! at the beginning of this manual, and in Chapter 5.**

The electrical system is of 12-volt negative earth type. Power for the lights and all electrical accessories is supplied by a lead/acid type battery, which is charged by the alternator.

This Chapter covers repair and service procedures for the various electrical components not associated with engine. Information on the battery, alternator and starter motor can be found in Chapter 5A.

It should be noted that, prior to working on any component in the electrical system, the battery negative terminal should first be disconnected, to prevent the possibility of electrical short-circuits and/or fires.

Caution: Before disconnecting the battery, refer to the information given in the preliminary Sections of this manual.

2 Electrical fault finding – general information

Note: *Refer to the precautions given in Safety first! and at the beginning of Chapter 5A before starting work. The following tests relate to testing of the main electrical circuits, and should not be used to test delicate electronic circuits (such as anti-lock braking systems), particularly where an electronic control module is used.*

General

1 A typical electrical circuit consists of an electrical component, any switches, relays, motors, fuses, fusible links or circuit breakers related to that component, and the wiring and connectors which link the component to both the battery and the chassis. To help to pinpoint a problem in an electrical circuit, wiring diagrams are included at the end of this manual.

2 Before attempting to diagnose an electrical fault, first study the appropriate wiring diagram, to obtain a more complete understanding of the components included in the particular circuit concerned. The possible sources of a fault can be narrowed down by noting whether other components related to the circuit are operating properly. If several components or circuits fail at one time, the problem is likely to be related to a shared fuse or earth connection.

3 Electrical problems usually stem from simple causes, such as loose or corroded connections, a faulty earth connection, a blown fuse, a melted fusible link, or a faulty relay (refer to Section 3 for details of testing relays). Visually inspect the condition of all fuses, wires and connections in a problem circuit before testing the components. Use the wiring diagrams to determine which terminal connections will need to be checked, in order to pinpoint the trouble-spot.

4 The basic tools required for electrical fault-finding include a circuit tester or voltmeter (a 12-volt bulb with a set of test leads can also be used for certain tests); a self-powered test light (sometimes known as a continuity tester); an ohmmeter (to measure resistance); a battery and set of test leads; and a jumper wire, preferably with a circuit breaker or fuse incorporated, which can be used to bypass suspect wires or electrical components. Before attempting to locate a problem with test instruments, use the wiring diagram to determine where to make the connections.

5 To find the source of an intermittent wiring fault (usually due to a poor or dirty connection, or damaged wiring insulation), a 'wiggle' test can be performed on the wiring. This involves wiggling the wiring by hand, to see if the fault occurs as the wiring is moved. It should be possible to narrow down the source of the fault to a particular section of wiring. This method of testing can be used in conjunction with any of the tests described in the following sub-Sections.

6 Apart from problems due to poor connections, two basic types of fault can occur in an electrical circuit - open-circuit, or short-circuit.

7 Open-circuit faults are caused by a break somewhere in the circuit, which prevents current from flowing. An open-circuit fault will prevent a component from working, but will not cause the relevant circuit fuse to blow.

8 Short-circuit faults are caused by a 'short' somewhere in the circuit, which allows the current flowing in the circuit to 'escape' along an alternative route, usually to earth. Short-circuit faults are normally caused by a breakdown in wiring insulation, which allows a feed wire to touch either another wire, or an earthed component such as the bodyshell. A short-circuit fault will normally cause the relevant circuit fuse to blow.

Finding an open-circuit

9 To check for an open-circuit, connect one lead of a circuit tester or voltmeter to either the negative battery terminal or a known good earth.

10 Connect the other lead to a connector in the circuit being tested, preferably nearest to the battery or fuse.

11 Switch on the circuit, bearing in mind that some circuits are live only when the ignition switch is moved to a particular position.

12 If voltage is present (indicated either by the tester bulb lighting or a voltmeter reading, as applicable), this means that the section of the circuit between the relevant connector and the battery is problem-free.

13 Continue to check the remainder of the circuit in the same fashion.

14 When a point is reached at which no voltage is present, the problem must lie between that point and the previous test point with voltage. Most problems can be traced to a broken, corroded or loose connection.

Finding a short-circuit

15 To check for a short-circuit, first disconnect the load(s) from the circuit (loads are the components which draw current from a circuit, such as bulbs, motors, heating elements, etc).

16 Remove the relevant fuse from the circuit, and connect a circuit tester or voltmeter to the fuse connections.

17 Switch on the circuit, bearing in mind that some circuits are live only when the ignition switch is moved to a particular position.

18 If voltage is present (indicated either by the tester bulb lighting or a voltmeter reading, as applicable), this means that there is a short-circuit.

19 If no voltage is present, but the fuse still blows with the load(s) connected, this indicates an internal fault in the load(s).

Finding an earth fault

20 The battery negative terminal is connected to 'earth' - the metal of the engine/transmission unit and the car body - and most systems are wired so that they only receive a positive feed, the current returning via the metal of the car body. This means that the component mounting and the body form part of that circuit. Loose or corroded mountings can therefore cause a range of electrical faults, ranging from total failure of a circuit, to a puzzling partial fault. In particular, lights may shine dimly (especially when another circuit sharing the same earth point is in operation), motors (eg wiper motors or the radiator cooling fan motor) may run slowly, and the operation of one circuit may have an apparently-unrelated effect on another. Note that on many vehicles, earth straps are used between certain components, such as the engine/transmission and the body, usually where there is no metal-to-metal contact between components, due to flexible rubber mountings, etc.

21 To check whether a component is properly earthed, disconnect the battery, and connect one lead of an ohmmeter to a known good earth point. Connect the other lead to the wire or earth connection being tested. The resistance reading should be zero; if not, check the connection as follows.

22 If an earth connection is thought to be faulty, dismantle the connection, and clean back to bare metal both the bodyshell and the wire terminal or the component earth connection mating surface. Be careful to remove all traces of dirt and corrosion, then use a knife to trim away any paint, so that a clean metal-to-metal joint is made. On reassembly, tighten the joint fasteners securely; if a wire terminal is being refitted, use serrated washers between the terminal and the bodyshell, to ensure a clean and secure connection. When the connection is remade, prevent the onset of corrosion in the future by applying a coat of petroleum jelly or silicone-based grease, or by spraying on (at regular intervals) a proprietary ignition sealer.

3 Fuses and relays –
general information

Fuses

1 Fuses are designed to break a circuit when a predetermined current is reached, in order to protect the components and wiring which could be damaged by excessive current flow. Any excessive current flow will be due to a fault in the circuit, usually a short-circuit (see Section 2).

2 The main fuses are located in the fusebox, below the left-hand side of the facia.

3 For access to the fuses, open the fusebox cover **(see illustration)**.

4 A blown fuse can be recognised from its melted or broken wire.

5 To remove a fuse, first ensure that the relevant circuit is switched off.

6 Using the plastic tool provided, pull the fuse from its location **(see illustration)**.

7 Before renewing a blown fuse, trace and rectify the cause, and always use a fuse of the correct rating. Never substitute a fuse of a higher rating, or make temporary repairs using wire or metal foil; more serious damage, or even fire, could result.

8 Note that the fuses are colour-coded as follows. Refer to the wiring diagrams for details of the circuits protected.

Colour	Rating
Orange	5A
Red	10A
Blue	15A
Yellow	20A
Clear or white	25A
Green	30A

9 Additional fuses may be located in the auxiliary fusebox, at the front left-hand corner of the engine compartment, next to the battery **(see illustration)**.

Relays

10 A relay is an electrically-operated switch, which is used for the following reasons:
a) A relay can switch a heavy current remotely from the circuit in which the current is flowing, allowing the use of lighter-gauge wiring and switch contacts.
b) A relay can receive more than one control input, unlike a mechanical switch.
c) A relay can have a timer function - for example, the intermittent wiper relay.

11 The relays are positioned in various locations around the vehicle, according to model and equipment fitted.

3.3 Fusebox cover opened to reveal fuses – fuse removal tool arrowed

12 If a circuit or system controlled by a relay develops a fault, and the relay is suspect, operate the system. If the relay is functioning, it should be possible to hear it 'click' as it is energised. If this is the case, the fault lies with the components or wiring of the system. If the relay is not being energised, then either the relay is not receiving a main supply or a switching voltage, or the relay itself is faulty. Testing is by the substitution of a known good unit, but be careful - while some relays are identical in appearance and in operation, others look similar but perform different functions.

13 To remove a relay, first ensure that the relevant circuit is switched off. The relay can then simply be pulled out from the socket, and pushed back into position **(see illustration)**.

4 Switches –
removal and refitting

Note: *Before removing any switch, using a screwdriver, release the securing clip, and remove the cover from the battery, then disconnect the battery negative lead. Reconnect the lead after refitting the switch.*

Ignition switch/steering column lock

1 Refer to Chapter 11.

3.9 Opening the auxiliary fusebox cover in the engine compartment

3.6 Using the plastic tool to remove a fuse

Steering column combination switch assembly

Removal

2 Remove the lower steering column shroud, as described in Chapter 12.

3 Lift the upper steering column shroud as necessary to expose the two stalk switch securing screws.

4 Unscrew the two switch securing screws, then withdraw the switch from the steering column, and disconnect the wiring connector.

Refitting

5 Refitting is a reversal of removal.

Lighting switch – models up to 1994

Removal

6 Remove the lower steering column shroud, as described in Chapter 12.

7 Carefully prise the switch from the shroud.

Refitting

8 Refitting is a reversal of removal.

Facia and centre console-mounted switches

Removal

9 Carefully prise the relevant switch from the facia or console, as applicable. If necessary, use a small screwdriver, but take care not to

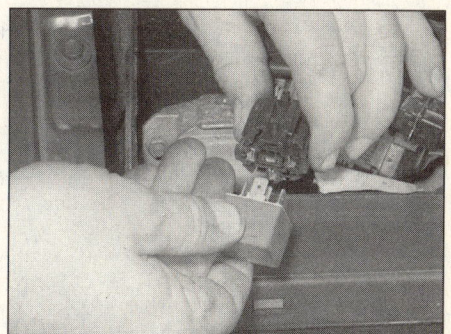

3.13 Removing a relay from the electronic control unit housing in the engine compartment

13

4.9 Removing a facia-mounted switch

4.15 Disconnecting a wiring plug from a steering wheel-mounted switch

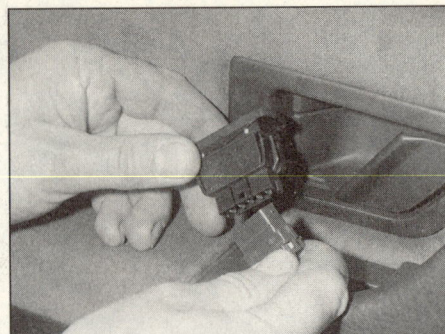
4.17 Removing an electric window switch from the door

damage the surrounding panels (see illustration).

10 Withdraw the switch and disconnect the wiring plug(s).

Refitting

11 Reconnect the wiring plug(s) and push the switch into position until the securing clips engage.

Steering wheel-mounted switches

Removal

12 On models without an air bag, carefully prise the centre pad from the steering wheel.

13 On models with an air bag, remove the air bag as described in Section 21.

14 Unscrew the two securing screws, and withdraw the relevant switch from the steering wheel.

15 Carefully disconnect the wiring connectors from the rear of the switch, and withdraw the switch assembly (see illustration).

Refitting

16 Refitting is a reversal of removal.

Door-mounted electric window switches

17 Proceed as described previously in this Section for the facia and centre console-mounted switches (see illustration).

Heater blower motor switch

18 The heater blower motor switch is integral with the heater control panel, and cannot be renewed separately. Details of heater control panel removal and refitting are given in Chapter 3.

Sunroof switch

Removal

19 Carefully prise the lens from the courtesy light, for access to the four roof console securing screws (see illustration).

20 Unscrew the securing screws, and lower the roof console, taking care not to strain the wiring connectors.

21 Disconnect the sunroof switch wiring connector, then push the switch out through the front of the roof console (see illustration).

Refitting

22 Refitting is a reversal of removal.

Stop light switch and parking brake 'on' warning light switch

23 Refer to Chapter 10.

Courtesy light switches

Body-mounted switch

24 The front door courtesy light switches are mounted in the front sill panels, and the rear door courtesy light switches are mounted in

the body centre pillars. Carefully prise the grommet from the switch.

25 Unscrew the securing screw, then withdraw the switch from the body panel, and disconnect the wiring. It is advisable to tape the wiring to the body panel to prevent it from falling into the body aperture.

26 Refitting is a reversal of removal, but make sure that the grommet is correctly fitted.

Light-mounted switches

27 The light-mounted switches are integral with the lights, and cannot be renewed separately.

Glovebox light switch

28 Proceed as described previously in this Section for the facia and centre console-mounted switches (see illustration).

Luggage compartment light switch – Hatchback models

Removal

29 The switch is integral with the luggage compartment rear trim panel. If the switch is damaged, the complete trim panel must be renewed.

30 To remove the trim panel, open the tailgate, then prise out the cover plugs, and unscrew the four screws securing the luggage compartment rear trim panel. Lift off the panel, and disconnect the wiring from the luggage compartment light switch.

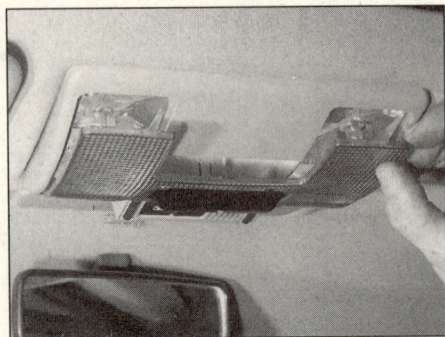
4.19 Prising the lens from the courtesy light

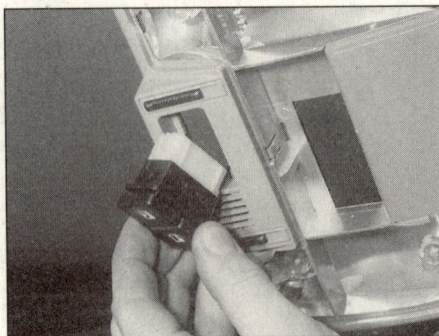
4.21 Removing the sunroof switch

4.28 Removing the glovebox light switch

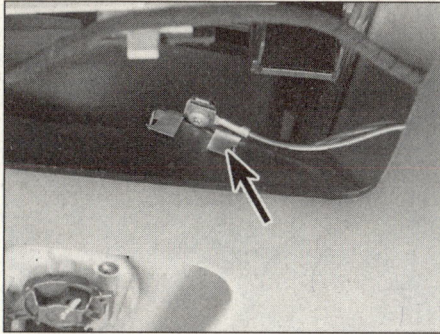
4.36 Luggage compartment light switch location (arrowed) – Estate model

5.3 Remove the headlight bulb cover . . .

5.4 . . . then disconnect the wiring from the bulb . . .

Refitting

31 Refitting is a reversal of removal.

Luggage compartment light switch – Estate models

Removal

32 The luggage compartment lights on Estate models are operated by a tilt-switch, mounted behind the tailgate trim panel.
33 Remove the securing screws, and withdraw the handle from the tailgate trim panel.
34 Work around the trim panel and remove the securing screws, then withdraw the trim panel.
35 The luggage compartment light switch is mounted on a bracket clipped to the right-hand side of the tailgate.
36 To remove the switch, unclip the bracket from the tailgate, then remove the securing screw, and disconnect the switch wiring **(see illustration)**.

Refitting

37 Refitting is a reversal of removal.

5 Bulbs (exterior lights) – renewal

1 Whenever a bulb is renewed, note the following points.
a) Using a screwdriver, release the securing clip, and remove the cover from the battery, then disconnect the battery negative lead.
b) Remember that, if the light has just been in use, the bulb may be extremely hot.
c) Always check the bulb contacts and holder, ensuring that there is clean metal-to metal contact between the bulb and its live(s) and earth. Clean off any corrosion or dirt before fitting a new bulb.
d) Wherever bayonet-type bulbs are fitted (see Specifications), ensure that the live contact(s) bear firmly against the bulb contact.
e) Always ensure that the new bulb is of the correct rating, and that it is completely clean before fitting it; this applies particularly to headlight/foglight bulbs (see below).

Headlight

2 Separate main and dipped beam bulbs are fitted to the headlights.
3 To renew a bulb, working in the engine compartment, at the rear of the light unit, twist the relevant bulb cover approximately an eighth-of-a-turn anticlockwise and remove it from the rear of the light **(see illustration)**.
4 Disconnect the wiring from the rear of the bulb **(see illustration)**.
5 Release the spring clip by compressing its ends, then withdraw the relevant bulb **(see illustrations)**.
6 When handling the new bulb, use a tissue or clean cloth, to avoid touching the glass with the fingers; moisture and grease from the skin can cause blackening and rapid failure of this type of bulb. If the glass is accidentally touched, wipe it clean using methylated spirit.
7 Fit the new bulb, ensuring that its locating tabs are correctly seated in the light cut-outs.
8 Secure the bulb in position with the spring clip, and reconnect the wiring plug, then refit the cover.

Front sidelight

9 Working in the engine compartment, at the rear of the sidelight assembly, press the securing tab to release the assembly, then push the assembly out from the front of the wing panel **(see illustrations)**.
10 Twist the bulbholder anticlockwise, and remove it from the rear of the light unit.

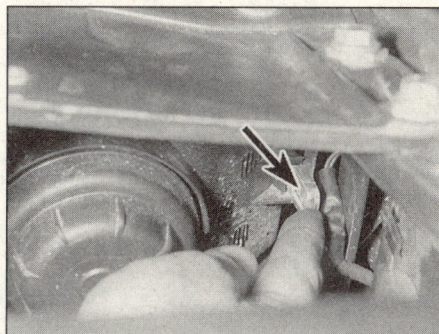
5.5a . . . release the spring clip . . .

5.5b . . . and withdraw the bulb

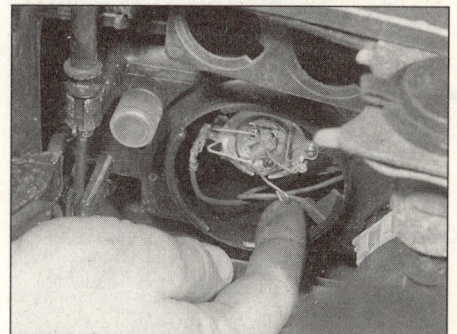
5.9a Press the securing tab (arrowed) . . .

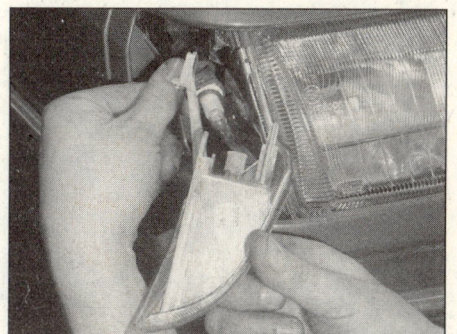
5.9b . . . then push the sidelight assembly from the wing panel

13

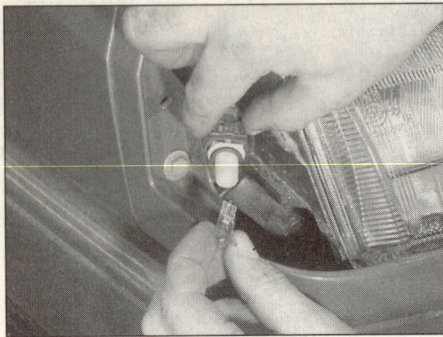
5.11 The bulb is a push-fit in the bulbholder

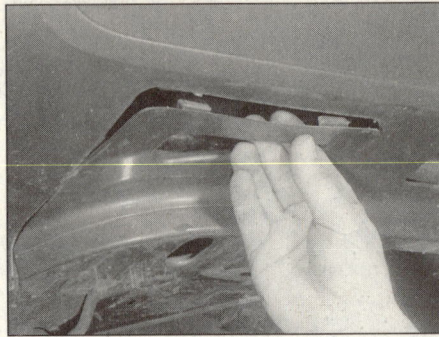
5.13 Unclip the cover for access to the spoiler-mounted lights (later model shown)

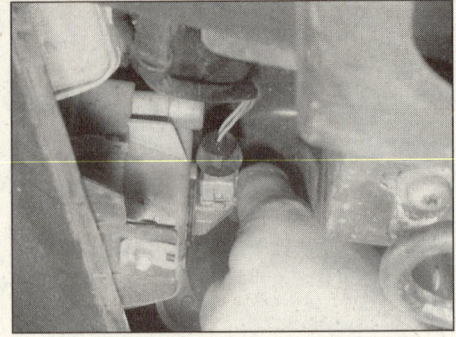
5.14 Twist the direction indicator bulbholder anti-clockwise . . .

11 The bulb is a push-fit in the bulbholder (see illustration).
12 Fit the new bulb using a reversal of the removal procedure, then refit the light unit, ensuring that the locating lugs on the light unit engage correctly with the headlight and body panel.

Front direction indicator light

13 Working underneath the front spoiler, unclip the cover for access to the spoiler-mounted lights (see illustration).
14 Twist the direction indicator bulbholder anti-clockwise and remove it from the light (see illustration).
15 The bulb is a bayonet-fit in the bulbholder (see illustration).

16 Fit the new bulb using a reversal of the removal procedure.

Front foglight

17 Working underneath the front spoiler, unclip the cover for access to the spoiler-mounted lights.
18 Release the securing clip and lower the cover from the bottom of the foglight (see illustration).
19 Release the spring clip, and withdraw the bulb from the light unit (see illustrations).
20 Release the clips, and open the wiring connector cover, then separate the two halves of the wiring connector and withdraw the bulb (see illustration).

21 When handling the new bulb, use a tissue or clean cloth, to avoid touching the glass with the fingers; moisture and grease from the skin can cause blackening and rapid failure of this type of bulb. If the glass is accidentally touched, wipe it clean using methylated spirit.
22 Fit the new bulb, ensuring that its locating tabs are correctly seated in the light cut-outs.
23 Further refitting is a reversal of removal.

Front direction indicator side repeater light

24 Push the light unit towards the front of the vehicle, then twist the bulbholder

5.15 . . . then remove the bulb from the bulbholder

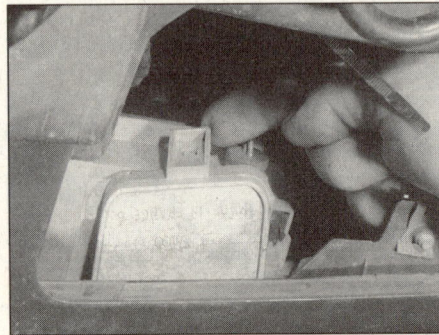
5.18 Release the clip and lower the foglight cover

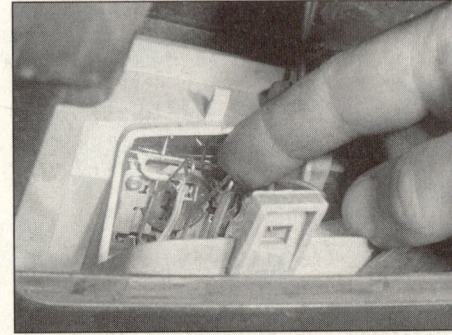
5.19a Release the spring clip . . .

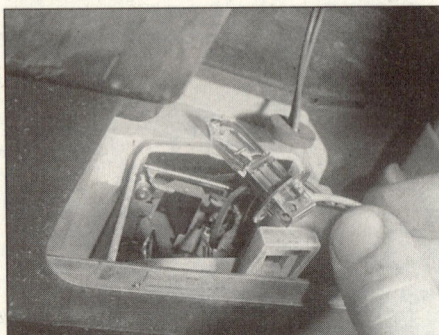
5.19b . . . and withdraw the bulb

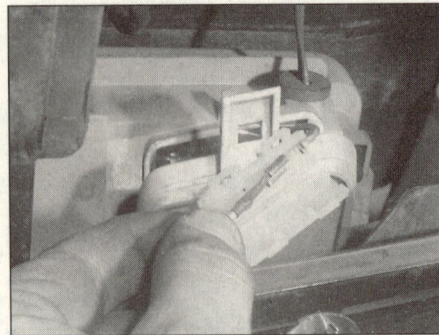
5.20 Release the clips and open the wiring connector cover

5.24a Push the light unit towards the front of the vehicle . . .

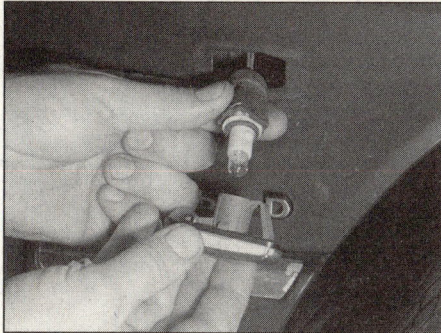

5.24b . . . then twist the bulbholder anti-clockwise

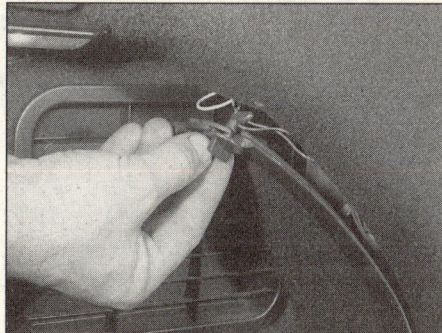

5.27 Unclip the rear wing-mounted light cover – Hatchback model

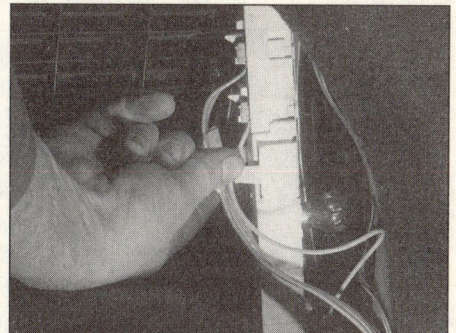

5.28 Squeeze the two lugs and remove the bulbholder – Hatchback model

anticlockwise to remove it from the light unit **(see illustration)**.
25 The bulb is a push-fit in the bulbholder.
26 Fit the new bulb using a reversal of the removal procedure.

Rear lights – Hatchback models
Rear wing-mounted lights
27 Working in the luggage compartment, unclip the light unit cover **(see illustration)**.
28 Squeeze the two securing lugs together, and remove the bulbholder from the light unit **(see illustration)**.
29 The bulbs are a bayonet-fit in the bulbholder **(see illustration)**.
30 Fit the new bulb using a reversal of the removal procedure.

Tailgate-mounted lights
31 Open the tailgate, then depress the securing lugs, and remove the bulb cover **(see illustration)**.
32 To remove the sidelight or foglight bulb, squeeze the securing tabs, and pull the bulbholder from the light unit. The bulbs are a bayonet-fit in the bulbholder **(see illustrations)**.
33 To remove the reversing light bulb, pull out the foam pad to reveal the bulbholder, then twist the bulbholder anticlockwise, and remove it from the light unit. The bulb is a bayonet-fit in the bulbholder.
34 Fit the new bulb using a reversal of the removal procedure.

Rear lights – Estate models
Rear wing-mounted lights
35 Working at the rear corner of the luggage compartment, turn the securing clip, and lower the cover panel from the side of the luggage compartment.
36 Unscrew the securing nut, and withdraw the storage box from the side of the luggage compartment **(see illustration)**.
37 Unscrew the two light unit securing nuts, then lift the light unit away from the rear wing.
38 Disconnect the wiring plug from the rear of the bulbholder **(see illustration)**.
39 Squeeze the two securing lugs together,

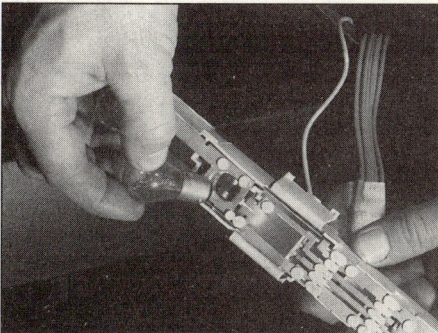

5.29 The bulbs are a bayonet-fit in the bulbholder – Hatchback model

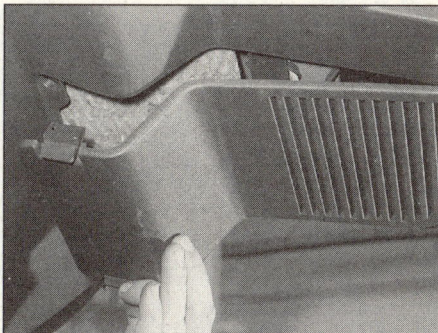

5.31 Remove the tailgate-mounted light cover – Hatchback model

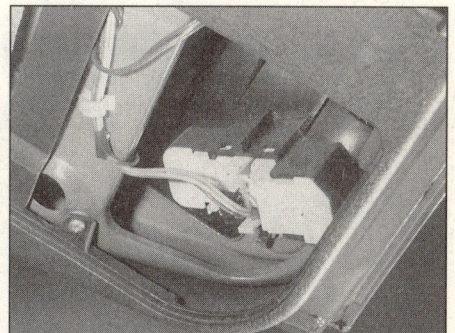

5.32a Squeeze the bulbholder securing tabs . . .

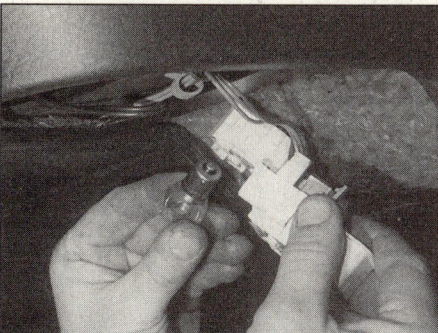

5.32b . . . then remove the bulb from the bulbholder – Hatchback model

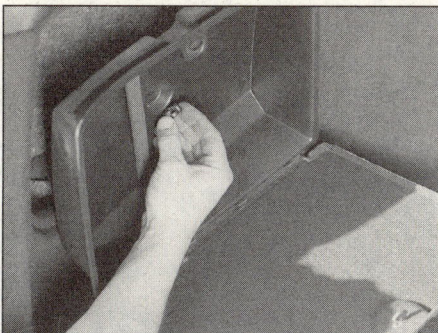

5.36 Removing the storage box from the side of the luggage compartment – Estate model

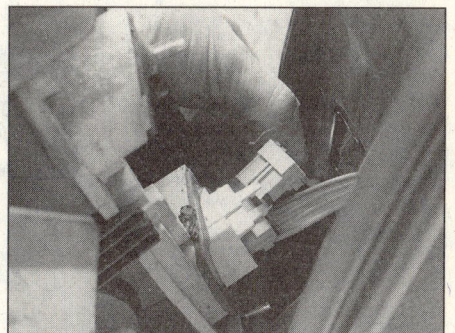

5.38 Disconnecting the wiring plug from the rear wing-mounted light bulbholder – Estate model

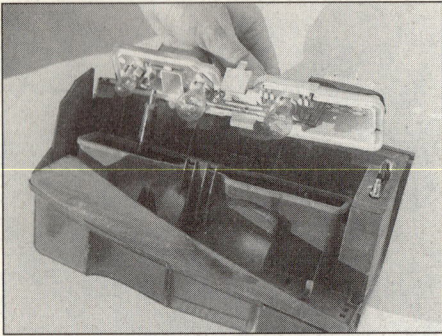

5.39 Removing the bulbholder from a rear wing-mounted rear light – Estate model

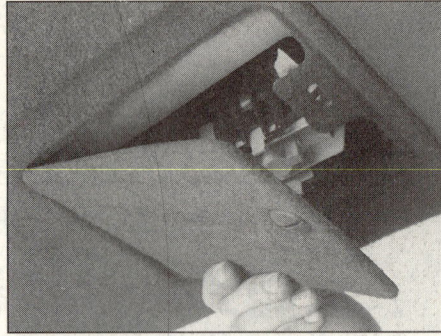

5.41 Remove the bulbholder cover from the tailgate trim panel . . .

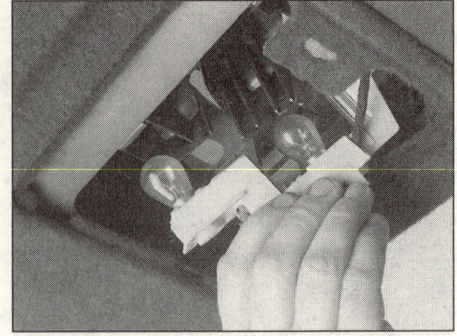

5.42 . . . then withdraw the bulbholder – Estate model

and withdraw the bulbholder from the rear of the light unit **(see illustration)**. The bulbs are a bayonet-fit in the bulbholder.
40 Fit the new bulb using a reversal of the removal procedure.

Tailgate-mounted lights

41 Open the tailgate, then release the clip, and remove the bulbholder cover from the tailgate trim panel **(see illustration)**.
42 Squeeze the two securing lugs together, and withdraw the bulbholder from the rear of the light unit **(see illustration)**. The bulbs are a bayonet-fit in the bulbholder.
43 Fit the new bulb using a reversal of the removal procedure.

Rear number plate light

44 Carefully prise the lens from the light unit,

using a small flat-bladed screwdriver, then pull out the bulb **(see illustrations)**.
45 Fit the new bulb using a reversal of the removal procedure.

6 Bulbs (interior lights) – renewal

General

1 Refer to Section 5, paragraph 1.

Front courtesy light

2 Carefully prise the lens from the courtesy light assembly **(see illustration)**.
3 Pull the relevant bulb from the bulbholder **(see illustration)**.

4 Push the new bulb into position, then refit the lens to the light assembly.

Rear courtesy and map reading lights

5 Carefully prise the light assembly from the roof, using a screwdriver if necessary **(see illustration)**.
6 Twist the relevant bulbholder anti-clockwise to remove it from the light. The bulbs are a push-fit in the bulbholders **(see illustration)**.
7 Fit the new bulb using a reversal of the removal procedure.

Kerb (door-mounted) lights

8 Carefully prise the light unit from the bottom edge of the door, then unclip the lens,

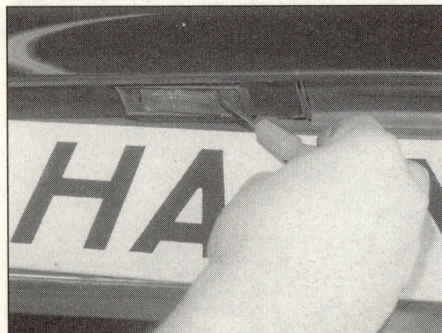

5.44a Prise the lens from the number plate light unit . . .

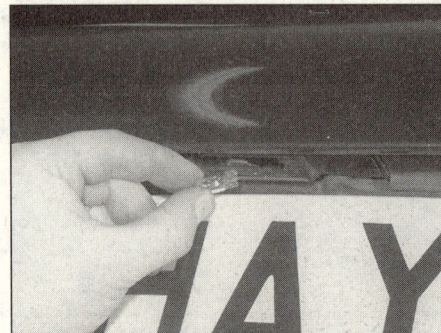

5.44b . . . then pull out the bulb – Hatchback models

6.2 Prise the lens from the courtesy light . . .

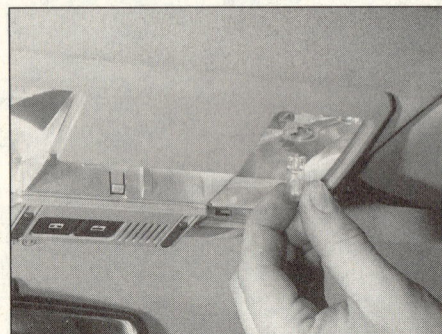

6.3 . . . then pull the bulb from the bulbholder

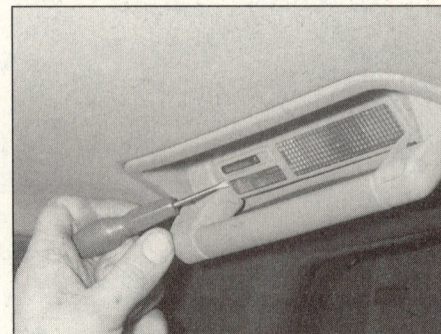

6.5 Prise the rear courtesy light assembly from the roof . . .

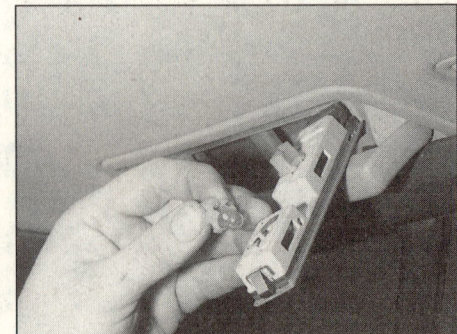

6.6 . . . then twist the bulbholder anti-clockwise

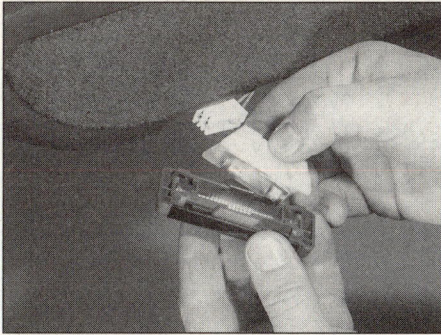
6.8 Unclipping the lens from the kerb light

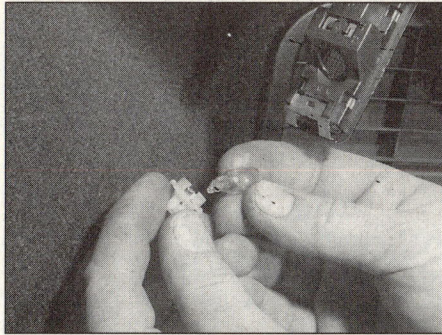
6.11 The luggage compartment light bulb is a push-fit in the bulbholder – Hatchback models

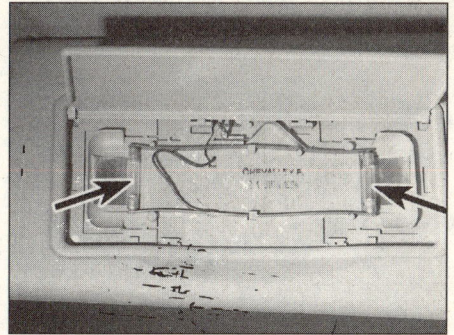
6.18 Vanity mirror lens removed to expose illumination bulbs (arrowed)

and pull the bulb from the contacts **(see illustration)**.

9 Fit the new bulb using a reversal of the removal procedure.

Luggage compartment light – Hatchback models

10 Carefully prise the light unit out from the trim panel.

11 Twist the bulbholder anti-clockwise and pull it from the light unit. The bulb is a push-fit in the bulbholder **(see illustration)**.

12 Fit the new bulb using a reversal of the removal procedure.

Glovebox light

13 Carefully prise the light assembly from the glovebox, using a screwdriver if necessary.

14 Twist the relevant bulbholder anti-clockwise to remove it from the light. The bulbs are a push-fit in the bulbholders.

15 Fit the new bulb using a reversal of the removal procedure.

Instrument panel illumination bulbs

16 Refer to Section 9.

Vanity mirror illumination bulbs

17 Open the mirror cover, then carefully prise off the lens.

18 Pull the relevant bulb from the contacts **(see illustration)**.

19 Fit the new bulb using a reversal of the removal procedure.

7 Exterior light units – removal and refitting

Note: *Before removing any light unit, using a screwdriver, release the securing clip, and remove the cover from the battery, then disconnect the battery negative lead. Reconnect the lead after refitting the light unit.*

Headlights

Removal

1 Remove the front grille panel as described in Chapter 12, section 20.

2 Working in the engine compartment,

disconnect the wiring plug from the rear of the headlight **(see illustration)**.

3 Where applicable, unclip the bonnet release cable from the top of the headlight **(see illustration)**.

4 Pull the headlight adjuster knobs from the top of the front body panel **(see illustration)**.

5 Unscrew the lower headlight securing bolt, located at the inboard lower corner of the light assembly. Note that the bolt may also secure other component brackets, such as the horn bracket – where applicable, note the orientation of the horn assembly, then disconnect the horn wiring plug(s) and withdraw the horn assembly **(see illustration)**.

6 Unscrew the two upper headlight securing bolts, then lift the headlight from the front of the vehicle **(see illustrations)**.

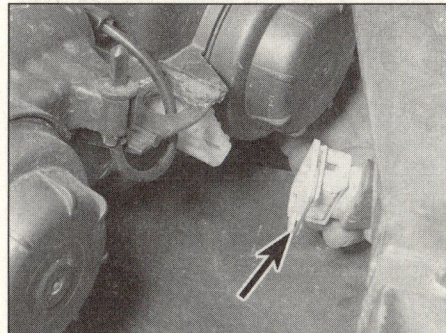
7.2 Disconnect the wiring plug (arrowed) from the rear of the headlight

7.3 Releasing the bonnet release cable from the top of the headlight

7.4 Pull out the headlight adjuster knobs

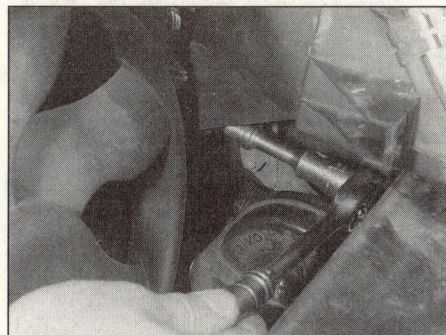
7.5 Unscrewing the lower headlight securing bolt

7.6a Unscrew the upper headlight securing bolts (arrowed) . . .

13

7.6b . . . then lift out the headlight

7.10 Front direction indicator/foglight assembly securing bolts (arrowed) – viewed with bumper removed

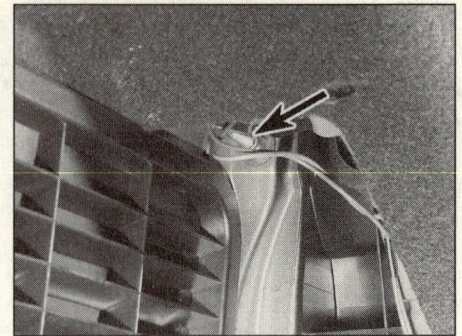

7.14a Unscrew the rear wing-mounted light upper securing nut . . .

Refitting

7 Refitting is a reversal of removal, but have the headlight beam alignment checked at the earliest opportunity.

Front sidelights

8 Removal and refitting is described as part of the bulb renewal procedure in Section 5.

Front direction indicator light/foglight assembly

Removal

9 Remove the front bumper as described in Chapter 12.
10 Unscrew the three bolts securing the light assembly to the bumper, and remove the light assembly **(see illustration)**.

Refitting

11 Refitting is a reversal of removal.

Rear lights - Hatchback models

Wing-mounted lights

12 Open the tailgate, then unclip the light unit cover.
13 Squeeze the two securing lugs together, and remove the bulbholder from the light unit.
14 Unscrew the two securing nuts and recover the washers, then withdraw the light unit from the outside of the wing panel **(see illustrations)**.
15 If desired, the plastic trim panels can be unclipped from the light unit **(see illustration)**.
16 Unscrew two nuts and remove light unit.(Recover washers).

17 Refitting is a reversal of removal.

Tailgate-mounted lights

18 Open the tailgate, then depress the securing lugs, and remove the bulb cover.
19 Squeeze the securing tabs, and pull the bulbholder from the light unit.
20 Unscrew the three securing nuts, then withdraw the light unit from the outside of the tailgate.
21 Refitting is a reversal of removal.

Rear lights – Estate models

Wing-mounted lights

22 The procedure is described as part of the bulb renewal procedure in Section 5.

Tailgate-mounted lights

23 Open the tailgate, then release the clip, and remove the bulbholder cover from the tailgate trim panel.
24 Squeeze the two securing lugs together, and withdraw the bulbholder from the rear of the light unit.
25 Prise the plastic cover from the side of the tailgate, for access to the outer light unit retaining clip.
26 Using a screwdriver inserted through the access hole in the side of the tailgate, depress the outer light unit retaining clip, then release the two inner retaining clips by inserting the screwdriver through the aperture in the tailgate trim panel **(see illustration)**.
27 Pull the light unit from the tailgate **(see illustration)**.

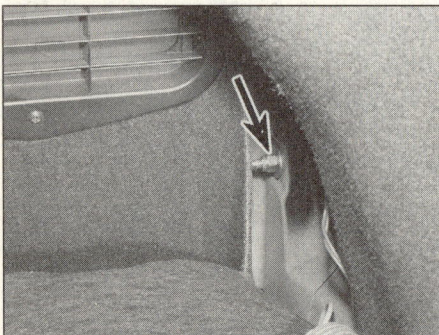

7.14b . . . and lower securing nut (arrowed) . . .

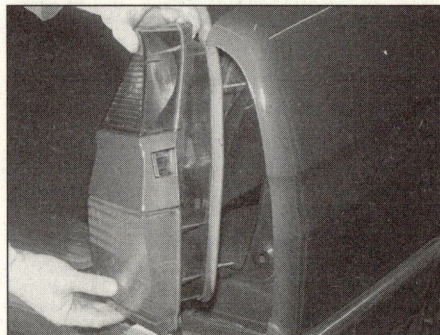

7.14c . . . then withdraw the light assembly – Hatchback model

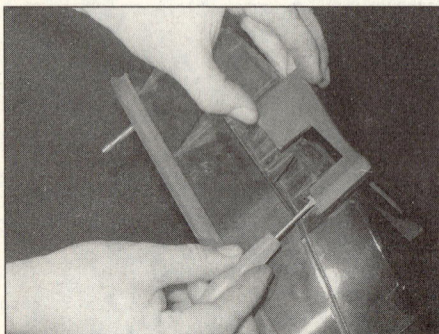

7.15 Unclipping a trim panel from a rear wing-mounted rear light – Hatchback model

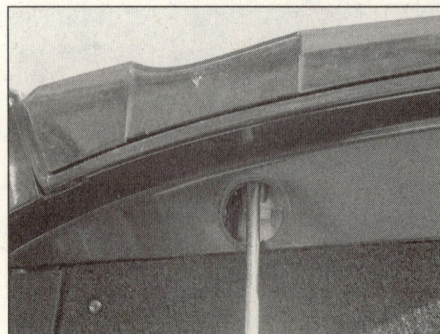

7.26 Depressing the tailgate-mounted rear light outer retaining clip – Estate model

7.27 Removing a tailgate-mounted rear light unit – Estate model

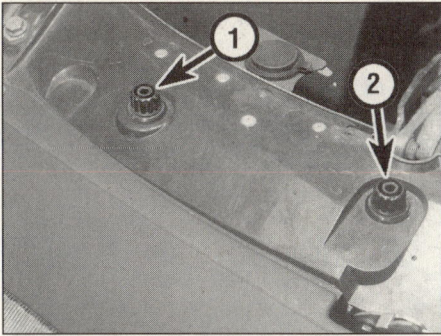

8.2 Headlight beam alignment adjuster screws

1 Vertical adjuster screw
2 Horizontal adjuster screw

28 Refitting is a reversal of removal, but make sure that the retaining clips engage fully with the tailgate.

8 Headlight beam alignment – general information

1 Accurate adjustment of the headlight beam is only possible using optical beam-setting equipment, and this work should therefore be carried out by a Citroën dealer or suitably-equipped workshop.

2 For reference, the headlights can be finely adjusted by rotating the adjuster screws fitted to the top of each light unit. The screws protrude through the top of the body front panel. The horizontal adjustment screw is mounted at the inner end of the headlight. The vertical adjustment screw is mounted at the centre of the headlight (see illustration).

9 Instrument panel and components – removal and refitting

Instrument panel - models up to 1994

Removal

1 Using a screwdriver, release the securing clip, and remove the cover from the battery, then disconnect the battery negative lead.

2 Working at each side of the instrument panel in turn, prise out the cover plate, and unscrew the instrument panel surround securing screw.

3 Carefully prise up the front of the instrument panel surround, and withdraw it forwards, taking care to release the three retaining lugs at the rear of the panel.

4 Unscrew the four instrument panel securing screws, then pull the instrument panel forwards from the facia, and disconnect the wiring plugs, noting their locations.

5 Where applicable, disconnect the speedometer cable from the rear of the instrument panel, then remove the instrument panel.

Refitting

6 Refit the panel following a reversal of the removal procedure.

Instrument panel - models from 1995-on

Removal

7 Remove the steering column shrouds as described in Chapter 12.

8 Carefully prise out the alarm bulb trim plate (or the blanking plate, as applicable) from the passenger's side end of the panel, to expose the upper switch/heater control/display trim panel securing screw (see illustration). Remove the screw.

9 Working on either side of the steering column, prise out the screw covers, then remove the remaining two upper switch/heater control/display trim panel securing screws (see illustrations).

10 Using a small flat-bladed screwdriver, carefully prise the heater control surround panel from the switch/heater control/display trim panel, then unscrew the four screws securing the heater control panel to the trim panel (see illustrations).

11 Unscrew the five lower securing screws from the bottom of the switch/heater control/display trim panel.

12 Lever the driver's side of the panel to release the securing clip, then withdraw the panel forward from the facia (see illustration).

9.8 Prise out the alarm bulb trim plate to expose the switch/heater control/display trim panel securing screw (arrowed)

9.9a Prise out the covers . . .

9.9b . . . then remove the remaining two upper switch/heater control/display trim panel securing screws (arrowed)

9.10a Prise out the heater control surround panel . . .

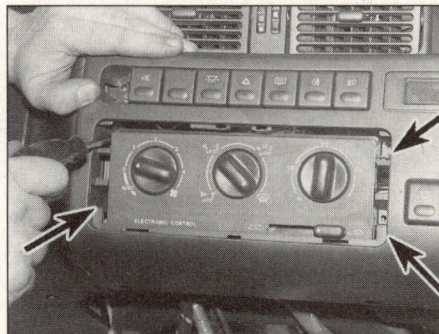

9.10b . . . then unscrew the four heater control panel securing screws (arrowed)

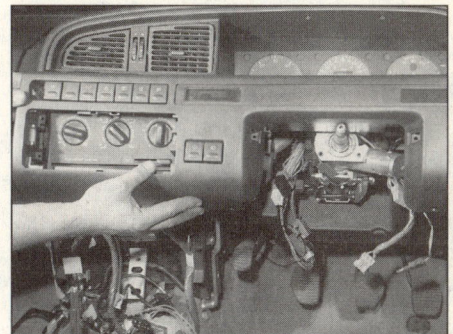

9.12 Withdraw the switch/heater control/display trim panel

13

9.13 Unscrew the two upper instrument panel shroud securing screws (arrowed)

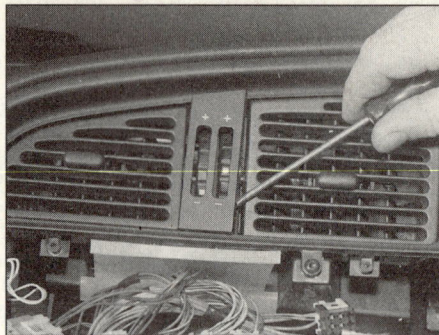
9.14a Prise off the trim plate . . .

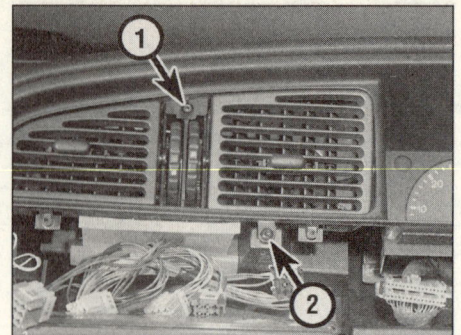
9.14b . . . to expose the remaining upper instrument panel shroud securing screw (1). Lower instrument panel securing screw (2) also shown

Working at the rear of the panel, disconnect the wiring plugs from the panel-mounted components, noting the locations of the plugs, and withdraw the panel.

13 Unscrew the two upper instrument panel shroud securing screws (see illustration).

14 Prise off the trim plate fitted between the centre heater vents, and unscrew the remaining upper instrument panel shroud securing screw (see illustrations).

15 Unscrew the lower instrument panel shroud securing screw, located beneath the driver's side centre heater vent, then withdraw the shroud from the facia (see illustration).

16 Unscrew the centre pins, and prise out the two lower instrument panel securing clips (see illustration).

17 Using a socket, and an extension with a universal joint, unscrew the two upper instrument panel securing nuts (see illustration). Take care not to allow the washers to drop down into the facia.

18 Carefully manipulate the washers from the upper instrument panel securing studs, then manipulate the instrument panel out from the facia, and disconnect the wiring plugs from the bottom of the panel, noting their locations (see illustration). Remove the instrument panel.

Refitting

19 Refit the panel following a reversal of the removal procedure. Refit the steering column shrouds with reference to Chapter 12.

9.15 Withdrawing the instrument panel shroud

9.16 Removing a lower instrument panel securing clip

9.17 Unscrewing an instrument panel upper securing nut

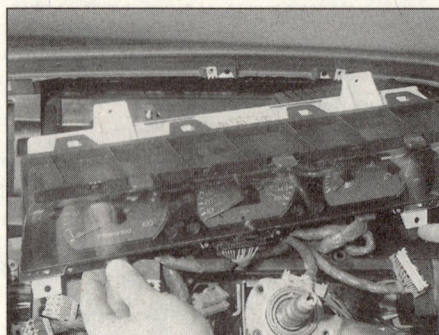
9.18a Withdraw the instrument panel . . .

9.18b . . . and disconnect the wiring plugs

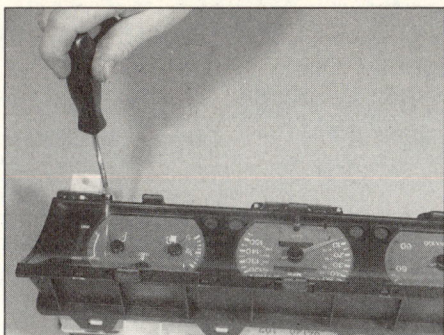

9.21a Removing an instrument panel lens securing screw

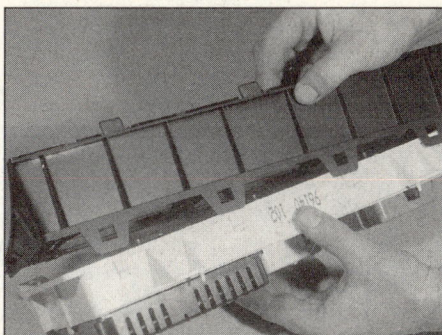

9.21b Removing the instrument panel lens

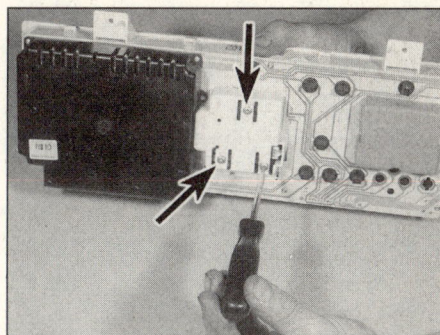

9.22 Remove the three rear speedometer securing screws (arrowed)

Speedometer

Removal

20 With the instrument panel removed as described previously in this Section, proceed as follows.

21 Remove the two securing screws, and release the securing clips, and withdraw the lens from the instrument panel **(see illustrations)**.

22 Working at the rear of the instrument panel, remove the three speedometer securing screws **(see illustration)**.

23 Working at the front of the panel, unscrew the two front speedometer securing screws, then lift the speedometer from the instrument panel and disconnect the wiring plug **(see illustrations)**.

Refitting

24 Refitting is a reversal of removal.

Tachometer ('rev counter')

Removal

25 Proceed as described in paragraphs 20 and 21.

26 Unclip the plastic cover from the rear of the tachometer **(see illustration)**.

27 Working at the rear of the instrument panel, unscrew the four securing nuts.

28 Working at the front of the instrument panel, unscrew the two securing screws, then lift the tachometer from the instrument panel.

Refitting

29 Refitting is a reversal of removal.

Fuel/temperature/oil level gauges

Removal

30 Proceed as described in paragraphs 20 and 21.

31 Working at the rear of the instrument panel, unscrew the four securing screws, and remove the plastic cover from the rear of the panel – take care, as the cover contains a circuit board, and the connector pins must be pulled out of the connector block without bending, as the cover is withdrawn **(see illustrations)**.

32 Working at the front of the instrument

panel, unscrew the two securing screws, and carefully pull the gauge assembly from the panel.

Refitting

33 Refitting is a reversal of removal, but take great care not to bend the connector pins when refitting the cover to the rear of the instrument panel.

Instrument panel illumination bulbs

Removal

34 Remove the instrument panel as described previously in this Section.

35 If the fuel/temperature/oil level gauge illumination bulbs are to be renewed, working at the rear of the instrument panel, unscrew

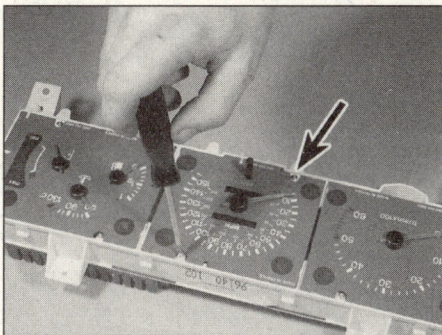

9.23a Unscrew the two front speedometer securing screws (arrowed) . . .

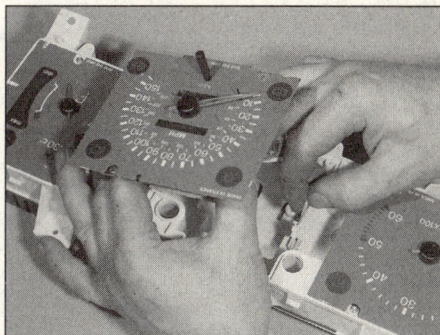

9.23b . . . and lift the speedometer from the instrument panel

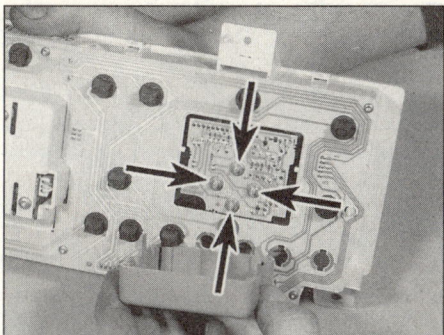

9.26 Unclip the plastic cover to expose the tachometer securing nuts (arrowed)

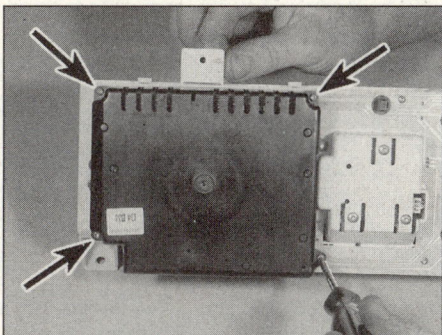

9.31a Unscrew the four securing screws (arrowed) . . .

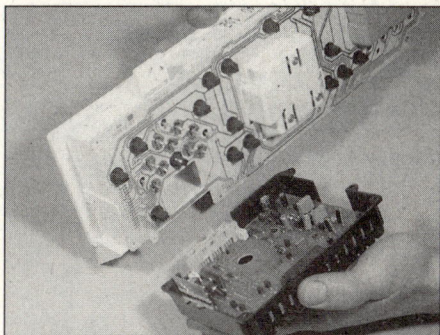

9.31b . . . and remove the plastic cover from the rear of the instrument panel

13

9.36 The instrument panel bulbs are a push-fit in the bulbholders

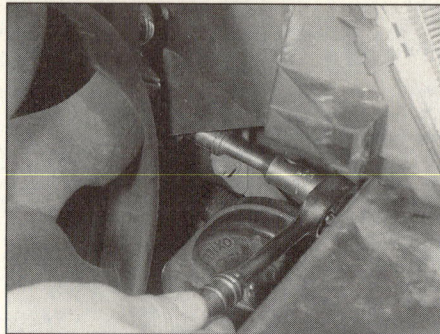
10.2 Unscrew the horn bracket securing bolt . . .

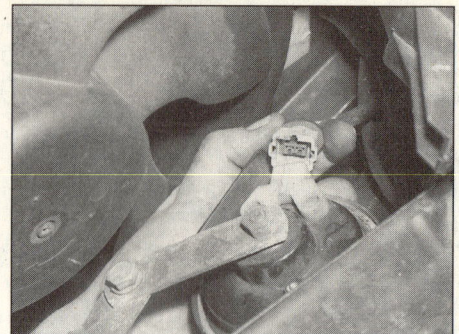
10.3 . . . then remove the bracket and disconnect the horn wiring plug

the four securing screws, and remove the plastic cover from the rear of the panel – take care, as the cover contains a circuit board, and the connector pins must be pulled out of the connector block without bending, as the cover is withdrawn.

36 Working at the rear of the instrument panel, twist the relevant bulbholder anti-clockwise, and withdraw it from the rear of the panel. The bulbs are a push-fit in the bulbholders **(see illustration)**.

Refitting

37 Fit the new bulb using a reversal of the removal procedure but, where applicable, take great care not to bend the connector pins when refitting the cover to the rear of the instrument panel.

10 Horn – removal and refitting

Removal

1 Remove the front grille panel as described in Chapter 12.
2 Reach through the front grille aperture, and unscrew the bolt securing the horn bracket (the bolt also acts as the lower headlight securing bolt) **(see illustration)**.
3 Lift out the horn and bracket assembly and disconnect the wiring plug **(see illustration)**.

Refitting

4 Refitting is a reversal of removal.

11 Speedometer transducer – removal and refitting

Removal

1 Using a screwdriver, release the securing clip, and remove the cover from the battery, then disconnect the battery negative lead.
2 Apply the parking brake, then jack up the front of the vehicle and support securely on axle stands (see *Jacking and Vehicle Support*).
3 Access to the transducer is limited, and is most easily obtained from behind the rear of the front suspension subframe **(see illustration)**.
4 Disconnect the wiring plug from the transducer.
5 Unscrew the transducer securing bolt and, where applicable, remove the heat shield.
6 Carefully withdraw the transducer from the transmission.

Refitting

7 Refitting is a reversal of removal, but check the condition of the O-ring on the transducer, and renew if necessary.

12 Wiper arm – removal and refitting

Removal

1 Operate the wiper motor, then switch it off so that the wiper arm returns to the at-rest position.

> **HAYNES HiNT**
> *Stick a piece of masking tape on the windscreen in line with the edge of the wiper blade, to use as an alignment aid on refitting.*

2 If removing a windscreen wiper arm, improved access can be obtained by removing the front scuttle trim panel as described in Chapter 12.
3 Unclip the wiper arm spindle nut cover, then slacken and remove the spindle nut. Where applicable, recover the washer **(see illustration)**.
4 Lift the blade off the glass, and pull the wiper arm off its spindle. Note that on some models, the wiper arms may be very tight on the spindle splines - it should be possible to lever the arm off the spindle, using a flat-bladed screwdriver (take care not to damage the scuttle cover panel) **(see illustration)**.

Refitting

5 Ensure that the wiper arm and spindle splines are clean and dry, then refit the arm to

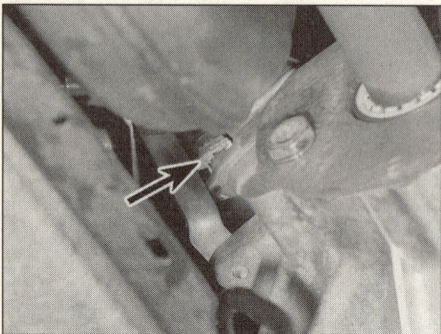
11.3 Speedometer transducer location (arrowed)

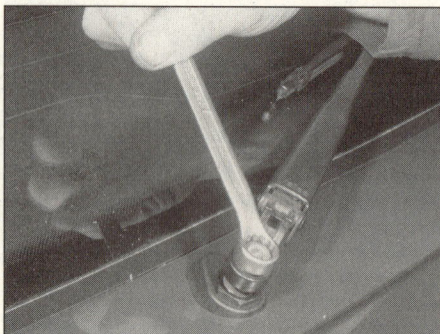
12.3 Unscrew the wiper arm spindle nut . . .

12.4 . . . then pull the arm from the spindle

13.4 Removing the washer fluid reservoir filler neck

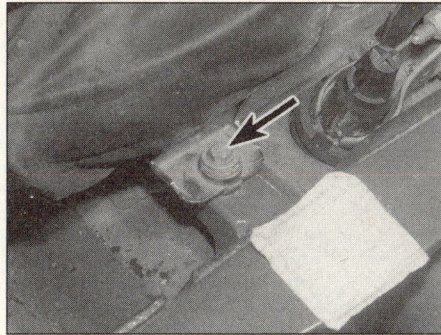

13.5a Unscrew bolt (arrowed) securing the wiper motor bracket

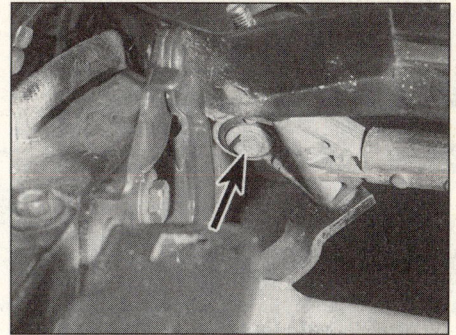

13.5b Wiper linkage-to-scuttle securing bolt (arrowed)

the spindle. Where applicable, align the wiper blade with the tape fitted on removal.

6 Refit the spindle nut (where applicable ensure that the washer is in place), tightening it securely, and clip the nut cover back into position.

13 Windscreen wiper motor and linkage – removal and refitting

Removal

1 Using a screwdriver, release the securing clip, and remove the cover from the battery, then disconnect the battery negative lead.

2 Remove the front scuttle trim panel as described in Chapter 12.

3 Remove the windscreen wiper arms, with reference to Section 12.

4 Where applicable, unscrew the two securing screws, and withdraw the washer fluid reservoir filler neck from the scuttle **(see illustration)**.

5 Unscrew the bolt securing the windscreen wiper motor bracket to the scuttle, then unscrew the two remaining bolts securing the ends of the linkage to the scuttle **(see illustrations)**.

6 Where applicable, pull the protective cover from the motor assembly, then disconnect the motor wiring plug **(see illustration)**.

7 Carefully prise the two linkage crank arms from their balljoints, using a screwdriver if necessary **(see illustration)**.

8 Make alignment marks between the linkage drive arm and the motor spindle, then unscrew the nut securing the linkage drive arm to the motor spindle, and prise the drive arm from the spindle **(see illustration)**.

9 Unscrew the three securing bolts, then separate the motor from the linkage and lift the motor from the scuttle **(see illustrations)**.

10 Carefully manipulate the linkage out from the scuttle, noting the orientation of the components **(see illustration)**.

Refitting

11 Refitting is a reversal of removal, bearing in mind the following points.
a) Ensure that the linkage is orientated as noted before removal.

13.6 Disconnecting the windscreen wiper motor wiring plug

13.7 Carefully prise the two linkage crank arms from their balljoints

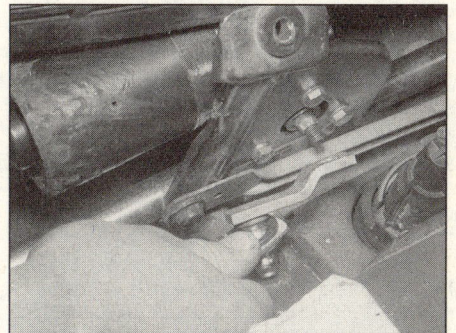

13.8 Unscrew the nut securing the linkage drive arm to the motor spindle

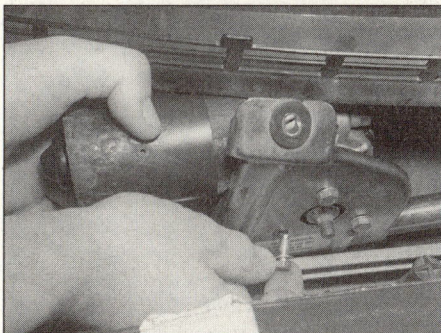

13.9a Unscrew the three securing bolts . . .

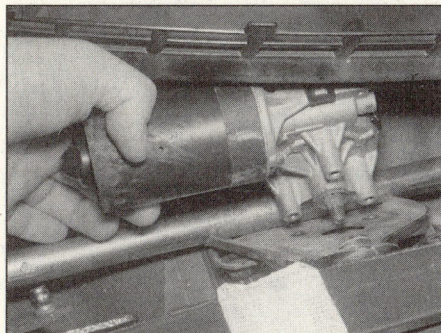

13.9b . . . and separate the motor from the linkage

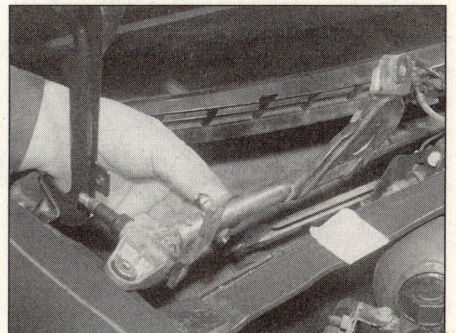

13.10 Manipulate the linkage out from the scuttle

13

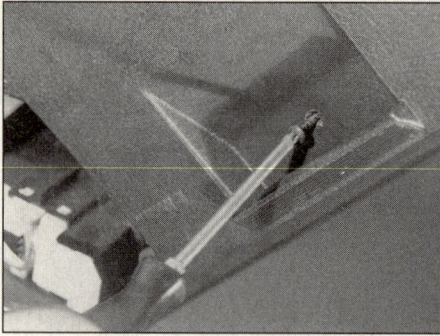
14.3a Remove the securing screws . . .

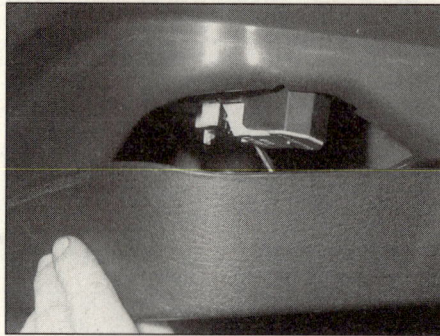
14.3b . . . and withdraw the tailgate trim panel – Hatchback model

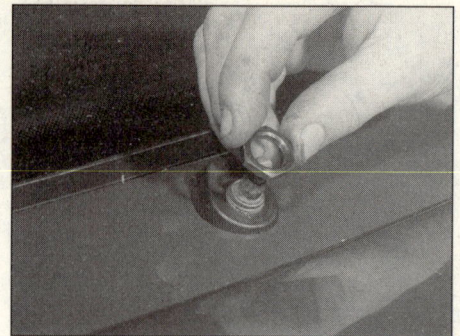
14.5 Unscrew the nut securing the motor spindle to the tailgate – Hatchback model

b) Where applicable, check the condition of the washer fluid bottle filler neck O-ring and renew if necessary.

c) Do not refit the windscreen wiper arms until the motor has been switched on and operated to set the linkage to the parked position.

14 Tailgate wiper motor – removal and refitting

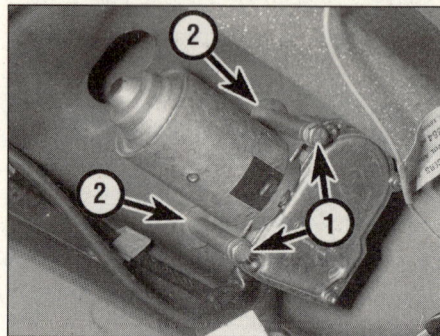
14.6a Unscrew the two motor securing bolts (1) – note the washer locations (2) . . .

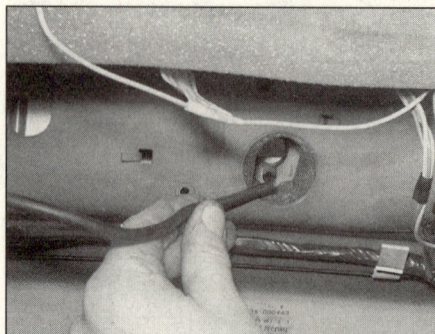
14.6b . . . and lower the motor from the tailgate – Hatchback model

Hatchback models

Removal

1 Using a screwdriver, release the securing clip, and remove the cover from the battery, then disconnect the battery negative lead.

2 Open the tailgate, and unclip the tailgate-mounted rear light covers.

3 Remove the securing screws, and withdraw the tailgate inner trim panel (see illustrations).

4 Working outside the tailgate, lift the cover, then unscrew the securing nut, and remove the tailgate wiper arm.

5 Unscrew the nut securing the motor spindle to the tailgate (see illustration).

6 Working inside the tailgate, unscrew the two securing bolts, then lower the motor from its location. Where applicable, recover the washers fitted between the motor and the tailgate (see illustrations).

7 Disconnect the washer fluid hose from the nozzle, and feed the hose through the foam attached to the motor (see illustration).

8 Disconnect the motor wiring plug, and withdraw the motor.

Refitting

9 Refitting is a reversal of removal. Where applicable, make sure that the washers are in position between the motor and the tailgate. Make sure that the wiper arm is fitted in its correct 'parked' position.

Estate models

Removal

10 Using a screwdriver, release the securing clip, and remove the cover from the battery, then disconnect the battery negative lead.

11 Remove the securing screws, and withdraw the handle from the tailgate trim panel.

12 Work around the trim panel and remove the securing screws, then withdraw the trim panel.

13 Working outside the tailgate, lift the cover, then unscrew the securing nut, and remove the tailgate wiper arm.

14 Working inside the tailgate, disconnect the washer fluid hose from the rear of the wiper motor (see illustration).

15 Disconnect the motor wiring plugs and, where applicable, unclip the wiring from the tailgate.

16 Unscrew the securing bolts, and withdraw the motor assembly from the tailgate.

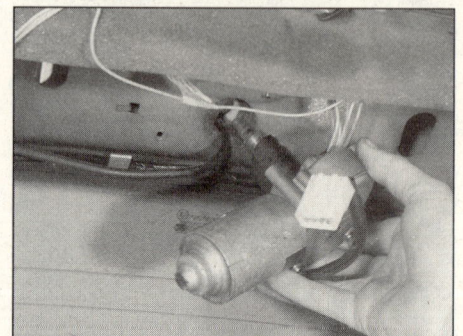
14.7 Disconnect the washer fluid hose from the nozzle – Hatchback model

Refitting

17 Refitting is a reversal of removal. Make sure that the wiper arm is fitted in its correct 'parked' position.

15 Windscreen/tailgate/headlight washer system components – removal and refitting

Washer fluid reservoir

Removal

1 The windscreen/tailgate washer fluid reservoir is located at the left-hand side of the

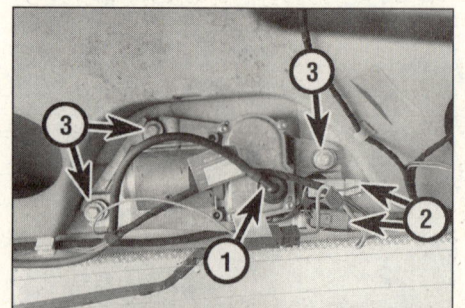
14.14 Tailgate wiper motor mounting details

1 Washer fluid hose connection
2 Wiring plugs
3 Securing bolts

15.3 Removing the washer fluid reservoir filler neck

15.6 Washer fluid reservoir front securing nut (arrowed)

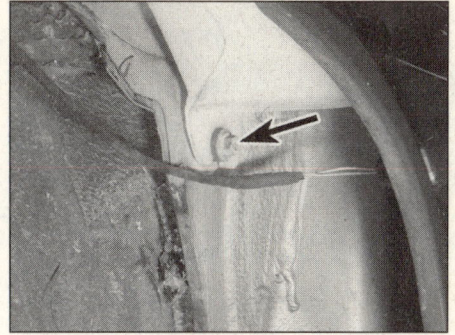

15.7 Washer fluid reservoir rear securing bolt (arrowed)

engine compartment. On models fitted with headlight washers, a separate reservoir is located at the right-hand side of the engine compartment. It is advisable to allow the reservoir to become empty before removal.

2 Using a screwdriver, release the securing clip, and remove the cover from the battery, then disconnect the battery negative lead.

3 Working in the scuttle at the rear of the engine compartment, unscrew the two securing screws, and pull out the reservoir filler neck **(see illustration)**.

4 Chock the rear wheels, then jack up the front of the vehicle and support securely on axle stands (see *Jacking and Vehicle Support*). Remove the relevant front roadwheel.

5 Remove the wheel arch liner with reference to Chapter 12.

6 Working in the engine compartment, unscrew the reservoir front securing nut (or bolt, as applicable), if necessary counterholding the bolt (or nut, as applicable) which fits from under the wheel arch (access is tricky). If the windscreen/tailgate washer fluid reservoir is being removed, it will be necessary to remove the air cleaner cover for access to the front reservoir securing nut or bolt **(see illustration)**.

7 Working under the wheel arch, unscrew the reservoir rear securing bolt **(see illustration)**.

8 Lower the reservoir, and disconnect the fluid pump wiring plugs and the fluid hoses,

noting their locations **(see illustrations)**. Be prepared for fluid spillage if the reservoir still contains fluid.

9 Withdraw the reservoir from under the wheel arch.

Refitting

10 Refitting is a reversal of removal, bearing in mind the following points.

a) Make sure that the rear end of the reservoir engages with the support bracket.

b) Do not fully tighten the reservoir securing nut(s) and bolt until the filler neck has been refitted (this will ensure correct alignment of the reservoir and filler neck).

c) Examine the filler neck O-ring and renew if necessary.

Washer fluid pump

Removal

11 Remove the fluid reservoir as described previously in this Section.

12 Disconnect the fluid hose(s) from the pump.

13 Pull the fluid pump from the grommet in the reservoir.

Refitting

14 Refitting is a reversal of removal, but check the condition of the rubber grommet in the reservoir, and renew if necessary. Refit the reservoir as described previously in this Section.

Windscreen washer nozzles

Removal

15 On models with heated windscreen washer nozzles, using a screwdriver, release the securing clip, and remove the cover from the battery, then disconnect the battery negative lead.

16 Remove the front scuttle trim panel as described in Chapter 12.

17 Where applicable, disconnect the washer nozzle heating element wiring plug, and release the plug from its retaining clip **(see illustration)**.

18 Disconnect the fluid hose from the washer nozzle.

19 Unscrew the washer nozzle securing nut, then withdraw the washer nozzle and the nozzle mounting plate **(see illustrations)**.

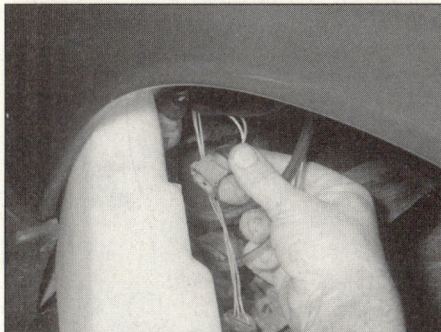

15.8a Lower the fluid reservoir . . .

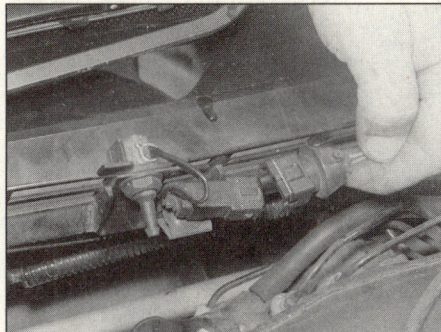

15.8b . . . and disconnect the fluid pump wiring plugs and fluid hoses

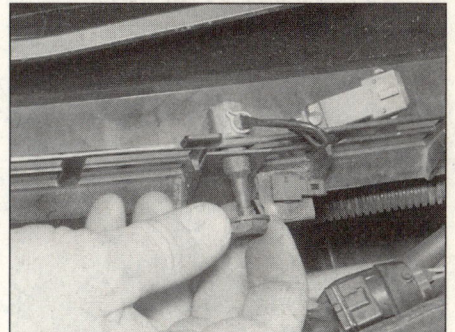

15.17 Disconnecting a windscreen washer nozzle heating element wiring plug

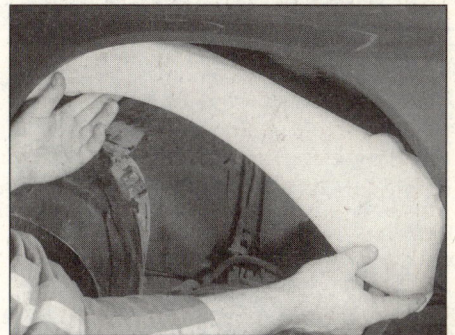

15.19a Unscrew the securing nut . . .

13

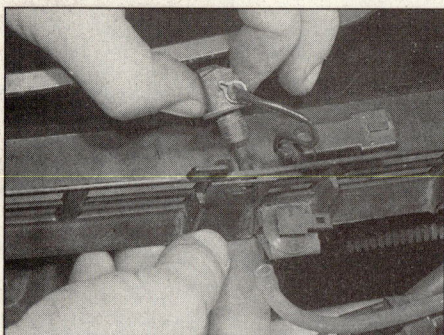
15.19b . . . then withdraw the washer nozzle and mounting plate

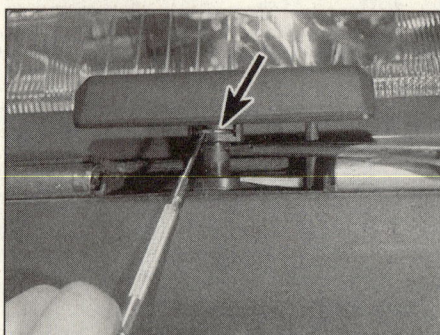
15.26 Prise the headlight washer nozzle from the bumper, and prise off the circlip (arrowed)

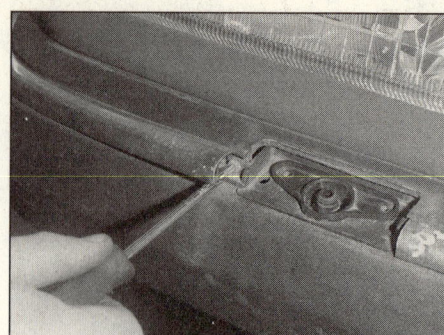
15.27a Unscrew the two feed tube securing screws . . .

Refitting

20 Refitting is a reversal of removal. Refit the scuttle trim panel with reference to Chapter 12.

Tailgate washer nozzle – Hatchback models

Removal

21 The washer nozzle is integral with the tailgate wiper motor spindle surround rubber.
22 Remove the tailgate wiper motor as described in Section 14.
23 Carefully pull the spindle surround rubber/washer nozzle from the outside of the tailgate.

Refitting

24 Push the spindle surround rubber/washer nozzle into position on the tailgate, then refit the wiper motor as described in Section 14.

Tailgate washer nozzle – Estate models

25 The nozzle is integral with the wiper arm. Refer to Section 12 for details of wiper arm removal.

Headlight washer nozzles

Removal

26 Carefully prise up the nozzle assembly from the bumper and, using a small screwdriver, prise off the circlip securing the nozzle assembly to the fluid feed tube. Pull off the nozzle assembly (see illustration).
27 If desired, the fluid feed tube can be removed as follows. Note that when removing the left-hand feed tube, it may be necessary to remove the bumper as described in Chapter 12, to allow the fluid hose to be reconnected to the fluid feed tube (see illustrations).
a) Unscrew the two feed tube surround plate securing screws, then carefully prise the feed tube assembly from the bumper.
b) Disconnect the washer fluid hose from the feed tube, then withdraw the assembly.

Refitting

28 Where applicable, refit the fluid feed tube using a reversal of the removal procedure. Where applicable, refit the bumper as described in Chapter 12.
29 Ensure that the circlip is correctly refitted to the feed tube, then carefully prise up the sprung section of the feed tube, using a small screwdriver, and push the nozzle assembly onto the feed tube until the circlip engages.
30 Release the assembly, and check that the nozzle assembly sits flush on the bumper.

16 Radio/cassette player – removal and refitting

Models up to 1994

Removal

Note: Two DIN removal tools will be required for this operation.
1 Using a screwdriver, release the securing clip, and remove the cover from the battery, then disconnect the battery negative lead.
2 Open the hinged cover to expose the radio/cassette player.
3 Insert the DIN removal tools into the holes on each side of the radio/cassette player, and push them until they click into place. If necessary, pull the tools gently to the left and right to release the locking tangs.
4 Gently pull the radio/cassette player from the facia, using the removal tools.
5 Disconnect the wiring plugs and the aerial lead, and withdraw the unit. If necessary, the oddments tray below the radio/cassette player can be pulled out to aid disconnection of the wiring.

Refitting

6 Reconnect the wiring plugs and the aerial

15.27b . . . then prise out the feed tube assembly . . .

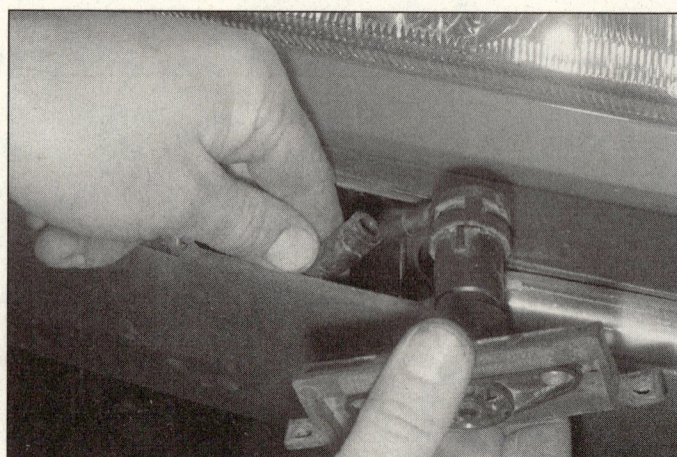
15.27c . . . and disconnect the washer fluid hose

16.10 Unclip the radio/cassette player surround panel – models from 1995

16.11 Removing the radio/cassette player – models from 1995

17.2 Prise the 'tweeter' loudspeaker cover from the facia . . .

lead, then push the unit into its housing until the securing clips engage.

7 On completion, reconnect the battery negative lead, refit the battery cover and, where applicable, enter the security code.

Models from 1995

Note: *Two DIN removal tools will be required for this operation.*

Removal

8 Using a screwdriver, release the securing clip, and remove the cover from the battery, then disconnect the battery negative lead.
9 Open the oddments tray cover above the radio/cassette player.
10 Release the securing clips at the top of the radio/cassette player surround panel, tilt the top of the panel forwards, and remove the panel **(see illustration)**.
11 Proceed as described in paragraphs 3 to 5,

ignoring the reference to the oddments tray **(see illustration)**.

Refitting

12 Proceed as described in paragraphs 6 and 7.

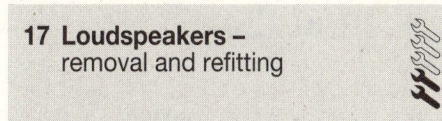

17 Loudspeakers – removal and refitting

Facia-mounted 'tweeter' speakers

Removal

1 Using a screwdriver, release the securing clip, and remove the cover from the battery, then disconnect the battery negative lead.
2 Carefully prise the loudspeaker cover from the facia **(see illustration)**.

3 Unscrew the two securing screws, then withdraw the loudspeaker from the facia, and disconnect the wiring **(see illustration)**.

Refitting

4 Refitting is a reversal of removal.

Front door-mounted loudspeakers

Removal

5 Proceed as described in paragraph 1.
6 Unclip the loudspeaker cover panel from the door, then remove the securing screws, withdraw the loudspeaker, and disconnect the wiring plugs **(see illustrations)**.

Refitting

7 Refitting is a reversal of removal.

Rear parcel shelf side support panel-mounted loudspeakers – Hatchback models

Removal

8 Remove the rear parcel shelf side support panel as described in Chapter 12.
9 Remove the securing screws and withdraw the loudspeaker **(see illustration)**.

Refitting

10 Refit the loudspeaker using a reversal of the removal procedure.

Rear loudspeakers – Estate models

Removal

11 Carefully prise the loudspeaker cover from the rear parcel shelf support panel **(see illustration)**.

17.3 . . . then unscrew the two loudspeaker securing screws (arrowed)

17.6a Unclip the loudspeaker cover panel . . .

17.6b . . . then remove the screws and withdraw the loudspeaker

17.9 Removing the rear loudspeaker – Hatchback models

17.11 Removing a rear loudspeaker cover – Estate model

13

18.1 Unscrew the securing screws and remove the sun visor

12 Unscrew the four securing screws, then lift out the loudspeaker and disconnect the wiring plug.

Refitting

13 Refitting is a reversal of removal.

18 Radio aerial – removal and refitting

Removal

1 Unscrew the two securing screws, and remove the central sun visor from the roof panel to expose the bottom of the aerial **(see illustration)**.
2 Using a small screwdriver, prise the cover from the bottom of the aerial **(see illustration)**.
3 Unscrew the aerial securing nut, and disconnect the aerial lead, then lift the aerial assembly from the top of the roof.

Refitting

4 Refitting is a reversal of removal, but ensure that the cover is securely pushed into position at the bottom of the aerial.

19 Anti-theft alarm system and engine immobiliser – general information

Certain vehicles are equipped with an anti-theft alarm system and/or and engine immobiliser system. Various types of system may be fitted depending on' vehicle specification, and market.

The anti-theft alarm system, is automatically activated by the central locking system (manually, or via the remote control, where applicable). The engine immobiliser system is operated by a coded key pad mounted in the centre console – the engine can only be started after entering the correct security code using the key pad. On some models, an engine immobiliser may operate in conjunction with the alarm system.

The alarm system has switches on the bonnet, tailgate and each of the doors.

Any faults with the system should be referred to a Citroën dealer.

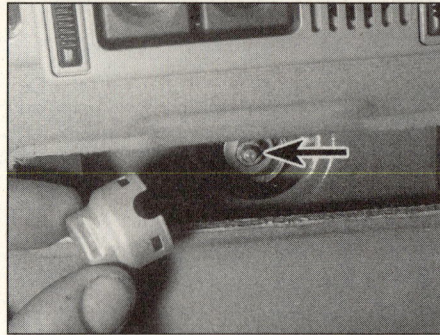

18.2 Prise the cover from the bottom of the aerial to expose the securing nut (arrowed)

20 Air bag system – general information, precautions and system de-activation

General information

A driver's side air bag is fitted as standard on some models, and is available as an option on others. The air bag is fitted to the steering wheel centre pad.

Similarly, a passenger's side air bag may be fitted as standard equipment, or as an option, depending on model.

The system is armed only when the ignition is switched on, however, a reserve power source maintains a power supply to the system in the event of a break in the main electrical supply. The system is activated by a 'g' sensor (deceleration sensor), incorporated in the electronic control unit. The electronic control unit may be incorporated in the steering wheel, or it may be mounted beneath the centre console. On some models, the driver's and passenger's airbag control systems are integrated with the front seat belt tensioners control system, and all are activated by a central control unit mounted beneath the centre console.

The air bags are inflated by gas generators, which force the bags out from their locations in the steering wheel, and the passenger's side facia, where applicable.

Precautions

⚠️ *Warning: The following precautions must be observed when working on vehicles equipped with an air bag system, to prevent the possibility of personal injury.*

General precautions

The following precautions **must** be observed when carrying out work on a vehicle equipped with an air bag.
a) *Do not disconnect the battery with the engine running.*
b) *Before carrying out any work in the vicinity of an air bag, removal of any of the air bag components, or any welding work on the vehicle, de-activate the system as described in the following sub-Section.*

c) *Do not attempt to test any of the air bag system circuits using test meters or any other test equipment.*
d) *If the air bag warning light comes on, or any fault in the system is suspected, consult a Citroën dealer without delay. Do not attempt to carry out fault diagnosis, or any dismantling of the components.*

Precautions to be taken when handling an air bag

a) *Transport the air bag by itself, bag upward.*
b) *Do not put your arms around the air bag.*
c) *Carry the air bag close to the body, bag outward.*
d) *Do not drop the air bag or expose it to impacts.*
e) *Do not attempt to dismantle the air bag unit.*
f) *Do not connect any form of electrical equipment to any part of the air bag circuit.*

Precautions to be taken when storing an air bag unit

a) *Store the unit in a cupboard with the air bag upward.*
b) *Do not expose the air bag to temperatures above 80ºC.*
c) *Do not expose the air bag to flames.*
d) *Do not attempt to dispose of the air bag - consult a Citroën dealer.*
e) *Never refit an air bag which is known to be faulty or damaged.*

De-activation of air bag system

The system must be de-activated as follows, before carrying out any work on the air bag components or surrounding area.
a) *Switch off the ignition.*
b) *Remove the ignition key.*
c) *Switch off all electrical equipment.*
d) *Using a screwdriver, release the securing clip, and remove the cover from the battery, then disconnect the battery negative lead.*
e) *Insulate the battery negative terminal and the end of the battery negative lead to prevent any possibility of contact.*
f) *Wait for at least ten minutes before carrying out any further work.*

System re-activation

When the battery has been reconnected, switch on the ignition as follows.
a) *Open the driver's window, then close the driver's door.*
b) *Reach in through the window, and switch on the ignition.*
c) *Observe the air bag warning light. When the ignition is first switched on, the air bag warning light (mounted on the instrument panel) should come on, and then go out 3 to 5 seconds later. If the warning light remains on, blinks continuously, or does not come on at all, consult a Citroën dealer without delay.*

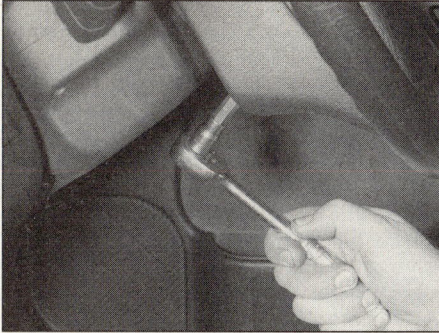

21.4 Unscrew the airbag unit securing screws . . .

21.5 . . . then ease the airbag unit from the steering wheel . . .

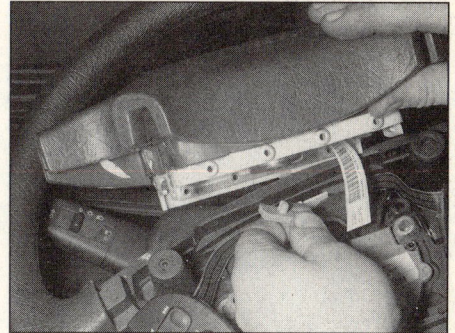

21.6 . . . then disconnect the wiring connector

21 Air bag system components – removal and refitting

Driver's side air bag unit

⚠ **Warning: Refer to the precautions given in Section 20 before attempting to carry out work on the air bag components.**

Removal

1 The air bag unit is an integral part of the steering wheel centre pad.

2 De-activate the air bag system as described in Section 20.

3 Move the steering wheel as necessary for access to the two air bag unit securing screws. The screws are located at the rear of the steering wheel centre pad.

4 Remove the two air bag unit securing screws (see illustration).

5 Gently ease the air bag unit from the steering wheel (see illustration).

6 Lift the air bag unit from the steering wheel, then disconnect the wiring connector from the rear of the air bag unit (see illustration).

7 Withdraw the air bag unit, and store it in a safe place, with reference to the precautions given in Section 20.

Refitting

8 Refitting is a reversal of removal, but make sure that the wiring connector is securely reconnected.

Driver's side air bag rotary switch

Removal

Note: *During this procedure, take note of the routing of all wiring to aid correct refitting.*

9 Remove the driver's air bag as described previously in this Section.

10 Remove the steering wheel as described in Chapter 11.

11 Working under the lower steering column shroud, unscrew the five screws securing the shroud, then lower the lower shroud from the steering column. Where applicable, unclip the steering lock illumination bulb from the lower shroud, and remove shroud.

12 Move the steering column height adjuster to fully lower the steering column.

13 Where applicable, depress the three securing tabs, then pull the rotary switch assembly from the centre of the steering column – take care not to strain the wiring (see illustrations).

14 Trace the wiring harness(es) back from the rotary switch, then separate the two halves of the wiring connector(s). Withdraw the rotary switch assembly, noting the routing of the wiring harness.

Refitting

15 Refitting is a reversal of removal, bearing in mind the following points.

a) Ensure that the wiring harness is routed correctly, as noted before removal.

b) Make sure that the direction indicator self-cancelling mechanism engages correctly.

c) Refit the steering wheel with reference to Chapter 11.

d) Refit the air bag as described previously in this Section.

Passenger's side air bag unit

16 At the time of writing, no information was available regarding the removal and refitting of the passenger's side air bag unit. If it is necessary to remove the air bag unit for any reason, the work should be entrusted to a Citroën dealer.

⚠ **Warning: Do not attempt to remove the passenger's side air bag yourself, without consulting a Citroën dealer.**

Electronic control unit

17 Various different types of control system have been used for airbag systems on XM models. On some models with a driver's side air bag, but no passenger's side air bag, the electronic control unit is integral with the steering wheel. On other models, the electronic control unit is mounted beneath the centre console, and may also control the passenger's side airbag and the seat belt tensioners, where fitted.

18 Removal and refitting of the steering wheel is described in Chapter 11.

19 If the control unit is mounted beneath the centre console, it should **not** be disturbed. If a fault is suspected, consult a Citroën dealer immediately – refer to the precautions given for working with airbags, in Section 20.

22 Cruise control system – general information

The cruise control system is vacuum-operated, and comprises the following components.

a) Vacuum pump.

b) Vacuum-operated throttle actuator.

c) Electronic control unit (ECU).

d) Road speed sensor.

e) Control switches mounted on the facia, steering wheel and brake and clutch pedals.

21.13a Depress the securing tabs . . .

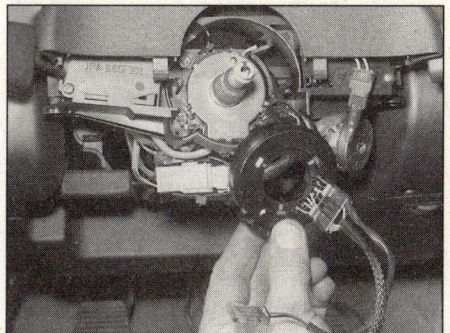

21.13b . . . then pull the rotary switch assembly from the steering column

13

The system allows a constant road speed to be maintained without the need to operate the throttle pedal. The desired 'cruising' speed can be set manually at speeds above 25 mph (40 km/h).

The speed is controlled by moving the throttle lever, by means of the throttle actuator. The throttle actuator is supplied with vacuum from the vacuum pump. The vacuum pump is controlled by the electronic control unit according to information provided by the road speed sensor and control switches.

The system does not affect the operation of the throttle pedal (although the pedal will move in accordance with the movement of the throttle actuator), and the system can be overridden at any time by depressing the accelerator, brake or clutch pedals, or using the main control switch.

23 Trip computer and auxiliary warning systems – general information

Certain models are fitted with a trip computer, which provides information on fuel consumption, average speed, range, and ambient temperature. The unit analyses information supplied by various sensors.

The trip computer display is mounted in the facia, below the instrument panel.

Similarly, some models are fitted with an auxiliary warning system, which provides visual warnings on a dot-matrix display panel, accompanied by an audible alarm chime. A variety of messages can be displayed using information from various existing sensors fitted to the vehicle (eg, washer fluid level sensor, front brake pad wear warning sensor, etc).

If problems are experienced with either the trip computer or the auxiliary warning system, consult a Citroën dealer or specialist for advice.

H29648
T.M.Marke

Key to symbols

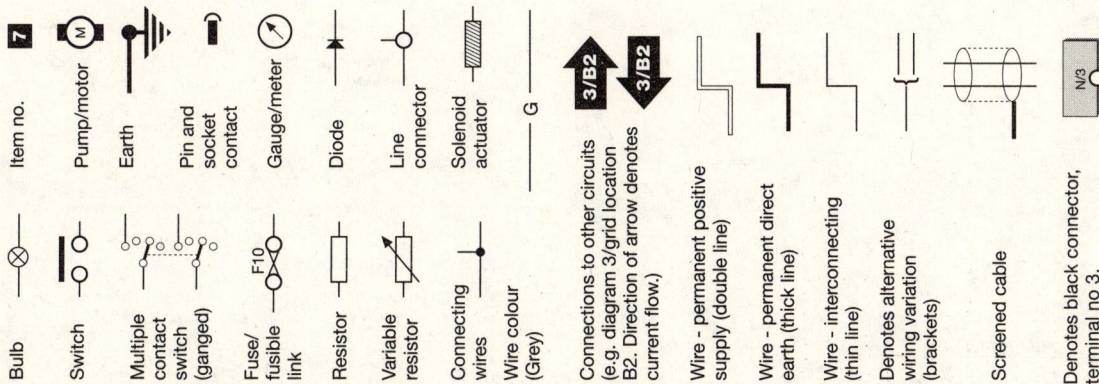

- Bulb
- Switch
- Multiple contact switch (ganged)
- Fuse/fusible link — F10
- Resistor
- Variable resistor
- Connecting wires
- Wire colour (Grey) — G
- Connections to other circuits (e.g. diagram 3/grid location B2. Direction of arrow denotes current flow.) — 3/B2
- Wire – permanent positive supply (double line)
- Wire – permanent direct earth (thick line)
- Wire – interconnecting (thin line)
- Denotes alternative wiring variation (brackets)
- Screened cable
- Denotes black connector, terminal no 3. — N/3
- Item no. — 7
- Pump/motor — M
- Earth
- Pin and socket contact
- Gauge/meter
- Diode
- Line connector
- Solenoid actuator

Typical passenger fusebox (in passenger compartment)

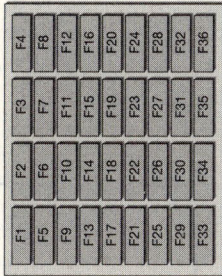

F4	F8	F12	F16	F20	F24	F28	F32	F36
F3	F7	F11	F15	F19	F23	F27	F31	F35
F2	F6	F10	F14	F18	F22	F26	F30	F34
F1	F5	F9	F13	F17	F21	F25	F29	F33

Typical Engine fusebox (in engine compartment)

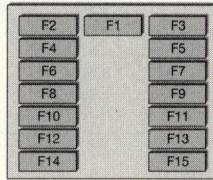

F2	F1	F3
F4		F5
F6		F7
F8		F9
F10		F11
F12		F13
F14		F15

Typical passenger fusebox

Fuse	Rating	Circuit protected
F1	10A	ABS
F2	10A	Stop lights
F3	10A	Heated rear window, A/C, heated seats
F4	5A	Language display
F5	30A	Radio, instrument panel, language display, heater temperature sensor
F6	5A	Rear foglights
F7	10A	Hydractive suspension
F8	5A	Language display
F9	30A	Controls for: heated rear window, front/rear wiper, A/C, electric windows, lights-on warning buzzer, flasher unit
F10	10A	Interior lighting
F11	10A	Interior lighting
F12	5A	Language display
F13	10A	Hazard warning lights, diagnostic socket
F14	5A	RH tail light, lighting rheostat
F15	5A	Temperature interface
F16	5A	Coolant temp. warning light, display supply
F17	20A	Heated rear seat
F18	5A	LH tail light
F19	15A	Reversing lights, instruments, diagnostic socket, compressor cut-off relay, lighting timer, temperature interface
F20	5A	Spare
F21	15A	Front foglights
F22/23	Shunt	Transit shunt (supply cut-off)
F24	20A	Heated front seats
F25	20A	Opening roof
F26	30A	Electric rear windows
F27	15A	Instruments, interior lighting, trip computer, clock, radio, alarm
F28	30A	RH electric front seat
F29	15A	Spare
F30	30A	LH electric front seat
F31	10A	Temperature interface - injection
F32	30A	Headlight washer, horn
F33	30A	Heated screen, heated mirrors
F34	30A	Central locking, Hydractive suspension
F35	30A	Front windows
F36	30A	Front/rear cigar lighters

Typical engine fusebox

Fuse	Rating	Circuit protected
F1	10A	Supply to EGR control unit
F2	30A	Supply to air conditioning
F3	30A	ABS
F4	-	Not used
F5	-	Not used
F6	-	Not used
F7	-	Not used
F8	10A	RH main beam
F9	-	Not used
F10	10A	LH main beam
F11	-	Not used
F12	10A	LH dipped beam
F13	-	Not used
F14	-	Not used
F15	10A	RH dipped beam

Diagram 1 : Information for wiring diagrams

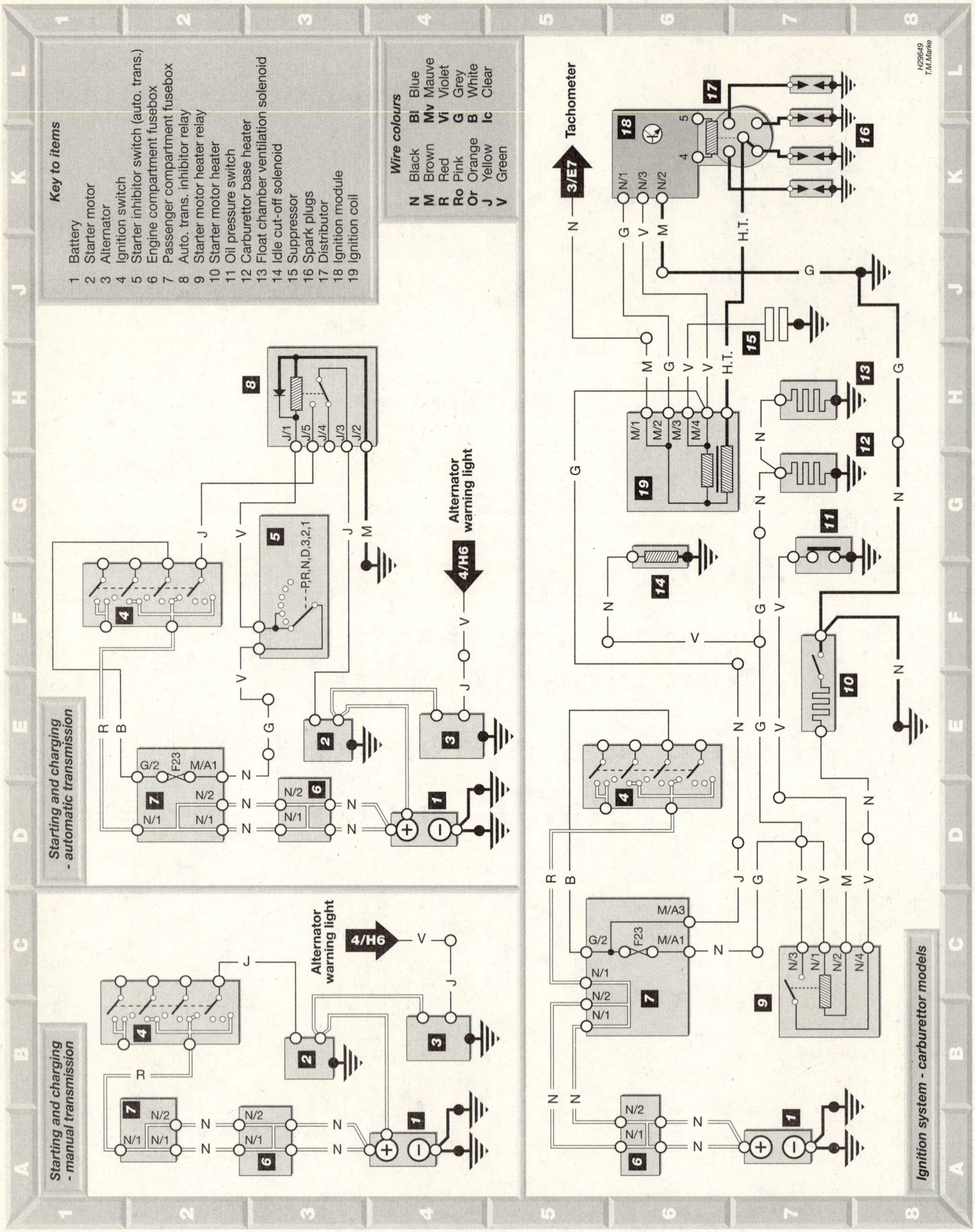

Key to items

1 Battery
2 Starter motor
3 Alternator
4 Ignition switch
5 Starter inhibitor switch (auto. trans.)
6 Engine compartment fusebox
7 Passenger compartment fusebox
8 Auto. trans. inhibitor relay
9 Starter motor heater relay
10 Starter motor heater
11 Oil pressure switch
12 Carburettor base heater
13 Float chamber ventilation solenoid
14 Idle cut-off solenoid
15 Suppressor
16 Spark plugs
17 Distributor
18 Ignition module
19 Ignition coil

Wire colours

N	Black	Bl	Blue
M	Brown	Mv	Mauve
R	Red	Vi	Violet
Ro	Pink	G	Grey
Or	Orange	B	White
J	Yellow	Ic	Clear
V	Green		

Starting and charging - automatic transmission

Starting and charging - manual transmission

Ignition system - carburettor models

Alternator warning light

Tachometer

H29649
T.M.Marke

Diagram 2 : Typical starting, charging and ignition system (carburettor models)

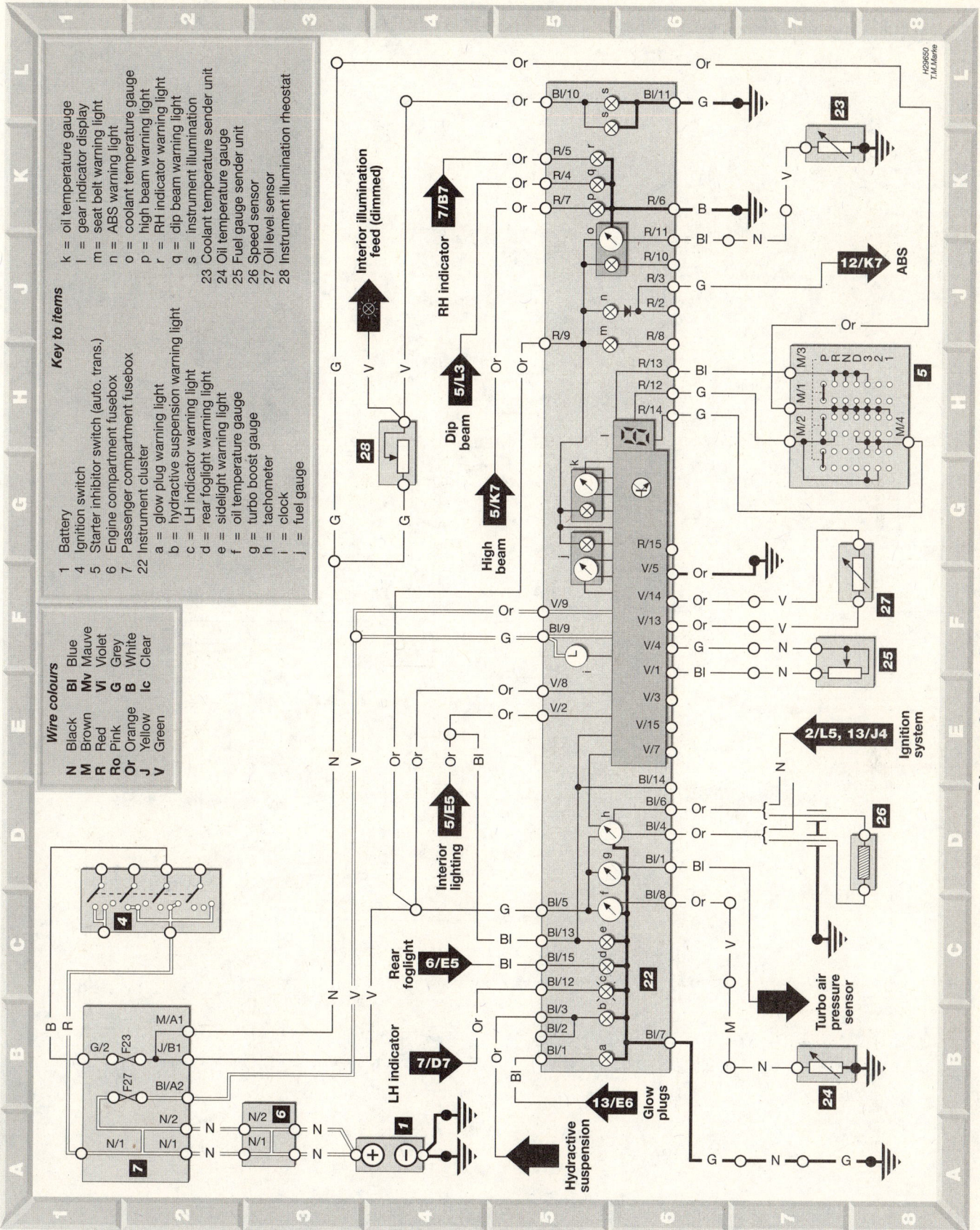

Key to items

1 Battery
4 Ignition switch
5 Starter inhibitor switch (auto. trans.)
6 Engine compartment fusebox
7 Passenger compartment fusebox
22 Instrument cluster
a = glow plug warning light
b = hydractive suspension warning light
c = LH indicator warning light
d = rear foglight warning light
e = sidelight warning light
f = oil temperature gauge
g = turbo boost gauge
h = tachometer
i = clock
j = fuel gauge
k = oil temperature gauge
l = gear indicator display
m = seat belt warning light
n = ABS warning light
o = coolant temperature gauge
p = high beam warning light
r = RH indicator warning light
q = dip beam warning light
s = instrument illumination
23 Coolant temperature sender unit
24 Oil temperature sender unit
25 Fuel gauge sender unit
26 Speed sensor
27 Oil level sensor
28 Instrument illumination rheostat

Wire colours

N Black   Bl   Blue
M Brown   Mv Mauve
R Red   Vi   Violet
Ro Pink   G   Grey
Or Orange   B   White
J Yellow   Ic   Clear
V Green

Interior illumination feed (dimmed)

RH indicator

7/B7

Dip beam 5/L3

High beam 5/K7

Interior lighting 5/E5

Rear foglight 6/E5

LH indicator 7/D7

Hydractive suspension

Glow plugs 13/E6

ABS 12/K7

Ignition system 2/L5, 13/J4

Turbo air pressure sensor

Or, G, V, N, Bl, M

Bl/10, Bl/11, R/5, R/4, R/7, R/6, R/11, R/10, R/3, R/2, R/9, R/8, R/13, R/12, R/14, R/15

V/9, V/5, V/14, V/13, V/4, V/1, V/8, V/3, V/15, V/7, V/2

Bl/14, Bl/6, Bl/4, Bl/1, Bl/8, Bl/5, Bl/13, Bl/15, Bl/12, Bl/3, Bl/2, Bl/7, Bl/9

M/3, M/1, M/2, M/4

P R N D 3 2 1

23, 5, 27, 25, 26, 24, 22, 28, 4, 6, 7, 1

M/A1, J/B1, Bl/A2, G/2, F23, F27, N/2, N/1, B, R

H29650 T.M.Marke

Diagram 3 : Typical warning lights and gauges

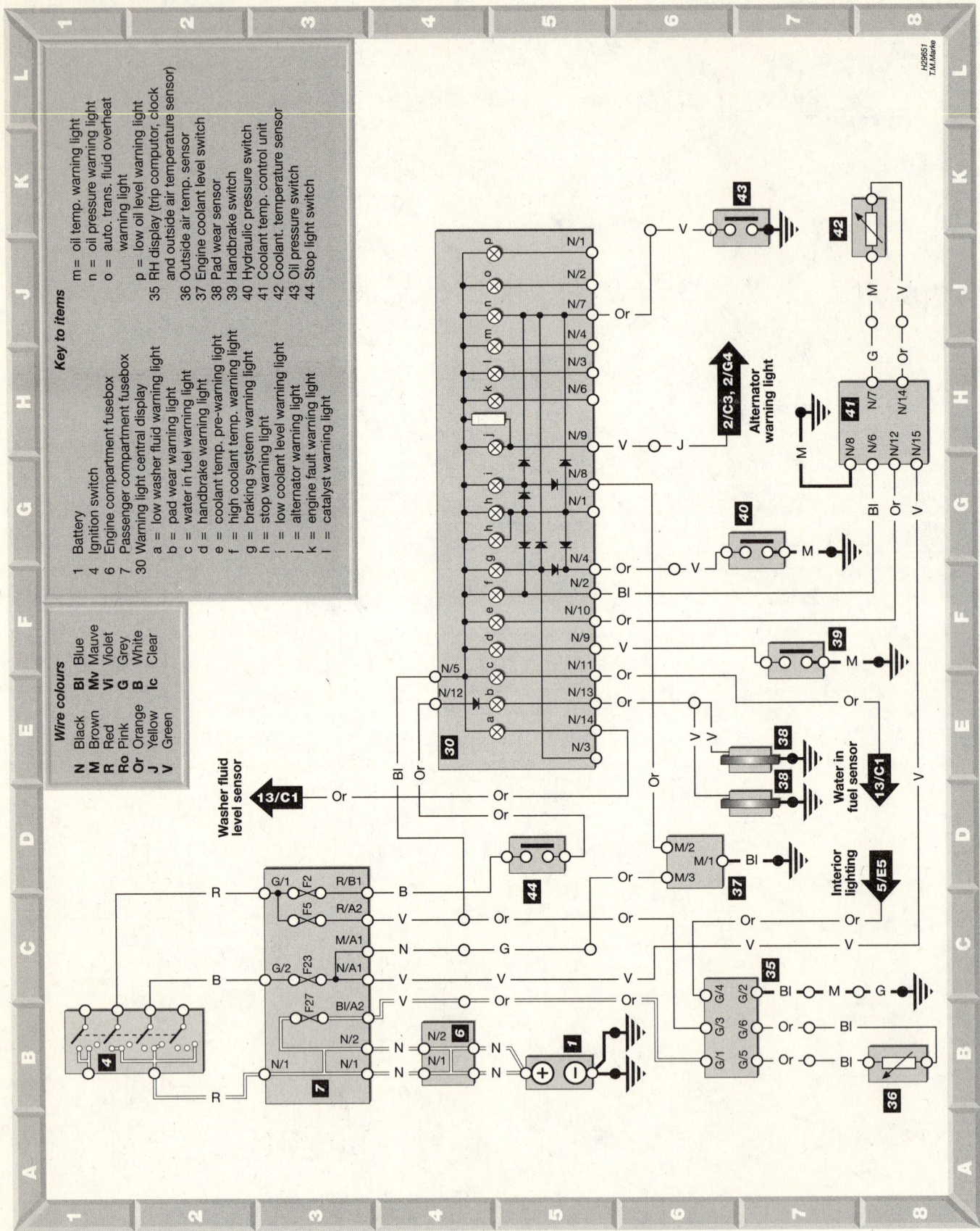

H29651
T.M.Marke

Key to items

1 Battery
4 Ignition switch
6 Engine compartment fusebox
7 Passenger compartment fusebox
30 Warning light central display
  a = low washer fluid warning light
  b = pad wear warning light
  c = water in fuel warning light
  d = handbrake warning light
  e = coolant temp. pre-warning light
  f = high coolant temp. warning light
  g = braking system warning light
  h = stop warning light
  i = low coolant level warning light
  j = alternator warning light
  k = engine fault warning light
  l = catalyst warning light
  m = oil temp. warning light
  n = oil pressure warning light
  o = auto. trans. fluid overheat warning light
  p = low oil level warning light
35 RH display (trip compurtor, clock and outside air temperature sensor)
36 Outside air temp. sensor
37 Engine coolant level switch
38 Pad wear sensor
39 Handbrake switch
40 Hydraulic pressure switch
41 Coolant temp. control unit
42 Coolant. temperature sensor
43 Oil pressure switch
44 Stop light switch

Wire colours

N	Black	Bl	Blue
M	Brown	Mv	Mauve
R	Red	Vi	Violet
Ro	Pink	G	Grey
Or	Orange	B	White
J	Yellow	Ic	Clear
V	Green		

Diagram 4 : Typical central display unit

H29652
T.M.Marke

Wire colours

N	Black	Bl	Blue
M	Brown	Mv	Mauve
R	Red	Vi	Violet
Ro	Pink	G	Grey
Or	Orange	B	White
J	Yellow	Ic	Clear
V	Green		

Key to items

1 Battery
4 Ignition switch
6 Engine compartment fusebox
50 Bulb failure unit
51 Light switch
52 LH number plate light
53 RH number plate light
54 LH sidelight
55 RH sidelight
56 LH stop/tail light
57 RH stop/tail light
58 LH tail light (in tailgate)
59 RH tail light (in tailgate)
60 Lights-on warning buzzer
61 LH front door switch
62 RH front door switch
63 LH rear door switch
64 RH rear door switch
65 LH headlight
66 RH headlight
67 Dip beam relay
68 LH switch (lighting, flasher, horn)

Dip beam warning light 3/H4

Main beam warning light 3/G5

Headlights

Sidelights

Interior lighting feed

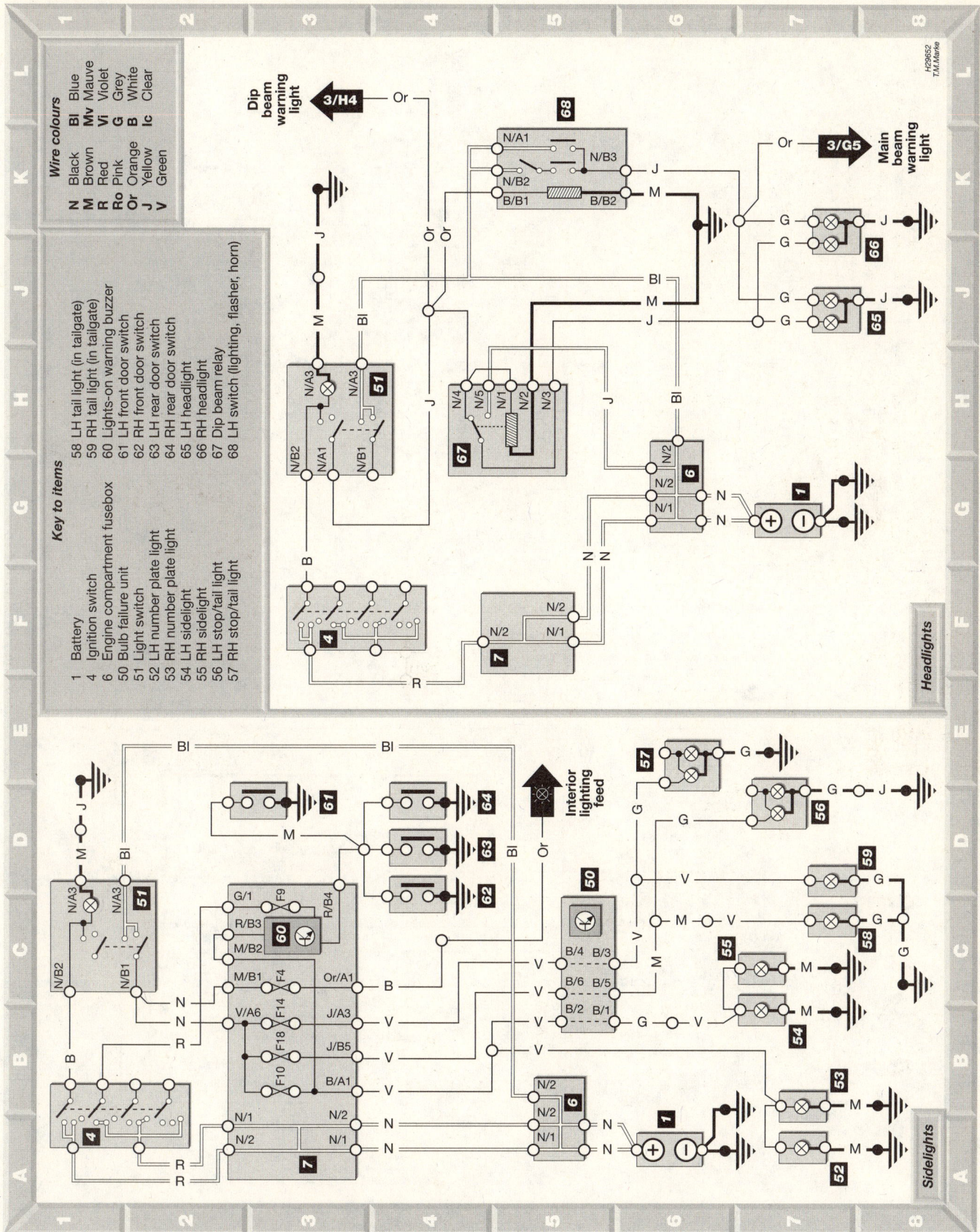

Diagram 5 : Typical exterior lighting - side and headlights

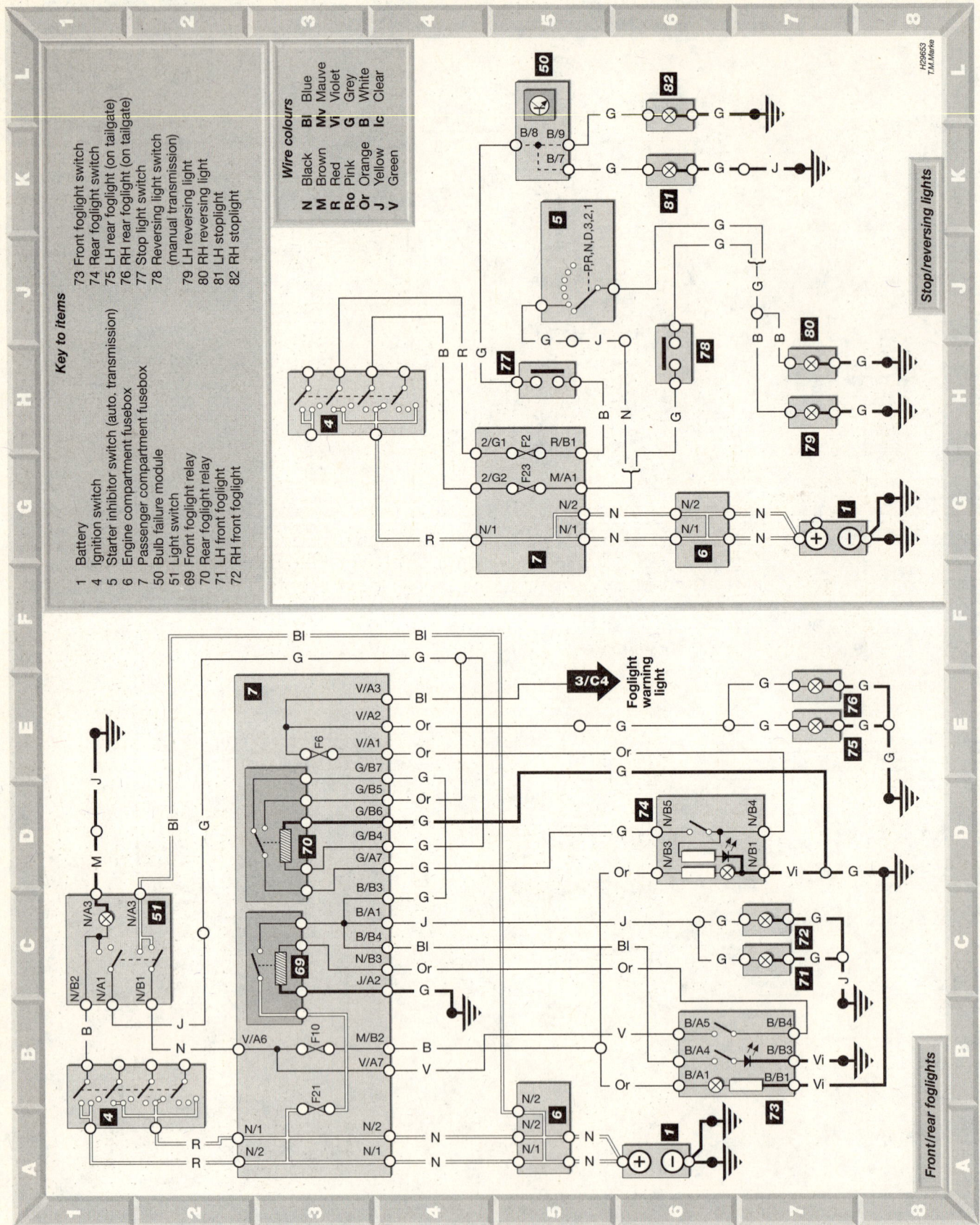

Key to items

1 Battery
4 Ignition switch
5 Starter inhibitor switch (auto. transmission)
6 Engine compartment fusebox
7 Passenger compartment fusebox
50 Bulb failure module
51 Light switch
69 Front foglight relay
70 Rear foglight relay
71 LH front foglight
72 RH front foglight
73 Front foglight switch
74 Rear foglight switch
75 LH rear foglight (on tailgate)
76 RH rear foglight (on tailgate)
77 Stop light switch
78 Reversing light switch (manual transmission)
79 LH reversing light
80 RH reversing light
81 LH stoplight
82 RH stoplight

Wire colours

N Black   Bl Blue
M Brown   Mv Mauve
R Red   Vi Violet
Ro Pink   G Grey
Or Orange   B White
J Yellow   Ic Clear
V Green

Stop/reversing lights

Front/rear foglights

Diagram 6 : Typical exterior lighting - front/rear foglights, stop and reversing lights

Horn

Cigar lighter and interior illumination

Key to items

1 Battery
2 Ignition switch
6 Engine compartment fusebox
7 Passenger compartment fusebox
51 Light switch
68 LH switch (lighting, flasher, horn)
85 Hazard warning switch
86 LH front direction indicator
87 LH indicator side repeater
88 RH front direction indicator
89 RH indicator side repeater
90 LH rear direction indicator
91 RH rear direction indicator
92 Direction indicator flasher unit
93 Horn
94 Cigar lighter
95 Ashtray illumination
96 Ashtray lid control light
97 Radio panel illumination
98 Luggage compartment light
99 Luggage compartment light switch
100 Glovebox light
101 Glovebox light switch

Wire colours

N Black   Bl Blue
M Brown   Mv Mauve
R Red   Vi Violet
Ro Pink   G Grey
Or Orange   B White
J Yellow   Ic Clear
V Green

Direction indicators and hazard warning lights

Interior lighting

LH indicator warning light

RH indicator warning light

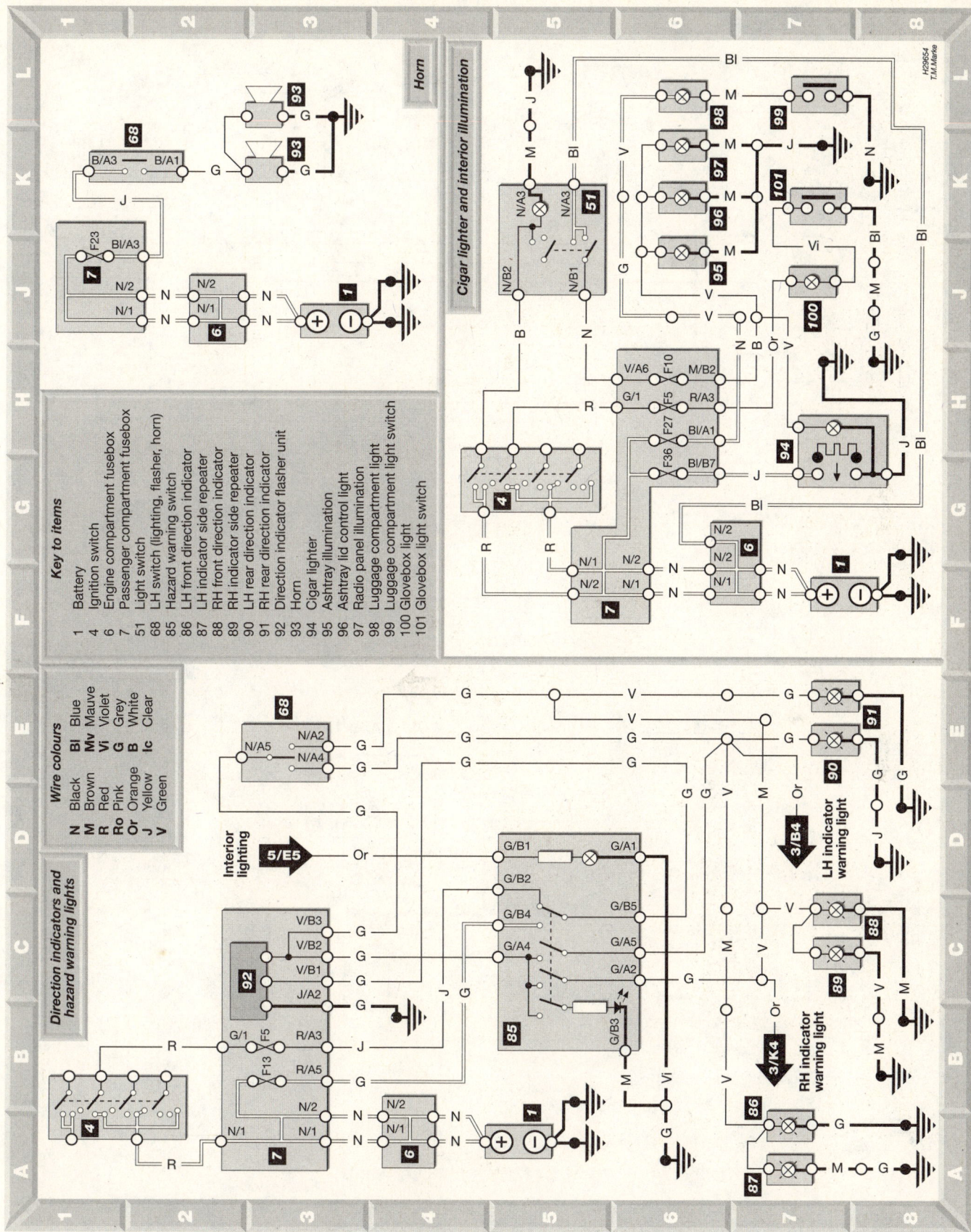

Diagram 7 : Typical exterior lighting – direction indicators, hazard warning and typical interior lighting

Front/rear wash/wipe

H29655
T.M.Marie

Wire colours

N	Black	BI	Blue
M	Brown	Mv	Mauve
R	Red	Vi	Violet
Ro	Pink	G	Grey
Or	Orange	B	White
J	Yellow	Ic	Clear
V	Green		

9/C3 Headlight washer

Washer level warning light 4/E3

Interior lighting 5/E5

Door Locking control unit

12/D2

12/E6

Interior lighting

Key to items

1 Battery
4 Ignition switch
6 Engine compartment fusebox
7 Passenger compartment fusebox
51 Light switch
61 LH front door switch
62 RH front door switch
63 LH rear door switch
64 RH rear door switch
105 Interior lighting timer unit
106 Centre roof light
107 Roof light switch
108 Front wiper relay
109 Rear wiper relay
110 Washer level sensor
111 Windscreen washer pump
112 Windscreen wiper motor
113 Front wiper switch
114 Rear wiper motor
115 Rear wiper switch
116 Rear washer pump

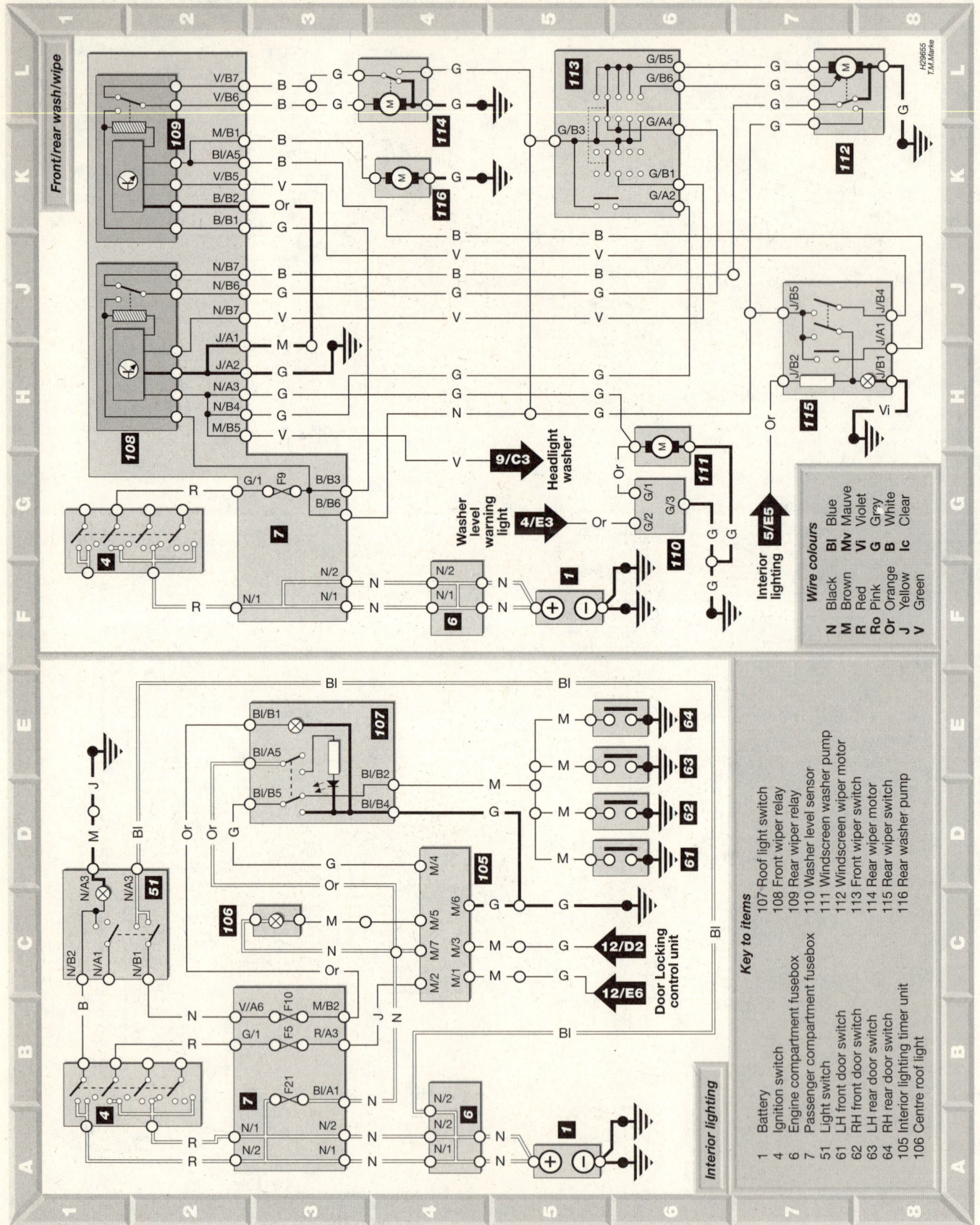

Diagram 8 : Typical interior lighting continued, front and rear wash/wipe

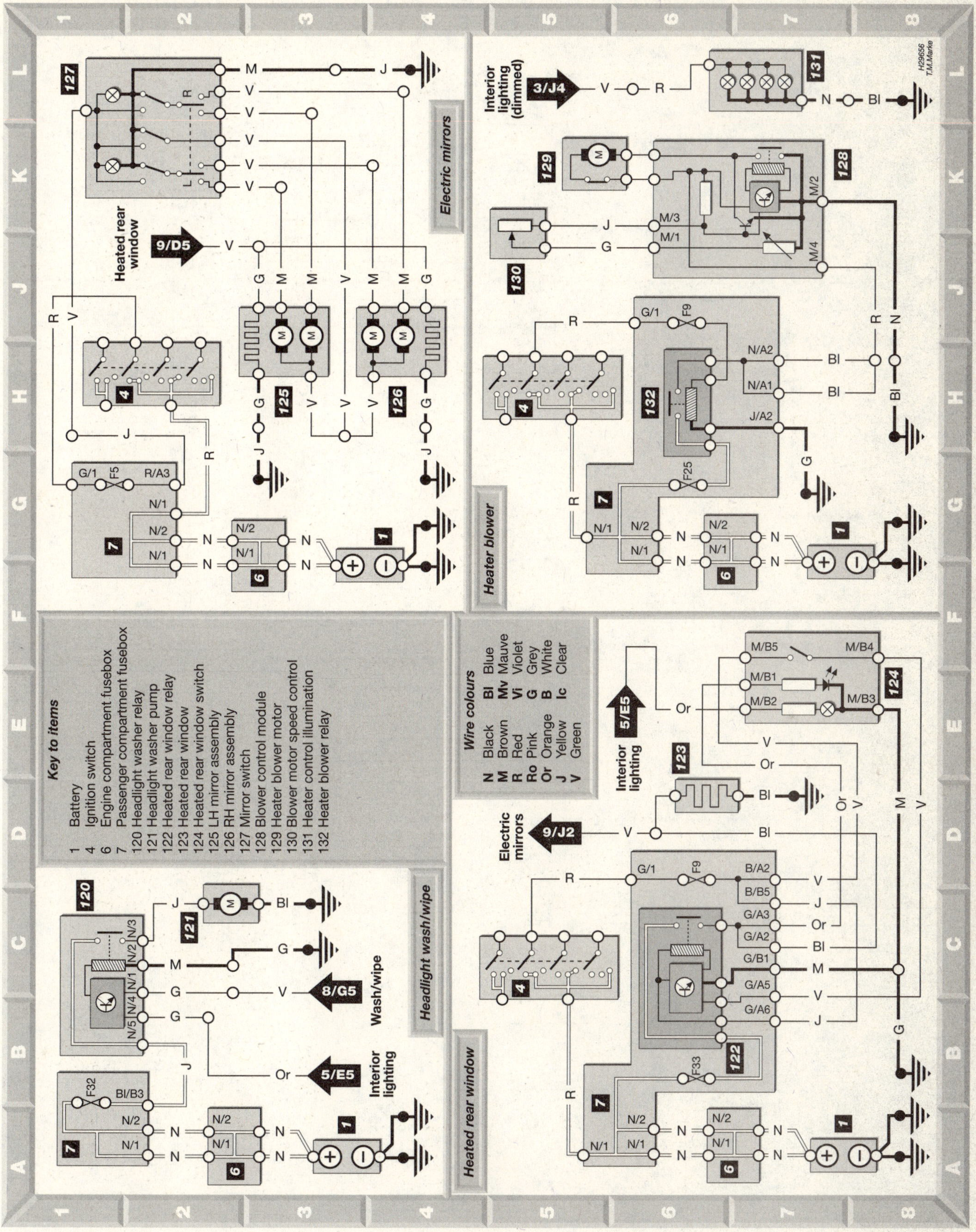

Electric mirrors

Heated rear window 9/D5

Interior lighting (dimmed) 3/J4

Heater blower

Key to items

1 Battery
4 Ignition switch
6 Engine compartment fusebox
7 Passenger compartment fusebox
120 Headlight washer relay
121 Headlight washer pump
122 Heated rear window relay
123 Heated rear window
124 Heated rear window switch
125 LH mirror assembly
126 RH mirror assembly
127 Mirror switch
128 Blower control module
129 Heater blower motor
130 Blower motor speed control
131 Heater control illumination
132 Heater blower relay

Wire colours

N	Black	Bl	Blue
M	Brown	Mv	Mauve
R	Red	Vi	Violet
Ro	Pink	G	Grey
Or	Orange	B	White
J	Yellow	Ic	Clear
V	Green		

Interior lighting 5/E5

Headlight wash/wipe

Wash/wipe 8/G5

Interior lighting 5/E5

Electric mirrors 9/J2

Heated rear window

H29656
T.M.Marke

Diagram 9 : Typical headlight washer, heater blower, heated rear window and electric mirrors

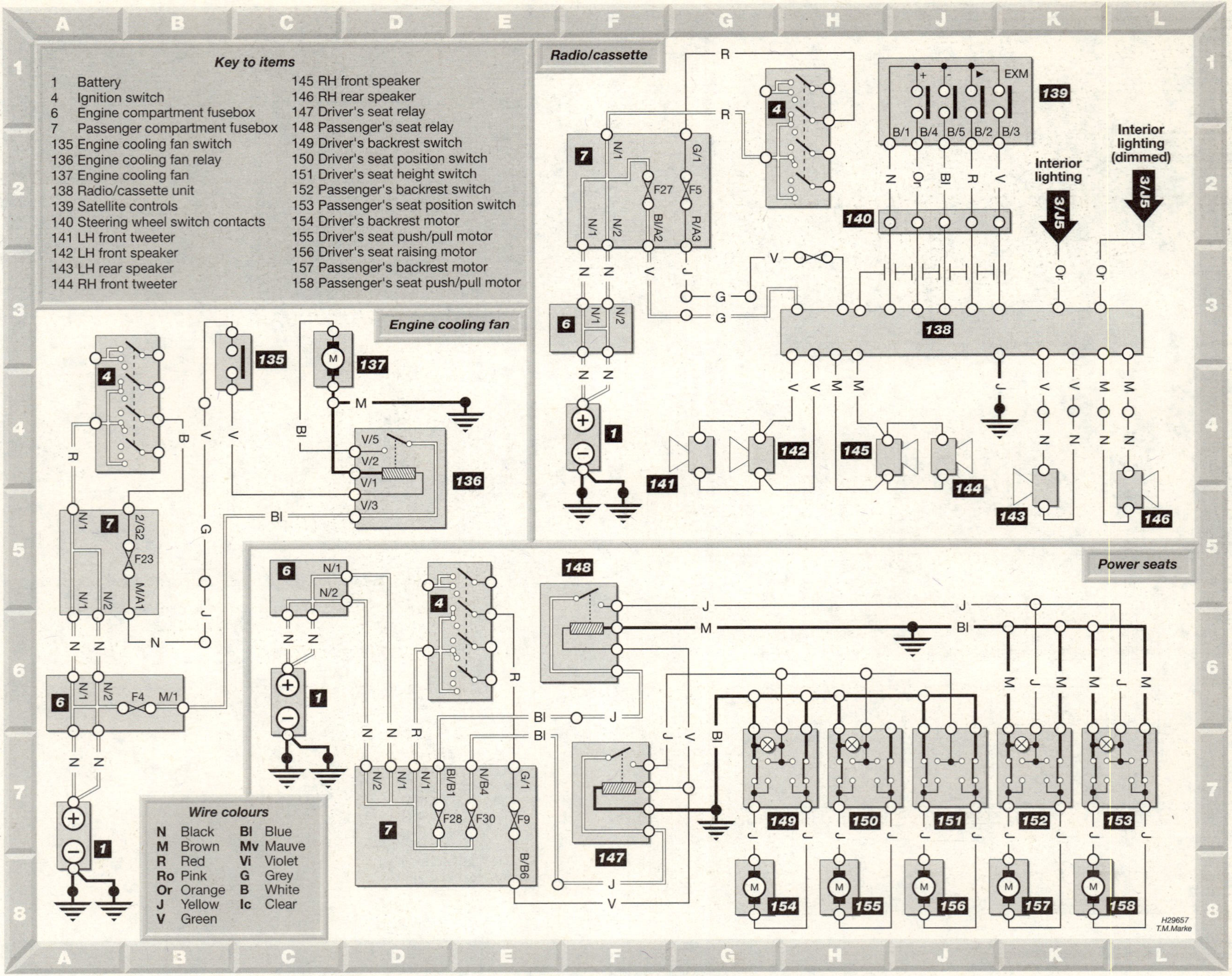

H29857
T.M.Marke

Interior lighting (dimmed)

3/J5

Interior lighting

3/J5

Power seats

Radio/cassette

Engine cooling fan

Key to items

1   Battery
4   Ignition switch
6   Engine compartment fusebox
7   Passenger compartment fusebox
135 Engine cooling fan switch
136 Engine cooling fan relay
137 Engine cooling fan
138 Radio/cassette unit
139 Satellite controls
140 Steering wheel switch contacts
141 LH front tweeter
142 LH front speaker
143 LH rear speaker
144 RH front tweeter

145 RH front speaker
146 RH rear speaker
147 Driver's seat relay
148 Passenger's seat relay
149 Driver's seat switch
150 Driver's seat position switch
151 Driver's seat height switch
152 Passenger's backrest switch
153 Passenger's seat position switch
154 Driver's backrest motor
155 Driver's seat push/pull motor
156 Driver's seat raising motor
157 Passenger's backrest motor
158 Passenger's seat push/pull motor

Wire colours

N	Black	**Bl**	Blue
M	Brown	**Mv**	Mauve
R	Red	**Vi**	Violet
Ro	Pink	**G**	Grey
Or	Orange	**B**	White
J	Yellow	**Ic**	Clear
V	Green		

Diagram 10 : Typical engine cooling fan, radio/cassette and power seats

Key to items

1 Battery
4 Ignition switch
6 Engine compartment fusebox
7 Passenger compartment fusebox
160 Front window control relay
161 Rear window control relay
162 Driver's door switch
163 Passenger's window switch (on driver's door)
164 Window control unit
165 Passenger's window switch (on passenger's door)
166 LH front window motor
167 RH front window motor
168 LH rear window switch (on console)
169 LH rear window switch (on door)
170 LH rear window motor
171 Rear window isolator
172 RH rear window switch (on console)
173 RH rear window switch (on door)
174 RH rear window motor

Wire colours

N	Black	Bl	Blue
M	Brown	Mv	Mauve
R	Red	Vi	Violet
Ro	Pink	G	Grey
Or	Orange	B	White
J	Yellow	Ic	Clear
V	Green		

5/E5 Interior illumination

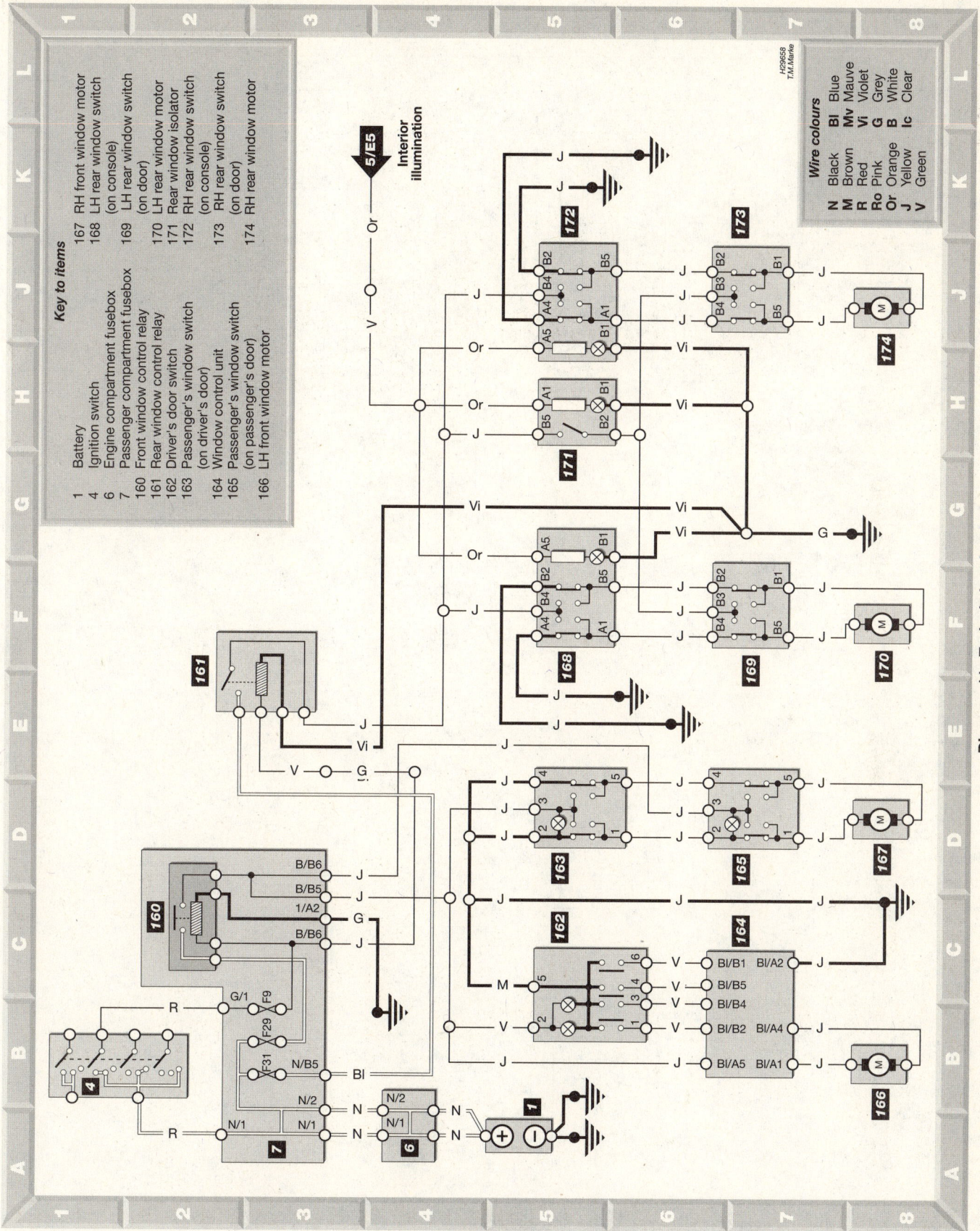

Diagram 11 : Typical electric windows

H29658
T.M.Marie

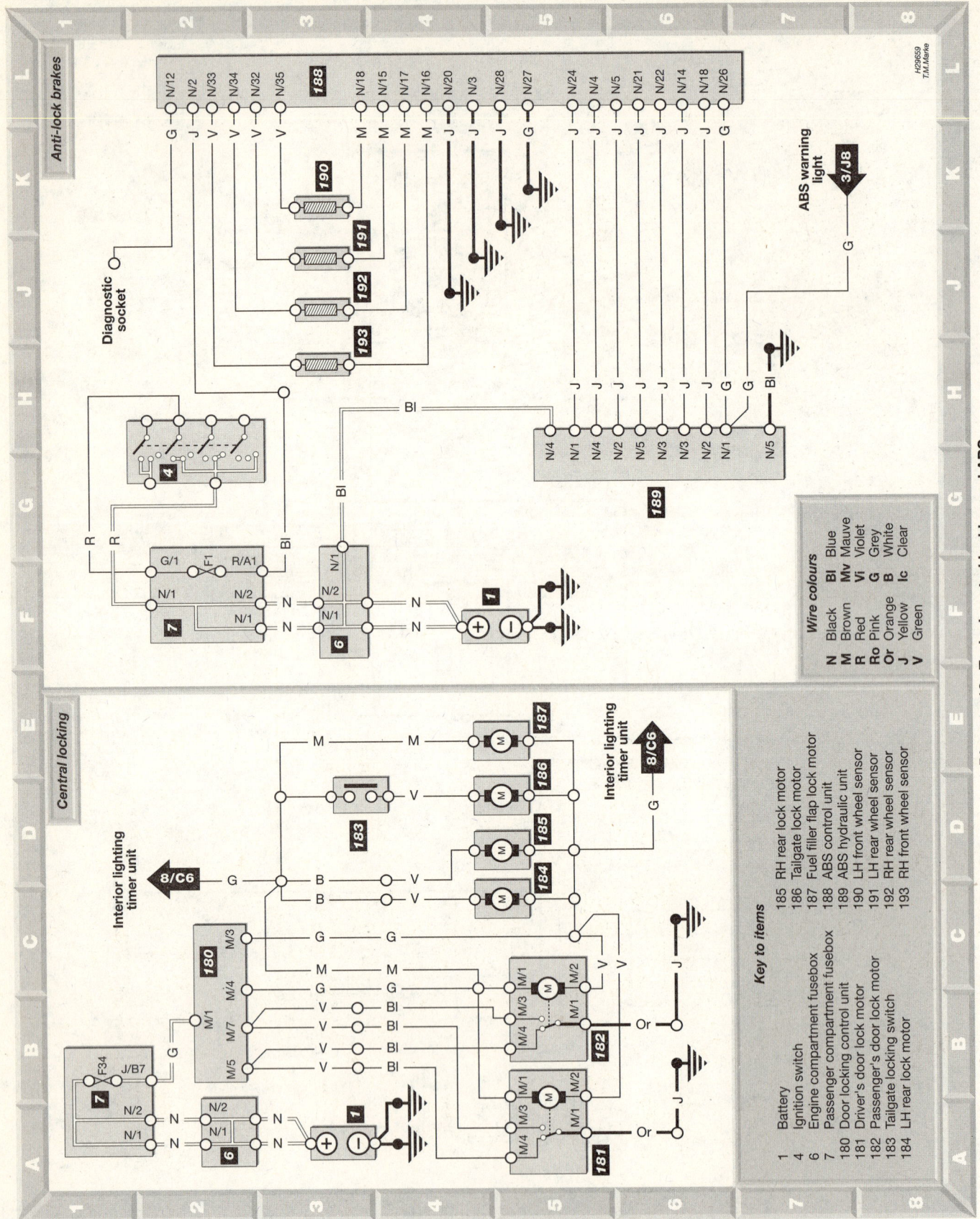

H29659
T.M.Marke

Anti-lock brakes

Diagnostic socket

ABS warning light

3/J8

Wire colours

N	Black	BI	Blue
M	Brown	Mv	Mauve
R	Red	Vi	Violet
Ro	Pink	G	Grey
Or	Orange	B	White
J	Yellow	Ic	Clear
V	Green		

Central locking

Interior lighting timer unit

Interior lighting timer unit

8/C6

8/C6

Key to items

1	Battery	185	RH rear lock motor
4	Ignition switch	186	Tailgate lock motor
6	Engine compartment fusebox	187	Fuel filler flap lock motor
7	Passenger compartment fusebox	188	ABS control unit
180	Door locking control unit	189	ABS hydraulic unit
181	Driver's door lock motor	190	LH front wheel sensor
182	Passenger's door lock motor	191	LH rear wheel sensor
183	Tailgate locking switch	192	RH rear wheel sensor
184	LH rear lock motor	193	RH front wheel sensor

Diagram 12 : Typical central locking and ABS

Fuel injection

H29660
T.M.Marke

Tachometer

3/E8

Wire colours

N	Black	Bl	Blue
M	Brown	Mv	Mauve
R	Red	Vi	Violet
Ro	Pink	G	Grey
Or	Orange	B	White
J	Yellow	Ic	Clear
V	Green		

Diesel - preheating

Water in fuel warning light

4/E8

Glow plug warning light

3/B6

Key to items

1	Battery
4	Ignition switch
6	Engine compartment fusebox
7	Passenger compartment fusebox
195	Preheater control unit
196	Glow plugs
197	Fuel pump solenoid
198	Pump advance solenoid/switch
199	Water in fuel sensor
200	Fuel injection control unit
201	Fuel injectors
202	Throttle switch
203	Air flow meter
204	Additional air control
205	Injection relay
206	Distributor
207	Ignition coil
208	Ignition module
209	Suppressor
210	Spark plugs
211	Fuel pump

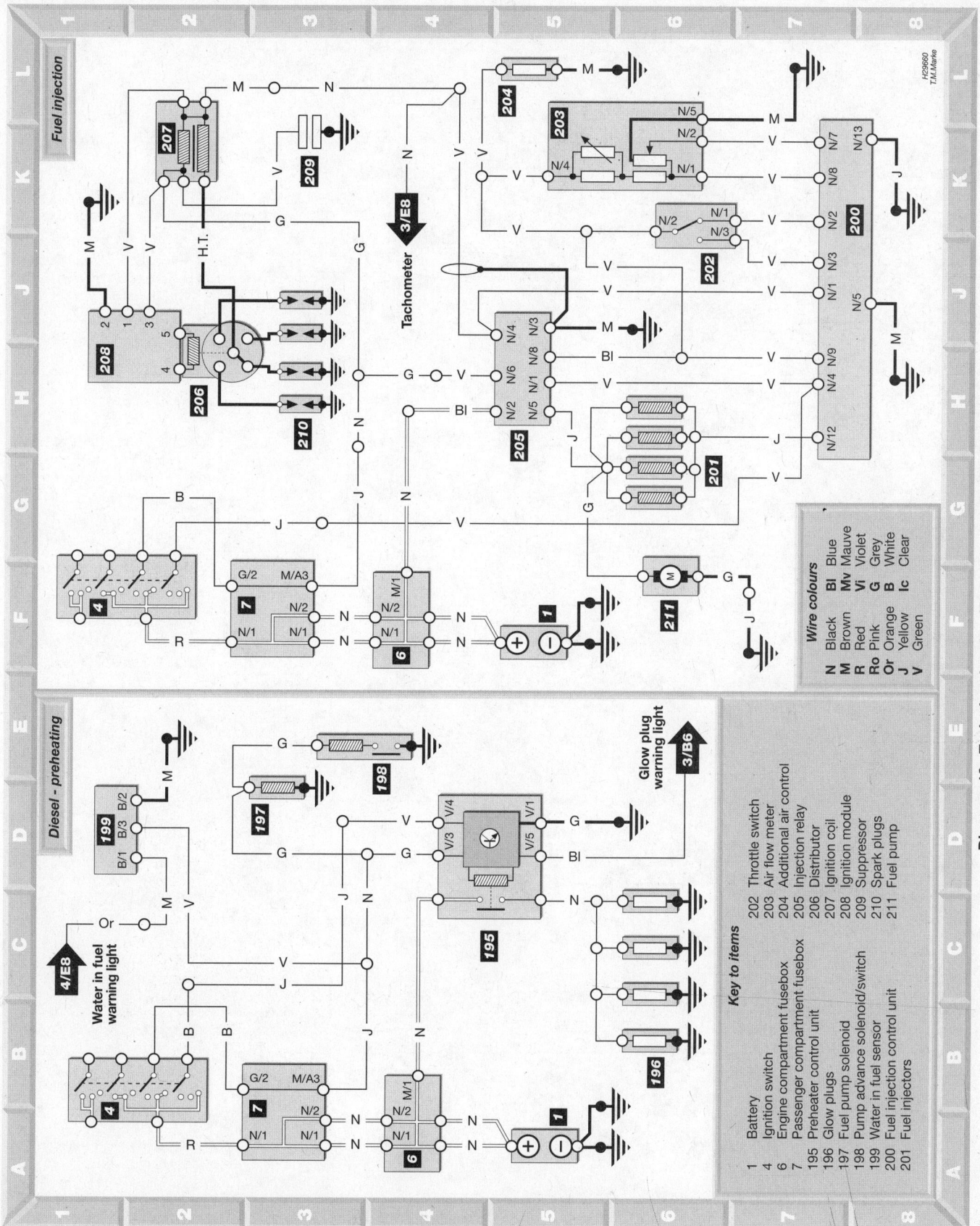

Diagram 13 : Typical preheating (Diesel) and fuel injection (petrol)

Notes

Reference REF•1

Dimensions and weights

Note: *All figures are approximate, and may vary according to model. Refer to manufacturer's data for exact figures.*

Dimensions

Overall width:
  Excluding mirrors .1790 mm
  Including mirrors . . . . .2070 mm

	Hatchback	Estate
Overall length	4710 mm	4960 mm
Overall height (unladen)	1380 mm	1470 mm
Wheelbase	2446 mm	2700 mm
Front track	1530 mm	1520 mm
Rear track	1450 mm	

Weights

Kerb weight (unladen):

	Manual	Auto
Hatchback models:		
Petrol engine models except turbo	1514 kg	1559 kg
Turbo petrol engine models	1534 kg	1568 kg
2.1 litre diesel engine models	1577 kg	1603 kg
2.5 litre diesel engine models	1587 kg	
Estate models:		
Petrol engine models except turbo	1577 kg	1604 kg
Turbo petrol engine models	1573 kg	1608 kg
2.1 litre diesel engine models	1623 kg	1654 kg
2.5 litre diesel engine models	1639 kg	

Weights (continued)

Maximum gross vehicle weight:

	Manual	Auto
Hatchback models:		
Petrol engine models except turbo	3220 kg	3265 kg
Turbo petrol engine models	3240 kg	3270 kg
2.1 litre diesel engine models	3480 kg	3505 kg
2.5 litre diesel engine models	3580 kg	
Estate models:		
Petrol engine models except turbo	3370 kg	3410 kg
Turbo petrol engine models	3400 kg	3425 kg
2.1 litre diesel engine models	3620 kg	3650 kg
2.5 litre diesel engine models	3740 kg	

Maximum towing weights:
  Trailer with brakes:
    Petrol engine models . . . . .1300 kg
    diesel engine models . . . . .1500 kg
  Maximum trailer nose weight:
    Hatchback models . . . . .110 kg
    Estate models . . . . .80 kg
Maximum roof rack load:
  Hatchback models . . . . .80 kg
  Estate models . . . . .100 kg

Conversion factors

Length (distance)

Inches (in)	x 25.4	= Millimetres (mm)	x 0.0394	= Inches (in)
Feet (ft)	x 0.305	= Metres (m)	x 3.281	= Feet (ft)
Miles	x 1.609	= Kilometres (km)	x 0.621	= Miles

Volume (capacity)

Cubic inches (cu in; in³)	x 16.387	= Cubic centimetres (cc; cm³)	x 0.061	= Cubic inches (cu in; in³)
Imperial pints (Imp pt)	x 0.568	= Litres (l)	x 1.76	= Imperial pints (Imp pt)
Imperial quarts (Imp qt)	x 1.137	= Litres (l)	x 0.88	= Imperial quarts (Imp qt)
Imperial quarts (Imp qt)	x 1.201	= US quarts (US qt)	x 0.833	= Imperial quarts (Imp qt)
US quarts (US qt)	x 0.946	= Litres (l)	x 1.057	= US quarts (US qt)
Imperial gallons (Imp gal)	x 4.546	= Litres (l)	x 0.22	= Imperial gallons (Imp gal)
Imperial gallons (Imp gal)	x 1.201	= US gallons (US gal)	x 0.833	= Imperial gallons (Imp gal)
US gallons (US gal)	x 3.785	= Litres (l)	x 0.264	= US gallons (US gal)

Mass (weight)

Ounces (oz)	x 28.35	= Grams (g)	x 0.035	= Ounces (oz)
Pounds (lb)	x 0.454	= Kilograms (kg)	x 2.205	= Pounds (lb)

Force

Ounces-force (ozf; oz)	x 0.278	= Newtons (N)	x 3.6	= Ounces-force (ozf; oz)
Pounds-force (lbf; lb)	x 4.448	= Newtons (N)	x 0.225	= Pounds-force (lbf; lb)
Newtons (N)	x 0.1	= Kilograms-force (kgf; kg)	x 9.81	= Newtons (N)

Pressure

Pounds-force per square inch (psi; lbf/in²; lb/in²)	x 0.070	= Kilograms-force per square centimetre (kgf/cm²; kg/cm²)	x 14.223	= Pounds-force per square inch (psi; lbf/in²; lb/in²)
Pounds-force per square inch (psi; lbf/in²; lb/in²)	x 0.068	= Atmospheres (atm)	x 14.696	= Pounds-force per square inch (psi; lbf/in²; lb/in²)
Pounds-force per square inch (psi; lbf/in²; lb/in²)	x 0.069	= Bars	x 14.5	= Pounds-force per square inch (psi; lbf/in²; lb/in²)
Pounds-force per square inch (psi; lbf/in²; lb/in²)	x 6.895	= Kilopascals (kPa)	x 0.145	= Pounds-force per square inch (psi; lbf/in²; lb/in²)
Kilopascals (kPa)	x 0.01	= Kilograms-force per square centimetre (kgf/cm²; kg/cm²)	x 98.1	= Kilopascals (kPa)
Millibar (mbar)	x 100	= Pascals (Pa)	x 0.01	= Millibar (mbar)
Millibar (mbar)	x 0.0145	= Pounds-force per square inch (psi; lbf/in²; lb/in²)	x 68.947	= Millibar (mbar)
Millibar (mbar)	x 0.75	= Millimetres of mercury (mmHg)	x 1.333	= Millibar (mbar)
Millibar (mbar)	x 0.401	= Inches of water (inH₂O)	x 2.491	= Millibar (mbar)
Millimetres of mercury (mmHg)	x 0.535	= Inches of water (inH₂O)	x 1.868	= Millimetres of mercury (mmHg)
Inches of water (inH₂O)	x 0.036	= Pounds-force per square inch (psi; lbf/in²; lb/in²)	x 27.68	= Inches of water (inH₂O)

Torque (moment of force)

Pounds-force inches (lbf in; lb in)	x 1.152	= Kilograms-force centimetre (kgf cm; kg cm)	x 0.868	= Pounds-force inches (lbf in; lb in)
Pounds-force inches (lbf in; lb in)	x 0.113	= Newton metres (Nm)	x 8.85	= Pounds-force inches (lbf in; lb in)
Pounds-force inches (lbf in; lb in)	x 0.083	= Pounds-force feet (lbf ft; lb ft)	x 12	= Pounds-force inches (lbf in; lb in)
Pounds-force feet (lbf ft; lb ft)	x 0.138	= Kilograms-force metres (kgf m; kg m)	x 7.233	= Pounds-force feet (lbf ft; lb ft)
Pounds-force feet (lbf ft; lb ft)	x 1.356	= Newton metres (Nm)	x 0.738	= Pounds-force feet (lbf ft; lb ft)
Newton metres (Nm)	x 0.102	= Kilograms-force metres (kgf m; kg m)	x 9.804	= Newton metres (Nm)

Power

Horsepower (hp)	x 745.7	= Watts (W)	x 0.0013	= Horsepower (hp)

Velocity (speed)

Miles per hour (miles/hr; mph)	x 1.609	= Kilometres per hour (km/hr; kph)	x 0.621	= Miles per hour (miles/hr; mph)

Fuel consumption*

Miles per gallon (mpg)	x 0.354	= Kilometres per litre (km/l)	x 2.825	= Miles per gallon (mpg)

Temperature

Degrees Fahrenheit = (°C x 1.8) + 32   Degrees Celsius (Degrees Centigrade; °C) = (°F - 32) x 0.56

It is common practice to convert from miles per gallon (mpg) to litres/100 kilometres (l/100km), where mpg x l/100 km = 282

Spare parts are available from many sources, including maker's appointed garages, accessory shops, and motor factors. To be sure of obtaining the correct parts, it will sometimes be necessary to quote the vehicle identification number. If possible, it can also be useful to take the old parts along for positive identification. Items such as starter motors and alternators may be available under a service exchange scheme - any parts returned should be clean.

Our advice regarding spare parts is as follows.

Officially appointed garages

This is the best source of parts which are peculiar to your car, and which are not otherwise generally available (eg, badges, interior trim, certain body panels, etc). It is also the only place at which you should buy parts if the vehicle is still under warranty.

Accessory shops

These are very good places to buy materials and components needed for the maintenance of your car (oil, air and fuel filters, light bulbs, drivebelts, greases, brake pads, touch-up paint, etc). Components of this nature sold by a reputable shop are usually of the same standard as those used by the car manufacturer.

Besides components, these shops also sell tools and general accessories, usually have convenient opening hours, charge lower prices, and can often be found close to home. Some accessory shops have parts counters where components needed for almost any repair job can be purchased or ordered.

Motor factors

Good factors will stock all the more important components which wear out comparatively quickly, and can sometimes supply individual components needed for the overhaul of a larger assembly (eg, brake seals and hydraulic parts, bearing shells, pistons, valves). They may also handle work such as cylinder block reboring, crankshaft regrinding, etc.

Tyre and exhaust specialists

These outlets may be independent, or members of a local or national chain. They frequently offer competitive prices when compared with a main dealer or local garage, but it will pay to obtain several quotes before making a decision. When researching prices, also ask what 'extras' may be added - for instance fitting a new valve and balancing the wheel are both commonly charged on top of the price of a new tyre.

Other sources

Beware of parts or materials obtained from market stalls, car boot sales or similar outlets. Such items are not invariably sub-standard, but there is little chance of compensation if they do prove unsatisfactory. in the case of safety-critical components such as brake pads, there is the risk not only of financial loss, but also of an accident causing injury or death.

Second-hand components or assemblies obtained from a car breaker can be a good buy in some circumstances, but this sort of purchase is best made by the experienced DIY mechanic.

Vehicle identification

Modifications are a continuing and unpublicised process in vehicle manufacture, quite apart from major model changes. Spare parts manuals and lists are compiled upon a numerical basis, the individual vehicle identification numbers being essential to correct identification of the component concerned.

When ordering spare parts, always give as much information as possible. Quote the car model, year of manufacture and registration, chassis and engine numbers as appropriate.

The *Vehicle Identification Number (VIN)* plate is riveted to the bonnet lock crossmember and is visible once the bonnet has been opened. The vehicle identification (chassis) number is also stamped into the panel at the right-hand side of the engine compartment **(see illustrations)**.

The spare/replacement parts identification number (RP number) is stamped onto the left-hand front door pillar, and can be viewed with the door open **(see illustration)**.

The *engine number* can be found at the front of the cylinder block **(see illustrations)**.

VIN plate location (arrowed)

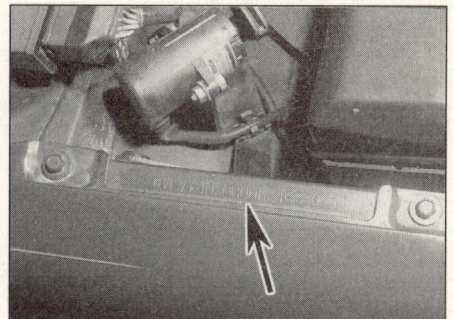

VIN number location (arrowed) on body panel

Spare/replacement parts identification (RP) number (arrowed)

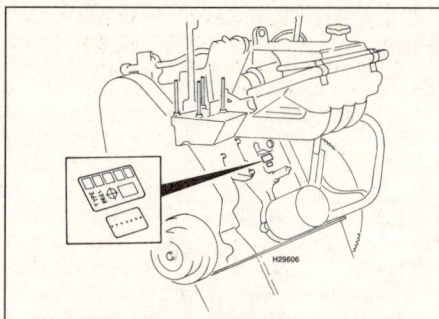

Engine number location (arrowed) – petrol engines

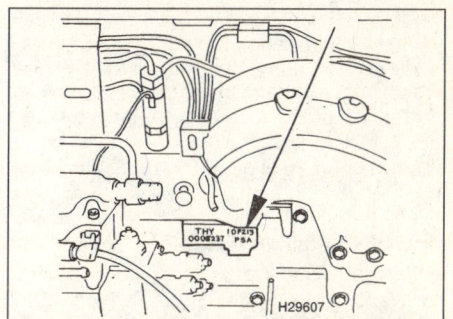

Engine number location (arrowed) – 2.5 litre diesel engine

Whenever servicing, repair or overhaul work is carried out on the car or its components, it is necessary to observe the following procedures and instructions. This will assist in carrying out the operation efficiently and to a professional standard of workmanship.

Joint mating faces and gaskets

When separating components at their mating faces, never insert screwdrivers or similar implements into the joint between the faces in order to prise them apart. This can cause severe damage which results in oil leaks, coolant leaks, etc upon reassembly. Separation is usually achieved by tapping along the joint with a soft-faced hammer in order to break the seal. However, note that this method may not be suitable where dowels are used for component location.

Where a gasket is used between the mating faces of two components, ensure that it is renewed on reassembly, and fit it dry unless otherwise stated in the repair procedure. Make sure that the mating faces are clean and dry, with all traces of old gasket removed. When cleaning a joint face, use a tool which is not likely to score or damage the face, and remove any burrs or nicks with an oilstone or fine file.

Make sure that tapped holes are cleaned with a pipe cleaner, and keep them free of jointing compound, if this is being used, unless specifically instructed otherwise.

Ensure that all orifices, channels or pipes are clear, and blow through them, preferably using compressed air.

Oil seals

Oil seals can be removed by levering them out with a wide flat-bladed screwdriver or similar tool. Alternatively, a number of self-tapping screws may be screwed into the seal, and these used as a purchase for pliers or similar in order to pull the seal free.

Whenever an oil seal is removed from its working location, either individually or as part of an assembly, it should be renewed.

The very fine sealing lip of the seal is easily damaged, and will not seal if the surface it contacts is not completely clean and free from scratches, nicks or grooves. If the original sealing surface of the component cannot be restored, and the manufacturer has not made provision for slight relocation of the seal relative to the sealing surface, the component should be renewed.

Protect the lips of the seal from any surface which may damage them in the course of fitting. Use tape or a conical sleeve where possible. Lubricate the seal lips with oil before fitting and, on dual-lipped seals, fill the space between the lips with grease.

Unless otherwise stated, oil seals must be fitted with their sealing lips toward the lubricant to be sealed.

Use a tubular drift or block of wood of the appropriate size to install the seal and, if the seal housing is shouldered, drive the seal down to the shoulder. If the seal housing is unshouldered, the seal should be fitted with its face flush with the housing top face (unless otherwise instructed).

Screw threads and fastenings

Seized nuts, bolts and screws are quite a common occurrence where corrosion has set in, and the use of penetrating oil or releasing fluid will often overcome this problem if the offending item is soaked for a while before attempting to release it. The use of an impact driver may also provide a means of releasing such stubborn fastening devices, when used in conjunction with the appropriate screwdriver bit or socket. If none of these methods works, it may be necessary to resort to the careful application of heat, or the use of a hacksaw or nut splitter device.

Studs are usually removed by locking two nuts together on the threaded part, and then using a spanner on the lower nut to unscrew the stud. Studs or bolts which have broken off below the surface of the component in which they are mounted can sometimes be removed using a stud extractor. Always ensure that a blind tapped hole is completely free from oil, grease, water or other fluid before installing the bolt or stud. Failure to do this could cause the housing to crack due to the hydraulic action of the bolt or stud as it is screwed in.

When tightening a castellated nut to accept a split pin, tighten the nut to the specified torque, where applicable, and then tighten further to the next split pin hole. Never slacken the nut to align the split pin hole, unless stated in the repair procedure.

When checking or retightening a nut or bolt to a specified torque setting, slacken the nut or bolt by a quarter of a turn, and then retighten to the specified setting. However, this should not be attempted where angular tightening has been used.

For some screw fastenings, notably cylinder head bolts or nuts, torque wrench settings are no longer specified for the latter stages of tightening, "angle-tightening" being called up instead. Typically, a fairly low torque wrench setting will be applied to the bolts/nuts in the correct sequence, followed by one or more stages of tightening through specified angles.

Locknuts, locktabs and washers

Any fastening which will rotate against a component or housing during tightening should always have a washer between it and the relevant component or housing.

Spring or split washers should always be renewed when they are used to lock a critical component such as a big-end bearing retaining bolt or nut. Locktabs which are folded over to retain a nut or bolt should always be renewed.

Self-locking nuts can be re-used in non-critical areas, providing resistance can be felt when the locking portion passes over the bolt or stud thread. However, it should be noted that self-locking stiffnuts tend to lose their effectiveness after long periods of use, and should be renewed as a matter of course.

Split pins must always be replaced with new ones of the correct size for the hole.

When thread-locking compound is found on the threads of a fastener which is to be re-used, it should be cleaned off with a wire brush and solvent, and fresh compound applied on reassembly.

Special tools

Some repair procedures in this manual entail the use of special tools such as a press, two or three-legged pullers, spring compressors, etc. Wherever possible, suitable readily-available alternatives to the manufacturer's special tools are described, and are shown in use. In some instances, where no alternative is possible, it has been necessary to resort to the use of a manufacturer's tool, and this has been done for reasons of safety as well as the efficient completion of the repair operation. Unless you are highly-skilled and have a thorough understanding of the procedures described, never attempt to bypass the use of any special tool when the procedure described specifies its use. Not only is there a very great risk of personal injury, but expensive damage could be caused to the components involved.

Environmental considerations

When disposing of used engine oil, brake fluid, antifreeze, etc, give due consideration to any detrimental environmental effects. Do not, for instance, pour any of the above liquids down drains into the general sewage system, or onto the ground to soak away. Many local council refuse tips provide a facility for waste oil disposal, as do some garages. If none of these facilities are available, consult your local Environmental Health Department, or the National Rivers Authority, for further advice.

With the universal tightening-up of legislation regarding the emission of environmentally-harmful substances from motor vehicles, most current vehicles have tamperproof devices fitted to the main adjustment points of the fuel system. These devices are primarily designed to prevent unqualified persons from adjusting the fuel/air mixture, with the chance of a consequent increase in toxic emissions. If such devices are encountered during servicing or overhaul, they should, wherever possible, be renewed or refitted in accordance with the vehicle manufacturer's requirements or current legislation.

OIL CARE
FOLLOW THE CODE
OIL BANK LINE
0800 66 33 66

Note: It is antisocial and illegal to dump oil down the drain. To find the location of your local oil recycling bank, call this number free.

The jack supplied with the vehicle tool kit should only be used for changing the roadwheels - see *Wheel changing* at the front of this manual. When carrying out any other kind of work, raise the vehicle using a hydraulic (or 'trolley') jack, and always supplement the jack with axle stands positioned under the vehicle jacking points.

To raise the front of the vehicle, position the jack head underneath body front crossmember. Use a block of wood between the jack head and the crossmember. Raise the vehicle to the required height and support it on axle stands positioned under the sides of the front subframe **(see illustrations)**.

To raise the rear of the vehicle, where applicable, remove the covers from the jacking points on at the rear of the sills, then position the jack head under the jacking bracket on the sill. Use a block of wood or a rag between the jack head and the jacking bracket. Lift the vehicle to the required height and support it on axle stands positioned underneath the rear suspension crosstube **(see illustrations)**.

The jack supplied with the vehicle locates with the jacking points on the sills. Where applicable, remove the covers from the jacking points, and ensure that the jack head is correctly engaged before attempting to raise the vehicle.

Never work under, around, or near a raised vehicle, unless it is adequately supported in at least two places with axle stands.

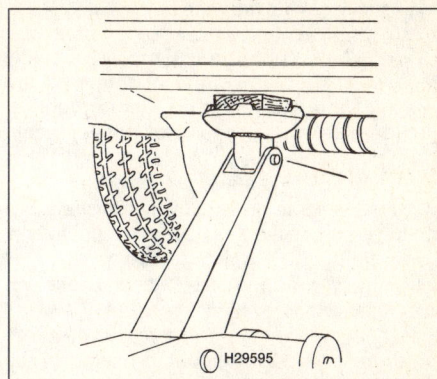

Jack up the front of the vehicle under the front body crossmember . . .

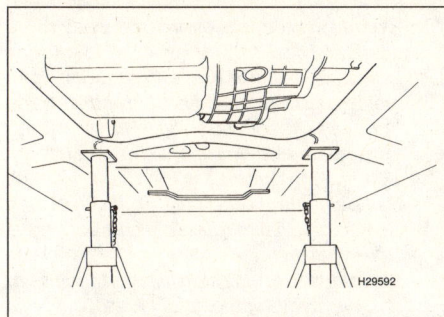

. . . and position axle stands under the sides of the front subframe

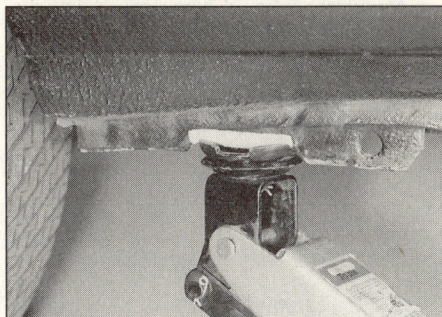

Jack up the rear of the vehicle under the flat sections of the jacking brackets on the sills . . .

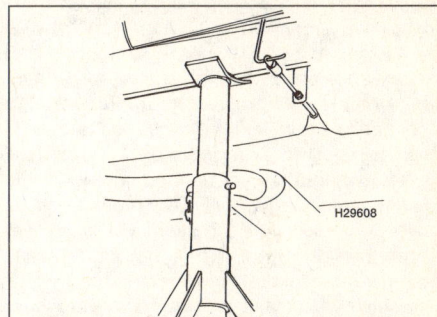

. . . and position axle stands under the rear suspension crosstube

Disconnecting the battery

Several systems fitted to the vehicle require battery power to be available at all times, to ensure that their continued operation (such as the clock and anti-theft alarm systems). Whenever the battery is to be disconnected therefore, first note the following, to ensure that there are no unforeseen consequences of this action:

a) First, on any vehicle with central locking, it is a wise precaution to remove the key from the ignition, and to keep it with you, so that it does not get locked in if the central locking should engage accidentally when the battery is reconnected.

b) If the battery is disconnected while the alarm system is armed or activated, the alarm will remain in the same state when the battery is reconnected. The same applies to the engine immobiliser system

(where fitted). Where applicable, ensure that you have the correct code for the immobiliser system before you disconnect the battery.

c) *If a security-coded audio unit is fitted, and the unit and/or the battery is disconnected, the unit will not function again on reconnection until the correct security code is entered. Details of this procedure, which varies according to the unit and model year, are given in the vehicle owner's handbook. Ensure you have the correct code before you disconnect the battery. If you do not have the code or details of the correct procedure, but can supply proof of ownership and a legitimate reason for wanting this information, a Citroën dealer may be able to help.*

Devices known as 'memory-savers' (or 'code-savers') can be used to avoid some of the above problems. Precise details vary according to the device used. Typically, it is plugged into the cigarette lighter, and is connected by its own wires to a spare battery; the vehicle's own battery is then disconnected from the electrical system, leaving the 'memory-saver' to pass sufficient current to maintain audio unit security codes and ECU memory values, and also to run permanently-live circuits such as the clock.

⚠ *Warning: Some of these devices allow a considerable amount of current to pass, which can mean that many of the vehicle's systems are still operational when the main battery is disconnected. If a memory-saver is used, ensure that the circuit concerned is actually dead before carrying out any work on it!*

Introduction

A selection of good tools is a fundamental requirement for anyone contemplating the ma 'tenance and repair of a motor vehicle. For the owner who does not possess any, their purchase will prove a considerable expense, offsetting some of the savings made by doing-it-yourself. However, provided that the tools purchased meet the relevant national safety standards and are of good quality, they will last for many years and prove an extremely worthwhile investment.

To help the average owner to decide which tools are needed to carry out the various tasks detailed in this manual, we have compiled three lists of tools under the following headings: *Maintenance and minor repair*, *Repair and overhaul*, and *Special*. Newcomers to practical mechanics should start off with the *Maintenance and minor repair* tool kit, and confine themselves to the simpler jobs around the vehicle. Then, as confidence and experience grow, more difficult tasks can be undertaken, with extra tools being purchased as, and when, they are needed. In this way, a *Maintenance and minor repair* tool kit can be built up into a *Repair and overhaul* tool kit over a considerable period of time, without any major cash outlays. The experienced do-it-yourselfer will have a tool kit good enough for most repair and overhaul procedures, and will add tools from the *Special* category when it is felt that the expense is justified by the amount of use to which these tools will be put.

Maintenance and minor repair tool kit

The tools given in this list should be considered as a minimum requirement if routine maintenance, servicing and minor repair operations are to be undertaken. We recommend the purchase of combination spanners (ring one end, open-ended the other); although more expensive than open-ended ones, they do give the advantages of both types of spanner.

- [ ] *Combination spanners:*
  Metric - 8 to 19 mm inclusive
- [ ] *Adjustable spanner - 35 mm jaw (approx.)*
- [ ] *Spark plug spanner (with rubber insert) - petrol models*
- [ ] *Spark plug gap adjustment tool - petrol models*
- [ ] *Set of feeler gauges*
- [ ] *Brake bleed nipple spanner*
- [ ] *Screwdrivers:*
  Flat blade - 100 mm long x 6 mm dia
  Cross blade - 100 mm long x 6 mm dia
  Torx - various sizes (not all vehicles)
- [ ] *Combination pliers*
- [ ] *Hacksaw (junior)*
- [ ] *Tyre pump*
- [ ] *Tyre pressure gauge*
- [ ] *Oil can*
- [ ] *Oil filter removal tool*
- [ ] *Fine emery cloth*
- [ ] *Wire brush (small)*
- [ ] *Funnel (medium size)*
- [ ] *Sump drain plug key (not all vehicles)*

Repair and overhaul tool kit

These tools are virtually essential for anyone undertaking any major repairs to a motor vehicle, and are additional to those given in the *Maintenance and minor repair* list. Included in this list is a comprehensive set of sockets. Although these are expensive, they will be found invaluable as they are so versatile - particularly if various drives are included in the set. We recommend the half-inch square-drive type, as this can be used with most proprietary torque wrenches.

The tools in this list will sometimes need to be supplemented by tools from the *Special* list:

- [ ] *Sockets (or box spanners) to cover range in previous list (including Torx sockets)*
- [ ] *Reversible ratchet drive (for use with sockets)*
- [ ] *Extension piece, 250 mm (for use with sockets)*
- [ ] *Universal joint (for use with sockets)*
- [ ] *Flexible handle or sliding T "breaker bar" (for use with sockets)*
- [ ] *Torque wrench (for use with sockets)*
- [ ] *Self-locking grips*
- [ ] *Ball pein hammer*
- [ ] *Soft-faced mallet (plastic or rubber)*
- [ ] *Screwdrivers:*
  Flat blade - long & sturdy, short (chubby), and narrow (electrician's) types
  Cross blade – long & sturdy, and short (chubby) types
- [ ] *Pliers:*
  Long-nosed
  Side cutters (electrician's)
  Circlip (internal and external)
- [ ] *Cold chisel - 25 mm*
- [ ] *Scriber*
- [ ] *Scraper*
- [ ] *Centre-punch*
- [ ] *Pin punch*
- [ ] *Hacksaw*
- [ ] *Brake hose clamp*
- [ ] *Brake/clutch bleeding kit*
- [ ] *Selection of twist drills*
- [ ] *Steel rule/straight-edge*
- [ ] *Allen keys (inc. splined/Torx type)*
- [ ] *Selection of files*
- [ ] *Wire brush*
- [ ] *Axle stands*
- [ ] *Jack (strong trolley or hydraulic type)*
- [ ] *Light with extension lead*
- [ ] *Universal electrical multi-meter*

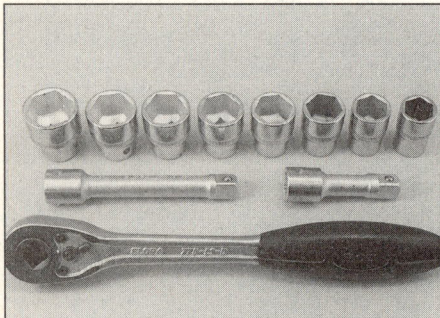

Sockets and reversible ratchet drive

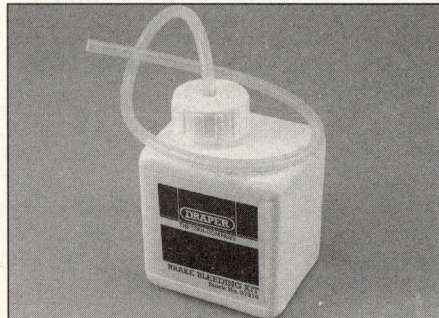

Brake bleeding kit

Torx key, socket and bit

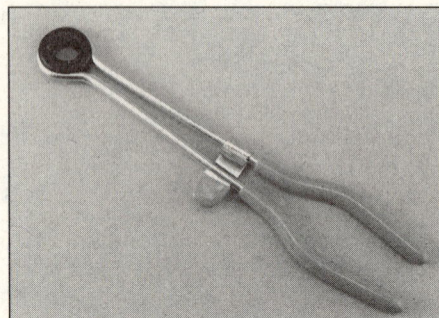

Hose clamp

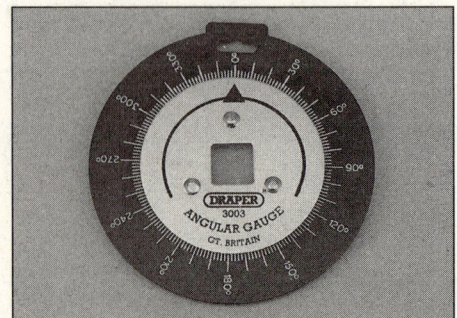

Angular-tightening gauge

Special tools

The tools in this list are those which are not used regularly, are expensive to buy, or which need to be used in accordance with their manufacturers' instructions. Unless relatively difficult mechanical jobs are undertaken frequently, it will not be economic to buy many of these tools. Where this is the case, you could consider clubbing together with friends (or joining a motorists' club) to make a joint purchase, or borrowing the tools against a deposit from a local garage or tool hire specialist. It is worth noting that many of the larger DIY superstores now carry a large range of special tools for hire at modest rates.

The following list contains only those tools and instruments freely available to the public, and not those special tools produced by the vehicle manufacturer specifically for its dealer network. You will find occasional references to these manufacturers' special tools in the text of this manual. Generally, an alternative method of doing the job without the vehicle manufacturers' special tool is given. However, sometimes there is no alternative to using them. Where this is the case and the relevant tool cannot be bought or borrowed, you will have to entrust the work to a dealer.

- [ ] Angular-tightening gauge
- [ ] Valve spring compressor
- [ ] Valve grinding tool
- [ ] Piston ring compressor
- [ ] Piston ring removal/installation tool
- [ ] Cylinder bore hone
- [ ] Balljoint separator
- [ ] Coil spring compressors (where applicable)
- [ ] Two/three-legged hub and bearing puller
- [ ] Impact screwdriver
- [ ] Micrometer and/or vernier calipers
- [ ] Dial gauge
- [ ] Stroboscopic timing light
- [ ] Dwell angle meter/tachometer
- [ ] Fault code reader
- [ ] Cylinder compression gauge
- [ ] Hand-operated vacuum pump and gauge
- [ ] Clutch plate alignment set
- [ ] Brake shoe steady spring cup removal tool
- [ ] Bush and bearing removal/installation set
- [ ] Stud extractors
- [ ] Tap and die set
- [ ] Lifting tackle
- [ ] Trolley jack

Buying tools

Reputable motor accessory shops and superstores often offer excellent quality tools at discount prices, so it pays to shop around.

Remember, you don't have to buy the most expensive items on the shelf, but it is always advisable to steer clear of the very cheap tools. Beware of 'bargains' offered on market stalls or at car boot sales. There are plenty of good tools around at reasonable prices, but always aim to purchase items which meet the relevant national safety standards. If in doubt, ask the proprietor or manager of the shop for advice before making a purchase.

Care and maintenance of tools

Having purchased a reasonable tool kit, it is necessary to keep the tools in a clean and serviceable condition. After use, always wipe off any dirt, grease and metal particles using a clean, dry cloth, before putting the tools away. Never leave them lying around after they have been used. A simple tool rack on the garage or workshop wall for items such as screwdrivers and pliers is a good idea. Store all normal spanners and sockets in a metal box. Any measuring instruments, gauges, meters, etc, must be carefully stored where they cannot be damaged or become rusty.

Take a little care when tools are used. Hammer heads inevitably become marked, and screwdrivers lose the keen edge on their blades from time to time. A little timely attention with emery cloth or a file will soon restore items like this to a good finish.

Working facilities

Not to be forgotten when discussing tools is the workshop itself. If anything more than routine maintenance is to be carried out, a suitable working area becomes essential.

It is appreciated that many an owner-mechanic is forced by circumstances to remove an engine or similar item without the benefit of a garage or workshop. Having done this, any repairs should always be done under the cover of a roof.

Wherever possible, any dismantling should be done on a clean, flat workbench or table at a suitable working height.

Any workbench needs a vice; one with a jaw opening of 100 mm is suitable for most jobs. As mentioned previously, some clean dry storage space is also required for tools, as well as for any lubricants, cleaning fluids, touch-up paints etc, which become necessary.

Another item which may be required, and which has a much more general usage, is an electric drill with a chuck capacity of at least 8 mm. This, together with a good range of twist drills, is virtually essential for fitting accessories.

Last, but not least, always keep a supply of old newspapers and clean, lint-free rags available, and try to keep any working area as clean as possible.

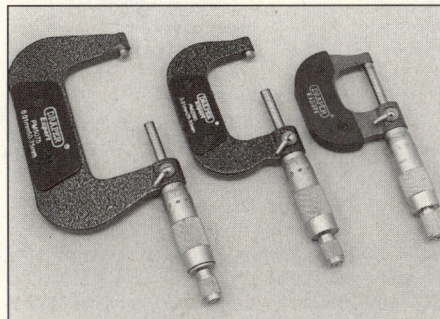

Micrometers

Dial test indicator ("dial gauge")

Strap wrench

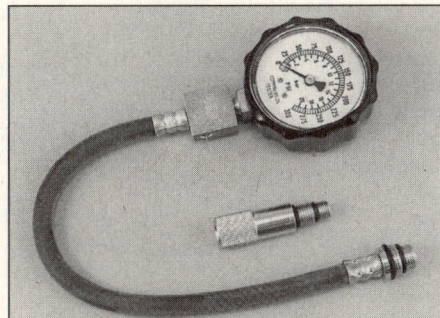

Compression tester

Fault code reader

This is a guide to getting your vehicle through the MOT test. Obviously it will not be possible to examine the vehicle to the same standard as the professional MOT tester. However, working through the following checks will enable you to identify any problem areas before submitting the vehicle for the test.

Where a testable component is in borderline condition, the tester has discretion in deciding whether to pass or fail it. The basis of such discretion is whether the tester would be happy for a close relative or friend to use the vehicle with the component in that condition. If the vehicle presented is clean and evidently well cared for, the tester may be more inclined to pass a borderline component than if the vehicle is scruffy and apparently neglected.

It has only been possible to summarise the test requirements here, based on the regulations in force at the time of printing. Test standards are becoming increasingly stringent, although there are some exemptions for older vehicles.

An assistant will be needed to help carry out some of these checks.

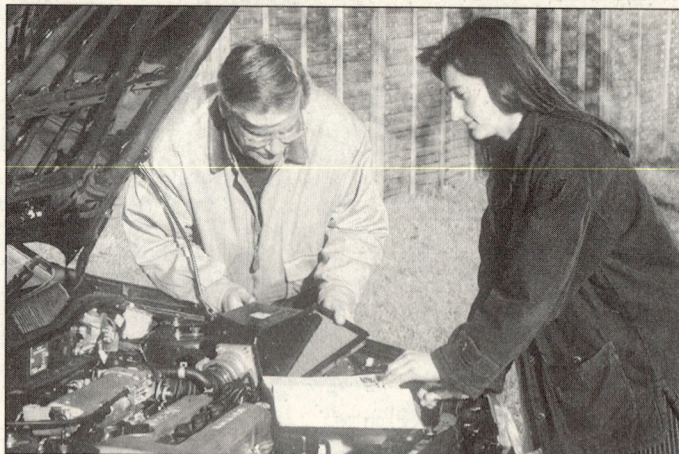

The checks have been sub-divided into four categories, as follows:

1 Checks carried out **FROM THE DRIVER'S SEAT**

2 Checks carried out **WITH THE VEHICLE ON THE GROUND**

3 Checks carried out **WITH THE VEHICLE RAISED AND THE WHEELS FREE TO TURN**

4 Checks carried out on **YOUR VEHICLE'S EXHAUST EMISSION SYSTEM**

1 Checks carried out **FROM THE DRIVER'S SEAT**

Handbrake

☐ Test the operation of the handbrake. Excessive travel (too many clicks) indicates incorrect brake or cable adjustment.

☐ Check that the handbrake cannot be released by tapping the lever sideways. Check the security of the lever mountings.

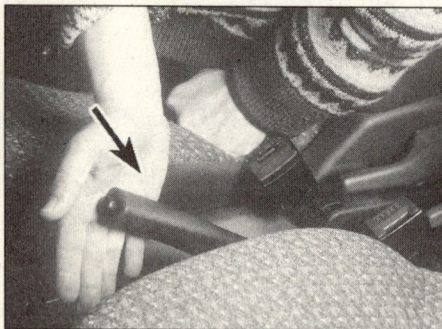

Footbrake

☐ Depress the brake pedal and check that it does not creep down to the floor, indicating a master cylinder fault. Release the pedal, wait a few seconds, then depress it again. If the pedal travels nearly to the floor before firm resistance is felt, brake adjustment or repair is necessary. If the pedal feels spongy, there is air in the hydraulic system which must be removed by bleeding.

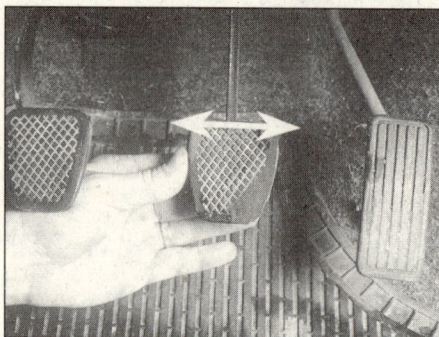

☐ Check that the brake pedal is secure and in good condition. Check also for signs of fluid leaks on the pedal, floor or carpets, which would indicate failed seals in the brake master cylinder.

☐ Check the servo unit (when applicable) by operating the brake pedal several times, then keeping the pedal depressed and starting the engine. As the engine starts, the pedal will move down slightly. If not, the vacuum hose or the servo itself may be faulty.

Steering wheel and column

☐ Examine the steering wheel for fractures or looseness of the hub, spokes or rim.

☐ Move the steering wheel from side to side and then up and down. Check that the steering wheel is not loose on the column, indicating wear or a loose retaining nut. Continue moving the steering wheel as before, but also turn it slightly from left to right.

☐ Check that the steering wheel is not loose on the column, and that there is no abnormal

movement of the steering wheel, indicating wear in the column support bearings or couplings.

Windscreen, mirrors and sunvisor

☐ The windscreen must be free of cracks or other significant damage within the driver's field of view. (Small stone chips are acceptable.) Rear view mirrors must be secure, intact, and capable of being adjusted.

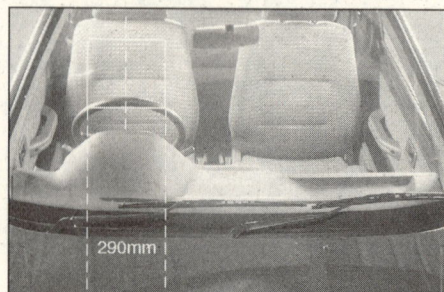

290mm

☐ The driver's sunvisor must be capable of being stored in the "up" position.

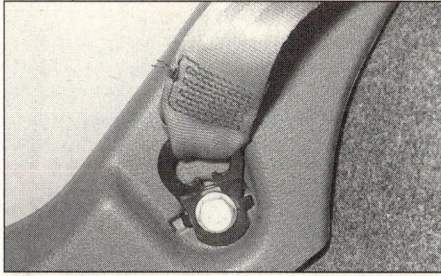

Seat belts and seats

Note: *The following checks are applicable to all seat belts, front and rear.*

☐ Examine the webbing of all the belts (including rear belts if fitted) for cuts, serious fraying or deterioration. Fasten and unfasten each belt to check the buckles. If applicable, check the retracting mechanism. Check the security of all seat belt mountings accessible from inside the vehicle.

☐ Seat belts with pre-tensioners, once activated, have a "flag" or similar showing on the seat belt stalk. This, in itself, is not a reason for test failure.

☐ The front seats themselves must be securely attached and the backrests must lock in the upright position.

Doors

☐ Both front doors must be able to be opened and closed from outside and inside, and must latch securely when closed.

2 Checks carried out WITH THE VEHICLE ON THE GROUND

Vehicle identification

☐ Number plates must be in good condition, secure and legible, with letters and numbers correctly spaced – spacing at (A) should be at least twice that at (B).

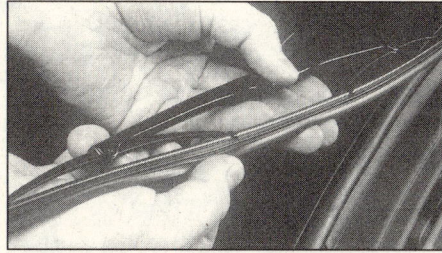

☐ The VIN plate and/or homologation plate must be legible.

Electrical equipment

☐ Switch on the ignition and check the operation of the horn.

☐ Check the windscreen washers and wipers, examining the wiper blades; renew damaged or perished blades. Also check the operation of the stop-lights.

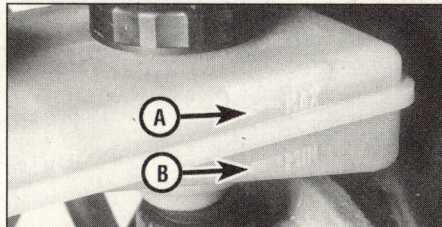

☐ Check the operation of the sidelights and number plate lights. The lenses and reflectors must be secure, clean and undamaged.

☐ Check the operation and alignment of the headlights. The headlight reflectors must not be tarnished and the lenses must be undamaged.

☐ Switch on the ignition and check the operation of the direction indicators (including the instrument panel tell-tale) and the hazard warning lights. Operation of the sidelights and stop-lights must not affect the indicators - if it does, the cause is usually a bad earth at the rear light cluster.

☐ Check the operation of the rear foglight(s), including the warning light on the instrument panel or in the switch.

☐ The ABS warning light must illuminate in accordance with the manufacturers' design. For most vehicles, the ABS warning light should illuminate when the ignition is switched on, and (if the system is operating properly) extinguish after a few seconds. Refer to the owner's handbook.

Footbrake

☐ Examine the master cylinder, brake pipes and servo unit for leaks, loose mountings, corrosion or other damage.

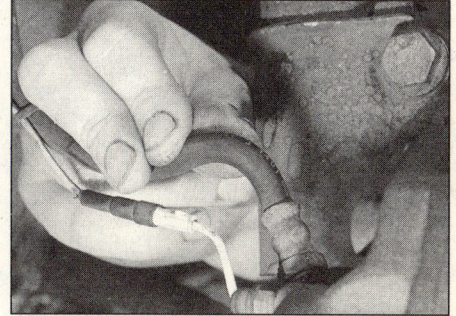

☐ The fluid reservoir must be secure and the fluid level must be between the upper (A) and lower (B) markings.

☐ Inspect both front brake flexible hoses for cracks or deterioration of the rubber. Turn the steering from lock to lock, and ensure that the hoses do not contact the wheel, tyre, or any part of the steering or suspension mechanism. With the brake pedal firmly depressed, check the hoses for bulges or leaks under pressure.

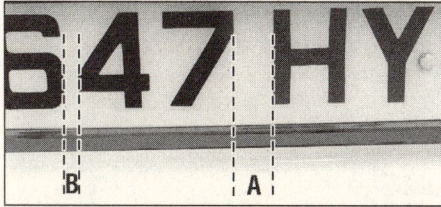

Steering and suspension

☐ Have your assistant turn the steering wheel from side to side slightly, up to the point where the steering gear just begins to transmit this movement to the roadwheels. Check for excessive free play between the steering wheel and the steering gear, indicating wear or insecurity of the steering column joints, the column-to-steering gear coupling, or the steering gear itself.

☐ Have your assistant turn the steering wheel more vigorously in each direction, so that the roadwheels just begin to turn. As this is done, examine all the steering joints, linkages, fittings and attachments. Renew any component that shows signs of wear or damage. On vehicles with power steering, check the security and condition of the steering pump, drivebelt and hoses.

☐ Check that the vehicle is standing level, and at approximately the correct ride height.

Shock absorbers

☐ Depress each corner of the vehicle in turn, then release it. The vehicle should rise and then settle in its normal position. If the vehicle continues to rise and fall, the shock absorber is defective. A shock absorber which has seized will also cause the vehicle to fail.

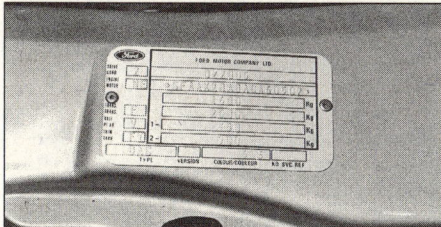

Exhaust system

☐ Start the engine. With your assistant holding a rag over the tailpipe, check the entire system for leaks. Repair or renew leaking sections.

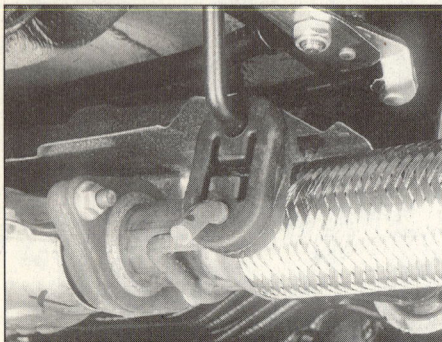

3 Checks carried out
WITH THE VEHICLE RAISED AND THE WHEELS FREE TO TURN

Jack up the front and rear of the vehicle, and securely support it on axle stands. Position the stands clear of the suspension assemblies. Ensure that the wheels are clear of the ground and that the steering can be turned from lock to lock.

Steering mechanism

☐ Have your assistant turn the steering from lock to lock. Check that the steering turns smoothly, and that no part of the steering mechanism, including a wheel or tyre, fouls any brake hose or pipe or any part of the body structure.
☐ Examine the steering rack rubber gaiters for damage or insecurity of the retaining clips. If power steering is fitted, check for signs of damage or leakage of the fluid hoses, pipes or connections. Also check for excessive stiffness or binding of the steering, a missing split pin or locking device, or severe corrosion of the body structure within 30 cm of any steering component attachment point.

Front and rear suspension and wheel bearings

☐ Starting at the front right-hand side, grasp the roadwheel at the 3 o'clock and 9 o'clock positions and rock gently but firmly. Check for free play or insecurity at the wheel bearings, suspension balljoints, or suspension mountings, pivots and attachments.
☐ Now grasp the wheel at the 12 o'clock and 6 o'clock positions and repeat the previous inspection. Spin the wheel, and check for roughness or tightness of the front wheel bearing.

☐ If excess free play is suspected at a component pivot point, this can be confirmed by using a large screwdriver or similar tool and levering between the mounting and the component attachment. This will confirm whether the wear is in the pivot bush, its retaining bolt, or in the mounting itself (the bolt holes can often become elongated).

☐ Carry out all the above checks at the other front wheel, and then at both rear wheels.

Springs and shock absorbers

☐ Examine the suspension struts (when applicable) for serious fluid leakage, corrosion, or damage to the casing. Also check the security of the mounting points.
☐ If coil springs are fitted, check that the spring ends locate in their seats, and that the spring is not corroded, cracked or broken.
☐ If leaf springs are fitted, check that all leaves are intact, that the axle is securely attached to each spring, and that there is no deterioration of the spring eye mountings, bushes, and shackles.

☐ The same general checks apply to vehicles fitted with other suspension types, such as torsion bars, hydraulic displacer units, etc. Ensure that all mountings and attachments are secure, that there are no signs of excessive wear, corrosion or damage, and (on hydraulic types) that there are no fluid leaks or damaged pipes.
☐ Inspect the shock absorbers for signs of serious fluid leakage. Check for wear of the mounting bushes or attachments, or damage to the body of the unit.

Driveshafts (fwd vehicles only)

☐ Rotate each front wheel in turn and inspect the constant velocity joint gaiters for splits or damage. Also check that each driveshaft is straight and undamaged.

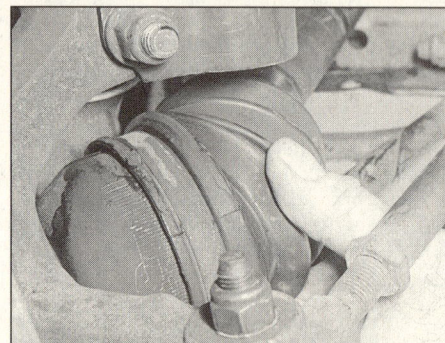

Braking system

☐ If possible without dismantling, check brake pad wear and disc condition. Ensure that the friction lining material has not worn excessively, (A) and that the discs are not fractured, pitted, scored or badly worn (B).

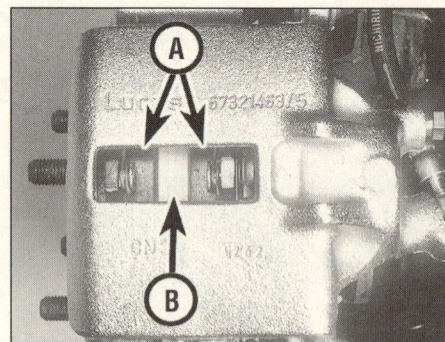

☐ Examine all the rigid brake pipes underneath the vehicle, and the flexible hose(s) at the rear. Look for corrosion, chafing or insecurity of the pipes, and for signs of bulging under pressure, chafing, splits or deterioration of the flexible hoses.
☐ Look for signs of fluid leaks at the brake calipers or on the brake backplates. Repair or renew leaking components.
☐ Slowly spin each wheel, while your assistant depresses and releases the footbrake. Ensure that each brake is operating and does not bind when the pedal is released.

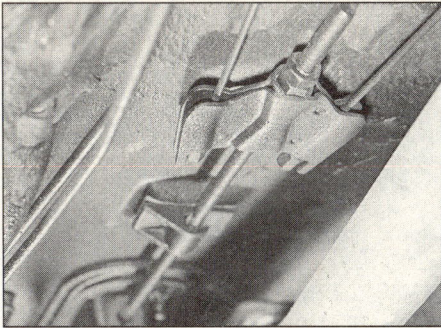

☐ Examine the handbrake mechanism, checking for frayed or broken cables, excessive corrosion, or wear or insecurity of the linkage. Check that the mechanism works on each relevant wheel, and releases fully, without binding.

☐ It is not possible to test brake efficiency without special equipment, but a road test can be carried out later to check that the vehicle pulls up in a straight line.

Fuel and exhaust systems

☐ Inspect the fuel tank (including the filler cap), fuel pipes, hoses and unions. All components must be secure and free from leaks.

☐ Examine the exhaust system over its entire length, checking for any damaged, broken or missing mountings, security of the retaining clamps and rust or corrosion.

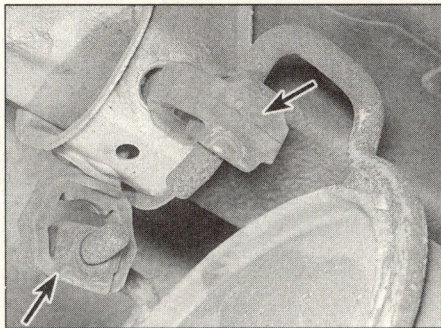

Wheels and tyres

☐ Examine the sidewalls and tread area of each tyre in turn. Check for cuts, tears, lumps, bulges, separation of the tread, and exposure of the ply or cord due to wear or damage. Check that the tyre bead is correctly seated on the wheel rim, that the valve is sound and properly seated, and that the wheel is not distorted or damaged.

☐ Check that the tyres are of the correct size for the vehicle, that they are of the same size and type on each axle, and that the pressures are correct.

☐ Check the tyre tread depth. The legal minimum at the time of writing is 1.6 mm over at least three-quarters of the tread width. Abnormal tread wear may indicate incorrect front wheel alignment.

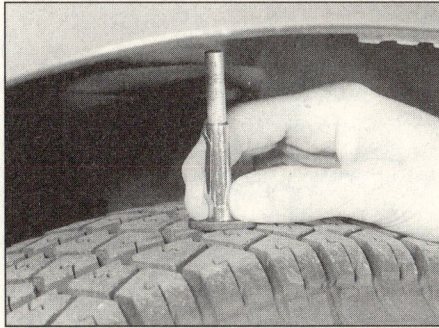

Body corrosion

☐ Check the condition of the entire vehicle structure for signs of corrosion in load-bearing areas. (These include chassis box sections, side sills, cross-members, pillars, and all suspension, steering, braking system and seat belt mountings and anchorages.) Any corrosion which has seriously reduced the thickness of a load-bearing area is likely to cause the vehicle to fail. In this case professional repairs are likely to be needed.

☐ Damage or corrosion which causes sharp or otherwise dangerous edges to be exposed will also cause the vehicle to fail.

4 Checks carried out on **YOUR VEHICLE'S EXHAUST EMISSION SYSTEM**

Petrol models

☐ Have the engine at normal operating temperature, and make sure that it is in good tune (ignition system in good order, air filter element clean, etc).

☐ Before any measurements are carried out, raise the engine speed to around 2500 rpm, and hold it at this speed for 20 seconds. Allow the engine speed to return to idle, and watch for smoke emissions from the exhaust tailpipe. If the idle speed is obviously much too high, or if dense blue or clearly-visible black smoke comes from the tailpipe for more than 5 seconds, the vehicle will fail. As a rule of thumb, blue smoke signifies oil being burnt (engine wear) while black smoke signifies unburnt fuel (dirty air cleaner element, or other carburettor or fuel system fault).

☐ An exhaust gas analyser capable of measuring carbon monoxide (CO) and hydrocarbons (HC) is now needed. If such an instrument cannot be hired or borrowed, a local garage may agree to perform the check for a small fee.

CO emissions (mixture)

☐ At the time of writing, for vehicles first used between 1st August 1975 and 31st July 1986 (P to C registration), the CO level must not exceed 4.5% by volume. For vehicles first used between 1st August 1986 and 31st July 1992 (D to J registration), the CO level must not exceed 3.5% by volume. Vehicles first

used after 1st August 1992 (K registration) must conform to the manufacturer's specification. The MOT tester has access to a DOT database or emissions handbook, which lists the CO and HC limits for each make and model of vehicle. The CO level is measured with the engine at idle speed, and at "fast idle". The following limits are given as a general guide:

> *At idle speed -*
> CO level no more than 0.5%
> *At "fast idle" (2500 to 3000 rpm) -*
> CO level no more than 0.3%
> (Minimum oil temperature 60ºC)

☐ If the CO level cannot be reduced far enough to pass the test (and the fuel and ignition systems are otherwise in good condition) then the carburettor is badly worn, or there is some problem in the fuel injection system or catalytic converter (as applicable).

HC emissions

☐ With the CO within limits, HC emissions for vehicles first used between 1st August 1975 and 31st July 1992 (P to J registration) must not exceed 1200 ppm. Vehicles first used after 1st August 1992 (K registration) must conform to the manufacturer's specification. The MOT tester has access to a DOT database or emissions handbook, which lists the CO and HC limits for each make and model of vehicle. The HC level is measured with the engine at "fast idle". The following is given as a general guide:

> *At "fast idle" (2500 to 3000 rpm) -*
> HC level no more than 200 ppm
> (Minimum oil temperature 60ºC)

☐ Excessive HC emissions are caused by incomplete combustion, the causes of which can include oil being burnt, mechanical wear and ignition/fuel system malfunction.

Diesel models

☐ The only emission test applicable to Diesel engines is the measuring of exhaust smoke density. The test involves accelerating the engine several times to its maximum unloaded speed.

Note: *It is of the utmost importance that the engine timing belt is in good condition before the test is carried out.*

☐ The limits for Diesel engine exhaust smoke, introduced in September 1995 are:

Vehicles first used before 1st August 1979:
Exempt from metered smoke testing, but must not emit "dense blue or clearly visible black smoke for a period of more than 5 seconds at idle" or "dense blue or clearly visible black smoke during acceleration which would obscure the view of other road users".

Non-turbocharged vehicles first used after 1st August 1979: 2.5m⁻¹

Turbocharged vehicles first used after 1st August 1979: 3.0m⁻¹

☐ Excessive smoke can be caused by a dirty air cleaner element. Otherwise, professional advice may be needed to find the cause.

Engine

- ☐ Engine fails to rotate when attempting to start
- ☐ Engine rotates, but will not start
- ☐ Engine difficult to start when cold
- ☐ Engine difficult to start when hot
- ☐ Starter motor noisy or excessively-rough in engagement
- ☐ Engine starts, but stops immediately
- ☐ Engine idles erratically
- ☐ Engine misfires at idle speed
- ☐ Engine misfires throughout the driving speed range
- ☐ Engine hesitates on acceleration
- ☐ Engine stalls
- ☐ Engine lacks power
- ☐ Engine backfires
- ☐ Oil pressure warning light illuminated with engine running
- ☐ Engine runs-on after switching off
- ☐ Engine noises

Cooling system

- ☐ Overheating
- ☐ Overcooling
- ☐ External coolant leakage
- ☐ Internal coolant leakage
- ☐ Corrosion

Fuel and exhaust systems

- ☐ Excessive fuel consumption
- ☐ Fuel leakage and/or fuel odour
- ☐ Excessive noise or fumes from exhaust system

Clutch

- ☐ Pedal travels to floor - no pressure or very little resistance
- ☐ Clutch fails to disengage (unable to select gears).
- ☐ Clutch slips (engine speed increases, with no increase in vehicle speed).
- ☐ Judder as clutch is engaged
- ☐ Noise when depressing or releasing clutch pedal

Manual transmission

- ☐ Noisy in neutral with engine running
- ☐ Noisy in one particular gear
- ☐ Difficulty engaging gears
- ☐ Jumps out of gear
- ☐ Vibration
- ☐ Lubricant leaks

Automatic transmission

- ☐ Fluid leakage
- ☐ Transmission fluid brown, or has burned smell
- ☐ General gear selection problems
- ☐ Transmission will not downshift (kickdown) with accelerator pedal fully depressed
- ☐ Engine will not start in any gear, or starts in gears other than Park or Neutral
- ☐ Transmission slips, shifts roughly, is noisy, or has no drive in forward or reverse gears

Driveshafts

- ☐ Vibration when accelerating or decelerating
- ☐ Clicking or knocking noise on turns (at slow speed on full-lock)

Braking system

- ☐ Vehicle pulls to one side under braking
- ☐ Noise (grinding or high-pitched squeal) when brakes applied
- ☐ Excessive brake pedal travel
- ☐ Brake pedal feels spongy when depressed
- ☐ Excessive brake pedal effort required to stop vehicle
- ☐ Judder felt through brake pedal or steering wheel when braking
- ☐ Brakes binding
- ☐ Rear wheels locking under normal braking

Suspension and steering

- ☐ Vehicle pulls to one side
- ☐ Wheel wobble and vibration
- ☐ Excessive pitching and/or rolling around corners, or during braking
- ☐ Wandering or general instability
- ☐ Excessively-stiff steering
- ☐ Excessive play in steering
- ☐ Lack of power assistance
- ☐ Tyre wear excessive

Electrical system

- ☐ Battery will not hold a charge for more than a few days
- ☐ Ignition/no-charge warning light remains illuminated with engine running
- ☐ Ignition/no-charge warning light fails to come on
- ☐ Lights inoperative
- ☐ Instrument readings inaccurate or erratic
- ☐ Horn inoperative, or unsatisfactory in operation
- ☐ Windscreen wipers inoperative, or unsatisfactory in operation
- ☐ Windscreen washers inoperative, or unsatisfactory in operation
- ☐ Electric windows inoperative, or unsatisfactory in operation
- ☐ Central locking system inoperative, or unsatisfactory in operation

Introduction

The vehicle owner who does his or her own maintenance according to the recommended service schedules should not have to use this section of the manual very often. Modern component reliability is such that, provided those items subject to wear or deterioration are inspected or renewed at the specified intervals, sudden failure is comparatively rare. Faults do not usually just happen as a result of sudden failure, but develop over a period of time. Major mechanical failures in particular are usually preceded by characteristic symptoms over hundreds or even thousands of miles. Those components which do occasionally fail without warning are often small and easily carried in the vehicle.

With any fault-finding, the first step is to decide where to begin investigations. Sometimes this is obvious, but on other occasions, a little detective work will be necessary. The owner who makes half a dozen haphazard adjustments or replacements may be successful in curing a fault (or its symptoms), but will be none the wiser if the fault recurs, and ultimately may have spent more time and money than was necessary. A calm and logical approach will be found to be more satisfactory in the long run. Always take into account any warning signs or abnormalities that may have been noticed in the period preceding the fault - power loss, high or low gauge readings, unusual smells, etc - and remember that failure of components such as fuses or spark plugs may only be pointers to some underlying fault.

The pages which follow provide an easy-reference guide to the more common problems which may occur during the operation of the vehicle. These problems and their possible causes are grouped under headings denoting various components or systems, such as Engine, Cooling system, etc. The Chapter and/or Section which deals with the problem is also shown in brackets. Whatever the fault, certain basic principles apply. These are as follows:

Verify the fault. This is simply a matter of being sure that you know what the symptoms are before starting work. This is particularly important if you are investigating a fault for someone else, who may not have described it very accurately.

Don't overlook the obvious. For example, if the vehicle won't start, is there fuel in the tank? (Don't take anyone else's word on this particular point, and don't trust the fuel gauge either!) If an electrical fault is indicated, look for loose or broken wires before digging out the test gear.

Cure the disease, not the symptom. Substituting a flat battery with a fully-charged one will get you off the hard shoulder, but if the underlying cause is not attended to, the new battery will go the same way. Similarly, changing oil-fouled spark plugs for a new set will get you moving again, but remember that the reason for the fouling (if it wasn't simply an incorrect grade of plug) will have to be established and corrected.

Don't take anything for granted. Particularly, don't forget that a 'new' component may itself be defective (especially if it's been rattling around in the boot for months), and don't leave components out of a fault diagnosis sequence just because they are new or recently-fitted. When you do finally diagnose a difficult fault, you'll probably realise that all the evidence was there from the start.

Engine

Engine fails to rotate when attempting to start

- [ ] Battery terminal connections loose or corroded (see *Weekly checks*)
- [ ] Battery discharged or faulty (Chapter 5A)
- [ ] Broken, loose or disconnected wiring in the starting circuit (Chapter 5A)
- [ ] Defective starter solenoid or switch (Chapter 5A)
- [ ] Defective starter motor (Chapter 5A)
- [ ] Starter pinion or flywheel ring gear teeth loose or broken (Chapters 2 and 5A)
- [ ] Engine earth strap broken or disconnected (Chapter 5A)

Engine rotates, but will not start

- [ ] Fuel cut-off switch energised (Chapter 4).
- [ ] Fuel tank empty
- [ ] Battery discharged (engine rotates slowly) (Chapter 5A)
- [ ] Battery terminal connections loose or corroded (*"Weekly checks"*)
- [ ] Ignition components damp or damaged - petrol models (Chapters 1 and 5B)
- [ ] Broken, loose or disconnected wiring in the ignition circuit - petrol models (Chapters 1A and 5B)
- [ ] Worn, faulty or incorrectly-gapped spark plugs - petrol models (Chapter 1A)
- [ ] Preheating system faulty - diesel models (Chapter 5C)
- [ ] Fuel injection system fault - petrol models (Chapter 4A)
- [ ] Stop solenoid faulty - diesel models (Chapter 4B)
- [ ] Air in fuel system - diesel models (Chapter 4B)
- [ ] Major mechanical failure (eg camshaft drive) (Chapter 2)

Engine difficult to start when cold

- [ ] Battery discharged (Chapter 5A)
- [ ] Battery terminal connections loose or corroded (see *Weekly checks*)
- [ ] Worn, faulty or incorrectly-gapped spark plugs - petrol models (Chapter 1A)
- [ ] Preheating system fault - diesel models (Chapter 5C)
- [ ] Fuel injection system fault - petrol models (Chapter 4A)
- [ ] Other ignition system fault - petrol models (Chapters 1A and 5B)
- [ ] Low cylinder compressions (Chapter 2)

Engine difficult to start when hot

- [ ] Air filter element dirty or clogged (Chapter 1)
- [ ] Fuel injection system fault - petrol models (Chapter 4A)
- [ ] Low cylinder compressions (Chapter 2)

Starter motor noisy or excessively-rough in engagement

- [ ] Starter pinion or flywheel ring gear teeth loose or broken (Chapters 2 and 5A)
- [ ] Starter motor mounting bolts loose or missing (Chapter 5A)
- [ ] Starter motor internal components worn or damaged (Chapter 5A)

Engine starts, but stops immediately

- [ ] Loose or faulty electrical connections in the ignition circuit - petrol models (Chapters 1A and 5B)
- [ ] Vacuum leak at the throttle body or inlet manifold - petrol models (Chapter 4A)
- [ ] Blocked injector/fuel injection system fault - petrol models (Chapter 4A)

Engine idles erratically

- [ ] Air filter element clogged (Chapter 1)
- [ ] Vacuum leak at the throttle body, inlet manifold or associated hoses - petrol models (Chapter 4A)
- [ ] Worn, faulty or incorrectly-gapped spark plugs - petrol models (Chapter 1A)
- [ ] Uneven or low cylinder compressions (Chapter 2)
- [ ] Camshaft lobes worn (Chapter 2)
- [ ] Timing belt incorrectly fitted (Chapter 2)
- [ ] Blocked injector/fuel injection system fault - petrol models (Chapter 4A)
- [ ] Faulty injector(s) - diesel models (Chapter 4B)

Engine misfires at idle speed

- [ ] Worn, faulty or incorrectly-gapped spark plugs - petrol models (Chapter 1A)
- [ ] Faulty spark plug HT leads - petrol models (Chapter 1A)
- [ ] Vacuum leak at the throttle body, inlet manifold or associated hoses - petrol models (Chapter 4A)
- [ ] Blocked injector/fuel injection system fault - petrol models (Chapter 4A)
- [ ] Faulty injector(s) - diesel models (Chapter 4B)
- [ ] Uneven or low cylinder compressions (Chapter 2)
- [ ] Disconnected, leaking, or perished crankcase ventilation hoses (Chapter 4C)

Engine misfires throughout the driving speed range

- [ ] Fuel filter choked (Chapter 1)
- [ ] Fuel pump faulty, or delivery pressure low (Chapter 4)
- [ ] Fuel tank vent blocked, or fuel pipes restricted (Chapter 4)
- [ ] Vacuum leak at the throttle body, inlet manifold or associated hoses - petrol models (Chapter 4A)
- [ ] Worn, faulty or incorrectly-gapped spark plugs - petrol models (Chapter 1A)
- [ ] Faulty spark plug HT leads - petrol models (Chapter 1A)
- [ ] Faulty injector(s) - diesel models (Chapter 4B)
- [ ] Faulty ignition coil - petrol models (Chapter 5A)
- [ ] Uneven or low cylinder compressions (Chapter 2)
- [ ] Blocked injector/fuel injection system fault - petrol models (Chapter 4A)

Engine (continued)

Engine hesitates on acceleration

- ☐ Worn, faulty or incorrectly-gapped spark plugs - petrol models (Chapter 1A)
- ☐ Vacuum leak at the throttle body, inlet manifold or associated hoses (Chapter 4)
- ☐ Blocked injector/fuel injection system fault - petrol models (Chapter 4A)
- ☐ Faulty injector(s) - diesel models (Chapter 4B)

Engine stalls

- ☐ Vacuum leak at the throttle body, inlet manifold or associated hoses - petrol models (Chapter 4A)
- ☐ Fuel filter choked (Chapter 1)
- ☐ Fuel pump faulty, or delivery pressure low - petrol models (Chapter 4A)
- ☐ Fuel tank vent blocked, or fuel pipes restricted (Chapter 4)
- ☐ Blocked injector/fuel injection system fault - petrol models (Chapter 4A)
- ☐ Faulty injector(s) - diesel models (Chapter 4B)

Engine lacks power

- ☐ Timing belt incorrectly fitted (Chapter 2)
- ☐ Fuel filter choked (Chapter 1)
- ☐ Fuel pump faulty, or delivery pressure low (Chapter 4)
- ☐ Uneven or low cylinder compressions (Chapter 2)
- ☐ Worn, faulty or incorrectly-gapped spark plugs - petrol models (Chapter 1A)
- ☐ Vacuum leak at the throttle body, inlet manifold or associated hoses - petrol models (Chapter 4A)
- ☐ Blocked injector/fuel injection system fault - petrol models (Chapter 4A)
- ☐ Faulty injector(s) - diesel models (Chapter 4B)
- ☐ Injection pump timing incorrect - diesel models (Chapter 4B)
- ☐ Brakes binding (Chapters 1 and 10)
- ☐ Clutch slipping (Chapter 6)
- ☐ Crushed or holed exhaust (Chapter 4C)

Engine backfires

- ☐ Timing belt incorrectly fitted (Chapter 2)
- ☐ Vacuum leak at the throttle body, inlet manifold or associated hoses - petrol models (Chapter 4A)
- ☐ Blocked injector/fuel injection system fault - petrol models (Chapter 4A)

Oil pressure warning light illuminated with engine running

- ☐ Low oil level, or incorrect oil grade (*"Weekly checks"*)
- ☐ Faulty oil pressure sensor (Chapter 5A)
- ☐ Worn engine bearings and/or oil pump (Chapter 2)
- ☐ High engine operating temperature (Chapter 3)
- ☐ Oil pressure relief valve defective (Chapter 2)
- ☐ Oil pick-up strainer clogged (Chapter 2)

Engine runs-on after switching off

- ☐ Excessive carbon build-up in engine (Chapter 2)
- ☐ High engine operating temperature (Chapter 3)
- ☐ Fuel injection system fault - petrol models (Chapter 4A)
- ☐ Faulty stop solenoid - diesel models (Chapter 4B)

Engine noises

Pre-ignition (pinking) or knocking during acceleration or under load

- ☐ Ignition timing incorrect/ignition system fault - petrol models (Chapters 1A and 5A)
- ☐ Incorrect grade of spark plug - petrol models (Chapter 1A)
- ☐ Incorrect grade of fuel (Chapter 1)
- ☐ Vacuum leak at the throttle body, inlet manifold or associated hoses - petrol models (Chapter 4A)
- ☐ Excessive carbon build-up in engine (Chapter 2)
- ☐ Blocked injector/fuel injection system fault - petrol models (Chapter 4A)

Whistling or wheezing noises

- ☐ Leaking inlet manifold or throttle body gasket - petrol models (Chapter 4A)
- ☐ Leaking exhaust manifold gasket or pipe-to-manifold joint (Chapter 4)
- ☐ Leaking vacuum hose (Chapters 4, 5 and 9)
- ☐ Blowing cylinder head gasket (Chapter 2)

Tapping or rattling noises

- ☐ Worn valve gear or camshaft (Chapter 2)
- ☐ Ancillary component fault (coolant pump, alternator, etc) (Chapters 3, 5, etc)

Knocking or thumping noises

- ☐ Worn big-end bearings (regular heavy knocking, perhaps less under load) (Chapter 2)
- ☐ Worn main bearings (rumbling and knocking, perhaps worsening under load) (Chapter 2)
- ☐ Piston slap (most noticeable when cold) (Chapter 2)
- ☐ Ancillary component fault (coolant pump, alternator, etc) (Chapters 3, 5A, etc)

Cooling system

Overheating

- ☐ Insufficient coolant in system (*Weekly Checks*)
- ☐ Thermostat faulty (Chapter 3)
- ☐ Radiator core blocked, or grille restricted (Chapter 3)
- ☐ Cooling fan faulty (Chapter 3)
- ☐ Inaccurate temperature gauge sender unit (Chapter 3)
- ☐ Airlock in cooling system (Chapter 3)
- ☐ Pressure cap faulty (Chapter 3)

Overcooling

- ☐ Thermostat faulty (Chapter 3)
- ☐ Inaccurate temperature gauge sender unit (Chapter 3)
- ☐ Cooling fan faulty (Chapter 3)

External coolant leakage

- ☐ Deteriorated or damaged hoses or hose clips (Chapter 1)

- ☐ Radiator core or heater matrix leaking (Chapter 3)
- ☐ Pressure cap faulty (Chapter 3)
- ☐ Coolant pump internal seal leaking (Chapter 3)
- ☐ Coolant pump-to-block seal leaking (Chapter 3)
- ☐ Boiling due to overheating (Chapter 3)
- ☐ Core plug leaking (Chapter 2)

Internal coolant leakage

- ☐ Leaking cylinder head gasket (Chapter 2)
- ☐ Cracked cylinder head or cylinder block (Chapter 2)

Corrosion

- ☐ Infrequent draining and flushing (Chapter 1)
- ☐ Incorrect coolant mixture or inappropriate coolant type (*"Weekly checks).*

Fuel and exhaust systems

Excessive fuel consumption

- ☐ Air filter element dirty or clogged (Chapter 1)
- ☐ Fuel injection system fault - petrol models (Chapter 4A)
- ☐ Faulty injector(s) - diesel models (Chapter 4B)
- ☐ Ignition timing incorrect/ignition system fault - petrol models (Chapters 1A and 5A)
- ☐ Tyres under-inflated ("Weekly checks")

Fuel leakage and/or fuel odour

- ☐ Damaged or corroded fuel tank, pipes or connections (Chapter 4)

Excessive noise or fumes from exhaust system

- ☐ Leaking exhaust system or manifold joints (Chapters 1 and 4C)
- ☐ Leaking, corroded or damaged silencers or pipe (Chapters 1 and 4C)
- ☐ Broken mountings causing body or suspension contact (Chapter 1)

Clutch

Pedal travels to floor - no pressure or very little resistance

- ☐ Broken clutch cable – cable-operated clutch (Chapter 6)
- ☐ Air in hydraulic system/faulty master or slave cylinder – hydraulic clutch (Chapter 6)
- ☐ Faulty clutch release system (Chapter 6)
- ☐ Broken clutch release bearing or fork (Chapter 6)
- ☐ Broken diaphragm spring in clutch pressure plate (Chapter 6)

Clutch fails to disengage (unable to select gears)

- ☐ Incorrect clutch cable adjustment – cable-operated clutch (Chapter 6)
- ☐ Air in hydraulic system/faulty master or slave cylinder – hydraulic clutch (Chapter 6)
- ☐ Faulty clutch release system (Chapter 6)
- ☐ Clutch disc sticking on gearbox input shaft splines (Chapter 6)
- ☐ Clutch disc sticking to flywheel or pressure plate (Chapter 6)
- ☐ Faulty pressure plate assembly (Chapter 6)
- ☐ Clutch release mechanism worn or incorrectly assembled (Chapter 6)
- ☐ Break up of clutch friction lining (Chapter 6)

Clutch slips (engine speed increases, with no increase in vehicle speed)

- ☐ Incorrect clutch cable adjustment – cable-operated clutch (Chapter 6)
- ☐ Faulty hydraulic release system – hydraulic clutch (Chapter 6)
- ☐ Clutch disc linings excessively worn (Chapter 6)
- ☐ Clutch disc linings contaminated with oil or grease (Chapter 6)
- ☐ Faulty pressure plate or weak diaphragm spring (Chapter 6)

Judder as clutch is engaged

- ☐ Clutch disc linings contaminated with oil or grease (Chapter 6)
- ☐ Clutch disc linings excessively worn (Chapter 6)
- ☐ Faulty or distorted pressure plate or diaphragm spring (Chapter 6).
- ☐ Worn or loose engine or gearbox mountings (Chapter 2)
- ☐ Clutch disc hub or gearbox input shaft splines worn (Chapter 6)

Noise when depressing or releasing clutch pedal

- ☐ Worn clutch release bearing (Chapter 6)
- ☐ Worn or dry clutch pedal bushes (Chapter 6)
- ☐ Faulty pressure plate assembly (Chapter 6)
- ☐ Pressure plate diaphragm spring broken (Chapter 6)
- ☐ Broken clutch disc cushioning springs (Chapter 6)

Manual transmission

Noisy in neutral with engine running

- ☐ Input shaft bearings worn (noise apparent with clutch pedal released, but not when depressed) (Chapter 7A)*
- ☐ Clutch release bearing worn (noise apparent with clutch pedal depressed, possibly less when released) (Chapter 6)

Noisy in one particular gear

- ☐ Worn, damaged or chipped gear teeth (Chapter 7A)*

Difficulty engaging gears

- ☐ Clutch fault (Chapter 6)
- ☐ Worn or damaged gearchange linkage/cable (Chapter 7A)
- ☐ Worn synchroniser units (Chapter 7A)*

Jumps out of gear

- ☐ Worn or damaged gearchange linkage (Chapter 7A)

- ☐ Worn synchroniser units (Chapter 7A)*
- ☐ Worn selector forks (Chapter 7A)*

Vibration

- ☐ Lack of oil (Chapter 1)
- ☐ Worn bearings (Chapter 7A)*

Lubricant leaks

- ☐ Leaking differential output oil seal (Chapter 7A)
- ☐ Leaking housing joint (Chapter 7A)*
- ☐ Leaking input shaft oil seal (Chapter 7A)*

Although the corrective action necessary to remedy the symptoms described is beyond the scope of the home mechanic, the above information should be helpful in isolating the cause of the condition, so that the owner can communicate clearly with a professional mechanic.

Automatic transmission

Note: *Due to the complexity of the automatic transmission, it is difficult for the home mechanic to properly diagnose and service this unit. For problems other than the following, the vehicle should be taken to a dealer service department or automatic transmission specialist. Do not be too hasty in removing the transmission if a fault is suspected, as most of the testing is carried out with the unit still fitted.*

Fluid leakage

☐ Automatic transmission fluid is usually dark in colour. Fluid leaks should not be confused with engine oil, which can easily be blown onto the transmission by airflow.

☐ To determine the source of a leak, first remove all built-up dirt and grime from the transmission housing and surrounding areas using a degreasing agent, or by steam-cleaning. Drive the vehicle at low speed, so airflow will not blow the leak far from its source. Raise and support the vehicle, and determine where the leak is coming from. The following are common areas of leakage:
a) *Fluid pan (Chapter 1 and 7B)*
b) *Dipstick tube (Chapter 1 and 7B)*
c) *Transmission-to-fluid cooler unions (Chapter 7B)*

Transmission fluid brown, or has burned smell

☐ Transmission fluid level low (Chapter 1)

Transmission will not downshift (kickdown) with accelerator pedal fully depressed

☐ Low transmission fluid level (Chapter 1)
☐ Incorrect selector cable adjustment (Chapter 7B)

General gear selection problems

☐ Chapter 7B deals with checking the selector cable on automatic transmissions. The following are common problems which may be caused by a faulty cable:
a) *Engine starting in gears other than Park or Neutral.*
b) *Indicator panel indicating a gear other than the one actually being used.*
c) *Vehicle moves when in Park or Neutral.*
d) *Poor gear shift quality or erratic gear changes.*

Engine will not start in any gear, or starts in gears other than Park or Neutral

☐ Incorrect starter/inhibitor (multi-function) switch adjustment (Chapter 7B)
☐ Incorrect selector cable adjustment (Chapter 7B)

Transmission slips, shifts roughly, is noisy, or has no drive in forward or reverse gears

☐ There are many probable causes for the above problems, but the home mechanic should be concerned with only one possibility - fluid level. Before taking the vehicle to a dealer or transmission specialist, check the fluid level and condition of the fluid as described in Chapter 1. Correct the fluid level as necessary, or change the fluid if needed. If the problem persists, professional help will be necessary.

Driveshafts

Vibration when accelerating or decelerating

☐ Worn inner constant velocity joint (Chapter 8)
☐ Bent or distorted driveshaft (Chapter 8)
☐ Worn intermediate bearing (Chapter 8)

Clicking or knocking noise on turns (at slow speed on full-lock)

☐ Worn outer constant velocity joint (Chapter 8)
☐ Lack of constant velocity joint lubricant, possibly due to damaged gaiter (Chapter 8)
☐ Worn intermediate bearing (Chapter 8)

Hydraulic system

Loss of hydraulic pressure

☐ Reservoir filters blocked (Chapter 9)
☐ Pump supply pipe leaking (Chapter 9)
☐ Pressure regulator unit faulty (Chapter 9)
☐ Pump faulty or drivebelt broken or slipping (Chapters 1 and 9)
☐ Pressure regulator pressure release screw loose (Chapter 9)

Excessive hydraulic pressure

☐ Pressure regulator unit faulty (Chapter 9)

Loss of suspension pressure

☐ Security valve faulty
☐ Height corrector faulty
☐ Suspension hydraulic unit faulty
☐ Vehicle ride height adjustment incorrect

Loss of brake pressure

☐ Compensator control valve faulty

Braking system

Note: *Before assuming that a brake problem exists, make sure that the tyres are in good condition and correctly inflated, that the front wheel alignment is correct, and that the vehicle is not loaded with weight in an unequal manner. Apart from checking the condition of all pipe and hose connections, any faults occurring on the anti-lock braking system should be referred to a Citroën dealer for diagnosis.*

Vehicle pulls to one side under braking

- ☐ Worn, defective, damaged or contaminated brake pads on one side (Chapters 1 and 10)
- ☐ Seized or partially-seized brake caliper piston (Chapters 1 and 10)
- ☐ A mixture of brake pad lining materials fitted between sides (Chapters 1 and 10)
- ☐ Brake caliper mounting bolts loose (Chapter 10)
- ☐ Worn/damaged steering/suspension components (Chapters 1 and 11)
- ☐ Damaged, cracked or warped brake disc (Chapter 10)

Noise (grinding or high-pitched squeal) when brakes applied

- ☐ Brake pad friction lining material worn down to metal backing (Chapters 1 and 10)
- ☐ Excessive corrosion of brake disc (may be apparent after the vehicle has been standing for some time (Chapters 1 and 10)
- ☐ Foreign object (stone chipping, etc) trapped between brake disc and shield (Chapters 1 and 10)

Excessive brake pedal effort required to stop vehicle

- ☐ Seized brake caliper piston (Chapter 10)
- ☐ Brake pads incorrectly fitted (Chapters 1 and 10)
- ☐ Incorrect grade of brake pads/shoes fitted (Chapters 1 and 10)
- ☐ Brake pad friction material contaminated (Chapters 1 and 10)

Judder felt through brake pedal or steering wheel when braking

- ☐ Excessive run-out or distortion of discs (Chapters 1 and 10)
- ☐ Brake pad friction material worn (Chapters 1 and 10)
- ☐ Brake caliper mounting bolts loose (Chapter 10)
- ☐ Wear in suspension or steering components or mountings (Chapters 1 and 11)

Brakes binding

- ☐ Seized brake caliper piston (Chapter 10)
- ☐ Incorrectly-adjusted parking brake mechanism (Chapter 10)

Rear wheels locking under normal braking

- ☐ Fault in hydraulic system (Chapter 9)

Suspension and steering

Note: *Before diagnosing suspension or steering faults, be sure that the trouble is not due to incorrect tyre pressures, mixtures of tyre types, or binding brakes.*

Vehicle pulls to one side

- ☐ Defective tyre (*Weekly checks*)
- ☐ Excessive wear in suspension or steering components (Chapters 1 and 11)
- ☐ Faulty suspension strut/hydraulic unit (Chapter 11)
- ☐ Incorrect front wheel alignment (Chapter 11)
- ☐ Accident damage to steering or suspension components (Chapter 1)

Wheel wobble and vibration

- ☐ Front roadwheels out of balance (vibration felt mainly through the steering wheel) (Chapters 1 and 11)
- ☐ Rear roadwheels out of balance (vibration felt throughout the vehicle) (Chapters 1 and 11)
- ☐ Roadwheels damaged or distorted (Chapters 1 and 11)
- ☐ Faulty or damaged tyre (*Weekly checks*)
- ☐ Worn steering or suspension joints, bushes or components (Chapters 1 and 11)
- ☐ Wheel bolts loose (Chapters 1 and 11)

Excessive pitching and/or rolling around corners, or during braking

- ☐ Hydraulic system fault (Chapter 9)
- ☐ Faulty suspension strut/hydraulic unit (Chapter 11)
- ☐ Worn or damaged anti-roll bar or mountings (Chapter 11)

Wandering or general instability

- ☐ Incorrect front wheel alignment (Chapter 11)
- ☐ Worn steering or suspension joints, bushes or components (Chapters 1 and 11)
- ☐ Roadwheels out of balance (Chapters 1 and 11)
- ☐ Faulty or damaged tyre (*Weekly checks*)
- ☐ Wheel bolts loose (Chapters 1 and 11)
- ☐ Hydraulic system fault (Chapter 19)

Excessively-stiff steering

- ☐ Seized steering linkage balljoint or suspension balljoint (Chapters 1 and 11)
- ☐ Incorrect front wheel alignment (Chapter 11)
- ☐ Steering gear or column damaged (Chapter 11)
- ☐ Hydraulic system fault (Chapter 19)

Excessive play in steering

- ☐ Worn steering column shaft joints (Chapter 11)
- ☐ Worn track rod balljoints (Chapters 1 and 11)
- ☐ Worn steering gear (Chapter 11)
- ☐ Worn steering or suspension joints, bushes or components (Chapters 1 and 11)

Tyre wear excessive

Tyres worn on inside or outside edges

- ☐ Tyres under-inflated (wear on both edges) (*Weekly checks*)
- ☐ Incorrect camber or castor angles (wear on one edge only) (Chapter 11)
- ☐ Worn steering or suspension joints, bushes or components (Chapters 1 and 11)
- ☐ Excessively-hard cornering
- ☐ Accident damage

Tyre treads exhibit feathered edges

- ☐ Incorrect toe-setting (Chapter 11)

Tyres worn in centre of tread

- ☐ Tyres over-inflated (*Weekly checks*)

Tyres worn on inside and outside edges

- ☐ Tyres under-inflated (*Weekly checks*)

Tyres worn unevenly

- ☐ Tyres/wheels out of balance (Chapter 1)
- ☐ Excessive wheel or tyre run-out (Chapter 1)
- ☐ Faulty suspension strut/hydraulic unit (Chapter 11)
- ☐ Faulty tyre (*Weekly checks*)

Electrical system

Note: For problems associated with the starting system, refer to the faults listed under Engine earlier in this Section.

Battery will not hold a charge more than a few days
- [ ] Battery defective internally (Chapter 5A)
- [ ] Battery terminal connections loose or corroded (*Weekly checks*)
- [ ] Auxiliary drivebelt worn or incorrectly adjusted (Chapter 1)
- [ ] Alternator not charging at correct output (Chapter 5A)
- [ ] Alternator or voltage regulator faulty (Chapter 5A)
- [ ] Short-circuit causing continual battery drain (Chapters 5 and 13)

Ignition/no-charge warning light remains illuminated with engine running
- [ ] Auxiliary drivebelt broken, worn, or incorrectly adjusted (Chapter 1)
- [ ] Internal fault in alternator or voltage regulator (Chapter 5A)
- [ ] Broken, disconnected, or loose wiring in charging circuit (Chapter 5A)

Ignition/no-charge warning light fails to come on
- [ ] Warning light bulb blown (Chapter 13)
- [ ] Broken, disconnected, or loose wiring in warning light circuit (Chapter 13)
- [ ] Alternator faulty (Chapter 5A)

Lights inoperative
- [ ] Bulb blown (Chapter 13)
- [ ] Corrosion of bulb or bulbholder contacts (Chapter 13)
- [ ] Blown fuse (Chapter 13)
- [ ] Faulty relay (Chapter 13)
- [ ] Broken, loose, or disconnected wiring (Chapter 13)
- [ ] Faulty switch (Chapter 13)

Instrument readings inaccurate or erratic
Instrument readings increase with engine speed
- [ ] Faulty voltage regulator (Chapter 13)
Fuel or temperature gauges give no reading
- [ ] Faulty gauge sender unit (Chapters 3 and 4)
- [ ] Wiring open-circuit (Chapter 13)
- [ ] Faulty gauge (Chapter 13)
Fuel or temperature gauges give continuous maximum reading
- [ ] Faulty gauge sender unit (Chapters 3 and 4)
- [ ] Wiring short-circuit (Chapter 13)
- [ ] Faulty gauge (Chapter 13)

Horn inoperative, or unsatisfactory in operation
Horn operates all the time
- [ ] Horn push either earthed or stuck down (Chapter 13)
- [ ] Horn cable-to-horn push earthed (Chapter 13)
Horn fails to operate
- [ ] Blown fuse (Chapter 13)
- [ ] Cable or cable connections loose, broken or disconnected (Chapter 13)
- [ ] Faulty horn (Chapter 13)
Horn emits intermittent or unsatisfactory sound
- [ ] Cable connections loose (Chapter 13)
- [ ] Horn mountings loose (Chapter 13)
- [ ] Faulty horn (Chapter 13)

Windscreen wipers inoperative, or unsatisfactory in operation
Wipers fail to operate, or operate very slowly
- [ ] Wiper blades stuck to screen, or linkage seized or binding (Chapters 1 and 13)
- [ ] Blown fuse (Chapter 13)
- [ ] Cable or cable connections loose, broken or disconnected (Chapter 13)
- [ ] Faulty relay (Chapter 13)
- [ ] Faulty wiper motor (Chapter 13)

Wiper blades sweep over too large or too small an area of the glass
- [ ] Wiper arms incorrectly positioned on spindles (Chapter 1)
- [ ] Excessive wear of wiper linkage (Chapter 13)
- [ ] Wiper motor or linkage mountings loose or insecure (Chapter 13)

Wiper blades fail to clean the glass effectively
- [ ] Wiper blade rubbers worn or perished (*Weekly checks*)
- [ ] Wiper arm tension springs broken, or arm pivots seized (Chapter 13)
- [ ] Insufficient windscreen washer additive to adequately remove road film (*Weekly checks*)

Windscreen washers inoperative, or unsatisfactory in operation
One or more washer jets inoperative
- [ ] Blocked washer jet (Chapter 1)
- [ ] Disconnected, kinked or restricted fluid hose (Chapter 13)
- [ ] Insufficient fluid in washer reservoir (*Weekly checks*)
Washer pump fails to operate
- [ ] Broken or disconnected wiring or connections (Chapter 13)
- [ ] Blown fuse (Chapter 13)
- [ ] Faulty washer switch (Chapter 13)
- [ ] Faulty washer pump (Chapter 13)
Washer pump runs for some time before fluid is emitted from jets
- [ ] Faulty one-way valve in fluid supply hose (Chapter 13)

Electric windows inoperative, or unsatisfactory in operation
Window glass will only move in one direction
- [ ] Faulty switch (Chapter 13)
Window glass slow to move
- [ ] Regulator seized or damaged, or in need of lubrication (Chapter 12)
- [ ] Door internal components or trim fouling regulator (Chapter 12)
- [ ] Faulty motor (Chapter 12)
Window glass fails to move
- [ ] Blown fuse (Chapter 13)
- [ ] Faulty relay (Chapter 13)
- [ ] Broken or disconnected wiring or connections (Chapter 13)
- [ ] Faulty motor (Chapter 12)

Central locking system inoperative, or unsatisfactory in operation
Complete system failure
- [ ] Blown fuse (Chapter 13)
- [ ] Faulty relay (Chapter 13)
- [ ] Broken or disconnected wiring or connections (Chapter 13)
- [ ] Faulty motor (Chapter 12)
Latch locks but will not unlock, or unlocks but will not lock
- [ ] Faulty master switch (Chapter 13)
- [ ] Broken or disconnected latch operating rods or levers (Chapter 12)
- [ ] Faulty relay (Chapter 13)
- [ ] Faulty motor (Chapter 12)
One solenoid/motor fails to operate
- [ ] Broken or disconnected wiring or connections (Chapter 13)
- [ ] Faulty operating assembly (Chapter 12)
- [ ] Broken, binding or disconnected latch operating rods or levers (Chapter 12)
- [ ] Fault in door latch (Chapter 12)

A

ABS (Anti-lock brake system) A system, usually electronically controlled, that senses incipient wheel lockup during braking and relieves hydraulic pressure at wheels that are about to skid.

Air bag An inflatable bag hidden in the steering wheel (driver's side) or the dash or glovebox (passenger side). In a head-on collision, the bags inflate, preventing the driver and front passenger from being thrown forward into the steering wheel or windscreen.

Air cleaner A metal or plastic housing, containing a filter element, which removes dust and dirt from the air being drawn into the engine.

Air filter element The actual filter in an air cleaner system, usually manufactured from pleated paper and requiring renewal at regular intervals.

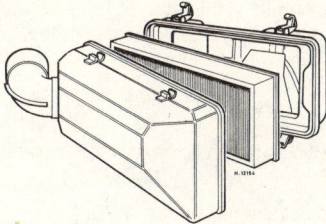

Air filter

Allen key A hexagonal wrench which fits into a recessed hexagonal hole.

Alligator clip A long-nosed spring-loaded metal clip with meshing teeth. Used to make temporary electrical connections.

Alternator A component in the electrical system which converts mechanical energy from a drivebelt into electrical energy to charge the battery and to operate the starting system, ignition system and electrical accessories.

Ampere (amp) A unit of measurement for the flow of electric current. One amp is the amount of current produced by one volt acting through a resistance of one ohm.

Anaerobic sealer A substance used to prevent bolts and screws from loosening. Anaerobic means that it does not require oxygen for activation. The Loctite brand is widely used.

Antifreeze A substance (usually ethylene glycol) mixed with water, and added to a vehicle's cooling system, to prevent freezing of the coolant in winter. Antifreeze also contains chemicals to inhibit corrosion and the formation of rust and other deposits that would tend to clog the radiator and coolant passages and reduce cooling efficiency.

Anti-seize compound A coating that reduces the risk of seizing on fasteners that are subjected to high temperatures, such as exhaust manifold bolts and nuts.

Asbestos A natural fibrous mineral with great heat resistance, commonly used in the composition of brake friction materials.

Asbestos is a health hazard and the dust created by brake systems should never be inhaled or ingested.

Axle A shaft on which a wheel revolves, or which revolves with a wheel. Also, a solid beam that connects the two wheels at one end of the vehicle. An axle which also transmits power to the wheels is known as a live axle.

Axleshaft A single rotating shaft, on either side of the differential, which delivers power from the final drive assembly to the drive wheels. Also called a driveshaft or a halfshaft.

B

Ball bearing An anti-friction bearing consisting of a hardened inner and outer race with hardened steel balls between two races.

Bearing The curved surface on a shaft or in a bore, or the part assembled into either, that permits relative motion between them with minimum wear and friction.

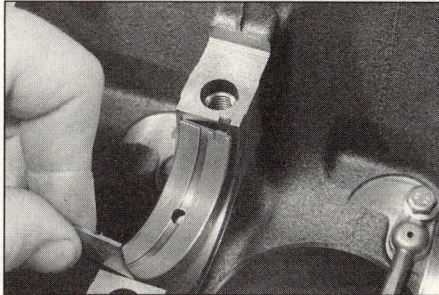

Bearing

Big-end bearing The bearing in the end of the connecting rod that's attached to the crankshaft.

Bleed nipple A valve on a brake wheel cylinder, caliper or other hydraulic component that is opened to purge the hydraulic system of air. Also called a bleed screw.

Brake bleeding Procedure for removing air from lines of a hydraulic brake system.

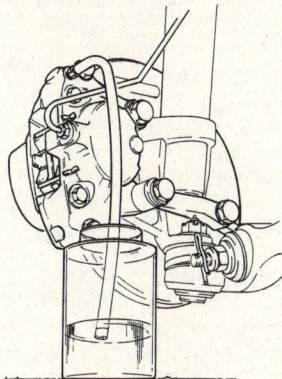

Brake bleeding

Brake disc The component of a disc brake that rotates with the wheels.

Brake drum The component of a drum brake that rotates with the wheels.

Brake linings The friction material which contacts the brake disc or drum to retard the vehicle's speed. The linings are bonded or riveted to the brake pads or shoes.

Brake pads The replaceable friction pads that pinch the brake disc when the brakes are applied. Brake pads consist of a friction material bonded or riveted to a rigid backing plate.

Brake shoe The crescent-shaped carrier to which the brake linings are mounted and which forces the lining against the rotating drum during braking.

Braking systems For more information on braking systems, consult the *Haynes Automotive Brake Manual*.

Breaker bar A long socket wrench handle providing greater leverage.

Bulkhead The insulated partition between the engine and the passenger compartment.

C

Caliper The non-rotating part of a disc-brake assembly that straddles the disc and carries the brake pads. The caliper also contains the hydraulic components that cause the pads to pinch the disc when the brakes are applied. A caliper is also a measuring tool that can be set to measure inside or outside dimensions of an object.

Camshaft A rotating shaft on which a series of cam lobes operate the valve mechanisms. The camshaft may be driven by gears, by sprockets and chain or by sprockets and a belt.

Canister A container in an evaporative emission control system; contains activated charcoal granules to trap vapours from the fuel system.

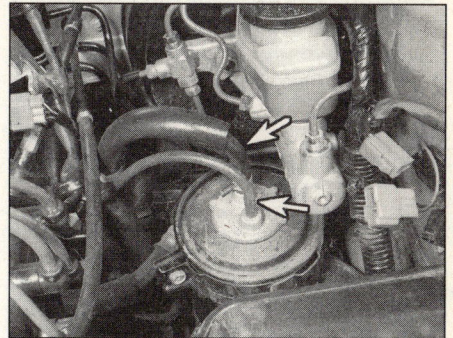

Canister

Carburettor A device which mixes fuel with air in the proper proportions to provide a desired power output from a spark ignition internal combustion engine.

Castellated Resembling the parapets along the top of a castle wall. For example, a castellated balljoint stud nut.

Castor In wheel alignment, the backward or forward tilt of the steering axis. Castor is positive when the steering axis is inclined rearward at the top.

Catalytic converter A silencer-like device in the exhaust system which converts certain pollutants in the exhaust gases into less harmful substances.

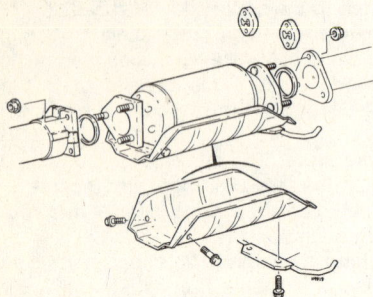

Catalytic converter

Circlip A ring-shaped clip used to prevent endwise movement of cylindrical parts and shafts. An internal circlip is installed in a groove in a housing; an external circlip fits into a groove on the outside of a cylindrical piece such as a shaft.

Clearance The amount of space between two parts. For example, between a piston and a cylinder, between a bearing and a journal, etc.

Coil spring A spiral of elastic steel found in various sizes throughout a vehicle, for example as a springing medium in the suspension and in the valve train.

Compression Reduction in volume, and increase in pressure and temperature, of a gas, caused by squeezing it into a smaller space.

Compression ratio The relationship between cylinder volume when the piston is at top dead centre and cylinder volume when the piston is at bottom dead centre.

Constant velocity (CV) joint A type of universal joint that cancels out vibrations caused by driving power being transmitted through an angle.

Core plug A disc or cup-shaped metal device inserted in a hole in a casting through which core was removed when the casting was formed. Also known as a freeze plug or expansion plug.

Crankcase The lower part of the engine block in which the crankshaft rotates.

Crankshaft The main rotating member, or shaft, running the length of the crankcase, with offset "throws" to which the connecting rods are attached.

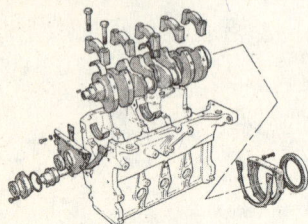

Crankshaft assembly

Crocodile clip See Alligator clip

D

Diagnostic code Code numbers obtained by accessing the diagnostic mode of an engine management computer. This code can be used to determine the area in the system where a malfunction may be located.

Disc brake A brake design incorporating a rotating disc onto which brake pads are squeezed. The resulting friction converts the energy of a moving vehicle into heat.

Double-overhead cam (DOHC) An engine that uses two overhead camshafts, usually one for the intake valves and one for the exhaust valves.

Drivebelt(s) The belt(s) used to drive accessories such as the alternator, water pump, power steering pump, air conditioning compressor, etc. off the crankshaft pulley.

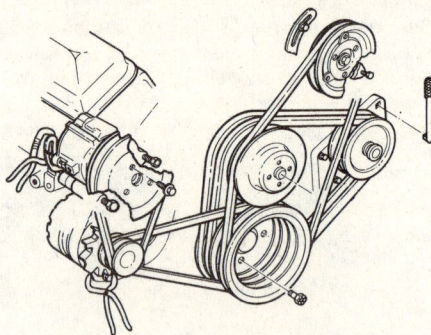

Accessory drivebelts

Driveshaft Any shaft used to transmit motion. Commonly used when referring to the axleshafts on a front wheel drive vehicle.

Drum brake A type of brake using a drum-shaped metal cylinder attached to the inner surface of the wheel. When the brake pedal is pressed, curved brake shoes with friction linings press against the inside of the drum to slow or stop the vehicle.

E

EGR valve A valve used to introduce exhaust gases into the intake air stream.

Electronic control unit (ECU) A computer which controls (for instance) ignition and fuel injection systems, or an anti-lock braking system. For more information refer to the *Haynes Automotive Electrical and Electronic Systems Manual*.

Electronic Fuel Injection (EFI) A computer controlled fuel system that distributes fuel through an injector located in each intake port of the engine.

Emergency brake A braking system, independent of the main hydraulic system, that can be used to slow or stop the vehicle if the primary brakes fail, or to hold the vehicle stationary even though the brake pedal isn't depressed. It usually consists of a hand lever that actuates either front or rear brakes mechanically through a series of cables and linkages. Also known as a handbrake or parking brake.

Endfloat The amount of lengthwise movement between two parts. As applied to a crankshaft, the distance that the crankshaft can move forward and back in the cylinder block.

Engine management system (EMS) A computer controlled system which manages the fuel injection and the ignition systems in an integrated fashion.

Exhaust manifold A part with several passages through which exhaust gases leave the engine combustion chambers and enter the exhaust pipe.

F

Fan clutch A viscous (fluid) drive coupling device which permits variable engine fan speeds in relation to engine speeds.

Feeler blade A thin strip or blade of hardened steel, ground to an exact thickness, used to check or measure clearances between parts.

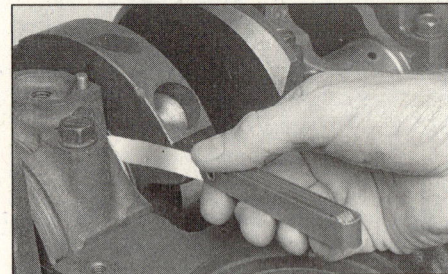

Feeler blade

Firing order The order in which the engine cylinders fire, or deliver their power strokes, beginning with the number one cylinder.

Flywheel A heavy spinning wheel in which energy is absorbed and stored by means of momentum. On cars, the flywheel is attached to the crankshaft to smooth out firing impulses.

Free play The amount of travel before any action takes place. The "looseness" in a linkage, or an assembly of parts, between the initial application of force and actual movement. For example, the distance the brake pedal moves before the pistons in the master cylinder are actuated.

Fuse An electrical device which protects a circuit against accidental overload. The typical fuse contains a soft piece of metal which is calibrated to melt at a predetermined current flow (expressed as amps) and break the circuit.

Fusible link A circuit protection device consisting of a conductor surrounded by heat-resistant insulation. The conductor is smaller than the wire it protects, so it acts as the weakest link in the circuit. Unlike a blown fuse, a failed fusible link must frequently be cut from the wire for replacement.

G

Gap The distance the spark must travel in jumping from the centre electrode to the side electrode in a spark plug. Also refers to the spacing between the points in a contact breaker assembly in a conventional points-type ignition, or to the distance between the reluctor or rotor and the pickup coil in an electronic ignition.

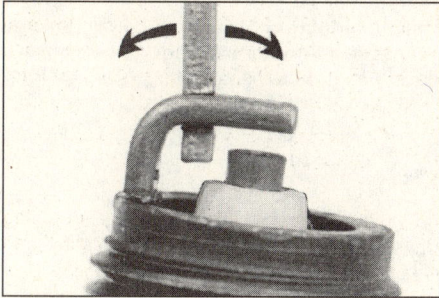

Adjusting spark plug gap

Gasket Any thin, soft material - usually cork, cardboard, asbestos or soft metal - installed between two metal surfaces to ensure a good seal. For instance, the cylinder head gasket seals the joint between the block and the cylinder head.

Gasket

Gauge An instrument panel display used to monitor engine conditions. A gauge with a movable pointer on a dial or a fixed scale is an analogue gauge. A gauge with a numerical readout is called a digital gauge.

H

Halfshaft A rotating shaft that transmits power from the final drive unit to a drive wheel, usually when referring to a live rear axle.

Harmonic balancer A device designed to reduce torsion or twisting vibration in the crankshaft. May be incorporated in the crankshaft pulley. Also known as a vibration damper.

Hone An abrasive tool for correcting small irregularities or differences in diameter in an engine cylinder, brake cylinder, etc.

Hydraulic tappet A tappet that utilises hydraulic pressure from the engine's lubrication system to maintain zero clearance (constant contact with both camshaft and valve stem). Automatically adjusts to variation in valve stem length. Hydraulic tappets also reduce valve noise.

I

Ignition timing The moment at which the spark plug fires, usually expressed in the number of crankshaft degrees before the piston reaches the top of its stroke.

Inlet manifold A tube or housing with passages through which flows the air-fuel mixture (carburettor vehicles and vehicles with throttle body injection) or air only (port fuel-injected vehicles) to the port openings in the cylinder head.

J

Jump start Starting the engine of a vehicle with a discharged or weak battery by attaching jump leads from the weak battery to a charged or helper battery.

L

Load Sensing Proportioning Valve (LSPV) A brake hydraulic system control valve that works like a proportioning valve, but also takes into consideration the amount of weight carried by the rear axle.

Locknut A nut used to lock an adjustment nut, or other threaded component, in place. For example, a locknut is employed to keep the adjusting nut on the rocker arm in position.

Lockwasher A form of washer designed to prevent an attaching nut from working loose.

M

MacPherson strut A type of front suspension system devised by Earle MacPherson at Ford of England. In its original form, a simple lateral link with the anti-roll bar creates the lower control arm. A long strut - an integral coil spring and shock absorber - is mounted between the body and the steering knuckle. Many modern so-called MacPherson strut systems use a conventional lower A-arm and don't rely on the anti-roll bar for location.

Multimeter An electrical test instrument with the capability to measure voltage, current and resistance.

N

NOx Oxides of Nitrogen. A common toxic pollutant emitted by petrol and diesel engines at higher temperatures.

O

Ohm The unit of electrical resistance. One volt applied to a resistance of one ohm will produce a current of one amp.

Ohmmeter An instrument for measuring electrical resistance.

O-ring A type of sealing ring made of a special rubber-like material; in use, the O-ring is compressed into a groove to provide the sealing action.

Overhead cam (ohc) engine An engine with the camshaft(s) located on top of the cylinder head(s).

Overhead valve (ohv) engine An engine with the valves located in the cylinder head, but with the camshaft located in the engine block.

Oxygen sensor A device installed in the engine exhaust manifold, which senses the oxygen content in the exhaust and converts this information into an electric current. Also called a Lambda sensor.

P

Phillips screw A type of screw head having a cross instead of a slot for a corresponding type of screwdriver.

Plastigage A thin strip of plastic thread, available in different sizes, used for measuring clearances. For example, a strip of Plastigage is laid across a bearing journal. The parts are assembled and dismantled; the width of the crushed strip indicates the clearance between journal and bearing.

Plastigage

Propeller shaft The long hollow tube with universal joints at both ends that carries power from the transmission to the differential on front-engined rear wheel drive vehicles.

Proportioning valve A hydraulic control valve which limits the amount of pressure to the rear brakes during panic stops to prevent wheel lock-up.

R

Rack-and-pinion steering A steering system with a pinion gear on the end of the steering shaft that mates with a rack (think of a geared wheel opened up and laid flat). When the steering wheel is turned, the pinion turns, moving the rack to the left or right. This movement is transmitted through the track rods to the steering arms at the wheels.

Radiator A liquid-to-air heat transfer device designed to reduce the temperature of the coolant in an internal combustion engine cooling system.

Refrigerant Any substance used as a heat transfer agent in an air-conditioning system. R-12 has been the principle refrigerant for many years; recently, however, manufacturers have begun using R-134a, a non-CFC substance that is considered less harmful to the ozone in the upper atmosphere.

Rocker arm A lever arm that rocks on a shaft or pivots on a stud. In an overhead valve engine, the rocker arm converts the upward movement of the pushrod into a downward movement to open a valve.

Rotor In a distributor, the rotating device inside the cap that connects the centre electrode and the outer terminals as it turns, distributing the high voltage from the coil secondary winding to the proper spark plug. Also, that part of an alternator which rotates inside the stator. Also, the rotating assembly of a turbocharger, including the compressor wheel, shaft and turbine wheel.

Runout The amount of wobble (in-and-out movement) of a gear or wheel as it's rotated. The amount a shaft rotates "out-of-true." The out-of-round condition of a rotating part.

S

Sealant A liquid or paste used to prevent leakage at a joint. Sometimes used in conjunction with a gasket.

Sealed beam lamp An older headlight design which integrates the reflector, lens and filaments into a hermetically-sealed one-piece unit. When a filament burns out or the lens cracks, the entire unit is simply replaced.

Serpentine drivebelt A single, long, wide accessory drivebelt that's used on some newer vehicles to drive all the accessories, instead of a series of smaller, shorter belts. Serpentine drivebelts are usually tensioned by an automatic tensioner.

Serpentine drivebelt

Shim Thin spacer, commonly used to adjust the clearance or relative positions between two parts. For example, shims inserted into or under bucket tappets control valve clearances. Clearance is adjusted by changing the thickness of the shim.

Slide hammer A special puller that screws into or hooks onto a component such as a shaft or bearing; a heavy sliding handle on the shaft bottoms against the end of the shaft to knock the component free.

Sprocket A tooth or projection on the periphery of a wheel, shaped to engage with a chain or drivebelt. Commonly used to refer to the sprocket wheel itself.

Starter inhibitor switch On vehicles with an automatic transmission, a switch that prevents starting if the vehicle is not in Neutral or Park.

Strut See MacPherson strut.

T

Tappet A cylindrical component which transmits motion from the cam to the valve stem, either directly or via a pushrod and rocker arm. Also called a cam follower.

Thermostat A heat-controlled valve that regulates the flow of coolant between the cylinder block and the radiator, so maintaining optimum engine operating temperature. A thermostat is also used in some air cleaners in which the temperature is regulated.

Thrust bearing The bearing in the clutch assembly that is moved in to the release levers by clutch pedal action to disengage the clutch. Also referred to as a release bearing.

Timing belt A toothed belt which drives the camshaft. Serious engine damage may result if it breaks in service.

Timing chain A chain which drives the camshaft.

Toe-in The amount the front wheels are closer together at the front than at the rear. On rear wheel drive vehicles, a slight amount of toe-in is usually specified to keep the front wheels running parallel on the road by offsetting other forces that tend to spread the wheels apart.

Toe-out The amount the front wheels are closer together at the rear than at the front. On front wheel drive vehicles, a slight amount of toe-out is usually specified.

Tools For full information on choosing and using tools, refer to the *Haynes Automotive Tools Manual*.

Tracer A stripe of a second colour applied to a wire insulator to distinguish that wire from another one with the same colour insulator.

Tune-up A process of accurate and careful adjustments and parts replacement to obtain the best possible engine performance.

Turbocharger A centrifugal device, driven by exhaust gases, that pressurises the intake air. Normally used to increase the power output from a given engine displacement, but can also be used primarily to reduce exhaust emissions (as on VW's "Umwelt" Diesel engine).

U

Universal joint or U-joint A double-pivoted connection for transmitting power from a driving to a driven shaft through an angle. A U-joint consists of two Y-shaped yokes and a cross-shaped member called the spider.

V

Valve A device through which the flow of liquid, gas, vacuum, or loose material in bulk may be started, stopped, or regulated by a movable part that opens, shuts, or partially obstructs one or more ports or passageways. A valve is also the movable part of such a device.

Valve clearance The clearance between the valve tip (the end of the valve stem) and the rocker arm or tappet. The valve clearance is measured when the valve is closed.

Vernier caliper A precision measuring instrument that measures inside and outside dimensions. Not quite as accurate as a micrometer, but more convenient.

Viscosity The thickness of a liquid or its resistance to flow.

Volt A unit for expressing electrical "pressure" in a circuit. One volt that will produce a current of one ampere through a resistance of one ohm.

W

Welding Various processes used to join metal items by heating the areas to be joined to a molten state and fusing them together. For more information refer to the *Haynes Automotive Welding Manual*.

Wiring diagram A drawing portraying the components and wires in a vehicle's electrical system, using standardised symbols. For more information refer to the *Haynes Automotive Electrical and Electronic Systems Manual*.

Note: *References throughout this index are in the form - "Chapter number" • "page number"*

Haynes Manuals – The Complete List

Title	Book No.
ALFA ROMEO	
Alfa Romeo Alfasud/Sprint (74 - 88) up to F	0292
Alfa Romeo Alfetta (73 - 87) up to E	0531
ALFA ROMEO	
Audi 80 (72 - Feb 79) up to T	0207
Audi 80, 90 (79 - Oct 86) up to D & Coupe (81 - Nov 88) up to F	0605
Audi 80, 90 (Oct 86 - 90) D to H & Coupe (Nov 88 - 90) F to H	1491
Audi 100 (Oct 82 - 90) up to H & 200 (Feb 84 - Oct 89) A to G	0907
Audi 100 & A6 Petrol & Diesel (May 91 - May 97) H to P	3504
Audi A4 (95 - Feb 00) M to V	3575
AUSTIN	
Austin/MG/Rover Maestro 1.3 & 1.6 (83 - 95) up to M	0922
Austin/MG Metro (80 - May 90) up to G	0718
Austin/Rover Montego 1.3 & 1.6 (84 - 94) A to L	1066
Austin/MG/Rover Montego 2.0 (84 - 95) A to M	1067
Mini (59 - 69) up to H	0527
Mini (69 - Oct 96) up to P	0646
Austin/Rover 2.0 litre Diesel Engine (86 - 93) C to L	1857
BEDFORD	
Bedford CF (69 - 87) up to E	0163
Bedford/Vauxhall Rascal & Suzuki Supercarry (86 - Oct 94) C to M	3015
BMW	
BMW 316, 320 & 320i (4-cyl) (75 - Feb 83) up to Y	0276
BMW 320, 320i, 323i & 325i (6-cyl) (Oct 77 - Sept 87) up to E	0815
BMW 3-Series (Apr 91 - 96) H to N	3210
BMW 3- & 5-Series (sohc) (81 - 91) up to J	1948
BMW 520i & 525e (Oct 81 - June 88) up to E	1560
BMW 525, 528 & 528i (73 - Sept 81) up to X	0632
CITROEN	
Citroën 2CV, Ami & Dyane (67 - 90) up to H	0196
Citroën AX Petrol & Diesel (87 - 97) D to P	3014
Citroën BX (83 - 94) A to L	0908
Citroën C15 Van Petrol & Diesel (89 - Oct 98) F to S	3509
Citroën CX (75 - 88) up to F	0528
Citroën Saxo Petrol & Diesel (96 - 98) N to S	3506
Citroën Visa (79 - 88) up to F	0620
Citroën Xantia Petrol & Diesel (93 - 98) K to S	3082
Citroën XM Petrol & Diesel (89 - 98) G to R	3451
Citroën ZX Diesel (91 - 93) J to L	1922
Citroën ZX Petrol (91 - 94) H to M	1881
Citroën 1.7 & 1.9 litre Diesel Engine (84 - 96) A to N	1379
COLT	
Colt/Mitsubishi 1200, 1250 & 1400 (79 - May 84) up to A	0600
FIAT	
Fiat 500 (57 - 73) up to M	0090
Fiat Cinquecento (93 - 98) K to R	3501
Fiat Panda (81 - 95) up to M	0793
Fiat Punto Petrol & Diesel (94 - Oct 99) L to V	3251
Fiat Regata (84 - 88) A to F	1167

Title	Book No.
Fiat Tipo (88 - 91) E to J	1625
Fiat Uno (83 - 95) up to M	0923
Fiat X1/9 (74 - 89) up to G	0273
FORD	
Ford Capri II (& III) 1.6 & 2.0 (74 - 87) up to E	0283
Ford Capri II (& III) 2.8 & 3.0 (74 - 87) up to E	1309
Ford Cortina Mk IV (& V) 1.6 & 2.0 (76 - 83) up to A	0343
Ford Escort (75 - Aug 80) up to V	0280
Ford Escort (Sept 80 - Sept 90) up to H	0686
Ford Escort & Orion (Sept 90 - 97) H to P	1737
Ford Escort Mk II Mexico, RS 1600 & RS 2000 (75 - 80) up to W	0735
Ford Fiesta (76 - Aug 83) up to Y	0334
Ford Fiesta (Aug 83 - Feb 89) A to F	1030
Ford Fiesta (Feb 89 - Oct 95) F to N	1595
Ford Fiesta Petrol & Diesel (Oct 95 - 97) N to R	3397
Ford Granada (Sept 77 - Feb 85) up to B	0481
Ford Granada & Scorpio (Mar 85 - 94) B to M	1245
Ford Ka (96 - 99) P to T	3570
Ford Mondeo Petrol (93 - 99) K to T	1923
Ford Mondeo Diesel (93 - 96) L to N	3465
Ford Orion (83 - Sept 90) up to H	1009
Ford Sierra 4 cyl. (82 - 93) up to K	0903
Ford Sierra V6 (82 - 91) up to J	0904
Ford Transit Petrol (Mk 2) (78 - Jan 86) up to C	0719
Ford Transit Petrol (Mk 3) (Feb 86 - 89) C to G	1468
Ford Transit Diesel (Feb 86 - 99) C to T	3019
Ford 1.6 & 1.8 litre Diesel Engine (84 - 96) A to N	1172
Ford 2.1, 2.3 & 2.5 litre Diesel Engine (77 - 90) up to H	1606
FREIGHT ROVER	
Freight Rover Sherpa (74 - 87) up to E	0463
HILLMAN	
Hillman Avenger (70 - 82) up to Y	0037
HONDA	
Honda Accord (76 - Feb 84) up to A	0351
Honda Civic (Feb 84 - Oct 87) A to E	1226
Honda Civic (Nov 91 - 96) J to N	3199
HYUNDAI	
Hyundai Pony (85 - 94) C to M	3398
JAGUAR	
Jaguar E Type (61 - 72) up to L	0140
Jaguar MkI & II, 240 & 340 (55 - 69) up to H	0098
Jaguar XJ6, XJ & Sovereign; Daimler Sovereign (68 - Oct 86) up to D	0242
Jaguar XJ6 & Sovereign (Oct 86 - Sept 94) D to M	3261
Jaguar XJ12, XJS & Sovereign; Daimler Double Six (72 - 88) up to F	0478
JEEP	
Jeep Cherokee Petrol (93 - 96) K to N	1943
LADA	
Lada 1200, 1300, 1500 & 1600 (74 - 91) up to J	0413
Lada Samara (87 - 91) D to J	1610
LAND ROVER	
Land Rover 90, 110 & Defender Diesel (83 - 95) up to N	3017
Land Rover Discovery Diesel (89 - 95) G to N	3016
Land Rover Series IIA & III Diesel (58 - 85) up to C	0529

Title	Book No.
Land Rover Series II, IIA & III Petrol (58 - 85) up to C	0314
MAZDA	
Mazda 323 (Mar 81 - Oct 89) up to G	1608
Mazda 323 (Oct 89 - 98) G to R	3455
Mazda 626 (May 83 - Sept 87) up to E	0929
Mazda B-1600, B-1800 & B-2000 Pick-up (72 - 88) up to F	0267
MERCEDES BENZ	
Mercedes-Benz 190, 190E & 190D Petrol & Diesel (83 - 93) A to L	3450
Mercedes-Benz 200, 240, 300 Diesel (Oct 76 - 85) up to C	1114
Mercedes-Benz 250 & 280 (68 - 72) up to L	0346
Mercedes-Benz 250 & 280 (123 Series) (Oct 76 - 84) up to B	0677
Mercedes-Benz 124 Series (85 - Aug 93) C to K	3253
MG	
MGB (62 - 80) up to W	0111
MG Midget & AH Sprite (58 - 80) up to W	0265
MITSUBISHI	
Mitsubishi Shogun & L200 Pick-Ups (83 - 94) up to M	1944
MORRIS	
Morris Ital 1.3 (80 - 84) up to B	0705
Morris Minor 1000 (56 - 71) up to K	0024
NISSAN	
Nissan Bluebird (May 84 - Mar 86) A to C	1223
Nissan Bluebird (Mar 86 - 90) C to H	1473
Nissan Cherry (Sept 82 - 86) up to D	1031
Nissan Micra (83 - Jan 93) up to K	0931
Nissan Micra (93 - 99) K to T	3254
Nissan Primera (90 - Aug 99) H to T	1851
Nissan Stanza (82 - 86) up to D	0824
Nissan Sunny (May 82 - Oct 86) up to D	0895
Nissan Sunny (Oct 86 - Mar 91) D to H	1378
Nissan Sunny (Apr 91 - 95) H to N	3219
OPEL	
Opel Ascona & Manta (B Series) (Sept 75 - 88) up to F	0316
Opel Ascona (81 - 88) *(Not available in UK see Vauxhall Cavalier 0812)*	3215
Opel Astra (Oct 91 - Feb 98) *(Not available in UK see Vauxhall Astra 1832)*	3156
Opel Calibra (90 - 98) *(See Vauxhall/Opel Calibra Book No. 3502)*	
Opel Corsa (83 - Mar 93) *(Not available in UK see Vauxhall Nova 0909)*	3160
Opel Corsa (Mar 93 - 97) *(Not available in UK see Vauxhall Corsa 1985)*	3159
Opel Frontera Petrol & Diesel (91 - 98) *(See Vauxhall/Opel Frontera Book No. 3454)*	
Opel Kadett (Nov 79 - Oct 84)	0634
Opel Kadett (Oct 84 - Oct 91) *(Not available in UK see Vauxhall Astra & Belmont 1136)*	3196
Opel Omega & Senator (86 - 94) *(Not available in UK see Vauxhall Carlton & Senator 1469)*	3157
Opel Omega (94 - 99) *(See Vauxhall/Opel Omega Book No. 3510)*	
Opel Rekord (Feb 78 - Oct 86) up to D	0543

Title	Book No
Opel Vectra (Oct 88 - Oct 95)	
(Not available in UK see Vauxhall Cavalier 1570)	**3158**
Opel Vectra Petrol & Diesel (95 - 98)	
(Not available in UK see Vauxhall Vectra 3396)	**3523**

PEUGEOT
Title	Book No
Peugeot 106 Petrol & Diesel (91 - 98) J to S	**1882**
Peugeot 205 (83 - 95) A to N	**0932**
Peugeot 305 (78 - 89) up to G	**0538**
Peugeot 306 Petrol & Diesel (93 - 99) K to T	**3073**
Peugeot 309 (86 - 93) C to K	**1266**
Peugeot 405 Petrol (88 - 96) E to N	**1559**
Peugeot 405 Diesel (88 - 96) E to N	**3198**
Peugeot 406 Petrol & Diesel (96 - 97) N to R	**3394**
Peugeot 505 (79 - 89) up to G	**0762**
Peugeot 1.7/1.8 & 1.9 litre Diesel Engine (82 - 96) up to N	**0950**
Peugeot 2.0, 2.1, 2.3 & 2.5 litre Diesel Engines (74 - 90) up to H	**1607**

PORSCHE
Title	Book No
Porsche 911 (65 - 85) up to C	**0264**
Porsche 924 & 924 Turbo (76 - 85) up to C	**0397**

PROTON
Title	Book No
Proton (89 - 97) F to P	**3255**

RANGE ROVER
Title	Book No
Range Rover V8 (70 - Oct 92) up to K	**0606**

RELIANT
Title	Book No
Reliant Robin & Kitten (73 - 83) up to A	**0436**

RENAULT
Title	Book No
Renault 5 (Feb 85 - 96) B to N	**1219**
Renault 9 & 11 (82 - 89) up to F	**0822**
Renault 18 (79 - 86) up to D	**0598**
Renault 19 Petrol (89 - 94) F to M	**1646**
Renault 19 Diesel (89 - 95) F to N	**1946**
Renault 21 (86 - 94) C to M	**1397**
Renault 25 (84 - 92) B to K	**1228**
Renault Clio Petrol (91 - May 98) H to R	**1853**
Renault Clio Diesel (91 - June 96) H to N	**3031**
Renault Espace Petrol & Diesel (85 - 96) C to N	**3197**
Renault Fuego (80 - 86) up to C	**0764**
Renault Laguna Petrol & Diesel (94 - 96) L to P	**3252**
Renault Mégane & Scénic Petrol & Diesel (96 - 98) N to R	**3395**

ROVER
Title	Book No
Rover 213 & 216 (84 - 89) A to G	**1116**
Rover 214 & 414 (89 - 96) G to N	**1689**
Rover 216 & 416 (89 - 96) G to N	**1830**
Rover 211, 214, 216, 218 & 220 Petrol & Diesel (Dec 95 - 98) N to R	**3399**
Rover 414, 416 & 420 Petrol & Diesel (May 95 - 98) M to R	**3453**
Rover 618, 620 & 623 (93 - 97) K to P	**3257**
Rover 820, 825 & 827 (86 - 95) D to N	**1380**
Rover 3500 (76 - 87) up to E	**0365**
Rover Metro, 111 & 114 (May 90 - 96) G to N	**1711**

SAAB
Title	Book No
Saab 90, 99 & 900 (79 - Oct 93) up to L	**0765**
Saab 900 (Oct 93 - 98) L to R	**3512**
Saab 9000 (4-cyl) (85 - 95) C to N	**1686**

SEAT
Title	Book No
Seat Ibiza & Cordoba Petrol & Diesel (Oct 93 - Oct 99) L to V	**3571**
Seat Ibiza & Malaga (85 - 92) B to K	**1609**

SKODA
Title	Book No
Skoda Estelle (77 - 89) up to G	**0604**
Skoda Favorit (89 - 96) F to N	**1801**
Skoda Felicia Petrol & Diesel (95 - 99) M to T	**3505**

SUBARU
Title	Book No
Subaru 1600 & 1800 (Nov 79 - 90) up to H	**0995**

SUZUKI
Title	Book No
Suzuki SJ Series, Samurai & Vitara (4-cyl) (82 - 97) up to P	**1942**
Suzuki Supercarry (86 - Oct 94) C to M	**3015**

TALBOT
Title	Book No
Talbot Alpine, Solara, Minx & Rapier (75 - 86) up to D	**0337**
Talbot Horizon (78 - 86) up to D	**0473**
Talbot Samba (82 - 86) up to D	**0823**

TOYOTA
Title	Book No
Toyota Carina E (May 92 - 97) J to P	**3256**
Toyota Corolla (Sept 83 - Sept 87) A to E	**1024**
Toyota Corolla (80 - 85) up to C	**0683**
Toyota Corolla (Sept 87 - Aug 92) E to K	**1683**
Toyota Corolla (Aug 92 - 97) K to P	**3259**
Toyota Hi-Ace & Hi-Lux (69 - Oct 83) up to A	**0304**

TRIUMPH
Title	Book No
Triumph Acclaim (81 - 84) up to B	**0792**
Triumph GT6 & Vitesse (62 - 74) up to N	**0112**
Triumph Spitfire (62 - 81) up to X	**0113**
Triumph Stag (70 - 78) up to T	**0441**
Triumph TR7 (75 - 82) up to Y	**0322**

VAUXHALL
Title	Book No
Vauxhall Astra (80 - Oct 84) up to B	**0635**
Vauxhall Astra & Belmont (Oct 84 - Oct 91) B to J	**1136**
Vauxhall Astra (Oct 91 - Feb 98) J to R	**1832**
Vauxhall/Opel Calibra (90 - 98) G to S	**3502**
Vauxhall Carlton (Oct 78 - Oct 86) up to D	**0480**
Vauxhall Carlton & Senator (Nov 86 - 94) D to L	**1469**
Vauxhall Cavalier 1600, 1900 & 2000 (75 - July 81) up to W	**0315**
Vauxhall Cavalier (81 - Oct 88) up to F	**0812**
Vauxhall Cavalier (Oct 88 - 95) F to N	**1570**
Vauxhall Chevette (75 - 84) up to B	**0285**
Vauxhall Corsa (Mar 93 - 97) K to R	**1985**
Vauxhall/Opel Frontera Petrol & Diesel (91 - Sept 98) J to S	**3454**
Vauxhall Nova (83 - 93) up to K	**0909**
Vauxhall/Opel Omega (94 - 99) L to T	**3510**
Vauxhall Vectra Petrol & Diesel (95 - 98) N to R	**3396**
Vauxhall/Opel 1.5, 1.6 & 1.7 litre Diesel Engine (82 - 96) up to N	**1222**

VOLKSWAGEN
Title	Book No
Volkswagen Beetle 1200 (54 - 77) up to S	**0036**
Volkswagen Beetle 1300 & 1500 (65 - 75) up to P	**0039**
Volkswagen Beetle 1302 & 1302S (70 - 72) up to L	**0110**
Volkswagen Beetle 1303, 1303S & GT (72 - 75) up to P	**0159**

Title	Book No.
Volkswagen Golf & Jetta Mk 1 1.1 & 1.3 (74 - 84) up to A	**0716**
Volkswagen Golf, Jetta & Scirocco Mk 1 1.5,1.6 & 1.8 (74 - 84) up to A	**0726**
Volkswagen Golf & Jetta Mk 1 Diesel (78 - 84) up to A	**0451**
Volkswagen Golf & Jetta Mk 2 (Mar 84 - Feb 92) A to J	**1081**
Volkswagen Golf & Vento Petrol & Diesel (Feb 92 - 96) J to N	**3097**
Volkswagen LT vans & light trucks (76 - 87) up to E	**0637**
Volkswagen Passat & Santana (Sept 81 - May 88) up to E	**0814**
Volkswagen Passat Petrol & Diesel (May 88 - 96) E to P	**3498**
Volkswagen Polo & Derby (76 - Jan 82) up to X	**0335**
Volkswagen Polo (82 - Oct 90) up to H	**0813**
Volkswagen Polo (Nov 90 - Aug 94) H to L	**3245**
Volkswagen Polo Hatchback Petrol & Diesel (94 - 99) M to S	**3500**
Volkswagen Scirocco (82 - 90) up to H	**1224**
Volkswagen Transporter 1600 (68 - 79) up to V	**0082**
Volkswagen Transporter 1700, 1800 & 2000 (72 - 79) up to V	**0226**
Volkswagen Transporter (air-cooled) (79 - 82) up to Y	**0638**
Volkswagen Transporter (water-cooled) (82 - 90) up to H	**3452**

VOLVO
Title	Book No.
Volvo 142, 144 & 145 (66 - 74) up to N	**0129**
Volvo 240 Series (74 - 93) up to K	**0270**
Volvo 262, 264 & 260/265 (75 - 85) up to C	**0400**
Volvo 340, 343, 345 & 360 (76 - 91) up to J	**0715**
Volvo 440, 460 & 480 (87 - 97) D to P	**1691**
Volvo 740 & 760 (82 - 91) up to J	**1258**
Volvo 850 (92 - 96) J to P	**3260**
Volvo 940 (90 - 96) H to N	**3249**
Volvo S40 & V40 (96 - 99) N to V	**3569**
Volvo S70, V70 & C70 (96 - 99) P to V	**3573**

YUGO/ZASTAVA
Title	Book No.
Yugo/Zastava (81 - 90) up to H	**1453**

AUTOMOTIVE TECHBOOKS
Title	Book No.
Automotive Brake Manual	**3050**
Automotive Carburettor Manual	**3288**
Automotive Diagnostic Fault Codes Manual	**3472**
Automotive Diesel Engine Service Guide	**3286**
Automotive Disc Brake Manual	**3542**
Automotive Electrical and Electronic Systems Manual	**3049**
Automotive Engine Management and Fuel Injection Systems Manual	**3344**
Automotive Gearbox Overhaul Manual	**3473**
Automotive Service Summaries Manual	**3475**
Automotive Timing Belts Manual – Austin/Rover	**3549**
Automotive Timing Belts Manual – Ford	**3474**
Automotive Timing Belts Manual – Peugeot/Citroën	**3568**
Automotive Timing Belts Manual – Vauxhall/Opel	**3577**
Automotive Welding Manual	**3053**
In-Car Entertainment Manual (3rd Edition)	**3363**

OTHER TITLES
Title	Book No.
Haynes Diesel Engine Systems & Data Book (91 -00)	**3548**
Haynes Petrol Models Data Book (94 - 00)	**3718**

CL09.04/00

Preserving Our Motoring Heritage

> The Model J Duesenberg Derham Tourster. Only eight of these magnificent cars were ever built – this is the only example to be found outside the United States of America

Almost every car you've ever loved, loathed or desired is gathered under one roof at the Haynes Motor Museum. Over 300 immaculately presented cars and motorbikes represent every aspect of our motoring heritage, from elegant reminders of bygone days, such as the superb Model J Duesenberg to curiosities like the bug-eyed BMW Isetta. There are also many old friends and flames. Perhaps you remember the 1959 Ford Popular that you did your courting in? The magnificent 'Red Collection' is a spectacle of classic sports cars including AC, Alfa Romeo, Austin Healey, Ferrari, Lamborghini, Maserati, MG, Riley, Porsche and Triumph.

A Perfect Day Out

Each and every vehicle at the Haynes Motor Museum has played its part in the history and culture of Motoring. Today, they make a wonderful spectacle and a great day out for all the family. Bring the kids, bring Mum and Dad, but above all bring your camera to capture those golden memories for ever. You will also find an impressive array of motoring memorabilia, a comfortable 70 seat video cinema and one of the most extensive transport book shops in Britain. The Pit Stop Cafe serves everything from a cup of tea to wholesome, home-made meals or, if you prefer, you can enjoy the large picnic area nestled in the beautiful rural surroundings of Somerset.

> John Haynes O.B.E., Founder and Chairman of the museum at the wheel of a Haynes Light 12.

> Graham Hill's Lola Cosworth Formula 1 car next to a 1934 Riley Sports.

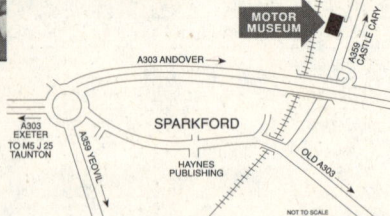

The Museum is situated on the A359 Yeovil to Frome road at Sparkford, just off the A303 in Somerset. It is about 40 miles south of Bristol, and 25 minutes drive from the M5 intersection at Taunton.

Open 9.30am - 5.30pm (10.00am - 4.00pm Winter) 7 days a week, *except Christmas Day, Boxing Day and New Years Day*
Special rates available for schools, coach parties and outings Charitable Trust No. 292048